Organic
Photochemistry

MOLECULAR AND SUPRAMOLECULAR PHOTOCHEMISTRY

Series Editors

V. RAMAMURTHY
Professor
Department of Chemistry
Tulane University
New Orleans, Louisiana

KIRK S. SCHANZE
Professor
Department of Chemistry
University of Florida
Gainesville, Florida

1. Organic Photochemistry, *edited by V. Ramamurthy and Kirk S. Schanze*
2. Organic and Inorganic Photochemistry, *edited by V. Ramamurthy and Kirk S. Schanze*
3. Organic Molecular Photochemistry, *edited by V. Ramamurthy and Kirk S. Schanze*
4. Multimetallic and Macromolecular Inorganic Photochemistry, *edited by V. Ramamurthy and Kirk S. Schanze*
5. Solid State and Surface Photochemistry, *edited by V. Ramamurthy and Kirk S. Schanze*
6. Organic, Physical, and Materials Photochemistry, *edited by V. Ramamurthy and Kirk S. Schanze*

ADDITIONAL VOLUMES IN PREPARATION

MOLECULAR AND SUPRAMOLECULAR PHOTOCHEMISTRY

Organic Photochemistry

VOLUME 1

edited by

V. Ramamurthy
Kirk S. Schanze

CRC Press
Taylor & Francis Group
Boca Raton London New York

CRC Press is an imprint of the
Taylor & Francis Group, an **informa** business

A TAYLOR & FRANCIS BOOK

First published 1997 by Marcel Dekker, Inc.

Published 2019 by CRC Press
Taylor & Francis Group
6000 Broken Sound Parkway NW, Suite 300
Boca Raton, FL 33487-2742

First issued in paperback 2019

No claim to original U.S. Government works

ISBN 13: 978-0-367-45584-2 (pbk)
ISBN 13: 978-0-8247-0012-6 (hbk)

**Visit the Taylor & Francis Web site at
http://www.taylorandfrancis.com**

**and the CRC Press Web site at
http://www.crcpress.com**

Series Introduction

Photochemistry has existed since the birth of the universe. Since ancient times, humans have recognized the connection between light and the quality of life as evidenced by the fact that the sun was an object of worship by almost every civilization. However, only during the second half of the 20th century was photochemistry recognized as an independent field, in which significant advances were just beginning to be made. During the third quarter of the century (1950–1975) this first generation of photochemists established the basic "rules" and mechanisms of the photochemistry of polyatomic molecules in condensed phases. Thus, photochemists described the primary excited states encountered in both organic and inorganic molecules, the patterns of reactivity expected from these states, and the fundamental photochemical reaction mechanisms expected for various functional groups. The significant advances that were made during this period were driven by a number of factors including theoretical developments such as the Woodward-Hoffman rules, theories of radiationless transitions and computational chemistry, in addition to remarkable progress in experimental techniques such as nanosecond and picosecond laser flash photolysis, matrix isolation and ESR spectroscopy. By the late 1970s, most of the basic principles of photochemistry were well established, and the field could then be characterized as a mature science.

Despite the maturity of the field, work in the area of photochemistry has continued at a relentless pace throughout the final quarter of the 20th century.

Currently, emphasis has shifted from applications of the basic principles of photochemistry to molecular systems of increasing complexity: through this shift, the field of supramolecular photochemistry was born. The new emphasis on supramolecular photochemistry is driven by further advances in ultrafast spectroscopy and synthetic methods that occurred during the 1980s and early 1990s. Significant progress has arisen from the application of photochemistry to probe supramolecular systems such as organized media and interfaces (micelles, vesicles, monolayers, multilayers, etc.), host–guest interactions (cyclodextrins, zeolites, silica, metal phosphonates, etc.) and colloidal semiconductors. Biological chemistry has also benefited from the advances in supramolecular photochemistry. Of greater significance is the research that has uncovered details of the primary events in photosynthesis. Other advances have come through the application of supramolecular photochemistry to understand the primary processes in vision and in the structure and function of DNA and proteins. For the past quarter of the 20th century, photochemistry has also found its way into a number of technologically and industrially important processes. Some significant applications include photography, photolithography, and xerography, and photochromic, electroluminescent, and analytical sensing devices.

The original series Organic Photochemistry has been renamed Molecular and Supramolecular Photochemistry and will consist of works that provide an accurate overview of the dynamic field of photochemistry. We sincerely hope that current and future photochemists will find these volumes to be valuable resources.

V. Ramamurthy
Kirk S. Schanze

Preface

This first volume of the Molecular and Supramolecular Photochemistry series features chapters by leading research scientists who work in the area of organic photochemistry. Although the volume by no means provides a comprehensive overview of this comparatively large research arena, the chapters do present a snapshot of the state of the science. Exploration of the photochemistry and photophysics of organic chromophores continues to be one of the main activities in organic photochemistry. This is highlighted by several chapters addressing this topic. For example, chapters on the photochemistry of particular organic functional groups, such as the sulfoxides, heterocycles, and polyalkynes, indicate this activity. Over time, even functionalities once thought to be intermediates have become chromophores in the hands of photochemists. One such example is found in the chapter on the photochemistry of carbocations. Much of the work described in these chapters is mechanistic, with particular emphasis placed on reactive intermediates and excited states involved in the photoinduced transformations. Detailed mechanistic insight is provided through applications of conventional organic techniques such as product and yield studies, combined with modern techniques such as nano- and picosecond pulsed laser spectroscopy.

Organic photochemists continue to demonstrate the utility of photochemical reactions in routine organic synthesis. With the change in the environmental scenario, it is quite likely that more photochemical reactions will become valuable in the hands of chemists who wish to pursue "green chemistry." [2+2] cycloaddition is of great value to build strained cyclobutane systems, and an up-to-date critical summary on regio- and steroselectivity on this reaction has been provided. Significant progress on photoinduced electron transfer reactions has taken place during the last decade. Many reactions prompted by electron transfer have been demonstrated to be of synthetic value. Two chapters in this volume provide in-depth coverage of this fascinating area of photochemistry.

Photochemists continue to keep an eye on the value of their activity with respect to long-range applications. Chemistry on squarine dyes and fullerenes summarized in three chapters illustrates the important connection between materials science and photochemistry.

Chemistry is moving from the molecular to the supramolecular level, and photochemists play an important role in this context. Certain aspects of activity in the area of supramolecular photochemistry are summarized by a chapter in this volume.

As the 21st century approaches, organic photochemistry continues to be a dynamic field. Significant new fundamental work on the photophysics and photochemistry of small organic molecules continues to emerge, as well as studies of supramolecular covalent and noncovalent assemblages of organic molecules of remarkable complexity. This first volume in the Molecular and Supramolecular Photochemistry series provides readers with an exciting glimpse of this evolving and highly active area of chemical research.

We would like to thank the contributors for the in-depth coverage of their topics, for their adherence to deadlines, and for serving as models for future and potential authors for this series.

V. Ramamurthy
Kirk S. Schanze

Contents

Contributors

Cornelia Bohne   Department of Chemistry, University of Victoria, Victoria, British Columbia, Canada

Mary K. Boyd   Department of Chemistry, Loyola University of Chicago, Chicago, Illinois

Cara L. Bradford   Department of Chemistry and Biochemistry, Brigham Young University, Provo, Utah

Suresh Das   Regional Research Laboratory, Council of Scientific and Industrial Research (CSIR), Trivandrum, India

Steven A. Fleming   Department of Chemistry and Biochemistry, Brigham Young University, Provo, Utah

J. Jerry Gao   Department of Chemistry and Biochemistry, Brigham Young University, Provo, Utah

Manapurathu V. George   Regional Research Laboratory, Council of Scientific and Industrial Research (CSIR), Trivandrum, India

Daniel D. Gregory   Iowa State University, Ames, Iowa

Yushen Guo   Iowa State University, Ames, Iowa

William S. Jenks   Iowa State University, Ames, Iowa

Mark H. Kleinman   Department of Chemistry, University of Victoria, Victoria, British Columbia, Canada

Kock-Yee Law   Wilson Center for Research and Technology, Xerox Corporation, Webster, New York

Woojae Lee   Iowa State University, Ames, Iowa

Yuzhuo Li   Department of Chemistry, Clarkson University, Potsdam, New York

Paul Margaretha   Institute of Organic Chemistry, University of Hamburg, Hamburg, Germany

Ganesh Pandey   Division of Organic Chemistry, National Chemical Laboratory, Pune, India

James W. Pavlik   Department of Chemistry and Biochemistry, Worcester Polytechnic Institute, Worcester, Massachusetts

Sang Chul Shim   Department of Chemistry, The Korea Advanced Institute of Science and Technology, Taejon, Korea

Ya-Ping Sun   Department of Chemistry, Clemson University, Clemson, South Carolina

Troy Tetzlaff   Iowa State University, Ames, Iowa

K. George Thomas   Regional Research Laboratory, Council of Scientific and Industrial Research (CSIR), Trivandrum, India

Organic
Photochemistry

1

The Photochemistry of Sulfoxides and Related Compounds

**William S. Jenks, Daniel D. Gregory,
Yushen Guo, Woojae Lee, and Troy Tetzlaff**
Iowa State University, Ames, Iowa

I. INTRODUCTION

The history of sulfoxide photochemistry dates back at least to the early 1960s, but this important functional group has received substantially less attention than some of the other chromophores whose chemistry was explored in those years [1,2]. On the other hand, the sulfoxide's chiral nature has brought its thermal chemistry into greater exposure [3-5]. Much of that is due to the relative ease of preparation of optically pure samples and their utility as chiral auxiliaries, directing the stereochemistry of subsequent synthetic steps. Also, the sulfoxide is an intermediate oxidation state of sulfur, which can be made achiral or to have different reactivity by easily achievable oxidations and reductions.

We have returned to examination of the sulfoxide with the purpose of developing a model based on the primary photochemical processes of the excited states. This review is generally organized in a way to reflect that approach. The reactions are broken down by the type of first step, but it will be seen that a variety of different products can arise from the subsequent chemistry of intermediates. Within each section, work is organized roughly in chronological fashion, though observations that are obviously tied to one another are discussed together. In addition to cataloging transformations and reporting proposed

1

mechanisms, we have, in places, suggested alternatives. In most such cases our job has been much easier by the benefit of time and a broader group of examples.

Unlike the carbonyl, the unconjugated sulfoxide is not associated with a long wavelength absorption of its own. Therefore most of what is reported herein has to do with conjugated sulfoxides. Aliphatic sulfoxides are treated separately, though there are many commonalities with aromatic sulfoxide chemistry. The aliphatic compounds have been examined by gas phase physical chemistry methods, which also justify a separate treatment.

We begin with a brief discussion of the sulfoxide chromophore itself. It will be seen that, just as the sulfoxide bond does not lend itself to simple description, the chromophore is fairly well understood phenomenologically but not yet well described in the shorthand photochemists have developed for other functional groups. Sections III through VII constitute the bulk of the paper and develop the chemistry of conjugated sulfoxides by reaction type. The photochemistry of aliphatic sulfoxides is discussed in Section VIII, with emphasis on gas phase reactions.

In some of the more complex molecules, there are instances in which the low energy chromophore is probably not localized on the sulfoxide, yet the sulfinyl (SO) group is involved in the observed transformations. Some attempt will be made to differentiate photochemical reactions of molecules that merely happen to contain a sulfinyl group and those for which the sulfoxide is the critical functional group and/or chromophore. Section IX is reserved for the chemistry of ketosulfoxides in which the chemistry is clearly carbonyl but the reaction involves of sulfinyl site. The photochemistry of sulfenic esters (R–S–O–R') appears throughout the text, particularly in Section III—the discussion of sulfoxide α-cleavage reactions. A short additional section on these sulfoxide isomers is also included at the end.

While we have attempted to be fairly comprehensive, inevitably there may be important contributions that are missed from time to time, and we apologize to any authors whose work we may have slighted. A point of notation: we have chosen to draw the sulfoxides as $R_2S{=}O$ but are aware that there are others who prefer different representations. In some cases, where stereochemistry is being specifically implied, we have used a single-bond hash or wedge for clarity. This is only to show spacial relationships and does not denote any change in oxidation state. Finally, because of the multitude of sulfur oxidation states, and the relative unfamiliarity of many photochemists with their nomenclature, Figure 1 is intended as a quick reference guide.

II. THE SULFOXIDE CHROMOPHORE

The sulfoxide chromophore, in the absence of conjugated aromatic groups, has a relatively high excitation energy; the absorption spectra of aliphatic sulfox-

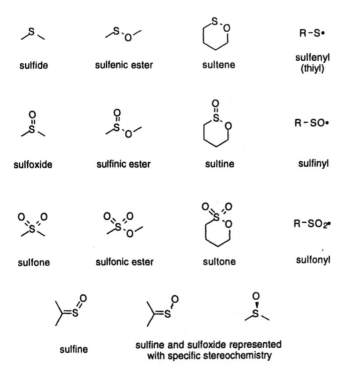

sulfide   sulfenic ester   sultene   sulfenyl (thiyl)

sulfoxide   sulfinic ester   sultine   sulfinyl

sulfone   sulfonic ester   sultone   sulfonyl

sulfine   sulfine and sulfoxide represented with specific stereochemistry

Figure 1 Illustrations of sulphur-containing functional groups.

ides have been examined by several authors [6–11]. Perhaps because of the more complex description of the sulfoxide bond, a meaningful simple orbital description (e.g., nπ*, $\pi\pi$*) has not been settled upon. Alkyl sulfoxides do not fluoresce to a detectable amount in solution, so the best estimate of the singlet energy derives from the extrapolated onset of absorption of a gas phase sample of dimethyl sulfoxide (DMSO) [12]. The value so obtained is 105 kcal/mol. A similar method was used to estimate a triplet energy of 83 kcal/mol. Xe was added to the sample as a heavy atom to induce $S_0 \rightarrow T_1$ absorption, which appeared as a shoulder on the original absorption spectrum. (Gas phase fluorescence of DMSO has been reported, but without the details of a spectrum [13]).

An attempt to observe the $S_0 \rightarrow T_1$ absorption for di-*tert*-butyl sulfoxide in solution was made using a similar, but flawed, method [14]. Very concentrated solutions of the sulfoxide in iodomethane showed an absorption at 455 nm (63 kcal/mol). This absorption was interpreted—in combination with chemistry that was sensitized by anthraquinone (E_T ca. 63 kcal/mol)—to indicate a triplet energy of that magnitude, even though it was known that the sulfoxide is thermally reactive with iodomethane. Given the rather higher triplet energy from the gas phase results of Gollnick, and the higher energy phosphorescence observed for various aromatic sulfoxides (*vide infra*), we are skeptical of this value and view it as likely an artifact.

Reliable calculations on the excited states of DMSO and related derivatives are not yet available. Nonetheless, both CNDO/2 [12] and RHF/6-31G(d,p) calculations agree that the occupied frontier molecular orbitals have qualitatively higher coefficients at oxygen than do the first few unoccupied orbitals. The canonical MO picture is consistent, at least, with the blue shift of the absorption spectrum observed on increasing the polarity of the solvent. The canonical HOMO at RHF/6-31G(d,p) is π-antibonding between S and O but is σ-bonding between C and S. The LUMO is localized more on the sulfur but is π-antibonding along SO and σ-antibonding along CS. The HOMO in the valence bond picture is the sulfur lone pair.

The absorption spectrum of the sulfoxide chromophore conjugated to aromatic nuclei has also been examined [6,7,9,15–17]. There are generally three bands for simple aryl alkyl sulfoxides between 200 and 300 nm. The middle band is most associated with the sulfoxide function because of the solvent effect, which parallels that of dialkyl sulfoxides. In some solvents, the central band, which is stronger than the benzene-like low energy absorption, overwhelms and hides the low energy band [10,17]. Figure 2 illustrates that an insulating CH_2 decouples the benzene and SO chromophores; save for the intensity of the sulfoxide absorption, dibenzyl sulfoxide appears as approximately the sum of bibenzyl and DMSO.

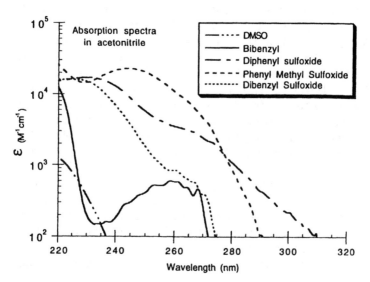

Figure 2   Absorption spectra of representative sulfoxides.

As with the smaller compounds, reliable computational descriptions of methyl phenyl sulfoxide excited states are not available. Ground state computations are easily accessible for molecules of this size. At the RHF/6-31G(d,p) level, the HOMO is π^* with regard to the SO bond but delocalized throughout the whole π-system. The next two descending orbitals are localized on the phenyl and SO, respectively. (The sulfur lone pair is the HOMO-2 when the valence bond orbitals are approximated by the Edmiston–Ruedenberg method.) While the LUMO is extensively delocalized, the LUMO+1 is entirely localized on the phenyl ring.

The lowest excited states of sulfoxides 1 and 2, the corresponding sulfides, and sulfones were studied by photoelectron spectroscopy (PES), voltammetry, absorption, and emission spectroscopy [18]. (Simpler sulfoxides, both saturated and aromatic, have also been studied by PES [19].) The phosphorescence spectrum of 1 and its small ΔE_{S-T} of about 6 kcal/mol are both typical of an aromatic $n\pi^*$ ketone. The same data for 2 are more consistent with a $\pi\pi^*$ triplet.

1 2 3 X = H, F, CH₃O 4 5 X = H, F, Cl, Br, CH₃O, CH₃

6 7 8 9

The 77 K luminescence of a set of non-carbonyl-containing sulfoxides 3–9 (and the corresponding sulfides and sulfones) has also been reported [20]. Fluorescence was noted only for 5-CH₃O, 8, and 9. Singlet-triplet gaps for those compounds were 19, 12, and 21 kcal/mol respectively. Phosphorescence among the sulfoxides was generally weak ($\Phi_p \leq 0.05$), and lifetimes were generally < 100 ms, though nonexponential. The triplet energies of the aryl alkyl sulfoxides are near 80 kcal/mol, with the diaryl sulfoxides in the range of 75–78 kcal/mol. The triplet energy of 9 is exceptionally low compared to the others due to the extended chromophore. It is not known whether the low phosphorescence yields are due to low triplet yields or inefficient luminescence. Based on these data, it was concluded that the lowest triplets were basically $\pi\pi^*$ states, but that there was a significant perturbation due to the sulfinyl group. Given that the triplet energy reported for DMSO[12] is only several kcal/mol higher, such mixing is probably to be expected.

III. α-CLEAVAGE

The primary photochemical process that is best established and probably most frequently invoked in sulfoxide photochemistry is homolytic C–S cleavage, or α-cleavage. Compared to the equivalent process in ketones, we shall see that the sulfoxide is, if anything, *more* susceptible. For instance, α-cleavage (under 1-photon conditions) is not observed for benzophenone, whereas diphenyl sulfoxide suffers this reaction, albeit with low quantum efficiency.

This section is broken up into three parts: 1.) benzylic and allylic systems, 2.) aryl alkyl and diaryl sulfoxides, and 3.) extrusions. It turns out that the isolated materials from these reactions are only infrequently the primary photochemical products. Thus we shall also see photochemistry of a few other functional groups, notably the sulfenic ester/sultene and the sulfine ($R_2C=S=O$). While secondary photochemistry can be a disadvantage in synthetic respects, the variety of rearrangements observed in sulfoxide photochemistry is all the richer for it.

A. Benzylic and Allylic Systems

By our usual definition, benzylic and allylic sulfoxides are not aromatic. Certainly, they are not conjugated to the aryl group. However, their behavior is more related to the aromatic sulfoxides than the other aliphatic compounds. Perhaps this is because the observed α-cleavage chemistry is as related to the benzyl chromophore as it is to the sulfoxide. Nonetheless, these reactions were among the first to be studied historically, and they set precedent for the thinking of authors working in conjugated systems.

The photochemistry of dibenzyl sulfoxide 10 was briefly reported in the mid 1960s [21,22]. It was said to decompose mainly to benzyl mercaptan (isolated as the disulfide 17) and benzaldehyde 16. Though no mechanism was suggested at the time, it is now clear that these products arise from a standard α-cleavage mechanism, followed by secondary photolysis of the sulfenic ester 13. The careful reader will note that the yield of bibenzyl (19) is very low in comparison to photolysis of dibenzyl ketone. Sulfinyl radicals rarely lose SO, though some net extrusions are discussed later.

$$\underset{10}{Ph\diagup\underset{\overset{\|}{O}}{S}\diagdown Ph} \xrightarrow{h\nu} \underset{11}{Ph\diagup\underset{\overset{\|}{O}}{S}\bullet} + \underset{12}{PhCH_2\bullet} \longrightarrow \underset{13}{Ph\diagup S\diagup O\diagdown Ph} \xrightarrow{h\nu}$$

$$\underset{14}{PhCH_2S\bullet} + \underset{15}{\bullet OCH_2Ph} \longrightarrow \underset{\underset{16}{37\%}}{PhCHO} \quad \underset{\underset{17}{14\%}}{PhCH_2SSCH_2Ph} \quad \underset{\underset{18}{4\%}}{PhCH_2OH} \quad \underset{\underset{19}{1\%}}{PhCH_2CH_2Ph}$$

Photolysis of dibenzoylstilbene episulfoxide 20 represented the first method for preparation of monothiobenzil 22 [23,24]. A mechanism involving formation of the sultene 21 was postulated, the authors favoring a concerted rearrangement over a discrete biradical. While there is no evidence to rule out a concerted mechanism, this seems likely to be an example of carbonyl β-cleavage, so a transient biradical is expected. The authors suggested that the sultene, postulated but not actually detected in the reaction mixture, would decompose thermally in the same fashion as related dioxetanes.

stereochemistry undetermined

Schultz and Schlessinger did several pioneering studies in sulfoxide photochemistry in the late 1960s and early 1970s. They appear in different sections of this review. Using naphthyl systems, they lowered the singlet and triplet energies significantly, compared to phenyl systems. As a result, their work involving triplet sensitization with benzophenone and its derivatives is more straightforward to interpret than other workers'. The first experiment they reported is just such a case.

Triplet sensitized photolysis of stereoisomers of 24 gave rise to the sulfines 26. They arise from disproportionation of the α-cleavage biradical and can be isolated under certain conditions [25]. Though triplet sensitization of the sulfine only causes interconversions of the stereoisomers, under the conditions of their original experiments, the nascent sulfine had sufficient absorption at the applied wavelengths also to undergo direct irradiation. Singlet chemistry of the sulfine yields the desulfurized ketone 28 [25,26]. The heterocycle 27 is suggested as an intermediate [27]. This remains the best rationalization for such desulfurization reactions. In asymmetric naphthyl systems 29, the analogous ketone products 30 are isolated, consistent with cleavage in the first step to form the more stable biradical [28].

Quite interestingly, direct irradiation of 24 gives a much different product mixture than does sensitization [29]. Interconversion of the sulfoxide isomers is severely curtailed, and two new products were introduced. Photolysis of trans 24 in the presence of piperylene as a triplet quencher resulted in high yields of the trans sultene 31. Secondary photolysis of 31 gradually converted it to the ether 33 with mixed stereochemistry. The loss of sulfur from this compound seems to imply S–O homolysis, but there is not detailed information on this process. This is apparently an example of how the spin state of the sulfinyl biradical can have a significant effect on its reactivity, with the singlet experiencing extremely rapid recombination to the sultene and the triplet giving disproportionation.

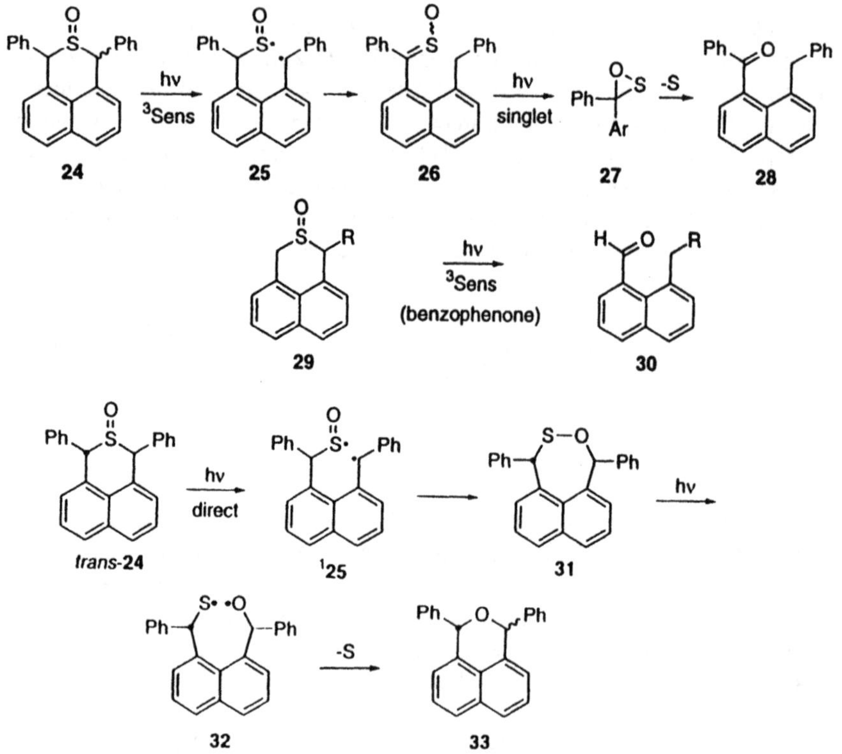

Desulfurization through α-cleavage was shown to be the major product for a series of compounds 34 illustrated below [30]. These compounds, both benzylic and allylic, consistently showed cleavage on the benzylic side but suffered E/Z isomerization in the enal products 36 and 37 when not constrained by substitution. Deuterium labeling was consistent with the internal hydrogen transfer proposed to reach the sulfine 35.

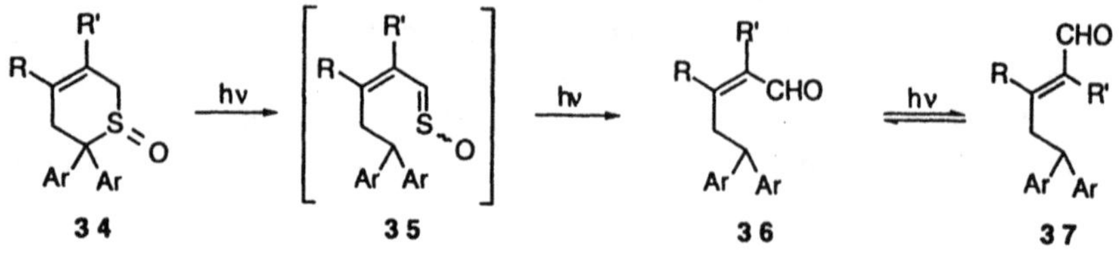

3 8

The photochemistry of this series of benzyl/allyl systems 34 was studied by chemically induced dynamic nuclear polarization (CIDNP) experiments [31]. Contrary to the assertion made here, the authors concluded that the formation of the (unisolated) sulfine 35 was an electrocyclic process, under either direct or sensitized irradiation, because no CIDNP signals were observed. On the other hand, CIDNP results for the acyclic 38 suggest that homolytic cleavage occurs on either side of the sulfoxide. Products were not identified.

B. Aryl Alkyl and Diaryl Systems

The first report of the photochemistry of a conjugated aromatic sulfoxide appears to be that of Kharasch and Khodair [32]. In a study more principally aimed at the sulfones, they showed that photolysis of diphenyl sulfoxide 5-H in benzene led to biphenyl 41 (53%), diphenyl sulfide (7%), and trace diphenyldisulfide 45. Biphenyl was also the major product of diphenyl sulfone photolysis, and it was shown to arise from attack of the photochemically generated phenyl radical on solvent (benzene). The same mechanism was clearly implied for the sulfoxide. The disulfide presumably comes from secondary photolysis of the unobserved phenyl benzenesulfenate 42. (The formation of the sulfide is addressed below in the section of deoxygenation.) Later workers showed that pyridine could be tolylated by photolysis of di-p-tolyl sulfoxide 5-Me with a product distribution quite consistent with other radical phenylations [33].

By 1970, Schultz and Schlessinger addressed the mechanism of sulfoxide stereomutation, a problem still under review for the more general cases (Section IV) [34]. Sensitized photolysis of cis or trans-48 gave a 4:1 photostationary state between the two. With high quantum yields, the authors favored the direct inversion implied in path *a* as the major stereomutation pathway, in competition with cleavage/recombination (path *b* via 49). The sultene 50 was a significant fraction of the photostationary mixture and had to be accounted for. The postulated mechanism was homolytic α-cleavage followed by recombination by path *c*. Actual observation of the sultene was an important result, since many subsequent invocations of the α-cleavage mechanism provided no direct evidence for sulfenic esters, though they are required to achieve the products whose formation is rationalized.

Triplet sensitization

A unique aspect of this work is that sensitized photolysis of 50 apparently leads to the sulfoxides [34]. Though S–O homolysis (*d*) is common, C–O homolysis (*c*) has not been observed in other cases. The authors suggested that homolysis along path *d* occurs but is merely nonproductive.

Another group that made many important contributions in the 1970s was that of Ian Still. This series of papers dealt with the photochemistry of derivatives of thiochromanone sulfoxide, which turned out to be remarkably sensitive to the pattern of substitution. Despite the expectation that the lowest excitations of such compounds would be related to the carbonyl function [18], the chemical "action" is clearly related to the sulfoxide.

Still established three types of reactivity [35]. The Type A reaction was observed for systems with electron donating substituents on the aryl ring (e.g., 52). In low yield, these compounds gave disulfide products 56 which appeared to derive from a surprising Aryl–S cleavage [36]. Labeling studies showed the phenolic oxygen in the product derived from the sulfoxidic oxygen in the starting material [35].

Type B reactivity was observed for systems in which substitution was found β to the sulfoxide [37]. Compound 57 is shown as a representative case. It was shown that 64 and 65 are probably derived from secondary photolysis of 63. Two mechanisms were proposed for the Type B transformations, each involving α-cleavage. First, alkyl-S cleavage can lead to the sultene 59. Further photolysis leads to S-O homolysis. The subsequent loss of atomic sulfur is the difficulty with this mechanism but may result from attack by other radicals in solution. Intramolecular hydrogen abstraction gives the major isolated product 65. The other proposed mechanism has aryl-S cleavage to give the sulfine 62, presumably followed by photochemical desulfurization [25,35].

Still later investigated a series of homologous naphthyl compounds. Most of them had only deoxygenation photochemistry, but an exception was found in 66, which showed type B chemistry, even without the carbonyl group of the thiochromanone series [38].

Moving the methyl groups α to the sulfoxide in the thiochromanones gave entirely different chemistry, Type C. The transformation of 68 to 71 is an example. This too could be rationalized by α-cleavage, but hydrogen abstraction mechanisms were also suggested [37]. This reaction is discussed further in Section V.

The structurally similar 72 was also photolyzed during this same time period [39]. The trans isomer is generally more reactive than the cis. It suffers epimerization, which is assumed to be α-cleavage and recombination. It also suffers from an unusual apparent disproportionation reaction, as the corresponding sulfone is observed as a significant fraction of the product mixture. However, the major identifiable product was found to be the dimer 77. Labeling of the sulfoxide with ^{18}O showed that each carboxylic group contained one sulfoxidic oxygen atom [39]. The proposed mechanism involves α-cleavage; loss of benzaldehyde from 73 gives a diradical 74, which can be thought of as a resonance form of the *ortho*-quinosulfine. The product distributions obtained with several sensitizers are also reported.

The loss of benzaldehyde can also be written as a concerted reaction, and there is no experimental evidence to distinguish the possibilities, aside from the preponderance of cleavage reactions for related substrates. However, a concerted loss of benzaldehyde would require either a simultaneous α-cleavage reaction or a direct pyramidal inversion to account for the isomerization reaction. The isolated product, 77, is presumed to derive from secondary photolysis of the mixed sulfenic/carboxylic anhydride 75.

The photolysis under direct and sensitized conditions of several simple dialkyl and alkyl aryl sulfoxides was reported by Shelton and Davis in 1973 [14]. Among these was phenyl *tert*-butyl sulfoxide 78. We will not discuss the results with sensitizers at length here, because we believe the mechanism of sensitization is unclear and some reactions had decidedly different product distributions as a function of sensitizer.

After extended photolysis in benzene through a Pyrex filter, *tert*-butyl alcohol and acetone were both observed in modest yield. They were postulated to derive from secondary chemistry of the unobserved sulfenic ester 79, which produces the *t*-butoxyl radical 81. Traces of phenyl disulfide 46 were also found, along with a significant amount of the sulfide 80, derived from deoxygenation.

In addition to the product studies, some direct physical evidence for α-cleavage was produced in the mid to late 1970s. First among these was a microsecond flash photolysis study that included dibenzyl sulfoxide, diphenyl sulfoxide, and di-*p*-tolyl sulfoxide [40]. The well-known benzyl absorption was observed on photolysis of dibenzyl sulfoxide, and new nearly identical transients (with maxima at 312 and 420 nm) were observed for the two diaryl sulfoxides. The new transients were sensitive to O_2 and assigned to the corresponding sulfinyl radicals but not otherwise characterized.

More convincing direct evidence for α-cleavage came from electron paramagnetic resonance (epr) detection of sulfinyl radicals. First observed by photolysis of phenylsulfinyl chloride at low temperature, the phenylsulfinyl radical 40 shows extensive benzylic-type delocalization, having larger coupling constants at the ortho and para protons than the meta [41]. Photolysis of diphenyl and related sulfoxides at low temperature in toluene was subsequently shown to generate appropriate sulfinyl signals, and the phenyl radicals were later visualized with nitrone spin traps [42,43]. While it is certainly true that epr detection does not prove that the radicals are the major species formed on photolysis, the combination of this with the overwhelming product analysis data now available is thoroughly convincing.

A series of papers on the photochemistry of a set of aromatic sulfoxides by German and Israeli authors relied very heavily on CIDNP results. A few of the conclusions reached in these papers may be over-reaching in retrospect, perhaps because of the limitations of the CIDNP technique, some unknown-until-later triplet energies, and the lack of product analysis in some cases. Nonetheless, this work gives very good evidence for the participation of homolytic α-cleavage.

First to appear was the polarized NMR spectrum obtained on benzophenone sensitized photolysis of 2-(methylsulfinyl)-biphenyl 85, which contained signals attributed to both methane and ethane [44]. This and all subsequent strong polarizations were derived from triplet radical pairs. A methane yield of 30% was found for 85. The reactions were carried out in C_6D_6, yet the observed methane was CH_4, not CH_3D, implying that the hydrogens came from other substrate molecules. No toluene was observed. Surprisingly, the methane observed in $CDCl_3$ was also CH_4, but that in CD_3OD was CH_3D. Methanol-d_1 was not used, as the main thrust of the work was NMR data.

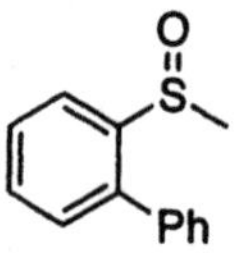

85 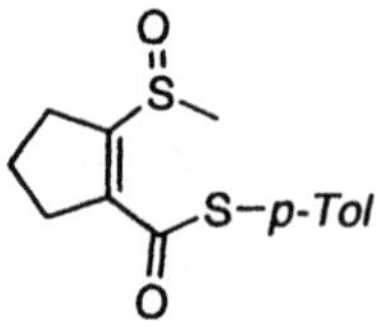   86 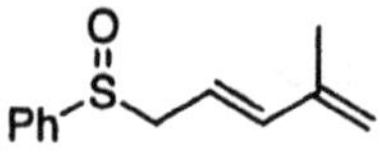   87

Several other ortho substituents (carbonyls, NHCOR) and CH_3S in any position led to similar spectra; these were significantly stronger than those of the unsubstituted parent (which was unlikely to have been triplet sensitized by benzophenone) and several other derivatives (ortho halogens, CH_3, CO_2H, RO, NH_2; para-CH_3, Br) [45]. Neither did this reactivity depend on the aromaticity of the sulfoxide, per se. The vinyl sulfoxide 86 showed a similar pattern of polarized products. Photolysis of the phenyl pentadienyl sulfoxide 87 gave rise to CIDNP signals for both the starting material and the 1,5 rearranged product. There was no indication of the 1,3 rearrangement product.

A very interesting photochemical transformation, illustrated below, of 1,4-dithiin sulfoxide 88 was also attributed to α-cleavage. Thermolysis of these compounds in polar solvents was known to yield a net extrusion of SO, forming thiophenes. Accompanying the extrusion was a ring-contracting rearrangement to 92, which dominated in nonpolar solvent. On photolysis, the extrusion is not observed, but a second ring-contracting rearrangement to 96 is [46,47].

The proposed mechanism proceeds through sultene 90 or 94, presumably arrived at through homolysis, though a concerted mechanism was also suggested. One of the sultenes was later isolated and characterized in low yield by another worker [48]. Most authors would now agree that a second photon is required to provide for sultene homolysis.

Kobayashi and Mutai also examined the photochemistry of the tosyl sulfilimines analogous to 88 (formed by substitution of TsN for O) [47]. Because S–C α-cleavage chemistry was observed, this is of some note. Previous studies of sulfilimine photochemistry, though limited, were usually done with S-dialkyl-N-tosyl or N-acyl compounds. In those compounds, the tosyl or acyl group was the chromophore, and the isolated products were more consistent with S–N cleavage to form a nitrene [49].

Sulfenic esters and sultenes are relatively stable thermodynamically, lying only several kcal/mol over the sulfoxides, all other things being equal [50]. However, in the laboratory, sulfenic esters are very difficult to handle without significant decomposition, in the absence of stabilizing substitutions [51]. Their absorption spectra often extend to the red of the isomeric sulfoxides, which contributes further to difficulty in their isolation by way of sulfoxide photochemistry. Thus it is exciting that two independent laboratories isolated stable sultene derivatives in the 1980s, derived from sulfoxide photolysis.

Capps et al. were able to prepare 2,1,4-oxathiazolidines 99 and 102 in good yield, and characterized 99 by x-ray crystallography [52]. The most reasonable mechanism for formation of these compounds involves α-cleavage to 98 or 101 and reclosure of the biradical on "the other side" of both centers. Margaretha showed that sulfinyl enone series 103 isomerizes in high quantum yield to the sultine 105 [53]. If cleavage α to the carbonyl occurs, it is to a minor extent and does not contribute significantly to the product mixture (90% 105a).

Margaretha's group followed up on these results some years later, studying the photolysis of 106 and related derivatives [54]. Like some of Still's work, it would appear that the lowest excited state ought to be associated with the

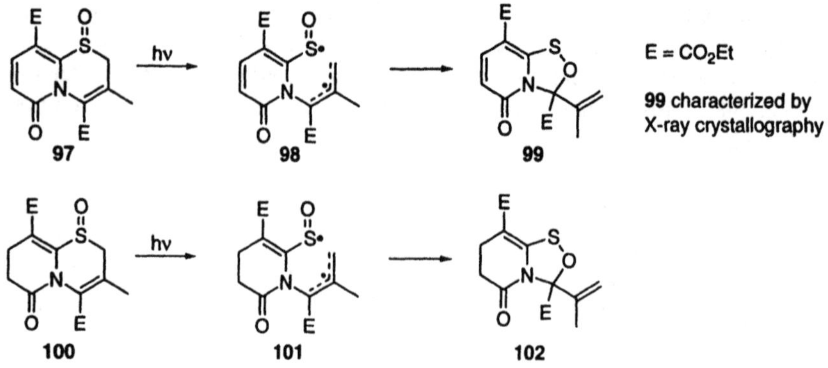
hv
E = CO_2Et
99 characterized by
X-ray crystallography
97
98
99
hv
100
101
102

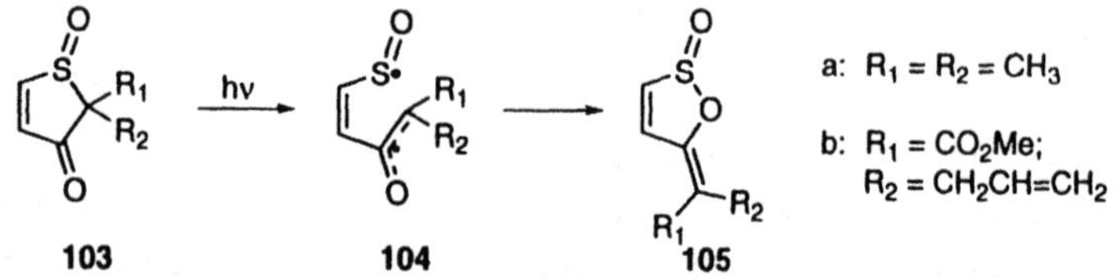
hv
a: $R_1 = R_2 = CH_3$
b: $R_1 = CO_2Me$;
$R_2 = CH_2CH{=}CH_2$
103
104
105

carbonyl, but nonetheless the action occurs at the sulfinyl group. The primary photochemical product is the sultene 107, which can be observed in good yield. Warming of 107 provided 109. The suggested mechanism for its formation started with an adaptation of the well-known [2,3] sigmatropic rearrangement of allylic sulfenates to give the unobserved 108. Loss of water was presumed to occur with acid (or other electrophilic) catalysis, along the lines of the Pummerer rearrangement.

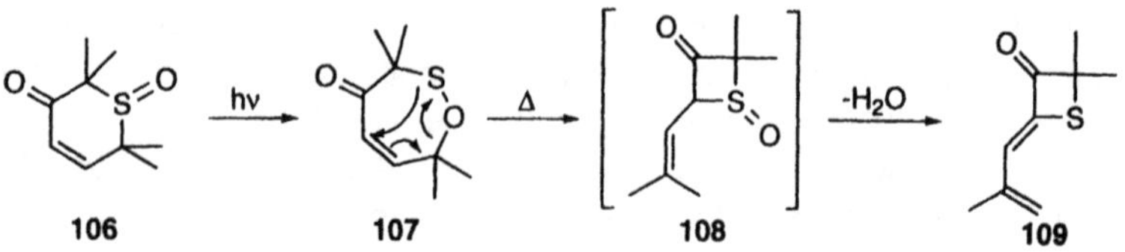
hv
Δ
$-H_2O$
106
107
108
109

Kropp and coworkers have been concerned with systems in which homolysis is followed by rapid electron transfer, such as the photolysis of certain alkyl halides [55]. Photolysis of the norbornyl phenyl sulfoxide 110 was examined. The notion was that sulfide photolysis is usually homolytic, but the sulfinyl radical is more electronegative than the sulfenyl (thiyl) radical, which might assist in electron transfer reactions. Thus, it was thought, ionic reactivity might be observed. In addition to inversion of the sulfur center and deoxygenation (*vide infra*), norbornane 111 and norbornene 112, both presumed to be due to homolytic reactions, were found. A trace of nortricyclene 113, the sought-after ionic product, was also observed. Photolysis in methanol led to a different mix of the two sulfoxide diastereomers and slightly more ionic-type products (4%).

In sulfoxide photochemistry, products that appear to derive from ionic intermediates (to the exclusion of radical intermediates) are rare indeed. Even here, the yields Kropp found were low. Perhaps they were only found because Kropp's group sought them. Regardless, the mass balances of many reactions of sulfoxides are low, and such homolysis/electron-transfer mechanisms that simply yield unobserved compounds cannot be ruled out. Results of photolysis of DMSO in water (*vide infra*) appear to follow just such a mechanism. Thus it should be something to consider for reactions in very polar media.

The 4-sulfinyl benzophenone 114 have been investigated as a polymerization initiator and by steady state and flash photolysis [56]. By analogy to halogenated benzophenones [57], aryl–S cleavage is expected in this reaction, although by analogy to the CIDNP work of Lüdersdorf et al. (*vide supra*) [31,44,45]. CH_3–S cleavage is also plausible.

In the absence of a triplet quencher (R = CH_3, acetonitrile), photolysis leads to isolation of 121 and the thiosulfonate 119 [56]. Addition of a triplet quencher increases the yield of 121 but severely reduces 119. An interaction of the excited state(s) with triethylamine observed in the transient absorption experiments is not chemically explained but could result in the formation of the aryl radical 120 and CH_3SO^- by $S_{RN}1$ type chemistry. The flash data are especially promising, as real kinetic treatments are greatly lacking in sulfoxide photochemistry. Nonetheless, there may be problems with the current interpretation, such as a reliance on an unrealistically long lifetime of S_1 for the benzophenone derivative 114 (ca. 3 ns).

The authors account for the higher yield of benzophenone in the presence of quencher with the following postulated pathway [56]:

An alternate, perhaps more plausible, explanation for the increase in quantum yield is within the pathways.

Furukuwa has shown that certain sulfoxides can be used as carbonyl synthons [58]. Photolysis of several derivatives of 123 in benzene gave near quantitative yields of the corresponding carbonyls and the aromatic heterocycle 125. The derivatives 130 and 131 were shown to be unreactive under the same conditions, which probably has to do with the absorption spectra. Photolysis of 132 also led to reaction but gave an inseparable mixture of products. The tosyl sulfilimines 126 have also been examined by the Furukawa group [59].

In that case, the authors tracked the formation and decay of the cyclic *N*-tosylsulfenamide 127 on the way to formation of the aldimines 129 in good yield.

These authors regard the sultene 124 as a reactive intermediate and attribute special effects to an S–S through space interaction. In the sulfilimine case, the formation and decay of 128 was observed, but formation of the cyclic compound was again attributed to "migration . . . via the S–S through space interaction." More likely, this is just a beautifully designed benzylic, cyclic system with a low energy chromophore. All of the results can be explained by the usual assumption of a homolytic α-cleavage followed by efficient closure. However, another very interesting question is whether the formation of the S–S-bond is simultaneous with or subsequent to elimination of the stable carbonyl or aldimine.

Finally, the Jenks group examined α-cleavage of some simplified systems in order to sort out some of the points that remained unsettled. Acyclic substrates were chosen to eliminate ambiguity regarding homolysis and/or electron transfer in the formation of the sulfenic ester. Reactions were generally carried out to low conversion. Photolysis of benzyl phenyl sulfoxide 133 in the solid provided either the sulfenic ester 134 or benzaldehyde and thiophenol, depending on the wavelength [60]. Use of longer wavelengths selectively photolyzes the ester, making its observation impossible. With excitation further to the blue (e.g., 254 nm), this is much less of a problem. Similar to the photolysis in solid, 134 was the near exclusive product in fairly viscous solvents like *tert*-butyl alcohol. In acetonitrile, small amounts of these escape products like the thiosulfonate 135 and bibenzyl 136 were observed. Photolysis in acetone as triplet sensitizer dramatically increased the yield of the escape products, which led the authors to conclude that the initial homolysis is largely a singlet process. Interestingly, photolysis of the sulfenic ester 134 did not provide any of the sulfoxide.

Subsequently, the Jenks group has worked with alkyl aryl sulfoxides where the α-cleavage would produce less favorable radicals: primary, secondary, and tertiary alkyls [61]. The essential results of this study are that secondary and tertiary S–C cleavage is strongly favored over Ph–S cleavage, but primary C–S cleavage is only competitive with Ph–S cleavage. This is qualitatively in agreement with Still's work, though the elaborate effects in the thiochromanone series are subtle. Sulfenic esters are major products of aryl (secondary- and tertiary-)alkyl sulfoxides, but alkenes formed by disproportionation of the sulfinyl/alkyl radical pair are also formed. The relative yields of the different alkenes serve readily to distinguish this mechanism of olefin formation from the well-known electrocyclic thermal elimination. Furthermore, the quantum yields for loss of starting material across the series range predictably down from above 0.2 for aryl benzyl systems to about 0.04 for aryl primary-alkyl (e.g., 3-CH_3) and diphenyl sulfoxide.

C. Extrusions

Photoextrusion of small stable molecules is a well-known phenomenon, with SO_2 being one of the standard leaving groups for that type of reaction. Photochemical extrusion of SO is also known, though less common, and is reviewed here. Loss of SO appears to be an uncommon process of sulfinyl radicals, at least near room temperature. This was foreshadowed by the earliest photolysis of dibenzyl sulfoxide (*vide supra*), in which sulfenate-derived products clearly dominate any loss of SO from the benzylsulfinyl radical, in marked contrast to the ketone case. The loss of SO from $CH_3SO\cdot$ is estimated to be endothermic by 50 kcal/mol [62,63], so it is clear that near simultaneous formation of a stable structure in the carbon portion of the molecule is a critical component in the design of extrusion reactions.

Being isoelectronic with O_2, the ground state of SO is a triplet. Also like O_2, there exists a very low lying excited singlet state. The state of SO when lost is thus a chemical variable of interest. Only the earliest report gives any significant clue as to whether these extrusions are concerted or stepwise—stepwise appears to be the answer—and the electronic state of the nascent SO is unknown.

The first example of SO photoextrusion to the best of our knowledge is that of Kellogg and Prins in 1974 with the dihydrothiophene derivatives 137 [64]. The analogous sulfones undergo stereospecific concerted thermal extrusion but were found photochemically to lead to some loss of stereochemistry in the diene, attributed to stepwise α-cleavage. The sulfoxides also quantitatively gave dienes 138 of mixed stereochemistry. This was taken to imply a biradical intermediate, presumably the initial allyl-sulfinyl biradical. No intermediates or other products (e.g., vinyl episulfoxides, sultenes) were observed. In contrast, episulfides *were* observed for photolysis of the analogous sulfides.

Two related systems have been examined in more recent years. Sulfoxide 139 was seen as an approach to the desired system 140 and indeed provided a low yield of that material (9%) on irradiation through Pyrex [65]. The major products were the isomeric bicycles 141 and 142. Direct or sensitized irradiation of 143 gives a mixture of 144 and 145 [66]. While the diene was shown to be a photochemical precursor to the arene, it was also shown that another oxidation process occurred, attributed to the presence of SO and/or its degradation products. Despite tertiary photochemistry of the arene, conditions were found that brought its yield to 80% [66].

Sulfoxide 146 and related derivatives were prepared in the late 1970s with the apparent expectation that rearrangement to the sultene would be observed. (The substituted compounds were apparently one diastereomer, but which one was not identified.) Photolysis of 146 (R=H) instead yielded the cyclopropane 149 in near quantitative yield. The initial photochemical step is almost certainly β-cleavage ketone chemistry, but it generates the same allylic-type biradical 147 as "true" sulfoxide α-cleavage. The authors suggest that formation of the sultene is followed by a second photolytic reaction resulting in loss of SO. Aside from the lack of evidence for such an intermediate, a homolytic C–O cleavage would generate the same intermediate as the first α-cleavage; only a concerted loss of SO from the sultene and the unprecedented stepwise cleavage of C–S then C–O represent alternate paths. With alkyl substitution at R, enone 150 was also observed. Given the energetics of loss of SO from a sulfinyl radical, biradical 148 may not be a discrete intermediate, and the products may arise from a direct S_R2-type process.

Diphenyl thiirene oxide 151 is a surprisingly thermally stable compound, whose chemistry was described in 1979 [67]. Photolysis near room temperature yields diphenyl acetylene in nearly quantitative yield. Whether this extrusion is concerted or stepwise and the state of the eliminated SO awaits further investigation.

IV. STEREOMUTATION

The sulfoxide center is stereogenic if it has two different carbon substituents. It has long been known that aromatic sulfoxides are subject to stereomutation under photochemical conditions, though the number of studies that treat it directly is limited. However, stereomutation should be an anticipated result of photolysis of aromatic sulfoxides. When there are diastereomeric interactions that favor one isomer over the other, the reaction may have synthetic value. This has been more notably demonstrated in a few dialkyl cases (*vide infra*).

The pioneering work in photostereomutation was done by Mislow and Hammond, with Mislow's group having carried out the most important work in the thermal stereomutations [68–72]. Direct irradiation of (S)-naphthyl tolyl sulfoxides 153 through a Pyrex filter yielded completely racemized starting material (70%) and unidentified products [73]. Direct irradiation of either isomer of thianthrene-S,S'-dioxide 154 resulted in isolation of the cis isomer.

Hammond, Mislow, and coworkers extended their work to alkyl tolyl sulfoxides, particularly with naphthalene sensitization, both inter- and intramolecular [74–76]. In general, direct irradiation caused substantially more decomposition than the sensitized cases. In the sensitized case, correlation was made

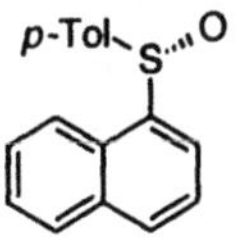

153 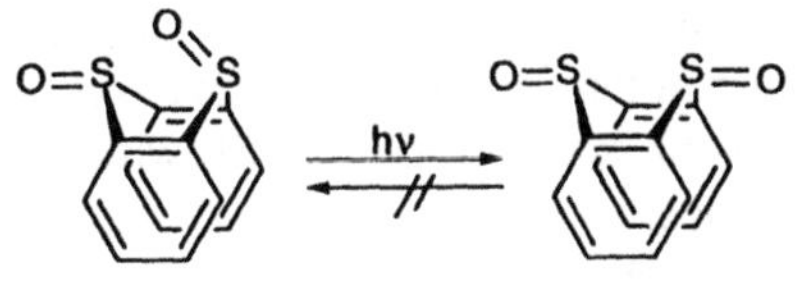   **trans-154**   **cis-154**

between the racemization reaction and relatively slow rate constants for quenching of naphthalene singlets [75,76]. It was recognized that the singlet and triplet states of naphthalene are both lower in energy than the respective states of the sulfoxides, and the suggestion was made that the active intermediate was an exciplex [75,76]. Recent work lends support to an exciplex and/or electron transfer hypothesis in which negative charge is transferred to the sulfoxide [77].

When there is a second stereogenic center in the molecule, stereomutation of the sulfoxide center yields a diastereomer, rather than an enantiomer. Kagan extended this concept to the supramolecular case and found that use of a chiral naphthalene derivative as sensitizer induces a modest enantiomeric excess in racemic sulfoxide substrates [78]. A unimolecular case was Kropp's phenyl norbornyl sulfoxide. Though other photochemistry accompanied the stereomutation (*vide supra*), a ratio of 2:3 for 110 to 155 was obtained on photolysis in THF. Another product in that mixture that is important to this discussion is the isomer 156, found in small yield.

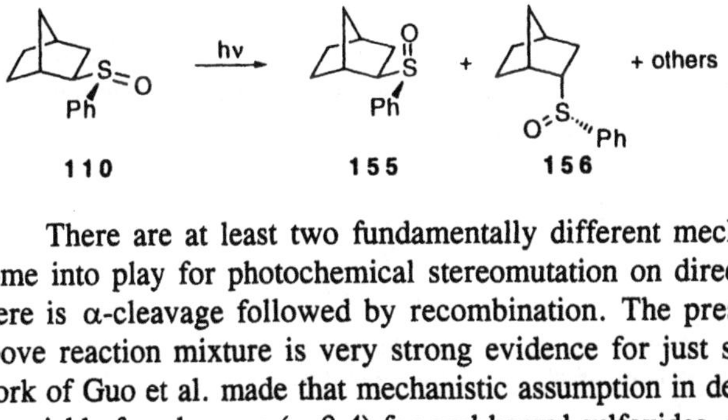

110   **155**   **156**

There are at least two fundamentally different mechanisms which may come into play for photochemical stereomutation on direct irradiation. First, there is α-cleavage followed by recombination. The presence of 156 in the above reaction mixture is very strong evidence for just such a process. The work of Guo et al. made that mechanistic assumption in determining the quantum yield of α-cleavage (~0.4) for aryl benzyl sulfoxides, as the quantum yield for loss of optical rotation was taken for the quantum yield of cleavage [60]. This is supported by the fact that homolytic cleavage is also the mechanism for thermal racemization of the same compound [71].

However, there may be another mechanism that is at least in competition with cleavage/recombination, as suggested by Schultz and Schlessinger (section IIIA, compound 48). This could involve inversion/racemization with little or no barrier in the excited state before otherwise "nonproductive" internal conversion down to the ground state (similar in principle to cis/trans isomerization of alkenes). Alternatively, internal conversion could lead to a hot ground state reaction. The thermal barriers for diaryl and aryl alkyl sulfoxide inversion are in the range of 35–40 kcal/mol [68,71,72], which is significantly less than the excited state energies.

Evidence for such an additional mechanism also comes from the work of Guo, in the form of the quantum yields for loss of optical activity and for loss of starting material for compounds 157, 158, and 3-CH$_3$ [61]. It is unlikely that the quantum yield for cleavage by methyl p-tolyl sulfoxide is higher than that for the benzyl compound, and that the sulfinyl methyl pair would be so over-whelmingly returned to the sulfoxide, as compared to escape products or sulfenic ester formation. Current evidence does not exist to establish the actual mechanism(s) firmly.

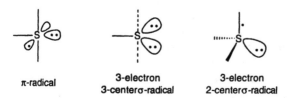

	157	158	3-CH$_3$
Φ(loss of optical rotation)	0.42	0.81	0.83
Φ(loss of starting material)	0.21	0.04	0.04

V. HYDROGEN ABSTRACTION

Probably in analogy to the well-known hydrogen abstraction reactions of car-bonyl and nitro compounds, hydrogen abstraction has been proposed as a pri-mary process of sulfoxide excited states. It is safe to say that while some of these suggestions appear quite reasonable, the actual evidence for hydrogen abstraction is much thinner than for α-cleavage.

π-radical	3-electron 3-center σ-radical	3-electron 2-center σ-radical

The trivalent intermediate implied by hydrogen abstraction is a 9-electron sulfuranyl radical with OH as one of the substituents (159). Sulfuranyl radicals are well known species whose reactivity generally involves loss of one ligand as a radical [79], though other types of reactivity are also known [80,81]. Per-sistent sulfuranyls and their epr spectra have been reported [82]. The evidence indicates that the geometry of trialkoxy- and dialkoxyalkylsulfuranyl radicals is a slightly distorted T-shape. The radical is a π-type, with the unpaired electron residing in an orbital derived from the sulfur p orbital perpendicular to the plane of the ligands [82,83]. The epr of an alkoxydialkyl sulfuranyl radical is more consistent with a σ-type radical in a similar T-shape geometry, best viewed as a trigonal bipyramidal structure with one electron removed from the usual 3-center 4-electron bond of the two apical centers with the sulfur [84]. Other sulfuranyl radicals, typically thiodialkyl systems, have epr spectra that have been interpreted as pyramidal structures with a two-center, three-electron (2c-3e) bond [81].

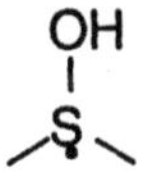

OH
|
S
/ \

159

Radicals such as 159 are postulated intermediates in the oxidation of various sulfides by HO· [85], though the existence of the parent case 159 as a real intermediate or a transition state has been a matter of some controversy, partly due to a lack of direct spectroscopic detection [86,87]. Calculations reported in 1993 find a minimum at the MP4/6-31G(2d) level, but fail to find a stable adduct at MP2/6-31G(d) [88]. The structure appears to be a pyramidal (2c-3e) radical with a very long SO bond (2.05 Å) with spin delocalized over the S and O. Experimental and theoretical values agree that 159 is bound, relative to the sulfide and HO·, by 15 kcal/mol or less. For purposes of this review, it will be assumed that formation of dialkylhydroxysulfuranyl is plausible. However, the π system of the aromatic substitution might influence the nature of the radical, and qualitatively different chemistry might be plausible from different types of systems. We now discuss specific reactions in which hydrogen abstraction has been suggested or implicated.

An early review of the photochemistry of sulfur compounds [1] presented some unpublished results that, to the best of our knowledge, were never found later in the primary literature. It was suggested that 160 underwent a reaction that was analogous to the Norrish II cleavage process.

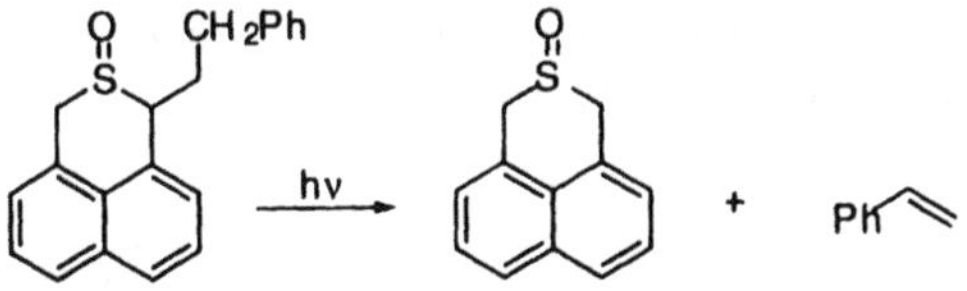

160   **161**   **162**

One reaction that might be attributed to hydrogen abstraction is the intramolecular redox reaction illustrated below [89]. Lüdersdorf and Praefcke reported that the benzoic acid 165 is isolated in modest yield on photolysis of the *ortho*-sulfinylaldehyde 163. When R = CH$_3$, the product is accompanied by significant deoxygenation. When R = *para*-tolyl, the major isolated product was 2-phenylbenzaldehyde, derived from an α-cleavage reaction followed by addition to solvent. An alternative formulation to explain the oxygen transfer reaction was put forward by the authors. It is a sequential two-step electrocyclic process, in which the first step is assumed to be photochemical. The well known photochemistry of ortho alkyl acetophenones involves formation of the photoenol in the triplet state, distinct from the singlet. The two possible intermediates drawn below (164a and b) can be seen as resonance forms of one another, so this may really be a point of semantics.

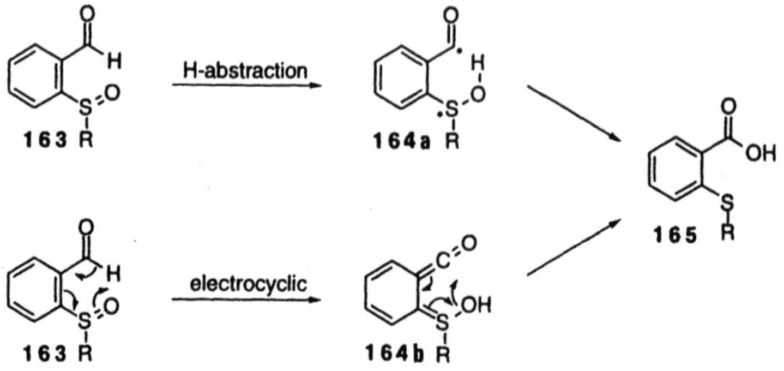

A very low chemical yield of *ortho*-(methylthio)-benzaldehyde was obtained from photolysis of *ortho*-methylsulfinylbenzyl alcohol [89], but other relevant experimental parameters (e.g., comparative quantum yields) were not reported. The analogous reaction does not occur with *ortho*-tolyl methyl sulfoxide. The reaction of 163 also proceeds under acetone sensitization.

An early report of sulfoxide photochemistry demonstrated that 2,2-dimethylthiochroman-*S*-oxide 166 was converted to 2-isopropylbenzothiophene 167 [90]. The suggested mechanism involved β-hydrogen abstraction as the primary photochemical event. Structure 166 is redrawn in a projection that shows the relationship between the O and the H that is abstracted. Most of the benzo group is removed for clarity. A conformational search on 166 using the PM3 model suggests that there are two nearly isoenergetic conformations, one of which is represented reasonably well by this drawing.

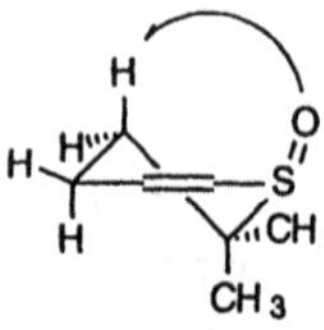

166
(benzo group deleted for clarity)

The proposed mechanism requires the formation of a strained tetravalent sulfur center (169), which ultimately leads to the ring contraction. Tetravalent sulfur is known to exist in equilibrium with sulfuranyl centers in certain cases [91], but the structural relationship between those and the present cases is tenuous. A second mechanism also accounts for the formation of 2-isopropyl-thiophene. Here, hydrogen abstraction is replaced by α-cleavage and disproportionation to form an unsaturated sulfenic acid (172). Formation of the five-membered ring from the olefin and sulfenic acid is well precedented under acid-catalyzed conditions [92,93]. (Both the current and the original mechanism have steps that are almost certainly catalyzed by adventitious acid.) Even an acetic anhydride-catalyzed thermal version of the same rearrangement [94] can be cast in terms similar to the closure of 172. It is also possible that 172 is subjected to S–O homolysis in a second photochemical step, leading to the exocyclic alkene by way of radical intermediates. Presumably acid-catalyzed rearrangement would give the final product.

A similar reaction, conversion of 173 to 177, was observed by Schultz and Schlessinger [95]. A labeling study showed that only the hydrogen that was proposed to be abstracted was lost. Again, an α-cleavage based mechanism can also account for the experimental results.

A structurally related case has been reported where rearrangement of the first ring-contracted alkene is not possible [35]. Still and coworkers found that photolysis of 68 produced a modest yield of the ring-contracted product 71. However, no methanol trapping of a cation (e.g., 184) was observed. This result supports the notion that ring closure is due to photo-homolysis of the sulfenic acid (e.g., 70, 172, 178), and that the methanol trapping may be observed as a result of acid-catalyzed rearrangement of the alkene. All of the observed products in the photolysis of 68 can be accounted for from the two arenesulfenic acids 70 and 181, both of which could arise from a proposed hydrogen abstraction or the corresponding α-cleavage.

Given that the reactions discussed in this section can be rationalized using either hydrogen abstraction reaction or α-cleavage, the Jenks group has done some investigation into systems in which predictable products might be unambiguously attributed to the former [61]. The reactivity of the putative sulfuranyl radical was speculated upon, and it was decided to concentrate on possible intramolecular oxidations brought about by two-step hydroxyl transfers. Electrocyclic reactions had to be unachievable. Given previous suggestions (and the notorious intramolecular charge transfer quenching of ketones [96]), systems were examined in which β- or γ-hydrogen abstraction might be favored. Some of these possible reactions are illustrated below.

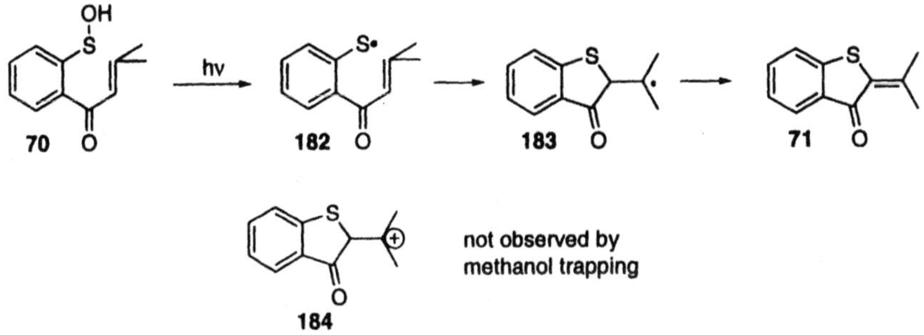

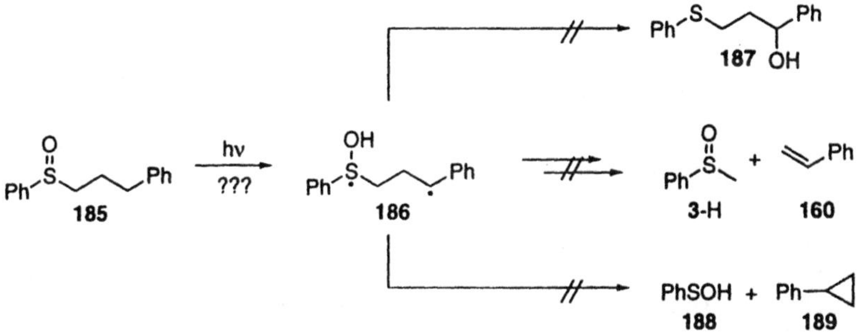

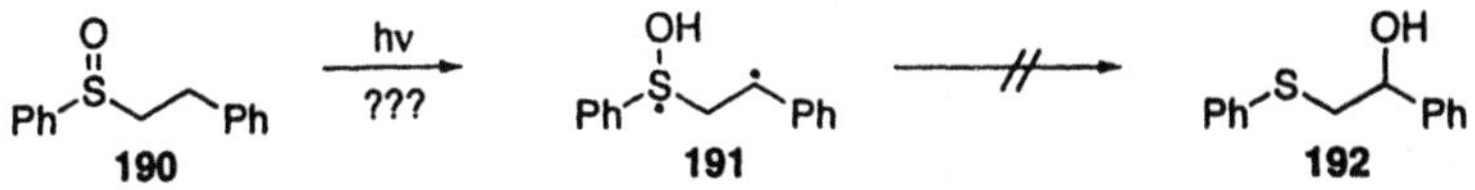

To date, all results of these studies have been negative. We are skeptical that hydrogen abstraction is truly a significant primary photochemical process of aromatic sulfoxides, but this remains an open question.

VI. DEOXYGENATION

A. Mechanistic Issues

The reduction of aromatic sulfoxides to sulfides is perhaps the single most interesting photochemical reaction of sulfoxides because it has no analog in ketone chemistry. Its mechanism is still unsettled. However, it will be stipulated for purposes of this discussion that there exists a mechanism for most of these reactions, which is not a disproportionation reaction. Authors who have investigated the reaction for aromatic sulfoxides have consistently reported that sulfone is not observed. (An exception to this has been noted, but no mechanistic study on that issue was performed [39]. Disproportionation has also been observed for concentrated solutions of dialkyl sulfoxides, as discussed in Section VIIIB.)

This discussion will center on the mechanism under conditions of direct irradiation because of the complications that arise on interpretation of some of the sensitization experiments [20,75–77]. However, we note that many of the systems discussed here show deoxygenation via sensitization with molecules that are traditionally thought of as triplet sensitizers.

Two mechanisms have appeared in the literature to account for sulfoxide deoxygenation. The first of these, to which we shall refer as "the dimer mechanism," was suggested nearly simultaneously by Shelton (explicitly) [14] and Posner (implicitly) [97]. It is outlined for a generic sulfoxide below. The excited sulfoxide triplet is trapped by another (ground state) sulfoxide molecule to produce a dimer (193) that contains an O–O bond. Posner suggested that the reaction produced singlet molecular oxygen, $^1O_2(^1\Delta_g)$, though he was not specific about the structure of a dimer or other intermediates. Shelton's mechanism assumed that ground state O_2 was produced.

The second mechanism in the literature was suggested several years later on the basis of CIDNP investigation of the photochemistry of aryl methyl sulfoxides [31,45]. It will be referred to as "the sulfinyl mechanism." The key step in the sulfinyl mechanism is the reaction of a free sulfinyl radical (formed by α-cleavage) with another radical to effect O atom transfer.

The methyl groups of the resulting aryl methyl sulfides 198 consistently showed an emissive CIDNP signal, while the aromatic protons were unpolarized. This was interpreted to signify that the sulfide was formed "as an intermediate result of the primary photochemical event, by the escape path of [the radical pair 195/84] [31]. The sulfinyl radical was said to (1) transfer its O–

atom, (2) lose its nuclear polarization, and (3) recombine with a spin polarized methyl radical to give the corresponding sulfide.

In spite of the CIDNP polarization pattern, we believe the sulfinyl mechanism can be dismissed. First, the SO bond in a sulfinyl radical is very strong. Using Benson's estimate for the heat of formation of the phenylsulfinyl radical (13 kcal/mol) [63] and standard values for the other relevant compounds [98], the S–O bond energy is ca. 102 kcal/mol, whereas the C–S bond is some 35 kcal/mol weaker. Transfer of an O atom from phenylsulfinyl to a methyl radical is endothermic by 11 kcal/mol, and to epoxidize ethylene endothermic by 40 kcal/mol. (The relevance of the latter example will become clear below.) Furthermore, from the α-cleavage work discussed previously, it is clear that the expected product from reaction to an arylsulfinyl radical and a carbon radical is a sulfenic ester or disproportionation product.

The Jenks laboratory has investigated two other mechanisms for sulfoxide deoxygenation, but all of the work has been on dibenzothiophene sulfoxide 9 [99,100]. It is conceivable, given other properties of this molecule [101], that 9 is an exceptional case. First, one additional considered possibility is that the sulfoxide undergoes a hydrogen abstraction, followed by hydroxyl transfer by 199 to the resultant solvent radical.

Another mechanism is proposed [100] in which the sulfoxide cleaves off an O atom directly out of the excited state. This produces the sulfide immediately. The O atom and sulfide exist in a solvent cage, so there will be competition between geminate recombination and escape of the O atom from the cage. Products isolated from solvent oxidation ought to reflect oxidations brought about by the O atom.

Here we consider the evidence for and against each of the three remaining mechanisms. Shelton and coworkers suggested the dimer mechanism based on the sulfur-containing products and because many of the reactions were run with traditional triplet sensitizers [14]. Posner found that, in nonaromatic solvents, several aromatic sulfoxides gave good yields of the corresponding sulfides [97]. Benzophenone ($E_T \sim 69$ kcal/mol [102]) was found to sensitize the deoxygenation of diphenyl sulfoxide ($E_T \sim 78$ kcal/mol [20]). It was also found that piperylene prevented the deoxygenation of diphenyl sulfoxide.

Some years later, Geneste and coworkers reported the photochemistry of several benzothiophene derivatives [103]. Photolysis of 2-methylbenzothiophene sulfoxide 200 leads to sulfide in modest yield and tars but is accompanied by none of the [2+2] olefin chemistry observed with other related derivatives. The dimer mechanism requires a dependence of the quantum yield on the sulfoxide concentration due to competition between unimolecular deactivation and bimolecular reaction. The quantum yield for deoxygenation of 200 was found to vary from about 0.03 to 0.08 over a concentration range of 2–9 mM. The data gave a linear double reciprocal plot with intercept/slope = 4.3. (For simple bimolecular quenching kinetics, this value is equivalent to $k_q\tau$). The fit of the quantum yield data to a double reciprocal plot is consistent with the dimer mechanism but was somewhat balanced by ambiguous results for cyclooctadiene quenching experiments in which no relationship between quantum yield and quencher could be established.

200   **201**   **202**

Evidence for the formation of singlet oxygen from Posner was based on chemical trapping. Photolysis of dibenzothiophene sulfoxide in a 90:10 mixture of cyclohexene and acetic acid provided a sample that tested positive for peroxides. After reduction with NaI, 2-cyclohexenol was obtained in 22–34% yield. The authors noted a lack of cyclohexanone and cyclohexene epoxide. This was rationalized as outlined below [97].

9   **203**   **204**   positive peroxide test

21 - 34% yield accounts for 42-64% of O atoms lost by deoxygenation.

205

Other authors have also tried to trap singlet oxygen. Geneste carried out unsuccessful trapping experiments with 200. Still and coworkers observed "inefficient" deoxygenation chemistry from a series of naphthyl sulfoxides derived from 201. (Unlike the thiochromanone derivatives, these molecules were otherwise photostable.) They explicitly repeated the cyclohexene trapping experiment and did not observe any positive peroxide tests [38]. Photolysis of thiochromanone derivative 202 also resulted in deoxygenation without evidence for $^1O_2$ [35].

Since the report by Posner, it has been found that cyclohexene itself is not a particularly good chemical trap for $^1O_2$, though it is a good physical quencher [104]. The more electron rich methylated homologues are much better chemical quenchers. Furthermore, 2-cyclohexenol is now detected after photolysis of 9, without the NaI reduction step, in yields comparable to those reported by Posner [99]. Minor amounts of the previously unreported cyclohexene epoxide were also found.

Further work from the Jenks laboratory on the deoxygenation has used dibenzothiophene exclusively, and it is possible that the results determined for that substrate are not general. However, this substrate has certain advantages, in that the formation of dibenzothiophene 203 is nearly quantitative and the sulfide is essentially photoinert [97].

Under conditions of direct photolysis, the dimer mechanism is not in effect for the deoxygenation of 9. The quantum yield, lower than that reported for 200, is invariant over the concentration range of 0.6 to 30 mM in benzene or THF. Also, the quantum yield of deoxygenation of 9 is independent of added diphenyl sulfoxide up to about 50 equivalents and no diphenyl sulfide is observed. Finally, the deoxygenation is effected even at concentrations as low as $< 10^{-6}$ M in ether/isopentane/alcohol glass at 77K [100].

The hydrogen abstraction mechanism would be expected to produce quantum yields that correlated qualitatively with the hydrogen donating ability of the solvent. In fact, the quantum yields for dexoygenation of 9 are extremely similar for solvents as varied in this quality as acetonitrile, benzene, toluene, 2-propanol, and tetrahydrofuran. A significant increase in quantum yield was noted for tetrahydrothiophene and cyclohexene, two solvents that would be expected to be more reactive with an electrophilic O atom. Quenching experiments show that a long-lived sulfoxide triplet is not involved [100].

Whatever the mechanism of sulfoxide reduction, oxidized solvent molecules were observed in the photolysis of 9. For example, benzene is converted to phenol, cyclohexane to cyclohexene and cyclohexanol, and cyclohexene to cyclohexene oxide and 2-cyclohexenol [99]. Oxygen atom accounting ranges from fairly poor (ca. 1/3) to quantitative, depending on the solvent substrate. The stepwise and fairly selective nature of the oxidizing agent are suggested by allylic rearrangements in alkene oxidations, loss of stereochemistry in epoxidations, and reasonable selectivity for tertiary centers over secondary over primary in hydroxylations [100]. These results also rule out $O(^1D)$ but are consistent with $O(^3P)$. Further evidence for $O(^3P)$ as the oxidizing agent comes from agreement between ratios of rate constants for the quenching of $O(^3P)$ [105], formed by photolysis of pyridine-N-oxide, and product competition analyses.

A completely different sort of photochemical deoxygenation reaction is suggested by the work of Kropp [55]. Though small quantities of the deoxygenation product of 209 were found on its photolysis in THF (accompanied by trace sulfone) or methanol, a 64% yield of the sulfide was obtained on photolysis in methanol containing 0.2 M NaOCH$_3$. Reductions of (ground state) sulfoxides are known with single electron transfer reagents [106], and it is implied that an electron transfer followed by proton transfer might generate a hydroxyl-substituted sulfuranyl radical, which in turn loses HO·. However, others have noted C–S α-cleavage reactions under photoinduced $S_{RN}1$ conditions [107].

B. Examples

Deoxygenation is most commonly a minor or merely competitive reaction pathway, with 9 being one of the exceptional cases in which it is an essentially quantitative reaction. The earliest examples of aromatic sulfoxide deoxygenation were those of Shelton and Posner, both with simple systems. Shelton showed that t-butyl phenyl sulfoxide 78 produced modest quantities of the sulfide on direct photolysis [14]. Methyl phenyl sulfoxide and diphenyl sulfoxide were other prototypical early examples, and a solvent dependence was suggested [97].

Among the compounds in Still's thiochromanone series, 202 and 68 showed significant deoxygenation [35–38]. Perhaps more remarkable than the observation of deoxygenation in these cases is the *lack* of its observation in the others. Still makes the sensible suggestion that deoxygenation is sort of a "background" process that is observed only when other processes are energetically disfavored [2]. Molecule 202 was cited as an example where other suggested mechanisms were blocked by substitution. However, this compound contains a tertiary center α to the sulfoxide, and cleavage might be expected to be an important process. Perhaps deoxygenation is observed for 68 and 202 because efficient recombination of the biradical formed by α-cleavage wastes many photons, allowing the less efficient primary process to be observed. Compounds 88, 185, 190, 211, and 212 are other compounds whose primary photochemistry is α-cleavage, but where deoxygenation is explicitly reported [31,45, 46,61].

VII. MISCELLANEOUS REACTIONS

There are several reports of reactions that do not fall among the classes of reactivity we have discussed in previous sections. In each of the transformations discussed below, though the sulfoxide is conjugated to the aromatic chromophore, it appears to behave as a "spectator," with its most significant contribution being the provision of an electron withdrawing group.

Common among these are reactions of olefins. Photolysis of benzothiophene-S-oxides 213 provides head-to-head dimers 214 in reasonable yields, depending on substitution [103,108]. Sensitization and quenching experiments indicated that dimerization proceeds through ^{3}213. As alluded to above, 2- and 2,3-substituted isomers also showed some deoxygenation. The dimers were stable to both thermal and photochemical methods of epimerization.

The sulfoxides 215 and 216 also show what is essentially alkene photochemistry [109]. Photostationary states of E/Z isomerization were obtained for
the analogous sulfides and sulfones as well. Interestingly, if the sulfoxide is
replaced by an ether, isomerization is followed by internal ketone hydrogen
abstraction from R and five-membered ring formation.

R = alkyl or aryl

Photolysis of methyl styryl sulfoxide in alcohols provides three types of
products, based on two types of chemistry [110]. One process is deoxygenation
to provide 220. Both it and the starting material are subject to the addition of
alcohols across the olefin. (It is worthwhile to note that conjugation to the low
energy chromophore was necessary for deoxygenation.) In methanol, the two
diastereomers of the alcohol/sulfoxide addition product are formed in 1:1 ratio with a total yield of 65% at complete conversion.

The *ortho*-nitrosulfoxide 221 is converted into the *ortho*-nitrososulfone
222 on exposure to Pyrex filtered light from a high pressure mercury lamp
[111,112]. Like the chemistry of *ortho*-nitroaldehydes or the *ortho*-nitrobenzyl
protecting group, the sulfoxide oxidation is attributed to reactivity of the nitro
chromophore, as opposed to the sulfoxide. This is reasonable based on the
excitation energies of nitroarenes and sulfinylarenes.

Archetypal photochemistry of the aromatic nitro group proceeds through
an nπ* triplet unless there is an electron dnoating group, which leads to a charge
transfer type of excited state [113]. Based on a lack of quenching by dienes or

oxygen, the authors suggested that this is singlet $\pi\pi^*$ chemistry. (However, it should be pointed out that a fast triplet reaction might also not be quenched.) The *ortho*-nitrosulfide does not undergo an analogous intramolecular oxygen transfer.

The photochemistry of *ortho*-sulfinyl benzoates has also been investigated [114]. Compound 223 is converted to *ortho*-(methylthio)-benzoic acid in 77% yield in benzene. The mechanism of this reaction was not well established, but it probably begins with cleavage of the carbonyl–O bond, as the same product was obtained from the thioester and selenoester analogs. In these latter cases, tolyl disulfide and tolyl diselenide were also isolated. It was speculated that the 1-photon product of 223 may actually be *ortho*-(methylsulfinyl)-benzaldehyde 161. As discussed in Section V, this compound is known to undergo photochemical intramolecular oxygen transfer.

Alkyl derivatives of 225 would not produce stabilized phenoxy radicals by carbonyl cleavages, and instead apparently suffer sulfoxide deoxygenation and aryl-S α-cleavage. In 226 the orientation of the ester has been reversed. The photochemical products are attributed to Photo-Fries chemistry and sulfoxide deoxygenation.

Another instance of the sulfoxide as an "interested spectator" in a photochemical reaction is the generation of α-sulfinyl carbenes [115]. Photolysis (or thermolysis) of the sulfinyl substituted pyrazoline 227 was shown by trapping experiments to generate the carbene 228, which undergoes a Wolff-type rearrangement to generate the sulfine. After loss of sulfur, the ketone 230 is isolated in near quantitative yield.

Finally, though not strictly a sulfoxide, the quantitative intramolecular oxygen migration reaction of benzotrithiole-2S-oxide 231 to 232 merits report [116,117]. This reaction is irreversible and not observed thermally. The reaction was proved not to be intermolecular by double label experiments and the benzotrithiole nucleus was found to be necessary by investigation of appropriate compounds. It was not quenched by dienes or oxygen, and the authors propose a singlet mechanism that passes through an oxadithiirene intermediate.

231   232

VIII. ALIPHATIC (SATURATED) SULFOXIDES

The photolysis of alkyl sulfoxides will be considered in two parts—gas phase and solution. Although, by necessity, the chemistry is closely related, the types of questions that have been asked and answered differ enough that it is more straightforward to handle them separately.

A. Gas Phase

Photolyses of several prototypical sulfoxides and related compounds have been carried out in recent years. Generally, 193 or 248 nm excitation has been used. Because of this, the lack of a solvent bath to absorb excess vibrational energy efficiently, and the generally unimolecular nature of gas phase photochemical reactions, the investigations have centered on the types of photodissociations that are observed and whether one or two carbon–sulfur cleavages occur in the primary photochemical event.

Weiner and coworkers have done substantial work with laser induced fluorescence (LIF) in the parent dimethyl sulfoxide and related systems, monitoring the rovibrational state distribution of nascent SO [13]. Following 193 nm photolysis of DMSO, the vibrational state of maximum population for the SO thus formed was $v = 2$. Two overall pathways were considered. The first could produce the methyl radicals in stepwise fashion or in a single concerted step.

$$DMSO + hv \ (193 \ nm) \rightarrow 2 \ CH_3 \cdot + SO \qquad \Delta H_{rxn} = 106 \ kcal/mol$$

$$DMSO + hv \ (193 \ nm) \rightarrow C_2H_6 + SO \qquad \Delta H_{rxn} = 25 \ kcal/mol$$

Given that the photon energy at 193 nm is equivalent to 148 kcal/mol, an inability to detect C_2H_6, and a relatively low, though inverted, vibrational energy of the SO, the concerted formation of ethane was ruled out. On the basis of the specific form of the vibrational distribution, a stepwise loss of two methyls was the preferred interpretation, and excitation into a repulsive electronic state was ruled out. However, the matter of two-body vs. three-body photodissociation was considered unsettled, especially because three-body fragmentation has been proposed for the 193 nm photolysis of acetone in the gas phase.

More detailed investigations [118] included a deeper look at the SO rovibrational spectrum and a resonance enhanced multiphoton ionization (REMPI) analysis of the methyl radicals. These resulted in total rovibrational energies for the fragments. The translational energy component was not measured. The quantum yield for formation of SO in its ground electronic state ($^3\Sigma$) was unity within experimental error. Evidence for two "types" of methyl radicals, as might be expected for stepwise decomposition, was not found, so the authors suggested that the three-body dissociation pathway was dominant [118].

In work that is yet to be published, Ng et al. have used time-of-flight mass spectrometry to detect the translational profiles of the fragments of DMSO photolysis; their analysis favors the two-body dissociation pathway by direct observation of $CH_3SO\cdot$ [119].

Multiphoton infrared excitation of the sulfoxide stretching chromophore (ca. 1100 cm^{-1}) of DMSO also leads to production of CH_3 and SO in their respective electronic ground states [120]. The direct formation of ethane was once again eliminated, this time by product analysis from double label experiments with DMSO and its D_6 isotopomer. Time resolved detection of IR absorptions was used to analyze products. Though, once again, $CH_3SO\cdot$ was not directly observed, these authors favor a stepwise decomposition because of a nonthermal rotational level distribution of the SO fragment.

Because of the two-body vs. three-body dissociation problem with DMSO, and because the system is now understood, we include here a brief discussion of the photolysis of thionyl chloride ($SOCl_2$).

Photofragment mass spectroscopy of thionyl chloride molecular beams irradiated at 193 and 248 nm was studied as early as 1984 [121]. The same three mechanistic possibilities, stepwise decomposition to 2 Cl· + SO (A), concerted decomposition to 2 Cl· + SO (B), and concerted elimination to Cl_2 and SO (C), were all considered.

$$SOCl_2 \rightarrow ClSO\cdot + Cl\cdot \tag{A}$$

$$SOCl_2 \rightarrow SO + 2\ Cl\cdot \tag{B}$$

$$SOCl_2 \rightarrow SO + Cl_2 \tag{C}$$

Neither Cl_2 nor $ClSO\cdot$ were detected as "molecular" ions, but their respective fragment ions (Cl^+ and SO^+) were assigned based on the time-of-flight (TOF) data. It was suggested that all three mechanisms occurred, with (A) being major at 248 nm photolysis and (B) major at 193 nm. However, it was asserted that the probable detailed mechanism for the three-body decomposition was via the formation of vibrationally very hot Cl_2, which decomposed almost immediately. The TOF fitting routine did not handle the three-body decay well, but it was shown that the data could not be fitted using combinations of the two two-body dissociations [121].

Thionyl chloride was reinvestigated with a high resolution system with a rotatable source [122]. It was found that the molecular channel C at 248 nm was dominated by the formation of SO in its lowest excited singlet state ($a^1\Delta$), 18 kcal/mol above the ground state. (Similar observations had previously been made for 193 nm photolysis [123]). The most abundant primary fragments, however, were due to channel A. With 193 nm excitation, channel C was observed, but unimportant. At this wavelength, the TOF data could not be accommodated without having channel B, the three-body dissociation, as the dominant pathway. An energy accounting by LIF study of the SO fragment [124] was consistent with these conclusions and suggested that the dissociation geometry is similar to the ground state, possibly implying a dissociative potential energy surface. A similar study on thionyl fluoride supports pathways analogous to A and C for photolysis at 193 nm [125].

Returning to true sulfoxides, the gas phase photochemistry of ethylene episulfoxide, trimethylene sulfoxide, and tetramethylene sulfoxide has been examined by organic product analysis and LIF analysis of the SO fragments. Interest is again in the timing of C–S cleavages and in the electronic state of the nascent SO. The simplest statement of the logic of the LIF studies with regard to cleavage timing is that sequential C–S cleavages are expected to generate a thermalized SO vibrational distribution, whereas an inverted vibrational distribution is taken as evidence for the concerted cleavage [126].

233   **234**   **235**

The vibrational population of SO formed as a result of 193 or 248 nm photolysis of 233 is inverted, with a maximum in $v = 1$ [126]. Thus it was concluded that the bond ruptures are concerted. The Franck–Condon/golden rule analysis used to fit the data includes calculation of the SO bond length immediately before cleavage, presumably in the appropriate excited state. These two values are different for 193 and 248 nm experiments, and it is suggested that this implies different precursor states. The analysis fails if ground state ethylene (or propylene, *vide infra*) energies are used, and it is concluded that the hydrocarbons are initially produced in an excited state. SO($^1\Delta$) was not directly observed in these measurements. However, it was noted that the signal for ground state SO($^3\Sigma^-$) grows as a function of time after the initial laser pulse. One explanation for the growth is relaxation of some of the ^1SO down to ground state [126].

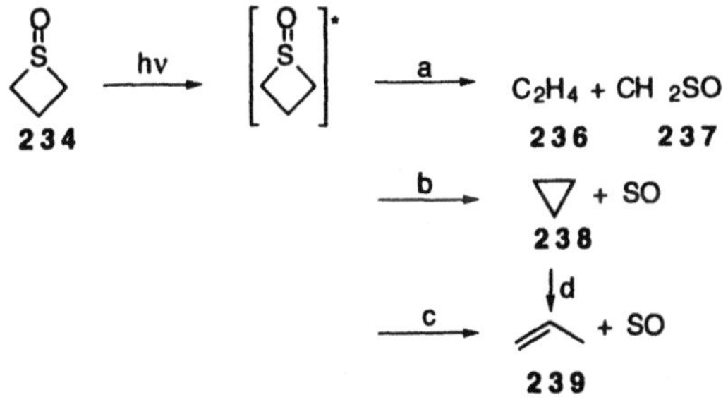

Product analysis was done for trimethylene sulfoxide 234 well in advance of the LIF experiments [127]. Photolysis was carried out from 214 nm to 254 nm and with Hg ($^3P_1$) sensitization, and the resulting quantities of C_3H_6 and C_2H_4 were determined. The quantites of each were comparable. Triplet sensitization favored the formation of C_3H_6 and SO, but quenching with O_2 did not eliminate it. It was assumed that path (d) was taken only by a fraction of the cyclopropane that was formed with sufficient internal energy, and this was borne out by experiments with diluent gases. No direct evidence on the state of SO was given, but it was speculated, based on cyclopropane/propene ratios, that direct photolysis along path (b) provided SO($^1\Delta$), while sensitized photolysis provided SO in the ground state.

LIF analysis of the nascent SO($^3\Sigma^-$) at either 193 or 248 nm showed a vibrational population inversion, which supports simultaneous cleavage, at least in that channel. The calculated pre-dissociation SO bond lengths were quite similar to those for 233, again suggesting two different states, depending on the excitation wavelength.

The photolysis of tetramethylene sulfoxide 235 has received the most attention of the series. Gas phase photolysis was carried out at multiple wavelengths, as with 234. The observed hydrocarbon products are illustrated below, derived from the suggested α-cleavage intermediate [128]. SO, CH_2SO, and 3-butenesulfenic acid were presumed to be formed but not observed. After breakage of one C-S bond (ca. 55 kcal/mol), significant energy remains in the system, and the product ratios are dependent on the quantity of that energy (variation of wavelength) and ability of the bath to take up the excess (variation of pressure). Sensitization with Hg($^3P_1$) provides only C_2H_4 and propene. With 147 nm excitation [129], tetramethylene sulfoxide yields mainly C_2H_4. (Similar results were also obtained as a neat liquid or solid.) The initially formed biradical was thought to decompose to tetramethylene, singlet and triplet trimethylene, and by stepwise loss of ethylene. Hot tetramethylene further decomposed to ethylene, as well as cyclobutane. Relative to the major product, 1-butene was more abundant in long wavelength photolyses.

$$235 \quad\quad\quad 240 \quad\quad\quad 241 \quad 236 \quad 238 \quad 239 \quad 242$$

Photolysis of 235 was also the subject of analysis by SINDO1 computational methodology, which lead to a fundamentally different conclusion about the mechanism for product formation [130]. Potential intermediates for all reasonable single and multiple cleavages were searched for. The lowest excited triplet (T_1) was found to dissociate along the C–S cleavage pathway to 240 without barrier. Further cleavage to tetramethylene and SO has a barrier of ~ 40 kcal/mol, and this was the lowest bond-breaking transition state found. Further product formation from the triplets was thus dismissed. On the other hand, only a barrier of a few kcal/mol stands between S_1 and three-body decomposition directly to $SO(^1\Delta)$ and 2 ethylenes. The low barrier cleavage out of S_2 is also symmetrical but yields tetramethylene and $SO(^1\Delta)$. (Similar pathways exist for T_1 and T_2, but they are discounted due to the barrierless α-cleavage process.) A transition state is found for concerted formation of trimethylene, and CH_2SO is found at 157 kcal/mol (182 nm $h\nu$), relative to the ground state, in a region where several states are close to one another. This is said to account for formation of the C_3 products. Qualitatively, C_3 products are increased (experimentally) at shorter wavelength photolysis and lower pressures.

Most recently, this compound has been examined by Weiner and LIF analysis of the diatomic products [126,131]. The quantum yield for $SO(^3\Sigma^-)$ was determined to be about 0.45 at either 193 or 248 nm. In contrast to the two smaller ring systems, the $SO(^3\Sigma^-)$ was found to be vibrationally thermalized at about 1250K with both excitation wavelengths, and the vibrational profile with 193 nm excitation was consistent only with tetramethylene as the other product after dissociation to two molecules. The thermalized nature of the SO spectra favors a stepwise process with the α-cleavage biradical 240 as an intermediate. If a stepwise process is invoked, then the tetramethylene fragment should be of a similar temperature. According to *ab initio* calculations by Doubleday [132], the ratio of cyclization to fragmentation should be about 0.27 at this temperature, consistent with experimental data [128] between 202 and 225 nm.

The 248 nm data are also consistent with stepwise cleavage, but the second step is found to lead to two molecules of ethylene, not tetramethylene. This would also be consistent with the variable wavelength product data. It is suggested that the difference in chemistry may arise due to excitation into a singlet state at 193 nm and a triplet state at 248 nm [126,131].

In addition to SO, HO· was detected in its ground vibrational state, even under collision free conditions, though the yield was not determined. The mechanism that was suggested for its formation was consistent with other α-cleavage chemistry in solution (*vide infra*).

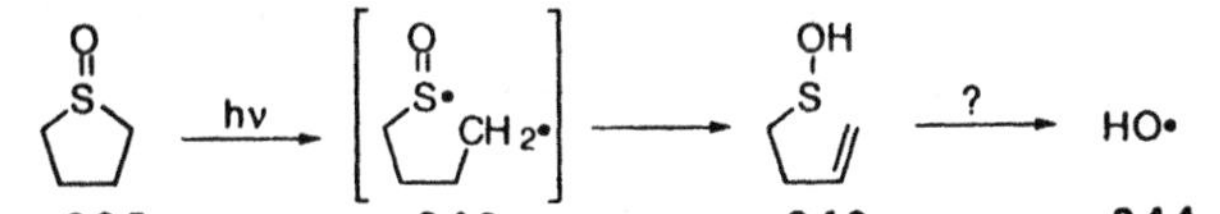

235 240 243 244

B. Solution

The photolysis of DMSO was studied in detail by Gollnick and Stracke [12,133,134], though it was mentioned preliminarily in previous studies [21,135]. Photolysis of solutions of DMSO in water, alcohols, or acetonitrile results in a variety of products. The reactions are done at high concentrations of DMSO (1–2 M), and it is reasonable to expect that sulfenic and sulfinic acids are further oxidized to the sulfonic acid by its presence. A careful accounting of the oxidation stoichiometry [12], however, shows that additional Me_2S is formed by another mechanism.

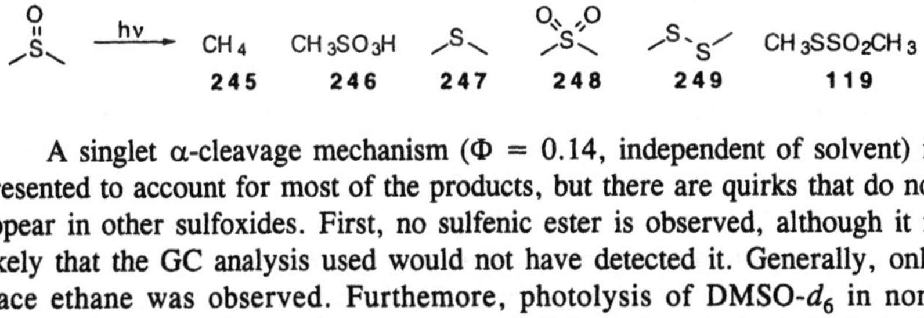

245 246 247 248 249 119

A singlet α-cleavage mechanism ($\Phi = 0.14$, independent of solvent) is presented to account for most of the products, but there are quirks that do not appear in other sulfoxides. First, no sulfenic ester is observed, although it is likely that the GC analysis used would not have detected it. Generally, only trace ethane was observed. Furthemore, photolysis of DMSO-d_6 in non-deuterated solvents (water, benzene, acetonitrile) results nearly exclusively in CHD_3, which eliminates the usual disproportionation reaction:

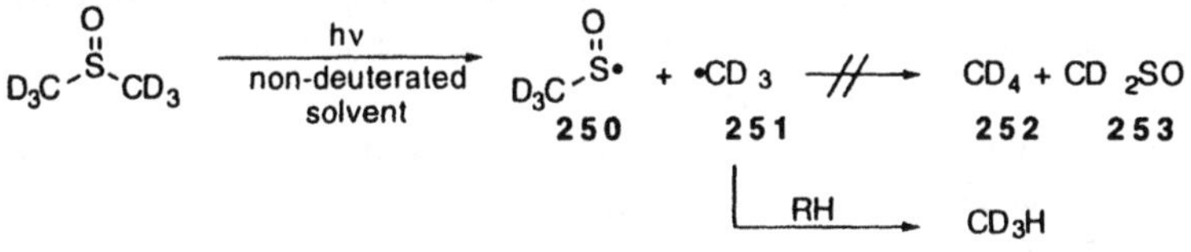

250 251 252 253 254

When the photolysis of DMSO is carried out in *O*-deuterated alcohols, the molar ratio of CH_4 to CH_3D varies from 1:1 (methanol) to 11:1 (isopropyl alcohol). These and the aqueous results point to a nonradical mechanism for the formation of methane. On the basis of pH dependence of the quantum yield, the pattern of the H/D ratio, the lack of ^{18}O exchange, and the energetics of heterolytic cleavage is argued that this pathway involves ordinary homolytic α-cleavage followed by electron transfer, i.e.,

$$H_3C \overset{\overset{O}{\parallel}}{S} CH_3 \xrightarrow[H_2O]{h\nu} H_3C \overset{\overset{O}{\parallel}}{S} \cdot + \cdot CH_3 \longrightarrow H_3CSO^+ + CH_3^- \xrightarrow{H_2O} H_3CSO_2H + CH_4$$

255   **115**   **256**   **251**   **258**   **245**

Competition between electron transfer and radical abstraction, along with the greater ability of water and methanol to solvate the ions help complete the rationalization. Methanesulfinic acid 258 is thought to participate in further redox reactions and is not isolated.

The rest of the dimethyl sulfide and the dimethyl sulfone is accounted for with a true photochemical dimerization reaction (i.e., not S–O cleavage and trapping of the O) supported by steady state kinetics [12,134]. The lack of dimethyl sulfide and O in the previously discussed gas phase studies supports the suggestion.

259   **260**   **261**   **262**   **263**

The photolysis of several other monofunctional sulfoxides was reported by Shelton and Davis [14]. Photolysis products in benzene with Pyrex filtration were the corresponding sulfide (without sulfone) and several products consistent with α-cleavage reactions. For instance, di-*tert*-butyl sulfoxide 259 produced *tert*-butyl alcohol, di-*tert*-butyl disulfide, a small amount of acetone, and a trace of thiosulfinate 269. Similar results were found for compounds 260-262. Photolysis of di-*n*-butyl sulfoxide 263 [136] also gave α-cleavage products: butyraldehyde, butanethiol, and the corresponding disulfide. Finally, (+)-(*S*)-butyl methyl sulfoxide was subjected to photolysis through a vycor filter, resulting in "extensive decomposition"; however, the recovered starting material had the same optical rotation as before photolysis, implying that stereomutation had not taken place [73].

The photochemistry of penicillin sulfoxide derivatives also clearly derives from α-cleavage. Though we have concentrated on direct irradiations until now, these results merit inclusion because of the historic importance of penicillin sulfoxides and because the sensitizer employed is acetone, whose triplet energy (ca. 79 kcal/mol) might be sufficient to sensitize these molecules by traditional energy transfer. The first preparation of an (R)-sulfoxide of a penicillin was by photolysis of 273 in acetone [137]. While there is some ambiguity in whether α-cleavage is involved in that particular stereomutation, photolysis of the slightly more complicated derivative 275 under identical conditions leads to all four possible diastereomers derived from cleavage between the sulfur atom and the tertiary carbon center [138]. α-Cleavage is clearly involved in the latter case.

Direct irradiation of 280 (λ = 277 nm) produced a mixture of 90% 280 and 10% of its diastereomer 281 [139]. The ratio did not change with further irradiation, but the mixture gradually decomposed due to an additional photochemical reaction. This was apparently not a true photostationary state, since irradiation of 281 produced a 77:23 mixture which also gradually decomposed. It is ambiguous whether this stereomutation is due to α-cleavage and recombination or another process not involving actual bond scission.

280   281

IX. KETOSULFOXIDES

An important group of reactions occur in (usually β-) ketosulfoxides, in which it is clear that the ketone is the chromophore and the sulfoxide merely an important actor in subsequent chemistry. A few of these have been seen already, such as the extrusion of SO from 146 [140]. The most common reaction of the β-ketosulfoxides is S–C cleavage, typically β to the ketone and α to the sulfoxide, as is common for other ketones with good β-leaving groups.

Interest in the photochemistry of these compounds began in the early 1970s with publications from three groups. Independent reports [141,142] within a year of each other gave acetophenone, dibenzoyl ethane, and methyl thiosulfonate as the major products from photolysis of 282 (R = H). Solvents and variations in the aromatic ring were shown to affect the product distribution, but the basic processes involved remained unchanged. Some combination of Type II and β-cleavage led to the observed acetophenone, whereas the other products were due to β-cleavage. No attempts were made to isolate the unstable sulfine 289.

Some years later, Wagner made a thorough investigation into the photochemistry of 282 (R = C_3H_7) and related compounds. Using quantum yields, H-atom abstraction trapping, quenching rate constants, and arguments from related structures, it was concluded that the rate of β-cleavage was ~6 × 10^9 s^{-1} and that there was no evidence for Type II reactivity [143]. The initial radical pair was shown to return to starting material about 60% of the time.

The Wagner group extended its work along the sulfinyl ketone series to γ-, δ-, and ε-sulfinyl compounds. No products with $\Phi > 0.001$ were observed for the γ-sulfinylketone. As with other phenones substituted in the same location, the lack of reactivity was attributed to rapid intramolecular charge transfer quenching of the triplet [144]. On the other hand, the δ-sulfinyl ketone produced normal Type II products, but with reduced quantum yield. The ε-sulfinyl compound underwent the internal hydrogen abstraction, but loss of the sulfinyl group was much faster than any other subsequent reaction; the sulfinyl radical leaving group was found to be faster than even Br· [144–147]. The observed products are 4-benzoyl-1-butene and the appropriate thiosulfonate.

Also reported in the early 1970s was the photochemistry of some alkyl α-substituted β-ketosulfoxides 290 [148]. Stereomutation at sulfur was observed in all cases where it could be detected due to diastereomeric interconversions. In addition to the products that are parallel to the previous example, ketone 293 and thioester 294 were observed. A mechanism in which Type I cleavage competed with β-cleavage was proposed. It included a cyclic sulfuranyl radical 297 in order to accomplish the oxygen migration. An alternative hypothesis has formation of a sulfenic ester 301, secondary photolysis, and typical chemistry of alkoxy radicals to get to the same intermediates. Sulfenic esters were not detected, but analysis was by GC or after column chromatography, neither of which would have been survived by such compounds.

Cy = cyclohexyl

Photolysis of the bicyclic ketosulfoxide 303 also resulted in stereomutation [148]. On extended photolysis, other products were observed. The only one identified was 305, which derives from the diketone. This is presumed to come about from desulfurization of the sulfine(s), as seen by Schultz and Schlessinger [25–28].

303   304   305

X. SULFENIC ESTERS

The photochemistry of sulfenic esters has come up repeatedly in this review as a result of their formation from sulfoxide α-cleavage. With the exception of 50, the result is S–O homolysis, just as one would expect intuitively. Indeed, photolysis of both simple and more complex sulfenic esters that will be discussed in this section appear to proceed along this pathway.

Nonetheless, it is worth mentioning that, given current experimental heats of formation, the S–O bond is not the weakest in most sulfenic esters. Take methyl methanesulenate for example. Hypothetical C–O cleavage leads to the same alkyl/sulfinyl radical pair obtained by S–C homolysis of the corresponding sulfoxide. All of the heats of formation for the radicals below are from a reliable database of experimental sources [98], except that of the methylsulfinyl radical, which is derived from Benson's estimate of the C–S bond energy of DMSO [63]. The estimate of the heat of formation of the sulfenic ester is based on Schlegel's calculation [50] and the known heat of formation of DMSO but is unimportant to the illustration. Given this data, homolysis of C–O is less costly than that of S–O by about 18 kcal/mol!

$$CH_3S\bullet \ + \ CH_3O\bullet \ \longleftarrow \ CH_3\text{–}S\text{–}O\text{–}CH_3 \ \longrightarrow \ CH_3SO\bullet \ + \ CH_3$$

ΔH_f (kcal/mol)   33.2   3.7   ca.–25   –16   34.8

Nonetheless, the experimental evidence overwhelmingly favors S–O homolysis in most instances. Sometimes O–C cleavage would go "unnoticed" because most of the reactions studied are cyclic systems derived from sulfoxides. However, Guo found that photolysis of benzyl benzenesulfenate did not lead to phenyl benzyl sulfoxide [60], even though the latter compound is known to racemize thermally through homolytic cleavage and reclosure of the sulfinyl/benzyl radical pair [71].

Very little is known about photolysis of simple alkyl sulfenic esters [149]. Photolysis of *tert*-butyl methanesulfenate at low temperature gives rise to an epr signal that was originally assigned to the methylsulfinyl radical [150]. The lack of signal from the *tert*-butyl radical and the assumption that S–O homolysis would be observed caused the authors to suggest a three-step mechanism that led to the appearance of these radicals. However, later workers showed that this assignment of the epr spectrum was incorrect and was in fact due to the sulfuranyl radical derived from addition of *tert*-butoxyl to the starting sulfenic ester [83]. The implication of this assignment to the photochemistry is still that

S–O homolysis is taking place, but the only reported product isolation in this system was "a high yield of isobutylene" [150], which might be consistent with a competition between the two homolyses.

In the 1990s, Pasto has taken advantage of the photochemistry of sulfenic esters to study, among other things, alkoxyl radicals. It was found that esters of 4-nitrobenzenesulfenic acid 306 were much easier to purify and handle than other aromatic sulfenates and had benefit of a chromophore extending well above 300 nm. A model computation with 4-nitrobenzenesulfenic acid placed the HOMO as a nonbonding orbital on sulfur and the LUMO as a π^* orbital of the aromatic ring [151]. Photolysis provides the sulfenyl (thiyl) and alkoxyl radicals. The latter often undergo β-scission reactions, generating alkyl radicals and carbonyl compounds. However, efficient generation of alkoxyl and/or alkyl radicals from nitrobenzenesulfenates requires the elimination of molecular oxygen from the sample. The cleavage apparently goes through a long-lived triplet state intermediate, which was shown to sensitize formation of singlet oxygen $O_2(^1\Delta_g)$ [151,152].

306

In the absence of other substrates, the corresponding alcohol, aldehyde/ketone, sulfide (alkyl + sulfenyl), and alkyl dimer are isolated [51]. The details of all of the studies conducted with this method are beyond the scope of this review, but a brief overview is appropriate.

In one study, various substituted allyl radicals were generated by sulfenate photolysis, and it was shown that coupling was controlled both by steric and by frontier molecular orbital considerations [153]. β-Scission is favored by α-substitution and the stability of the putative alkyl radical. A particularly clever device used in another study was the thermal equilibrium between allylic sulfoxides and sulfenates used to generate allyloxy and other C_3H_5O radicals [154].

By photolyzing a solution of the allyl aryl sulfoxide at slightly elevated temperature at appropriate wavelengths, allyloxy radicals were produced from photolysis of the steady state concentration of the sulfenate. Another study examined the regioselectivity of addition of the alkyl and sulfenyl radicals across olefins and allenes [155]. Control of these elements allowed potentially useful synthetic transformations to be designed [156], particularly as the sulfenate may be viewed as an O–H abstraction "synthon".

$$ROH + ArSCl \longrightarrow ROSAr \xrightarrow{h\nu} RO\bullet + \bullet SAr$$
$$\;\;\;314\;\;\;\;315\;\;\;\;\;\;\;\;\;\;\;\;316\;\;\;\;\;\;\;\;\;\;\;\;\;\;317\;\;\;\;307$$

XI. CONCLUDING REMARKS

It should be evident from the body of work described in this review that a good start has been made on the understanding of sulfoxide photochemistry, yet many fundamental questions remain unanswered. The role of α-cleavage is well established in a number of reactions, for instance, but nothing is known about the rates of these reactions, and little is known about sulfinyl radical reactivity outside of recombination and disproportionation of biradicals and radical pairs [62]. The role, if any, of something as simple as hydrogen abstraction as a primary photochemical process remains an open question! The existence of a mechanism of stereomutation by "inversion" in addition to the obvious cleavage/closure process is hinted at but not well established. Very little has been discussed here about anything outside of unimolecular reactions. There are hints here and there throughout this work (e.g., reaction of phenyl norbornyl sulfoxide in basic methanol and the sensitized racemizations of Hammond) that electron transfer reactions occur with important chemical consequences.

Sulfoxide photochemistry is unlikely suddenly to catch fire as a field of interest due to intense synthetic interest in certain reactions. Nonetheless, we should keep a watchful eye for useful, as well as interesting, chemistry. The work of Pasto in sulfenic esters clearly shows that a well-chosen substitution can tame otherwise difficult compounds, and it would not be surprising to see other dramatic effects on sulfoxide photochemistry, which has been scoped out with "simpler," relatively unsubstituted systems. The scope of rearrangements here is wide because of the flexible reactivity of some of the intermediates; the clever soul among us should not hesitate to design transformations around the chemistry described herein.

ACKNOWLEDGMENTS

The work carried out at Iowa State was supported by the Petroleum Research Foundation, The Research Corporation, and the National Science Foundation, and the authors gratefully thank these institutions. The support of the department of chemistry at Iowa State University, both financial and intangible, has also been invaluable.

REFERENCES

1. Block, E. *Quarterly Reports on Sulfur Chem.* 1969, *4*, 315–319.
2. Still, I. W. J. In *The Chemistry of Sulfones and Sulfoxides;* S. Patai; Z. Rappaport, and C. J. M. Stirling, eds.; John Wiley & Sons: New York, 1988, pp. 873–887.
3. Fuji, K. *Chem. Rev.* 1993, *93*, 2037–2066.
4. Drabowicz, J.; Kielbainski, P.; Mikolajaczyk, M. In *The Syntheses of Sulphones, Sulphoxides and Cyclic Sulphides*; S. Patai and Z. Rappoport, Ed.; John Wiley and Sons: New York, 1994, pp. 111–388.
5. Carreño, M. C. *Chem. Rev.* 1995, *95*, 1717–1760.
6. Leandri, G.; Mangini, A.; Passerini, R. *J. Chem. Soc.* 1957, 1386–1395.
7. Frolov, Y. L.; Sinegovskaya, L. M.; Gusarova, N. K.; Efremova, G. G.; Trofimov, B. A. *Bull. Acad. Sci. USSR Div. Chem. Sci.* 1978, 901–904.
8. Koch, H. P. *J. Chem. Soc.* 1950, 2892–2894.
9. Cumper, C. W. N.; Read, J. F.; Vogel, A. I. *J. Chem. Soc.* (A) 1966, 239–242.
10. Mislow, K.; Green, M. M.; Laur, P.; Melillo, J. T.; Simmons, T.; Ternay, A. L. J. *J. Am. Chem. Soc.* 1965, *87*, 1958–1976.
11. Mislow, K.; Green, M. M.; Laur, P.; Chisolm, D. R. *J. Am. Chem. Soc.* 1965, *87*, 665–666.
12. Gollnick, K.; Stracke, H.-U. *Pure Appl. Chem.* 1973, 219–245.
13. Chen, X.; Asmar, F.; Wang, H.; Weiner, B. R. *J. Phys. Chem.* 1991, *95*, 6415–6417.
14. Shelton, J. R.; Davis, K. E. *Int. J. Sulfur Chem.* 1973, *8*, 217–228.
15. Baliah, V.; Varadachari, R. *Indian J. Chem.* 1989, *28A*.
16. Bordwell, F. G.; Boutan, P. J. *J. Am. Chem. Soc.* 1957, *79*, 717–722.
17. Mangini, A.; Pallotti, M.; Tiecco, M.; Dondoni, A.; Vivarelli, P. *Int. J. Sulfur Chem. A* 1972, *2*, 69–78.
18. Loutfy, R. O.; Still, I. W. J.; Thompson, M.; Leong, T. S. *Can J. Chem.* 1979, *57*, 638–644.
19. Bock, H.; Solouki, B. *Chem. Ber.* 1974, *107*, 2299–2318.
20. Jenks, W. S.; Lee, W.; Shutters, D. *J. Phys. Chem.* 1994, *98*, 2282–2289.
21. Sato, T.; Yamada, E.; Akiyama, T.; Inoue, H.; Hata, K. *Bull. Chem. Soc. Japan* 1965, *38*, 1225.
22. Sato, T.; Goto, Y.; Tohyama, T.; Hayashi, S.; Hata, K. *Bull. Chem. Soc. Japan* 1967, *40*, 2975–2976.
23. Dittmer, D. C.; Levy, G. C.; Kuhlmann, G. E. *J. Am. Chem. Soc.* 1967, *89*, 2793–2794.
24. Dittmer, D. C.; Kuhlmann, G. E.; Levy, G. C. *J. Org. Chem.* 1970, *35*, 3676–3679.

25. Schultz, A. G.; Schlessinger, R. H. *J. Chem. Soc. Chem. Commun.* 1969, 1483–1484.
26. Schultz, A. G.; DeBoer, C. D.; Schlessinger, R. H. *J. Am. Chem. Soc.* 1968, *90*, 5314–5315.
27. Schlessinger, R. H.; Schultz, A. G. *Tetrahedron Lett.* 1969, 4513–4516.
28. Schultz, A. G.; Schlessinger, R. H. *J. Chem. Soc., Chem. Commun.* 1970, 1051–1052.
29. Schultz, A. G.; Schlessinger, R. H. *Tetrahedron Lett.* 1973, 3605–3608.
30. Praefcke, K.; Weichsel, C. *Liebigs Ann. Chem.* 1980, 333–343.
31. Khait, I.; Lüdersdorf, R.; Muszkat, K. A.; Praefcke, K. *J. Chem. Soc., Perkins Trans 2* 1981, 1417–1429.
32. Kharasch, N.; Khodair, A. I. A. *J. Chem. Soc., Chem. Commun.* 1967, 98–100.
33. Nakabayashi, T.; Horii, T.; Kawamura, S.; Hamada, M. *Bull. Chem. Soc. Japan* 1977, *50*, 2491–2492.
34. Schultz, A. G.; Schlessinger, R. H. *J. Chem. Soc. Chem. Commun.* 1970, 1294–1295.
35. Still, I. W. J.; Arora, P. C.; Chauhan, M. S.; Kwan, M.-H.; Thomas, M. T. *Can. J. Chem.* 1976, *54*, 455–470.
36. Still, I. W. J.; Cauhan, M. S.; Thomas, M. T. *Tetrahedron Lett.* 1973, 1311–1314.
37. Still, I. W. J.; Thomas, M. T. *Tetrahedron Lett.* 1970, 4225–4228.
38. Still, I. W. J.; Arora, P. C.; Hasan, S. K.; Kutney, G. W.; Lo, L. Y. T.; Turnbull, K. *Can. J. Chem.* 1981, *59*, 199–209.
39. Larson, B. S.; Kolc, J.; Lawesson, S.-O. *Tetrahedron* 1971, *27*, 5163–5176.
40. Thyrion, F. C. *J. Phys. Chem.* 1973, *77*, 1478–1482.
41. Gilbert, B. C.; Kirk, C. M.; Norman, O. C.; Laue, H. A. H. *J. Chem. Soc. Perkins 2* 1977, 497–501.
42. Gilbert, B. C.; Gill, B.; Sexton, M. D. *J. Chem. Soc. Chem. Commun.* 1978, 78–79.
43. Chatgilialoglu, C.; Gilbert, B. C.; Gill, B.; Sexton, M. D. *J. Chem. Soc. Perkin Trans. 2* 1980, 1141–1150.
44. Muszkat, K. A.; Praefcke, K.; Khait, I.; Lüdersdorf, R. *J. Chem. Soc. Chem. Commun.* 1979, 898–899.
45. Lüdersdorf, R.; Khait, I.; Muszkat, K. A.; Praefcke, K.; Margaretha, P. *Phosph. and Sulfur* 1981, *12*, 37–54.
46. Kobayashi, K.; Mutai, K. *Tetrahedron Lett.* 1981, *22*, 5201–5204.
47. Kobayashi, K.; Mutai, K. *Phosph. and Sulf.* 1985, *25*, 43–51.
48. Gajurel, C. L. *Indian J. Chem. (B)* 1986, *25*, 319–320.
49. Oae, S.; Furukawa, N. *Sulfilimines and Related Derivatives;* American Chemical Society: Washington, D.C., 1983; Vol. 179, pp. 340.
50. See, for instance, Wolfe, S.; Schlegel, H. B. *Gazz. Chim. Ital.* 1990, *120*, 285–290.
51. See, for instance, Pasto, D. J.; Hermine, G. L. *J. Org. Chem.* 1990, *55*, 5815–5816.
52. Capps, N. K.; Davies, G. M.; Hitchcock, P. B.; McCabe, R. W.; Young, D. W. *J. Chem. Soc. Chem. Commun.* 1983, 199–200.

53. Kowalewski, R.; Margaretha, P. *Angew Chem. Int. Ed. Engl.* 1988, *27*, 1374–1375.
54. Kowalewski, R.; Margaretha, P. *Helv. Chim. Acta* 1993, *76*, 1251–1257.
55. Kropp, P. J.; Fryxell, G. E.; Tubergen, M. W.; Hager, M. W.; Harris, G. D., Jr., McDermott, T. P., Jr.; Tornero-Velez, R. *J. Am. Chem. Soc.* 1991, *113*, 7300–7310.
56. Fouassier, J. P.; Lougnot, D. J. *Polymer* 1995, *36*, 5005–5010.
57. Wagner, P. J.; Sedon, J; Waite, C.; Gudmundsdottir, A. *J. Am. Chem. Soc.* 1994, *116*, 10284–10285.
58. Furukuwa, N.; Fujii, T.; Kimura, T.; Fujihara, H. *Chem. Lett.* 1994, 1007–1010.
59. Fujii, T.; Kimura, T.; Furukawa, N. *Tetrahedron Lett.* 1995, *36*, 1075–1078.
60. Guo, Y.; Jenks, W. S. *J. Org. Chem.* 1995, *60*, 5480–5486.
61. Guo, Y.; Jenks, W. S. *J. Org. Chem.* 1997, in press.
62. Chatgilialoglu, C. In *The Chemistry of Sulfones and Sulfoxides*; S. Patai; Z. Rappoport; and C. J. M. Stirling, Eds.; John Wiley and Sons: New York, 1988; pp. 1081–1087.
63. Benson, S. W. *Chem. Rev.* 1978, *78*, 23–35.
64. Kellogg, R. M.; Prins, W. L. *J. Org. Chem.* 1974, *39*, 2366–2374.
65. Kato, H.; Arikawa, Y.; Hashimoto, M.; Masuzawa, M. *J. Chem. Soc., Chem. Commun.* 1983, 938–938.
66. Thiemann, C.; Thiemann, T.; Li, Y.; Sawada, T.; Nagano, Y.; Tashiro, M. *Bull. Chem. Soc. Jpn.* 1994, *67*, 1886–1893.
67. Carpino, L. A.; Chen, H.-W. *J. Am. Chem. Soc.* 1979, *101*, 390–394.
68. Rayner, D. R.; Miller, E. G.; Bickert, P.; Gordon, A. J.; Mislow, K. *J. Am. Chem. Soc.* 1966, *88*, 3138–3139.
69. Miller, E. G.; Rayner, D. R.; Mislow, K. *J. Am. Chem. Soc.* 1966, *88*, 3139–3140.
70. Bickart, P.; Carson, F. W.; Jacobus, J.; Miller, E. G.; Mislow, K. *J. Am. Chem. Soc.* 1968, *90*, 4869–4876.
71. Miller, E. G.; Rayner, D. R.; Thomas, H. T.; Mislow, K. *J. Am. Chem. Soc.* 1968, *90*, 4861–4868.
72. Rayner, D. R.; Gordon, A. J.; Mislow, K. *J. Am. Chem. Soc.* 1968, *90*, 4854–4860.
73. Mislow, K.; Axelrod, M.; Rayner, D. R.; Gottardt, H; Coyne, L. M.; Hammond, G. S. *J. Am. Chem. Soc.*, 1965, *87*, 4958–4959.
74. Hammond, G. S.; Gottardt, H.; Coyne, L. M.; Axelrod, M.; Rayner, D. R.; Mislow, K. *J. Am. Chem. Soc.* 1965, *87*, 4959–4960.
75. Cooke, R. S.; Hammond, G. S. *J. Am. Chem. Soc.* 1968, *90*, 2958–2959.
76. Cooke, R. S.; Hammond, G. S. *J. Am. Chem. Soc.* 1970, *92*, 2739–2745.
77. Charlesworth, P.; Lee, W.; Jenks, W. S. *J. Phys. Chem.* 1996, *100*, 15152–15155.
78. Balavoine, G.; Jugé, S.; Kagan, H. B. *Tetrahedron Lett.* 1973, 4159–4162.
79. See, for instance, Franz, J. A.; Roberts, D. H.; Ferris, K. F. *J. Org. Chem.* 1987, *52*, 2256–2262.
80. Chatgilialoglu, C. In *The Chemistry of Sulfenic Acids and Their Derivatives;* S. Patai, Ed.; John Wiley and Sons, New York, 1990; pp. 549–569.

81. Anklam, E.; Margaretha, P. *Res. Chem. Intermed.* 1989, *11*, 127–155.
82. Perkins, C. W.; Clarkson, R. B.; Martin, J. C. *J. Am. Chem. Soc.* 1986, *108*, 3206–3210.
83. Gara, W. B.; Roberts, B. P.; Gilbert, B. C.; Kirk, C. M.; Norman, R. O. C. *J. Chem. Res. (M)* 1977, 1748–1760.
84. Perkins, C. W.; Martin, J. C.; Arduengo, A. J.; Lau, W.; Alegria, A.; Kochik, J. *J. Am. Chem. Soc.* 1980, *102*, 7753–7759.
85. Saltelli, A.; Hjorth, J. *J. Atmos. Chem.* 1995, *21*, 187–221.
86. Gu, M.; Turecek, F. *J. Am. Chem. Soc.* 1992, *114*, 7146–7151.
87. Hynes, A. J.; Wine, P. H.; Semmes, D. H. *J. Phys. Chem.* 1986, *90*, 4148–4156.
88. McKee, M. L. *J. Phys. Chem.* 1993, *97*, 10971–10976.
89. Lüdersdorf, R.; Praefcke, K. *Z. Naturforsch* 1976, *31B*, 1658–1661.
90. Archer, R. A.; Kitchell, B. S. *J. Am. Chem. Soc.* 1966, *88*, 3462–3463.
91. Perkins, C. W.; Martin, J. C. *J. Am. Chem. Soc.* 1986, *108*, 3211–3214.
92. Chen, C. H.; Donatelli, B. A. *J. Org. Chem.* 1976, *41*, 3053–3054.
93. Morin, R. B.; Jackson, B. G.; Mueller, R. A.; Lavagnino, E. R.; Scanlon, W. B.; Andrews, S. L. *J. Am. Chem. Soc.* 1963, *85*, 1896–1897.
94. Morin, R. B.; Spry, D. O.; Mueller, R. A. *Tetrahedron Lett.* 1969, 849–852.
95. Schultz, A. G.; Schlessinger, R. H. *Tetrahedron Lett.* 1973, 4787–4890.
96. Wagner, P. J.; Kelso, P. A.; Kemppainen, A. E.; Haug, A.; Graber, D. R. *Mol. Photochem.* 1970, *2*, 81–85.
97. Gurria, G. M.; Posner, G. H. *J. Org. Chem.* 1973, *38*, 2419–2420.
98. Stein, S. E.; Lias, S. G.; Liebman, J. F.; Levin, R. D.; Kafafi, S. A. In U.S. Department of Commerce, NIST: Gaithersburg, MD, 1994.
99. Wan, Z.; Jenks, W. S. *J. Am. Chem. Soc.* 1995, *117*, 2667–2668.
100. Gregory, D. D.; Wan, Z.; Jenks, W. S. *J. Am. Chem. Soc.* 1997, in press.
101. Jenks, W. S.; Matsunaga, N; Gordon, M. *J. Org. Chem.* 1996, *61*, 1275–1283.
102. Murov, S. L.; Carmichael, I.; Hug, G. L. *Handbook of Photochemistry*; 2d ed.; Marcel Dekker: New York, NY, 1993.
103. El Amoudi, M. S. E. F.; Geneste, P.; Olivé, J.-L. *J. Org. Chem.* 1981, *46*, 4258–4262.
104. Ando, W.; Takata, T. In *Singlet Oxygen;* A. A. Frimer, Ed.; CRC Press: Boca Raton, 1983; Vol. III.
105. Bücher, G.; Scaiano, J. C. *J. Phys. Chem.* 1994, *98*, 12471–12473.
106. Grossert, J. S. In *The Chemistry of Sulphoxides and Sulphones*; S. Patai; Z. Rappoport and C. J. M. Stirling, Ed.; John Wiley and Sons: New York, 1988; pp. 925–967.
107. Cheng, C.; Stock, L. M. *J. Org. Chem.* 1991, *56*, 2436–2443.
108. El Amoudi, M. S. E. F.; Geneste, P.; Olivé, J. L. *Tetrahedron Lett.* 1978, 999–1000.
109. Furukawa, N.; Fukumua, M.; Nishio, T.; Oae, S. *Phosphorus and Sulfur* 1978, *5*, 191–196.
110. Miyamoto, N.; Nozaki, H. *Tetrahedron* 1973, *29*, 3819–3824.
111. Tanikaga, R.; Higashio, Y.; Kaji, A. *Tetrahedron Lett.* 1970, 3273–3276.
112. Tanikaga, R.; Kaji, A. *Bull. Chem. Soc. Japan* 1973, *46*, 3814–3817.

113. Gilbert, A.; Baggott, P. J. *Essentials of Molecular Photochemistry;* CRC Press: Boca Raton, 1991.
114. Lüdersdorf, R.; Martens, J.; Pakzad, B.; Praefcke, K. *Liebigs Ann. Chem.* 1976, 1992-2017.
115. Franck-Neuman, M.; Lohmann, J. J. *Tetrahedron Lett.* 1979, 2397-2400.
116. Yomoji, N.; Takahashi, S.; Chida, S.; Ogawa, S.; Sato, R. *J. Chem. Soc., Perkin Trans. 1* 1993, 1995-2000.
117. Yomoji, N.; Satoh, S.; Ogawa, S.; Sato, R. *Tetrahedron Lett.* 1993, *34*, 673-676.
118. Chen, X.; Wang, H.; Weiner, B. R.; Hawley, M.; Nelson, H. H. *J. Phys. Chem.* 1993, *97*, 12269-12274.
119. Zhao, A.-Q.; Cheung Y.-S.; Heck, D. P.; Ng, C. Y.; Tetzlaff, T.; Jenks, W. S. *J. Chem. Phys.* 1997, in press.
120. Gross, H.; He, Y.; Quack, M.; Schmid, A.; Seyfang, G. *Chem. Phys. Lett.* 1993, *213*, 122-130.
121. Kawasaki, M.; Kasatani, K.; Sato, H.; Ohtoshi, H.; Tanaka, I. *Chem. Phys.* 1984, *91*, 285-291.
122. Baum, G.; Effenhauser, C. S.; Felder, P.; Huber, J. R. *J. Phys. Chem.* 1992, 756-764.
123. Kanamori, H.; Tiemann, E.; Hirota, E. *J. Chem. Phys.* 1988, *89*, 621-624.
124. Wang, H.; Chen, X.; Weiner, B. R. *J. Phys. Chem.* 1993, *97*, 12260-12268.
125. Wang, H.; Chen, X.; Weiner, B. R. *Chem. Phys. Lett.* 1993, *216*, 537-543.
126. Wu, F.; Chen, X.; Weiner, B. R. *S. P. I. E.* 1995, *254 B*, 355-364.
127. Dorer, F. H.; Salomon, K. E. *J. Phys. Chem.* 1980, *84*, 3024-3028.
128. Dorer, F. H.; Salomon, K. E. *J. Phys. Chem.* 1980, *84*, 1302-1305.
129. Scala, A. A.; Colon, I.; Rourke, W. *J. Phys. Chem.* 1981, *85*, 3603-3607.
130. Jug, K.; Neumann, F.; Schluff, H. P. *J. Org. Chem.* 1993, 58, 6634-6640.
131. Wu, F.; Chen, X.; Weiner, B. R. *J. Phys. Chem.* 1995, *99*, 17380-17385.
132. Doubleday, C., Jr. *J. Am. Chem. Soc.* 1993, *115*, 11968-11983.
133. Gollnick, K.; Stracke, H. U. *Tetrahedron Lett.* 1971, 207-210.
134. Gollnick, K.; Stracke, H. U. *Tetrahedron Lett.* 1971, 203-206.
135. Horner, L.; Dörges, J. *Tetrahedron Let.* 1963, 757-759.
136. Petrova, R. G.; Freidlina, R. K. *Bull. Akad. Sci. USSR Div. Chem. Soc.* (Engl. Transl.) 1966, 1797-1798.
137. Archer, R. A.; De Marck, P. V. *J. Am. Chem. Soc.* 1969, *91*, 1530-1532.
138. Spry, D. O. *J. Am. Chem. Soc.* 1970, *92*, 5006-5008.
139. Kishi, M.; Komeno, T. *Tetrahedron Lett.* 1971, *28*, 2641-2644.
140. Ito, S.; Mori, J. *Bull. Chem. Soc. Japan* 1978, *51*, 3403-3404.
141. Majeti, S. *Tetrahedron Lett.* 1971, 2523-2526.
142. Nozaki, H.; Shirafuji, T.; Kuno, K.; Yamamoto, Y. *Bull. Chem. Soc. Japan* 1972, *45*, 856-859.
143. Wagner, P. J.; Lindstrom, M. J. *J. Am. Chem. Soc.* 1987, *109*, 3062-3067.
144. Wagner, P. J.; Lindstrom, M. J. *J. Am. Chem. Soc.* 1987, *109*, 3057-3062.
145. Wagner, P. J.; Sedon, J. H.; Lindstrom, M. J. *J. Am. Chem. Soc.* 1978, *100*, 2579-2580.
146. Wagner, P. J. *Colloq. Int. C.N.R.S.* 1978, *278*, 169-188.

147. Wagner, P. J.; Lindstrom, M. J.; Sedon, J. H.; Ward, D. R. *J. Am. Chem. Soc.* 1981, *103*, 3842–3849.
148. Ganter, C.; Moser, J.-F. *Helv. Chim. Acta.* 1971, *54*, 2228–2251.
149. Horspool, W. In *The Chemistry of Sulfenic Acids and Their Derivatives;* S. Patai, Ed.; John Wiley and Sons: New York, 1990; pp. 517–547.
150. Kawamura, T.; Krusic, P. J.; Kochi, J. K. *Tetrahedron Lett.* 1972, 4075–4078.
151. Pasto, D. J.; Cottard, F.; Horgan, S. *J. Org. Chem.* 1993, *58*, 4110–4112.
152. Pasto, D. J.; Cottard, F. *J. Am. Chem. Soc.* 1994, *116*, 8973–8977.
153. Pasto, D. J.; L'Hermine, G. *Tetrahedron* 1993, *49*, 3259–3272.
154. Pasto, D. J.; Cottard, F.; Picconatto, C. *J. Org. Chem.* 1994, *59*, 7172–7177.
155. Pasto, D. J.; Cottard, F. *J. Org. Chem.* 1994, *59*, 4642–4646.
156. Pasto, D. J.; Cottard, F. *Tetrahedron Lett.* 1994, *35*, 4303–4306.

2

The Photochemistry of Pyrazoles and Isothiazoles

James W. Pavlik
Worcester Polytechnic Institute, Worcester, Massachusetts

I. INTRODUCTION

The phototransposition chemistry of N-substituted pyrazoles and of isothiazoles has been of considerable interest [1] since the first reports that N-methylpyrazole 1 undergoes photoconversion to N-methylimidazole 2 [2] and that irradiation of isothiazole 3 leads to the formation of thiazole 4 [3].

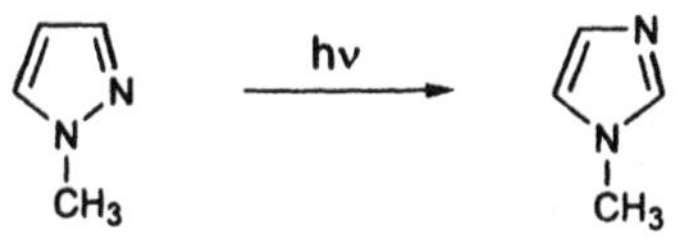

1   **2**

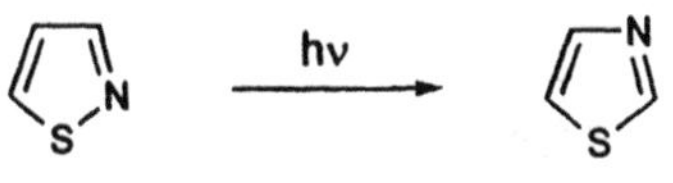

3   **4**

Mechanistic pathways suggested for these transformations assumed that both reactions involve a 2,3-interchange. Such an assumption was not based on experimental fact, however, since due to the absence of suitable labels, it was not possible to determine where C-4 and C-5 of the product originated in the reactant. Thus the reactions could have involved a greater number of atom transpositions and thus required different mechanistic conclusions.

A. Permutation Pattern Analysis

Barltrop and Day suggested that such ambiguities could be removed by determining the permutation pattern for the reaction before postulating a mechanistic pathway [4–6]. For such a five-membered cyclic compound there are just 12 different ways of transposing the five ring atoms resulting in the 12 permutation patterns shown in Table 1. According to this symbolism, which was first suggested by Barltrop and Day, the outer pentagon represents the order in which the ring atoms are bonded in the reactant, while the inner pattern represents the order in which the ring atoms are bonded in the transposed product. The permutation pattern thus provides a map of the transposition by showing where each ring atom in the product originated in the reactant. This provides a precise definition of all bonds that have been broken and formed during the reaction. This information is important because it greatly restricts the mechanisms possible for the transposition. Permutation pattern analysis has been used throughout this essay.

II. *N*-SUBSTITUTED PYRAZOLE PHOTOCHEMISTRY

A. 1-Methylpyrazoles: P_4, P_6, and P_7 Pathways

Schmidt and coworkers rationalized the transposition of 1-methylpyrazole 1 to 1-methylimidazole 2 in terms of a ring contraction and ring expansion mecha-

Table 1   Permutation patterns in Five-Membered Rings

P_1 P_2 P_3 P_4 P_5 P_6

P_7 P_8 P_9 P_{10} P_{11} P_{12}

nism [2], involving the intermediacy of 2-(N-methylimino)-2H-azirine 5 formed by cyclization of a photochemically generated biradical species. Although the intermediacy of acylazirines has been adequately demonstrated in the analogous isoxazole–oxazole phototransposition [7], such iminoazirines have not been detected in a pyrazole → imidazole reaction, and thus this mechanistic suggestion has never been experimentally confirmed.

A more recent investigation of this reaction revealed that in addition to undergoing phototransposition to 2 in 70% yield, 1 also undergoes photocleavage to 3-(N-methylamino)propenenitrile 6 in 21% yield [8]. Barltrop, Day, and Wakamatsu have also observed that other pyrazoles with hydrogen at position 3 also undergo photoring cleavage to enaminonitriles [9]. Although some enaminonitriles undergo E-Z photoisomerization [10], others have been observed to undergo photocyclization to imidazoles [11]. In the present case, enaminonitrile 6 was observed to undergo both E-Z photoisomerization as well as photocyclization to 1-methylimidazole 2 [8]. This confirms that the photo-cleavage-photocyclization pathway *via* an enaminonitrile (1 → 6 → 2) is one pathway for the phototransposition of 1-methylpyrazole 1 to 1-methylimidazole 2. This photocyclization reaction was observed, however, to be a very inefficient process. Thus under conditions of short-duration irradiation, most, if not all, of the 1-methylimidazole 2 formed must arise via another pathway not involving an enaminonitrile intermediate.

The operation of an additional pathway was confirmed by Beak and co-workers, who reported that 1,3,5-trimethylpyrazole 7 phototransposes to 1,2,4-trimethylimidazole 8 and to 1,2,5-trimethylimidazole 9 [12]. Although 9 can result from the 2,3-interchange pathway, product 8 cannot be rationalized by this route. Rather, this product was suggested to arise via a transposition pathway (Scheme 1) that includes initial electrocyclic ring closure, [1,3]-sigmatropic shift of nitrogen, and rearomatization of the resulting 2,5-diazabicyclo-[2.1.0]pentene to provide 8.

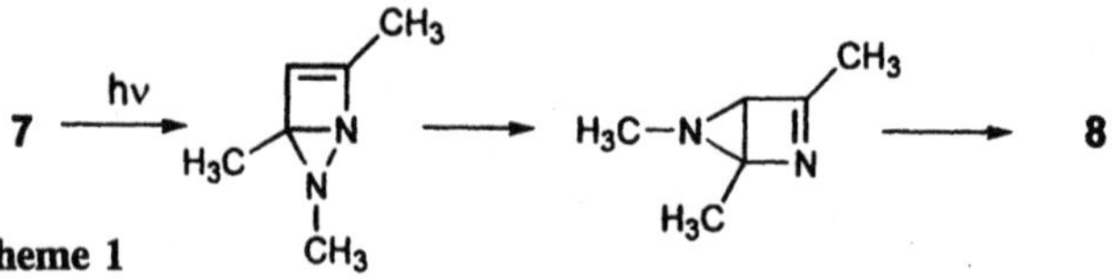

Scheme 1

Barltrop, Day, and colleagues later observed that 3-cyano-1,5-dimethyl-pyrazole 10 undergoes phototransposition to three primary products, 11 and 12, which can be rationalized by the 2,3-interchange pathway and the one-step nitrogen walk mechanism respectively, and 13, which cannot arise by either of these pathways but was suggested to arise via a double nitrogen walk mechanism shown in Scheme 2 [10]. Such a double walk process had formerly been implicated in the phototransposition chemistry of cyanothiophenes [13] and cyanopyrroles [5,6]. It should be pointed out that, unlike pyrazoles 1 and 7, all ring positions in 10 are uniquely substituted and therefore product identification allows the permutation patterns to be assigned. Thus 11, 12, and 13 are

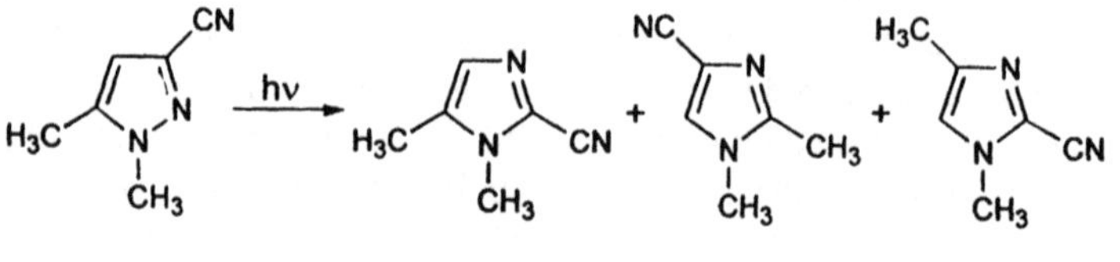

10   **11**   **12**   **13**

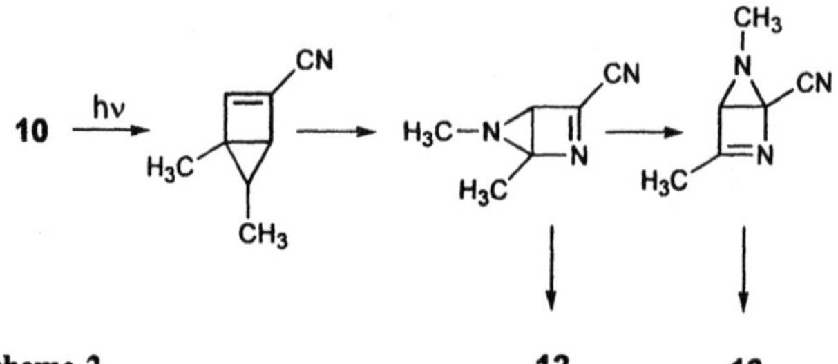

Scheme 2   **12**   **13**

formed from 10 via P_4, P_6, and P_7 permutation processes respectively. Transposition of 10 to 11 via the P_4 pathway also confirms that the reaction involves only a 2,3-interchange and that additional ring atom transpositions are not involved. Scheme 3 summarizes the various mechanistic pathways implicated in the 1-methylpyrazole to 1-methylimidazole transposition and identifies the permutation pattern associated with each route.

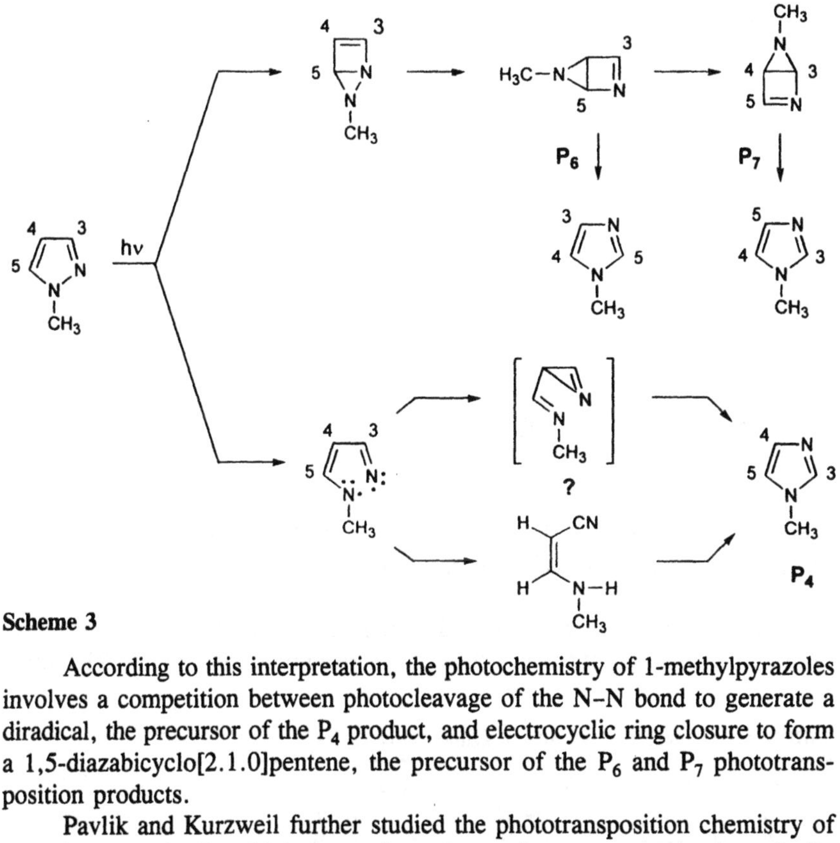

Scheme 3

According to this interpretation, the photochemistry of 1-methylpyrazoles involves a competition between photocleavage of the N–N bond to generate a diradical, the precursor of the P_4 product, and electrocyclic ring closure to form a 1,5-diazabicyclo[2.1.0]pentene, the precursor of the P_6 and P_7 phototransposition products.

Pavlik and Kurzweil further studied the phototransposition chemistry of 1-methylpyrazoles in which the various ring carbon atoms are systematically labeled with a second methyl group. The primary products shown in Table 2 and the results of deuterium labeling studies shown in Table 3 confirm that these dimethylpyrazoles undergo only pyrazole to imidazole phototransposition by as many as three distinct permutation pathways. Thus, whereas 1,5-dimethylpyrazole 16 transposes by the P_4, P_6. and P_7 pathways to yield a mixture of 1,5-, 1,2-, and 1,4-dimethylimidazoles 19, 17, and 18 respectively, 1,3-dimethylpyrazole 14 transposes only via the P_4 and P_6 pathways to form 1,2- and 1,4-dimethylimidazoles 17 and 18. Finally, 1,4-dimethylpyrazole 15 trans-

Table 2 Phototransposition of Dimethylpyrazoles. Primary Products

poses to a single product, 1,4-dimethylimidazole 18, presumably via a P_4 pathway, since of the two possible permutations, i.e., P_4 or P_{11}, the P_4 pathway is common to the other two reactions.

In addition, both 1,4- and 1,5-dimethylpyrazoles 15 and 16 were also observed to undergo photoring cleavage to enaminonitriles 20 and 21, which

Table 3 Deuterium Labeling of Dimethylpyrazoles

were also observed to undergo inefficient photocyclization to the P_4 imidazoles 18 and 19 respectively [8].

These phototransposition products also show that the ring methyl group does not act as an inert positional label but that it influences the number of phototransposition pathways operating in each reaction. In order to study the transposition process in 1-methylpyrazole 1 with minimum substituent perturbation, Pavlik and Kurzweil also studied the phototransposition chemistry of 3,4-dideuterio-1-methylpyrazole $1\text{-}3,4d_2$ [8]. This labeling pattern allows distinction between the three permutation pathways, since Scheme 3 shows that conversion of 1 to 2 via the P_4, P_6, or P_7 pathways is accompanied by transposition of C-5 of the reactant to ring position 5, 2, or 4 in the product.

After less than 10% photoconversion of $1\text{-}3,4d_2$, quantitative ^{1}HNMR analysis of the dideuterio-1-methylimidazole $2\text{-}d_2$ revealed that the C-5 proton of the reactant had transposed to ring positions 5, 2, and 4 of the dideuterio-1-methylimidazole ring, confirming that $1\text{-}3,4d_2$ undergoes phototransposition via the P_4, P_6, and P_7 pathway in a ratio of 4.8:6.5:1.0. Thus the 1-methylpyrazole 1 to 1-methylimidazole 2 phototransposition is considerably more complicated than originally suggested by Schmidt.

Recent experimental evidence supports the electrocyclic ring closure–heteroatom migration mechanism for the formation of the P_6 and P_7 N-methyl-imidazole products. Thus Pavlik and Kebede have observed that whereas irradiation of 1-methyl-5-phenylpyrazole 22 in methanol results in the formation of the anticipated P_4, P_6, and P_7 imidazoles 23–25 and to the enaminonitrile 26 in the chemical and quantum yields shown, irradiation of 22 in neat furan leads only to the pyrazole-furan [4 + 2] adduct 27 [14]. The formation of this product is consistent with furan trapping of a photochemically generated 1,5-diazabicyclopentene species 28. This is the first report of the trapping of a bicyclic species photochemically generated from a pyrazole [15]. It is interesting to note that since only pyrazole to imidazole transpositions are observed, the [1,3] sigmatropic shift of nitrogen must take place away from the azetine nitrogen to form the 2,5-diazabicyclo species but not in the opposite direction toward the azetine nitrogen to yield an isomeric 1,5-diazabicyclic species and eventually a pyrazole to pyrazole transformation.

	22	23	24	25	26
%	-23.2	16.0	30.3	45.6	trace
Φ	0.0410	0.0080	0.140	0.0220	-

B. Details of the P_6 and P_7 Pathways

Whereas these studies have established the gross structural changes associated with the P_6 and P_7 transportation pathways, MNDO calculations have provided insights into the electronic details of the structural changes occurring on the ground- and excited-state transposition coordinates [16].

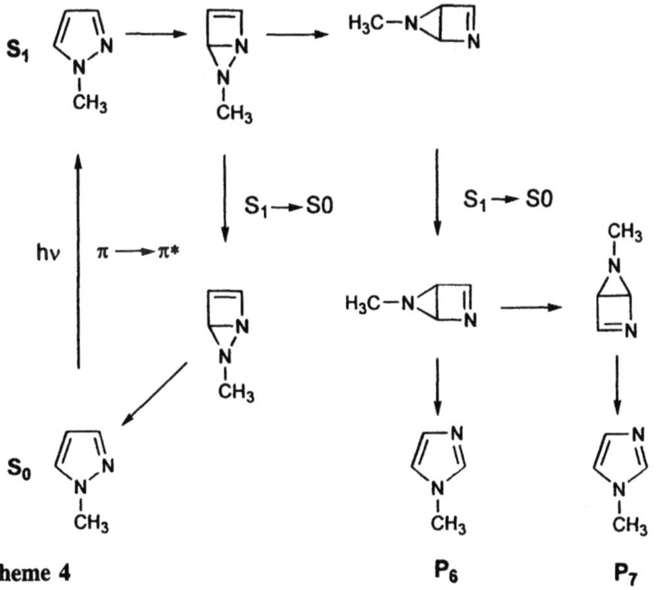

Scheme 4 P_6 P_7

According to these calculations, $\pi \rightarrow \pi^*$ (Scheme 4) excitation converts the planar ground state of 1-methylpyrazole to the Franck–Condon excited singlet. As this planar excited singlet relaxes, calculations show that the ring begins to undergo disrotatory deformation, resulting in an energy minimized S_1 1-methylpyrazole in which the 1-methylnitrogen is ~12° out of the plane of the ring. From this point the molecule undergoes facile electrocyclic ring closure to yield the initial 1,5-diazabicyclic species.

At this point on the reaction coordinate the calculated energy gap between the S_1 and S_0 surfaces is relatively small. It was concluded, however, that nonradiative crossover to the S_0 surface would lead only to aromatization to the reactant pyrazole and would thus be an energy wasting process. The first [1,3]-sigmatropic shift of nitrogen was thus predicted to occur on the S_1 surface to yield the 2,5-diazabicyclopentene excited singlet state. Theoretical calculations led to the same conclusion in the case of [1,3]-shifts of sulfur and oxygen in the phototransposition reactions of thiophene [17] and oxazole [18] respectively. It was further concluded that at this point on the reaction coordinate the molecule crosses over to the S_0 surface and that the ground state diazabicyclo species either aromatizes to the P_6 imidazole or undergoes a second nitrogen migration to an isomeric 2,5-diazabicyclopentene and ultimately the P_7 imidazole. These calculations thus suggest that the ratio of the P_6 to the P_7 imidazoles is determined by the activation barriers for rearomatization and the second sigmatropic shift of nitrogen.

In the case of 1,5-dimethylpyrazole 16, the calculated activation barriers for these two pathways were found to favor formation of the P_6 imidazole by ~8 kcal mol^{-1}. Variable temperature studies of the phototransposition chemistry of 16 were consistent with this conclusion. Thus although the total amount of the P_6 and P_7 products, 17 and 18 respectively, remain reasonably constant over the temperature range studied, the P_6 to P_7 ratio changed from 1.6 at 30°C to 23 at –30°C [19].

C. The P_4 Pathway

Although the electrocyclic ring closure–heteroatom migration pathway adequately rationalizes the formation of the P_6 and P_7 imidazoles, until recently the P_4 pathway has been less well understood. As in the case of 1,4-dimethylpyrazole 15, 1-phenylpyrazole 29 and the methyl substituted-1-phenylpyrazoles

Table 4 Phototransposition of N-Phenylpyrazoles

30–32 have been shown to undergo regiospecific P_4 phototransposition to the 1-phenylimidazoles 33–36 via the excited singlet states of the heterocycles [20]. Interestingly, unlike 1-methylpyrazole 1, which is nonplanar in the first excited singlet state [16], computational studies show that the pyrazole ring in 1-phenylpyrazole 29 remains planar upon excitation [20]. This is consistent with its lack of reactivity via the electrocyclic ring closure pathway. Calculations do show, however, that the angle of twist between the phenyl and pyrazole rings changes from 26° in S_0 to 88° in the energy minimized S_1 state. This brings the p-orbitals of the phenyl ring into coplanarity with the σ-orbital of the N–N bond in the pyrazole ring. This would assist in breaking the N–N bond by resonance stabilization of the resulting N-phenyl radical as well as by stabilization of the incipient radical in the transition state leading to it. Accordingly, these calculations also reveal that excitation is also accompanied by an increase in the N–N bond length from 1.35 Å in S_0 to 1.78 Å in S_1. This corresponds to a change in the N–N bond order from 1.12 in S_0 to 0.38 in S_1 and indicates that by the time the molecule reaches the energy minimized perpendicular S_1 state, the N–N bond is essentially broken and the molecule is well along the P_4 reaction coordinate.

Pavlik and Kebede have also reported that 1-methyl-4-phenylpyrazole 37 phototransposes regiospecifically to the P_4 product 1-methyl-4-phenylimidazole 38. Interestingly, in addition to the expected photocleavage product, (E,Z)-3-(N-methylamino)-2-phenylpropenenitrile 39, a second photocleavage product, (E,Z)-2-(N-methylamino)-1-phenylethenylisocyanide 40 was also formed in 68% yield. Upon direct photolysis, (Z)-40 undergoes $Z \rightarrow E$ photoisomerization and photocyclization to the P_4 imidazole with quantum yields of 0.216 and 0.116 respectively [21]. Thus unlike the photocleavage-photocyclization pathway via an enaminonitrile intermediate, which is very inefficient, the analogous pathway via an isocyanide intermediate is a major reaction sequence on the P_4 transposition pathway.

	37	38	39	40
%	-20.5	17.2	12.0	68.1
Φ	0.200±.003	0.037±.002	0.008±.001	0.114±.001

Although isocyanides have been spectroscopically detected as intermediates in the analogous P_4 isoxazole to oxazole phototransposition [22], they have not been previously observed in the pyrazole to imidazole phototransposition. Subsequent studies by Pavlik and Kebede have shown that the intermedicacy of isocyanides is not restricted to this single reaction but they are involved in all other pyrazole → imidazole phototranspositions investigated.

From these studies a general mechanistic scheme for the P_4 transition pathway is emerging. As shown in Scheme 5, photocleavage of the N–N bond is suggested to result in the formation of a species I-1 that can be described as either a biradical or a zwitterion. Rotation around the original C_3–C_4 bond and transfer of a proton or hydrogen atom from C-3 to N-1 (Path A) would lead directly to the enaminonitrile photocleavage product EN. Alternatively, it has also been suggested that the initially formed zwitterion could undergo a 1,2-shift of C_4 from C_3 to the adjacent electron deficient nitrogen (Path B) to give the nitrile ylide I-2. This species could undergo direct cyclization to the P_4 imidazole or to proton transfer resulting in the formation of the isocyanide IC, a known precursor of the P_4 imidazole.

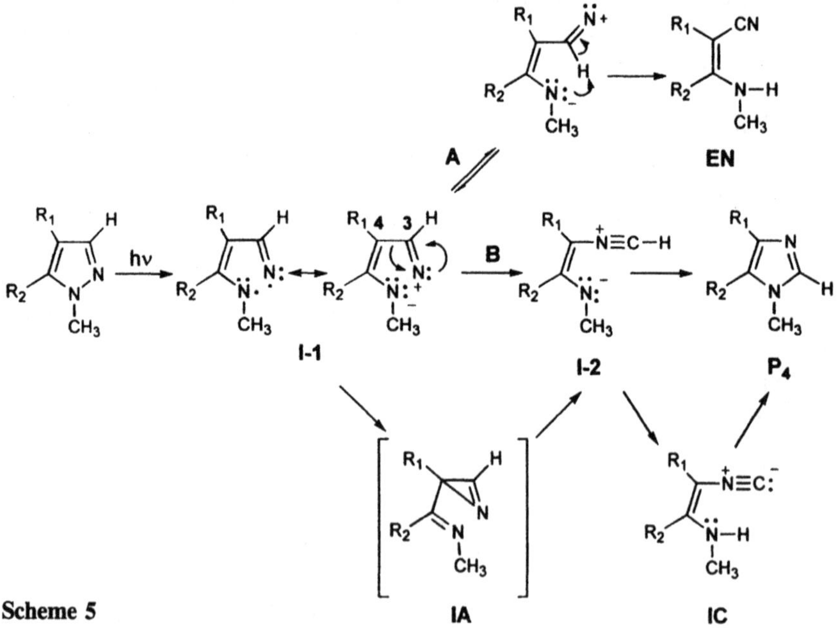

Scheme 5

It should also be noted that either the diradical or the zwitterion could also undergo cyclization to the iminoazirine IA, as originally suggested by Schmidt. Interestingly, electrocyclic ring opening of this azirine would be expected to yield nitrile ylide I-2 [23]. Thus the net effect of the ring closure and opening of the azirine is a 1,2-shift of C-3 from C_4 to N_2. If the azirine IA is formed on the ground state, calculations suggest that the most likely reaction pathway on this surface leads back to the reactant pyrazole [24,25]. Alternatively, calculations also reveal that excitation to the S_1 surface is accompanied by an increase in the carbon–carbon bond length in the azirine from 1.49 to 1.74 Å. Accordingly, if an azirine-like structure is reached on the reaction coordinate before crossing over to the S_0 surface, the excited species would be expected to undergo facile conversion to the nitrile ylide rather than resulting in an isolable product.

III. ISOTHIAZOLE-THIAZOLE PHOTOCHEMISTRY

A. Methylisothiazoles and Methylthiazoles

The photoisomerization of isothiazole 3 to thiazole 4 was the first reported phototransposition in the isothiazole–thiazole heterocyclic system [3]. Methylisothiazoles have also been shown to undergo transposition. In neutral solvents, Lablache-Combier and coworkers reported that 3-methylisothiazole 41 and 4-methylisothiazole 42 each transpose to a single product, 2-methylthiazole 44 and 4-methylthiazole 45 respectively. According to this report, however, 5-methylisothiazole 43 transposes to three primary products, 5-methylthiazole 46, 3-methylisothiazole 41, and 4-methylisothiazole 42 [26].

41   44

42   45

43   46

These workers suggested that these methylisothiazoles 41–43 transpose via tricyclic zwitterionic intermediates (Scheme 6) analogous to the intermediates invoked to rationalize the phototransposition reactions of 2-phenylthiophene [27]. Although this tricyclic zwitterion mechanism accounts for the isothiazole to thiazole and the isothiazole to isothiazole phototranspositions reported for 5-methylisothiazole 43, it also predicts that 3- and 4-methylisothiazoles 41 and 42 should also phototranspose to other isothiazole isomers.

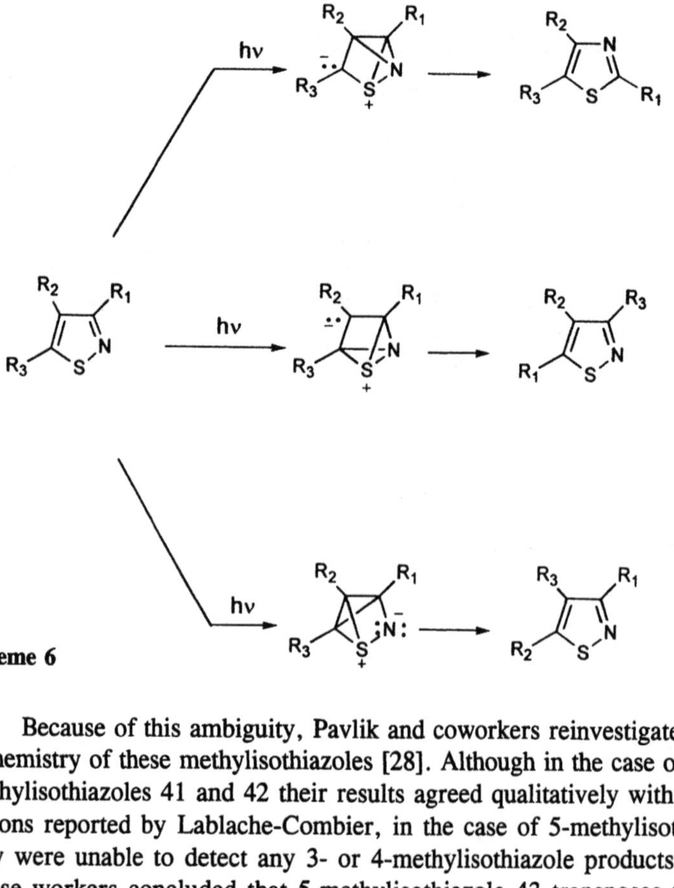

Scheme 6

Because of this ambiguity, Pavlik and coworkers reinvestigated the photochemistry of these methylisothiazoles [28]. Although in the case of 3- and 4-methylisothiazoles 41 and 42 their results agreed qualitatively with the observations reported by Lablache-Combier, in the case of 5-methylisothiazole 43 they were unable to detect any 3- or 4-methylisothiazole products 41 or 42. These workers concluded that 5-methylisothiazole 43 transposes to a single product, 5-methylthiazole 46.

Permutation pattern analysis indicates that these products are formed via a P_4 permutation pathway, since this is the only pattern common to all cases. This suggests that the methylisothiazole to methylthiazole transposition is a 2,3-

Table 5 Phototransposition of Methylthiazolium Ions

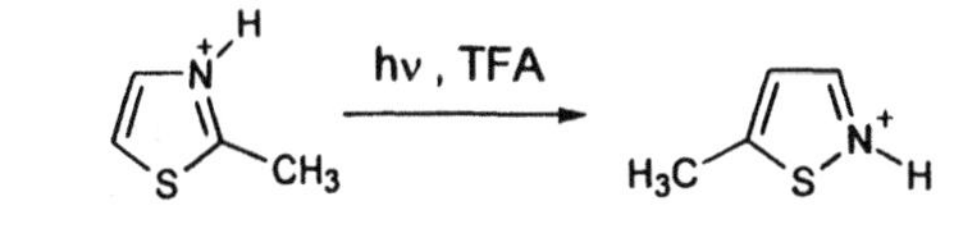

interchange process mechanistically analogous to that observed for pyrazole to imidazole photochemistry.

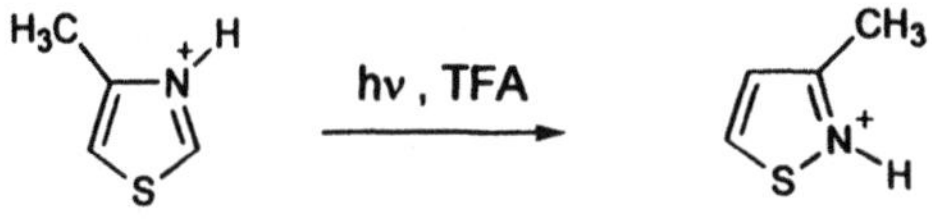

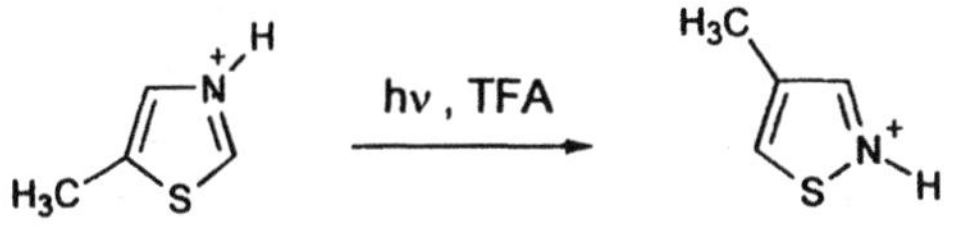

Although methylthiazoles do not undergo phototransposition upon irradiation in a variety of neutral solvents, Pavlik and coworkers have reported that the methylthiazolium cations $44H^+$–$46H^+$, formed by dissolving methylthiazoles 44–46 in trifluoroacetic acid (TFA), do undergo phototransposition (Table 5) to methylisothiazolium ions $41H^+$–$43H^+$ via a P_5 permutation process.

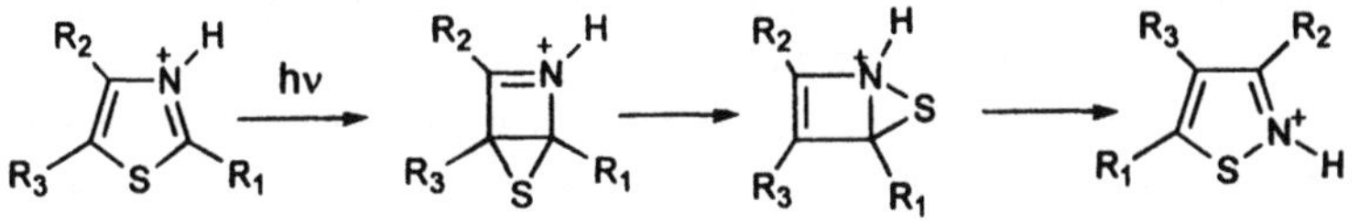

Scheme 7

The 3,5-interchange demanded by the P_5 permutation is consistent with the electrocyclic ring closure heteroatom migration mechanism (Scheme 7) in which the heteroatom migrates regiospecifically toward the positively charged azetine nitrogen. This regiospecificity is in marked contrast to that observed in neutral 1-methylimidazoles, which undergo only P_6 imidazole to imidazole transpositions. Thus in the case of neutral imidazoles, the heteroatom migrates only away from the azetine ring nitrogen.

B. Phenylisothiazoles and Phenylthiazoles

1.   Interconversions in Benzene Solution

Vernin and colleagues conducted extensive studies of the phototransposition chemistry of phenylisothiazoles 47–49 and phenylthiazoles 50–52 in benzene solution [29–34]. They concluded that while most of the observed products can be explained by the electrocyclic ring closure heteroatom migration mechanism, other observed products could not arise via this pathway. In particular, these workers reported that 5-phenylthiazole 52 phototransposes to all three isomeric phenylisothiazoles, viz., 4-phenylisothiazole 48, 5-phenylisothiazole 49, and 3-phenylisothiazole 47. Although the former product requires formation via a P_5 or P_7 pathway, and is therefore predicted by the electrocyclic ring closure heteroatom migration pathway, the latter two products would require P_4 or P_{11} or P_{10} or P_{12} permutation pathways, which are not possible by this mechanism. Accordingly, to account for the formation of these two products, Vernin et al. concluded that no one mechanistic pathway could explain the formation of all products [31]. They therefore suggested that the mechanism originally suggested by Kellogg to allow for the interchange of any two ring atoms in the phototransposition chemistry of phenylthiophenes [35] was also in operation in the reactions of phenylthiazoles.

| 52 | 48 | 49 | 47 |

Pavlik and coworkers reinvestigated the phototransposition chemistry of these compounds. Their results differ from those reported by Vernin in several ways. Most importantly, careful capillary column gas chromatographic analysis, including coinjections with authentic samples of the three isomeric phenylisothiazoles 47–49, confirmed that 5-phenylthiazole 52 transposes to 4-phenylisothiazole 48 but that 5-phenylisothiazole 49 and 3-phenylisothiazole 47 are *not* formed in this reaction. As a consequence, Pavlik and colleagues concluded that it was not necessary to evoke the Kellogg mechanism to explain the formation of any products in these photoreactions [36].

| 50 | 48 |

According to the observed photochemical products and the results of deuterium labeling studies, the six isomeric phenylisothiazoles and phenylthiazoles can be organized into a tetrad of four isomers that interconvert mainly via P_5, P_6, and P_7 transposition pathways and a dyad in which 5-phenylthiazole 52 transposes via P_5 and P_7 pathways to 4-phenylisothiazole 48 (Scheme 8), the only isomer that did not yield a transposition product upon irradiation in benzene solution. With one minor exception, no interconversions between the tetrad and dyad were observed. In that case, in addition to transposing to members of the tetrad, 5-phenylisothiazole 49 also transposed to 5-phenylthiazole 52, the first member of the dyad, in less than 1% yield. This conversion was assumed to occur via a P_4 permutation process.

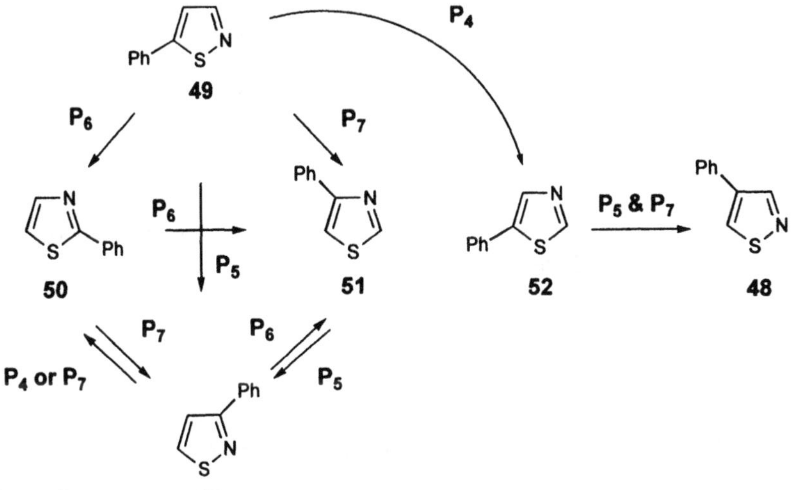

Scheme 8

The interconversions within the tetrad are consistent with the electrocyclic ring closure heteroatom migration mechanistic pathway shown in Scheme 9. Thus photochemical excitation of any member of the tetrad was suggested to result in electrocyclic ring closure, leading directly to the azathiabicyclo-[2.1.0]pentene intermediates. The four bicyclic intermediates, and hence the four members of the tetrad, are interconvertible via 1,3-sigmatropic shifts of the sulfur around the four sides of the azetine ring. Thus sulfur walk followed by rearomatization allows sulfur insertion into the four different sites in the carbon–nitrogen sequence.

In the dyad, because of the symmetry of the 5-phenylthiazole 52 ring, insertion of sulfur between ring positions C-1 and C-4 or C-1 and C-2 leads to the same compound, 5-phenylthiazole 52, while insertion of the sulfur atom

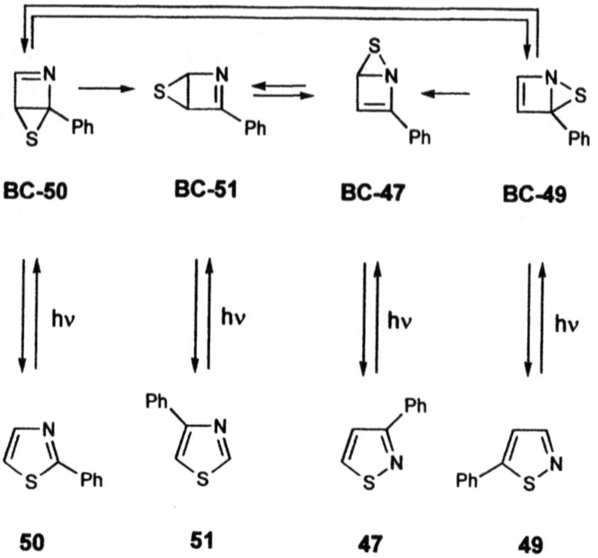

Scheme 9

between N-3 and C-4 or between N-3 and C-2 leads to 4-phenylisothiazole 48. These two compounds thus constitute a dyad.

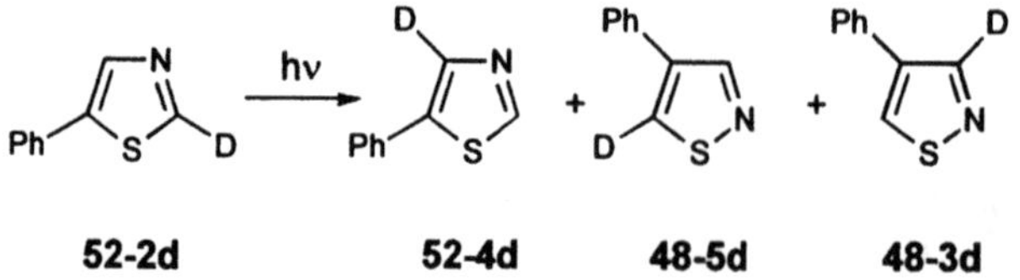

This symmetry is removed in the case of 2-deuterio-5-phenylthiazole 52-2d, and hence deuterium labeling expands the dyad into a tetrad. As expected from this analysis, irradiation of 2-deuterio-5-phenylthiazole 52-2d resulted in the formation of 4-deuterio-5-phenylthiazole 52-4d, the product of a P_6 permutation process not observed in the absence of deuteration, and to 5-deuterio-4-phenylisothiazole 48-5d and 3-deuterio-4-phenylisothiazole 48-3d, formed via P_5 and P_7 pathways respectively. Again, these observations are entirely consistent with the electrocyclic ring closure heteroatom migration mechanism shown in Scheme 10.

2. Photodeuteration Studies

Maeda and Kojima also studied the phototransposition chemistry of phenylisothiazoles 47-49 and phenylthiazoles 50-52 in benzene or ether saturated with

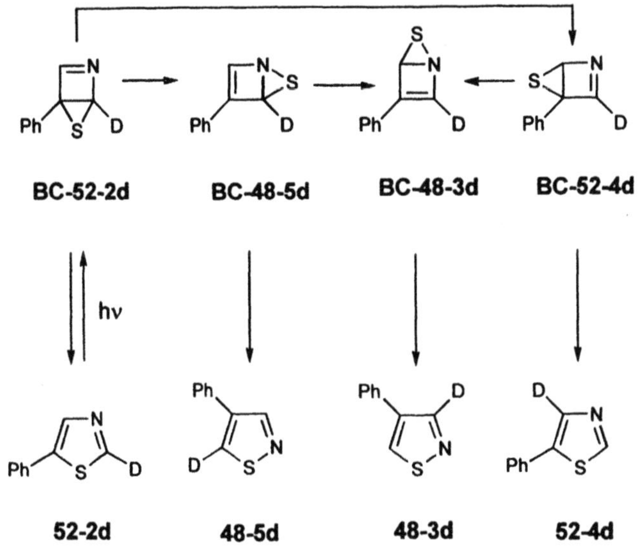

Scheme 10

D$_2$O [37–40]. These workers reported that irradiation of 2-phenylthiazole 50, 4-phenylthiazole 51, or 5-phenylisothiazole 49 under these conditions resulted in the formation of 3-phenylisothiazole 47-4d with deuterium incorporation into ring position 4. In the case of 2-phenylthiazole 50, 4-phenylthiazole 51 was also formed but without deuterium incorporation. Finally, they reported that none of the reactants underwent photodeuteration prior to isomerization.

Maeda and Kojima reasoned that the bicyclic intermediates in the electrocyclic ring closure heteroatom migration mechanism (Scheme 9) would not be expected to react with D$_2$O to incorporate deuterium. Furthermore, they argued that if these intermediates did react with D$_2$O, deuterium should also be incorporated into 4-phenylthiazole 51 during its formation from 2-phenylthiazole 50. These workers thus concluded that deuterium incorporation demanded a carbanion intermediate. They therefore rejected the mechanism involving the interconversion of bicyclic intermediates and suggested that all three reactants isomerize to 3-phenylisothiazole 45 via a common tricyclic zwitterionic intermediate TC-1 (Scheme 11), which incorporates deuterium.

Reinvestigation of these reactions, however, gave substantially different results. Thus Pavlik and coworkers observed that 3-phenylisothiazole 47 is formed with deuterium incorporation at C-4 upon irradiation of either 2-phenylthiazole 50 or 5-phenylisothiazole 49 but with approximately twice the extent of incorporation from the latter reactant. Furthermore, although Maeda

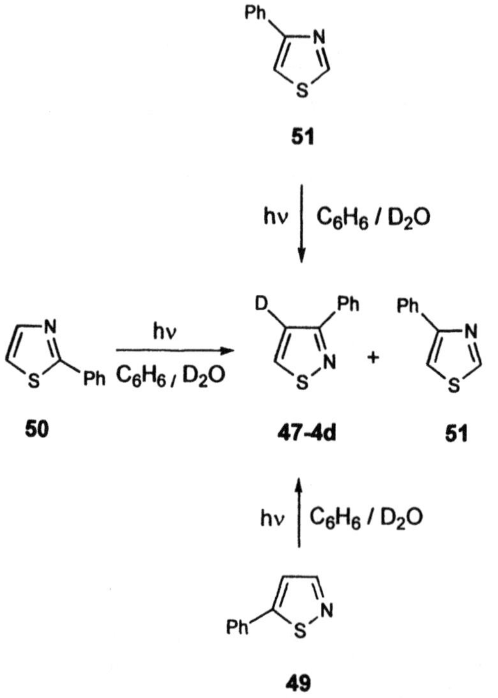

51

hv | C₆H₆ / D₂O

50 →(hv C₆H₆/D₂O)→ **47-4d** + **51**

hv | C₆H₆ / D₂O

49

and Kojima reported that none of the recovered phenylthiazoles had undergone deuteration when irradiated in benzene-D₂O, reinvestigation revealed that 4-phenylthiazole 51 undergoes photodeuteration at C-2 when irradiated under these conditions much more rapidly than it transposes to 3-phenylisothiazole 47.

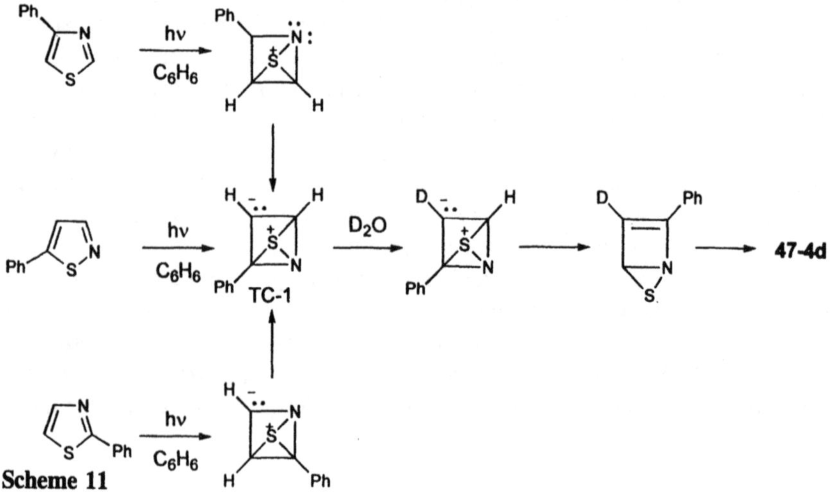

Scheme 11

Thus upon prolonged irradiation the initially formed 2-deuterio-4-phenylthiazole 51-2d transposes to 5-deuterio-3-phenylisothiazole 47-5d without additional deuterium incorporation [36].

These results show that within the tetrad (Scheme 9) the intermediate that incorporates deuterium is on both the 2-phenylthiazole 50 and 5-phenyl-isothiazole 49 to 3-phenylisothiazole 47 pathway but is *not* on the 4-phenylthiazole 51 to 3-phenylisothiazole 47 pathway. Accordingly, these photodeuteration results, which Maeda and Kojima concluded *required* the intermediary of tricyclic zwitterion intermediates, actually *preclude* their involvement in the phototransposition pathway.

In order to rationalize the photodeuteration results, Pavlik and coworkers postulated [36] that bicyclic intermediate BC-49 undergoes deuteration with simultaneous sulfur migration to yield BC-49-4d$^+$. This accounts for the lack of photodeuteration of 5-phenylisothiazole 49 and for the greater deuterium incorporation in 3-phenylisothiazole 47 formed from 5-phenylisothiazole 49 than from 2-phenylthiazole 50. From 5-phenylisothiazole 49, all molecules of 3-phenylisothiazole 47 must arise via BC-49, i.e. 49 + $h\nu$ → BC-49 → BC-47 → 47 (Scheme 9). From 2-phenylthiazole 50, however, 3-phenylisothiazole 47 can be formed by the pathway 50 + $h\nu$ → BC-50 → BC-49 → BC-47 → 47 with deuterium incorporation or by the pathway 50 + $h\nu$ → BC-50 → BC-51 → BC-47 → 47 without deuterium incorporation.

BC-49   **BC-49-4d$^+$**   **47-4d**

These results reveal that the photochemistry of isothiazoles and pyrazoles is not as mechanistically different as was once concluded. Thus both hetero-cycles transpose by the electrocyclic ring closure heteroatom migration mechanistic pathway. In the case of pyrazoles, heteroatom migration in the initially formed 1,5-diazabicyclopentene occurs regiospecifically (Scheme 12A) away from the azetine ring nitrogen, resulting in the formation of a 2,5-diaza-bicyclopentene isomer that can aromatize to an imidazole and a P$_6$ transposition. In the case of phenylisothiazoles, heteroatom migration in the initially formed 1-aza-5-thiabicyclopentene (Scheme 12B) occurs in both directions, resulting in the formation of 2-aza-5-thiabicyclopentene or 1-aza-5-thiabicy-clopentene isomers that upon aromatization result in the formation of an iso-meric P$_6$ thiazole or P$_5$ isothiazole respectively. The regiospecificity of the diazabicyclopentene migration could be because of the greater stability of a 2,5-diazabicyclopentene relative to a 1,5-diazabicyclopentene due to the greater strength of the C–N bond in the 2,5-isomer relative to the N–N bond in the

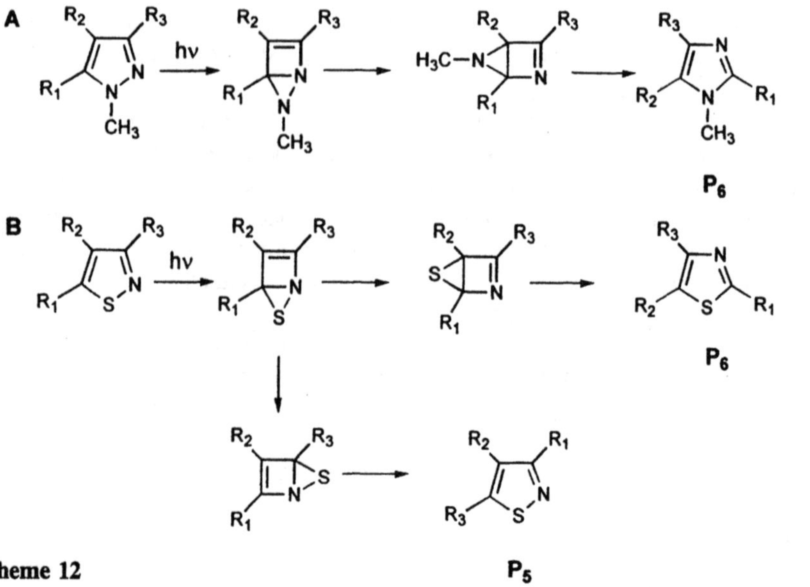

Scheme 12 P_5

1,5-diaza species. In the azathiabicyclopentene case, the difference between the S–N and S–C bond strengths is not expected to be as great and is therefore not anticipated to exert such a pronounced effect on the regiochemistry of the migration.

3. The P_4 Pathway

A more significant difference between pyrazole and isothiazole photochemistry, however, appears to be the minor role of the P_4 pathway in isothiazole transposition. Thus although the P_4 process is a major transposition pathway in pyrazole chemistry, it has been observed as only a minor pathway upon irradiation of phenylisothiazoles in benzene solution [36]. In fact, 4-phenylisothiazole 48, the compound most expected to react via the P_4 pathway, was the only isomer that did not yield a transposition product upon irradiation in benzene solution. This is not due to the photostability of the compound. Indeed, based on their quantum yields for consumption, 4-phenylisothiazole 48 is the most reactive of the six isomers. Nevertheless, even after consumption of 85% of 4-phenylisothiazole 48, no phototransposition product could be detected.

In marked contrast, when the reaction was carried out in methanol containing a small amount of NH_3, the P_4 transposition product, 4-phenylthiazole 51, was obtained in 90% yield along with a small amount of the deprotonated photocleavage product 53⁻, which was trapped by reaction with benzylbromide to yield the (E/Z)-benzylthioether 54. If the reaction is carried out in metha-

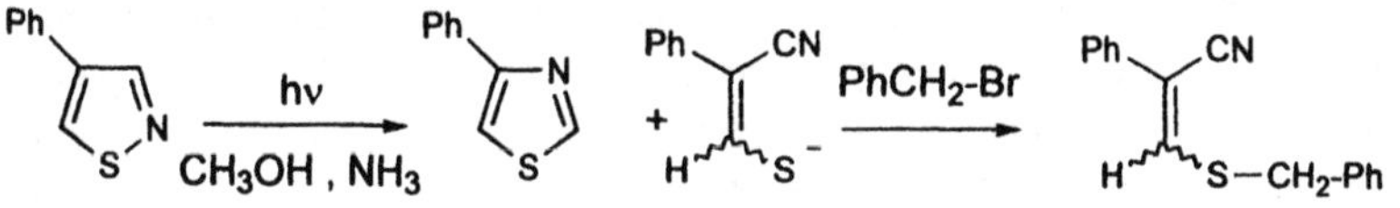

48   **51**   **53⁻**   **54**

nol containing a small amount of aqueous HCl, the phototransposition is completely quenched, and the only product observed is the photocleavage product 53H [41].

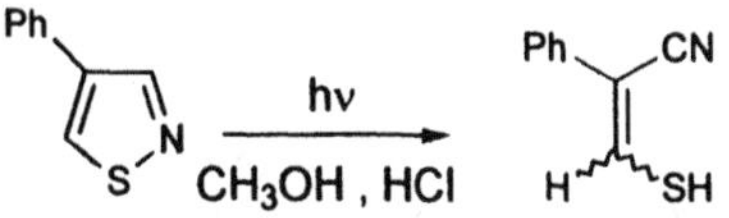

48   **53 H**

These results suggest that the biradical formed from photocleavage of the S–N bond (Scheme 13) partitions between hydrogen transfer from C-3 to S to yield the thionitrile photocleavage product 53H, which is deprotonated in the basic media, and isomerization to another intermediate, I, which is photochemically and/or thermally converted to the P_4 product, 4-phenylthiazole 51. Interestingly, the infrared spectrum of the crude benzylthioether 54 exhibited an absorption band at 2205 cm^{-1}, as expected for the nitrile functional group in 54, and another band at 2102 cm^{-1}, suggesting the presence of an isocyanide in the product mixture.

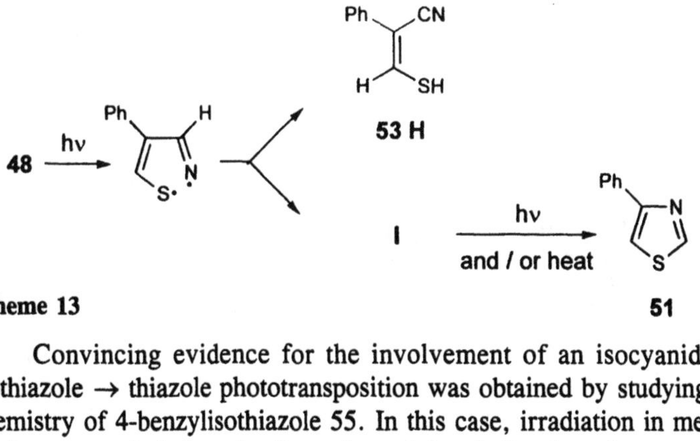

Scheme 13   **51**

Convincing evidence for the involvement of an isocyanide in the P_4 isothiazole → thiazole phototransposition was obtained by studying the photochemistry of 4-benzylisothiazole 55. In this case, irradiation in methanol containing ammonia led to the formation of the nitrile photocleavage product 56, which was trapped by reaction with benzylbromide to yield benzylthioether 57, and to the formation of the isocyanide photocleavage product 58, which was trapped by reaction with benzyl bromide followed by hydrolysis to yield the formamide 59. Experimental evidence also revealed that isocyanide 58 is thermally, but not photochemically, converted to the P_4 product, 4-benzylthiazole 60 [41].

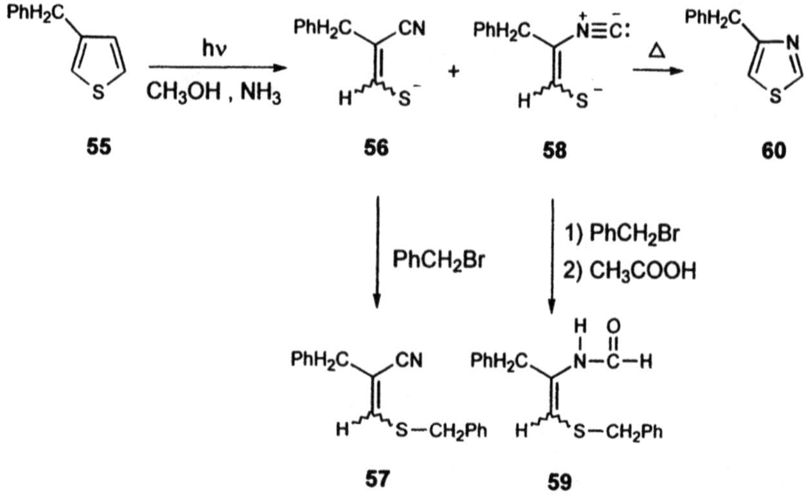

The P_4 isothiazole → thiazole transposition is thus mechanistically quite similar to the analogous pyrazole → imidazole isomerization. Scheme 14 shows that both reactions involve competition between ring cleavage to yield a nitrile cleavage product and to transposition involving an isocyanide intermediate.

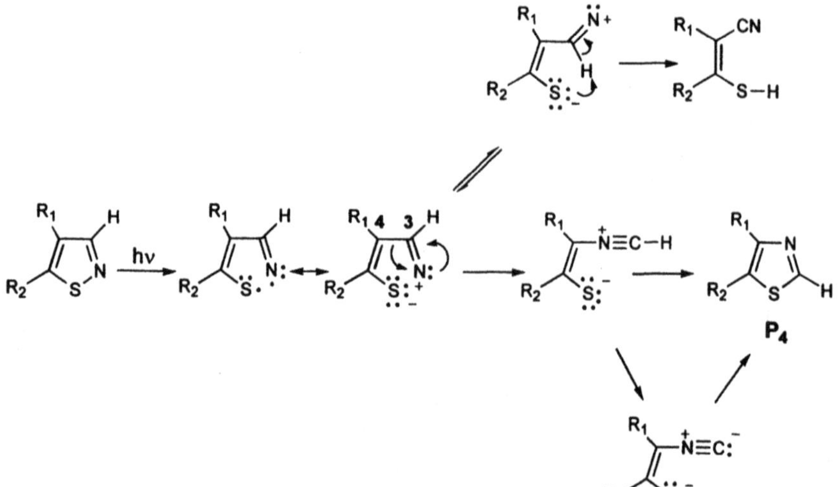

Scheme 14

IV. CONCLUSIONS

Contrary to earlier conclusions, more recent studies are showing that the photochemistry of pyrazoles and isothiazoles can be rationalized by a single unifying mechanistic scheme. Thus excitation of either heterocycle is followed by a competition between electrocyclic ring closure and subsequent heteroatom migration leading to P_5, P_6, and/or P_7 photoproduct formation, and to cleavage of the bond between the two heteroatoms to yield a diradical or zwitterionic species. This latter intermediate is the precursor of the nitrile photocleavage product and of the P_4 phototransposition product. Mounting evidence indicates that isocyanides are imortant intermediates on the P_4 transposition pathway.

ACKNOWLEDGMENTS

The author gratefully acknowledges the diligent and dedicated collaboration of the graduate students whose names appear as coauthors in the references.

REFERENCES

1. Lablache-Combier, A. *Photochemistry of Heterocyclic Compounds;* Buchardt, O., Ed.; Wiley: New York, 1976; p. 123. Padwa, A. *Rearrangements in Ground and Excited States*; de Mayo, P., Ed.; Academic Press: New York, 1980; Vol. 3, p. 501. Lablache-Combier, A. *CRC Handbook of Photochemistry and Photobiology;* Horspool, W. M.; Song, P.-S., Eds.; CRC Press: New York, 1994; pp. 803, 1063.

2. Tiefenthaler, H.; Dörscheln, W.; Goth, H.; Schmid, H. *Helv. Chim. Acta.* 1967, *50*, 2244–2258.

3. Catteau, J. P.; LaBlache-Combier, A.; Pollet, A. *J. Chem. Soc., Chem. Commun.* 1969, 1018.

4. Barltrop, J. A.; Day, A. C. *J. Chem. Soc., Chem. Commun.* 1975, 177–179.

5. Barltrop, J. A.; Day, A. C.; Moxon, P. D.; Ward, R. W. *J. Chem. Soc., Chem. Commun.* 1975, 786–787.

6. Barltrop, J. A.; Day, A. C.; Ward, R. W. *J. Chem. Soc., Chem. Commun.* 1978, 131–133.

7. (a) Kurtz, D. W.; Schechter, H. *J. Chem. Soc., Chem. Commun.* 1966, 689–690. (b) Ullman, E. G.; Singh, B. *J. Am. Chem. Soc.* 1966, *88*, 1844–1845; 1967, *89*, 6911–6916. (c) Singh, A.; Zweig, A.; Gallivan, J. B. *J. Am. Chem. Soc.*, 1972, *94*, 1199–1206. (d) Nishiwaki, T.; Nakano, A.; Matsuoka, H. *J. Chem. Soc.* C 1970, 1825–1829. (e) Nishiwaki, T.; Fujiyama, F. *J. Chem. Soc., Perkin Trans.* 1 1972, 1456–1459. (f) Wamhoff, H. *Chem. Ber.* 1972, *105*, 748–752. (g) Sato, T.; Yamamoto, K.; Fukui, K. *Chem. Lett.* 1973, 111–114. (h) Boeth, H.; Gragneux, A. R.; Eugster, C. H.; Schmid, H. *Helv. Chim. Acta* 1967, *50* , 137–142. (i) Padwa, A.; Chen, E.; Kua, A. *J. Am. Chem. Soc.* 1975, *97*, 6484–6491. (j) Dietliker, K.; Gilgen, P.; Heimgartner, H.; Schmid, H. *Helv. Chim. Acta* 1976, *59*, 2074–2099.

8. Pavlik. J. W.; Kurzweil, E. M. *J. Org. Chem.* 1991, *56*, 6313–6320.

9. Wakamatsu, S.; Barltrop, J. A.; Day, A. C. *Chem. Lett.*, 1982, 667–670.

10. Barltrop, J. A.; Day, A. C.; Mac, A. G.; Shahrisa, A.; Wakamatsu, S. *J. Chem. Soc., Chem. Commun.* 1981, 604–606.

11. (a) Ferris, J. P.; Orgel, L. E. *J. Am. Chem. Soc.* 1966, *88*, 1074. (b) Ferris, J. P.; Sanchez, R. A.; Orgel, L. E. *J. Mol. Biol.* 1968, *33*, 693–704. (c) Ferris, J. P.; Kuder, J. E. *J. Am. Chem. Soc.* 1970, *92*, 2527–2533. (d) Ferris, J. P.; Antonucci, F. R.; Trimmer, R. W. *J. Am. Chem. Soc.*, 1973, *95*, 919–920. (e) Ferris, J. P.; Antonucci, F. R. *J. Am. Chem. Soc.*, 1974, *96*, 2104–2110. (f) Koch, T. H.; Rodehorst, R. M. *J. Am. Chem. Soc.* 1974, *96*, 6706–6710. (g) Ferris, J. P.; Prabhu, K. V.; Strong, R. L. *J. Am. Chem. Soc.* 1975, *97*, 2835–2839. (h) Ferris, J. P.; Trimmer, R. W. *J. Org. Chem.* 1976, *41*, 19–24. (i) Ferris, J. P.; Narang, R. S.; Newton, T. A.; Rao, V. R. *J. Org. Chem.* 1979, *44*, 1273–1278.

12. (a) Beak, P.; Miesel, J. L.; Messer, W. R. *Tetrahedron Lett.* 1967, 5315–5317. (b) Beak, P.; Messer, W. R. *Tetrahedron* 1969, *25*, 3287–3295.

13. (a) Barltrop, J. A.; Day, A. C.; Irving, E. *J. Chem. Soc., Chem. Commun.* 1979, 881–883. (b) Barltrop, J. A.; Day, A. C.; Irving, E. *J. Chem. Soc., Chem. Commun.* 1979, 966–968.

14. Pavlik, J. W.; Kebede, N. Manuscript submitted for publication.

15. Bicyclic species generated by photolysis of pyrroles have been trapped by reaction with furan. See: Barltrop, J. A.; Day, A. C.; Ward, R. W. *J. Chem. Soc., Chem. Commun.* 1978, 131–133.

16. Connors, R. E.; Pavlik, J. W.; Burns, D. S.; Kurzweil, E. M. *J. Org. Chem.*, 1991, *56*, 6321–6326.

17. Matsushita, T.; Tanaka, H.; Nishimoto, K.; Osamura, Y. *Theor. Chim. Acta.*, 1983, *63*, 55–68.

18. Tanaka, H.; Matsushita, T.; Nishimoto, K. *J. Am. Chem. Soc.*, 1983, *105*, 1753–1760.

19. Connors, R. E.; Burns, D. S.; Kurzweil, E. M.; Pavlik, J. W. *J. Org. Chem.* 1992, *57*, 1937–1940.

20. Pavlik, J. W.; Connors, R. E.; Burns, D. S.; Kurzweil, E. M. *J. Am. Chem. Soc.* 1993, *115*, 7645–7652.

21. Pavlik, J. W.; Kebede, N.; Bird, N. P.; Day, A. C.; Barltrop, J. A. *J. Org. Chem.* 1995, *60*, 8138–8139.

22. (a) Ferris, J. P.; Antonucci, F. R. *J. Am. Chem. Soc.* 1974, *96*, 2014–2019. (b) Ferris, J. P.; Antonucci, R. R.; Trimmer, R. W. *J. Am. Chem. Soc.* 1973, *95*, 919–920. (c) Ferris, J. P.; Trimmer, R. W. *J. Org. Chem.* 1976, *41*, 13–19.

23. (a) Padwa, A. *Acc. Chem. Res.* 1976, *9*, 371–378. (b) Griffin, G. W.; Padwa, A. In *Photochemistry of Heterocyclic Compounds*; Buchardt, O., Ed.; John Wiley and Sons: New York, 1976.

24. 3-Phenyl-2-(N-phenylimino)-2H-azirine is photochemically converted to 1,2-diphenylimidazole but rearranges thermally to yield 1,4-diphenylpyrazole. See Ref. 25.

25. (a) Padwa, A.; Smolanoff, J.; Tremper, A. *Tetrahedron Lett.* 1974, 29–32. (b) Padwa, A.; Smolanoff, J.; Tremper, A. *J. Am. Chem. Soc.* 1975, *97*, 4682–4691.

26. LaBlache-Combier, A.; Pollet, A. *Tetrahedron* 1972, *28*, 3141–3151.
27. Wynberg, H.; van Driel, H.; Kellogg, R. M.; Butler, J. *J. Am. Chem. Soc.* 1967, *89*, 3487–3494.
28. Pavlik, J. W.; Pandit, C. R.; Samuel, C. J.; Day, A. C. *J. Org. Chem.* 1993, *58*, 3407–3410.
29. Vernin, G.; Poite, J. C.; Metzger, J.; Aune, J. P.; Dou, J. M. *Bull. Soc. Chim. Fr.* 1971, 1101–1104.
30. Vernin, G.; Jauffred, R.; Richard, C.; Dou, H. J. M.; Metzger, J. *J. Chem. Soc. Perkin Trans.* 2 1972, 1145–1150.
31. Vernin, G.; Riou, C.; Dou, H. J. M.; Bouscasse, L.; Metzger, J.; Loridan, G. *Bull. Soc. Chim. Fr.* 1973, 1743–1751.
32. Riou, C.; Poite, J. C.; Vernin, G.; Metzger, J. *Tetrahedron* 1974, *30*, 879–898.
33. Riou, C.; Vernin, G.; Dou, H. J. M.; Metzger, J. *Bull. Soc. Chem. Fr.* 1972, 2673–2678.
34. Vernin, G.; Poite, J. C.; Dou, H. J. M.; Metzger, J. *Bull. Soc. Chem. Fr.* 1972, 3157–3167.
35. Kellog, R. R. *Tetrahedron Lett.* 1972, 1423–1429.
36. Pavlik, J. W.; Tongcharoensirikul, P.; Bird, N. P.; Day, A. C.; Barltrop, J. A. *J. Am. Chem. Soc.* 1994, *116*, 2292–2300.
37. Kojima, M.; Maeda, M. *J. Chem. Soc., Chem. Commun.* 1970, 386–387.
38. Maeda, M.; Kojima, M. *Tetrahedron Lett.* 1973, 3523–3526.
39. Maeda, M.; Kawahara, A.; Kai, M.; Kojima, M. *Heterocycles*, 1978, *3*, 389–393.
40. Maeda, M.; Kojima, M. *J. Chem. Soc., Perkin* 1 1978, 685–692.
41. Pavlik, J. W.; Trongcharoensirikul, P. Submitted for publication.

3

Photochemistry of (S-Hetero)cyclic Unsaturated Carbonyl Compounds

Paul Margaretha

Institute of Organic Chemistry, University of Hamburg, Hamburg, Germany

I. INTRODUCTION

The photochemistry of organosulfur compounds in general has been the subject both of annual reports [1,2] and of timely scattered reviews [3–5]. Light-induced reactions of specific chromophores, such as sulfoxides and sulfones [6], thiocarbonyl compounds [7], and thiols [8] have been discussed in more detail, as has the photochemistry of thiophenes and certain other specific ring systems [9,10].

The formal substitution of a C–S–C unit into a molecule in place of a C–CH$_2$–C fragment results in several chemically significant changes. These include 1.) longer bond lengths for the C–S bonds (181 vs. 153 pm), 2.) smaller bond angles for the C–S–C group (95° vs. 109°), and 3.) weaker C–S bonds (60 vs. 83 kcal/mol). In addition, the change from C–O–C to C–S–C introduces a lower lone-pair ionization potential at the heteroatom. As will be apparent in the chapters that follows, these structural factors can have important effects on photochemical behavior. These chapters deal with the photochemistry of unsaturated cyclic carbonyl compounds containing one S-atom in the ring, with a focus on how these important structural factors affect the interactions of the sulfur atom, double bond, and carbonyl group.

II. ABSORPTION SPECTRA

The absorption maximum around $\lambda \approx 210$ nm for simple dialkyl sulfides [11], corresponding to a n-σ* transition, experiences an important bathochromic shift for β-ketosulfides ($\lambda_{max} \approx 240$ nm) [12] due to a conjugative effect between the two functional groups. This interaction is illustrated in the spectra of 2H,6H-thiin-3-ones [13] and also of their corresponding 1-oxides [14], with the appearance of an additional absorption band around 275 nm that does not appear in the spectrum of e.g. 2H,6H-pyran-3-ones [15] (Scheme 1).

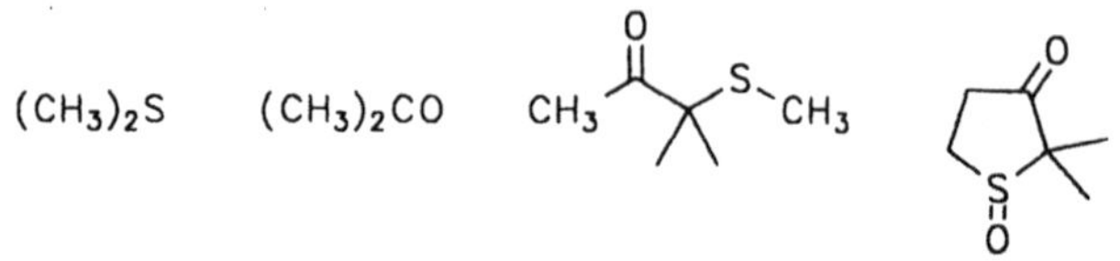

	$(CH_3)_2S$	$(CH_3)_2CO$		
n$\sigma^\bullet$	213 (3.0)	—	244 (2.7)	232 (2.5)
n$\pi^\bullet$	—	279 (1.1)	296 (2.6)	297 (2.0)

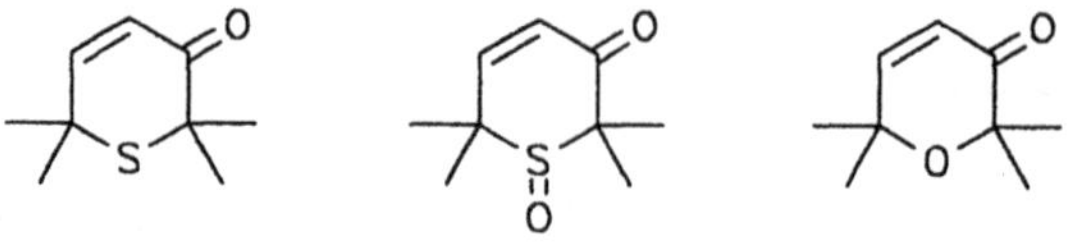

$\pi\pi$*	223 (3.8)	223 (3.8)	221 (4.0)
nσ*	277 (2.6)	271 (2.3)	—
nπ*	350 (2.3)	349 (2.0)	346 (1.6)

Scheme 1

In contrast to aliphatic esters or lactones which only show end absorption around 230 nm, thiolesters exhibit a strong absorption band with $\lambda_{max} \approx$ 235 nm, corresponding to a π-π* transition [16]. Conjugation of the C(O)S group with an additional C=C double bond shifts the maximum to 265–270 nm as illustrated for 2(3H)-thiophenones [17] or 2(5H)-thiophenones [18] (Scheme 2). These latter compounds exhibit an additional n-σ* absorption band at $\lambda \approx$ 320 nm.

The auxochromic effect of an S-atom linked directly to a C=C double bond is reflected in the spectrum of ethylthioethene (λ_{max} = 229 nm) [19]. A

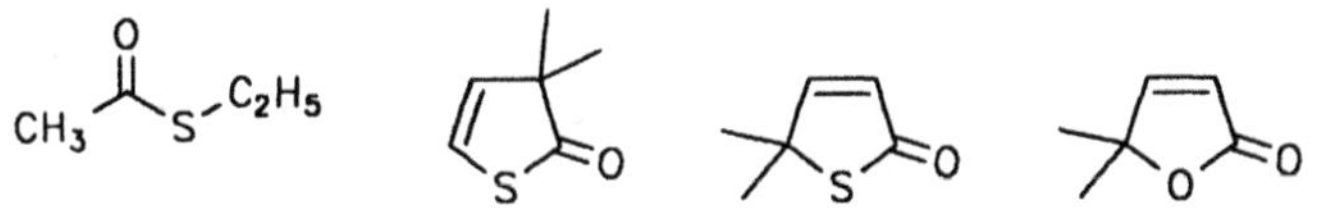

$\pi\pi*$	235 (3.1)	267 (3.2)	267 (3.3)	232 (3.9)
$n\sigma*$			319 (1.6)	

Scheme 2

similar effect is caused by an alkylthio group linked to $C(\beta)$ of an α,β-unsaturated ketone [20], as illustrated in the substitution of the CH_2 group in 5,5-dimethylcyclopent-2-enone by different heteroatoms, e.g. S [21], S(O) [22], O [23], or N-alkyl [21], inducing a bathochromic shift of the π-$\pi*$ absorption band (Scheme 3).

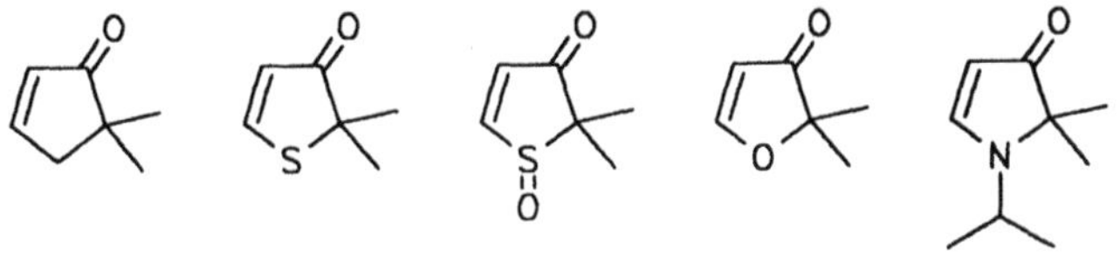

$\pi\pi*$	222 (4.0)	311 (3.8)	276 (3.8)	254 (3.8)	322 (4.0)
$n\pi*$	335 (1.7)	352 (2.3)	350 (2.0)	333 (1.7)	n.o.

Scheme 3

A superposition of these last two effects is observed in the spectrum of 1-thiocoumarin [24], as compared to that of coumarin [25] or 1-quinolone [26] (Scheme 4), although assignment of each band in these spectra becomes more complex.

From all these examples it becomes evident that S-heterocyclic unsaturated carbonyl compounds are convenient compounds for preparative irradiations,

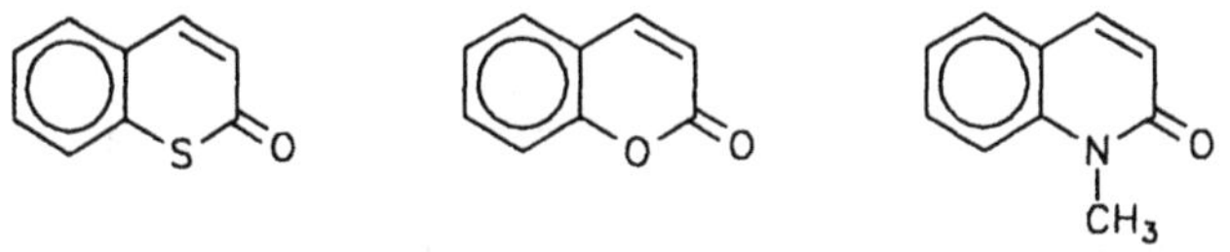

240 (4.1)	274 (4.0)	275 (3.9)
286 (3.6)	308 (3.7)	324 (3.7)
300 (3.5)		
340 (3.1)		

Scheme 4

as these reactions can be carried out at wavelengths > 305 nm and often even
> 350 nm, thus minimizing the risk of consecutive photochemical conversions
of the primary photoproducts.

III. COMPOUNDS WHEREIN THE S-ATOM IS NOT LINKED TO A sp²-HYBRIDIZED C-ATOM

Phenacyl sulfides are known to undergo three main processes from their triplet state: 1.) charge transfer from the heteroatom to the carbonyl group [27–
29], 2.) cleavage of the bond between the C-atom next to the carbonyl C-atom
and the S-atom (so-called β-cleavage) [28,29], and 3.) γ-hydrogen abstraction
by the carbonyl O-atom with subsequent cleavage of the same C–S bond as
above in the 1,4-biradical [30,31]. This last sequence is used to obtain thio-
aldehydes which can be trapped by enes or dienes (Scheme 5).

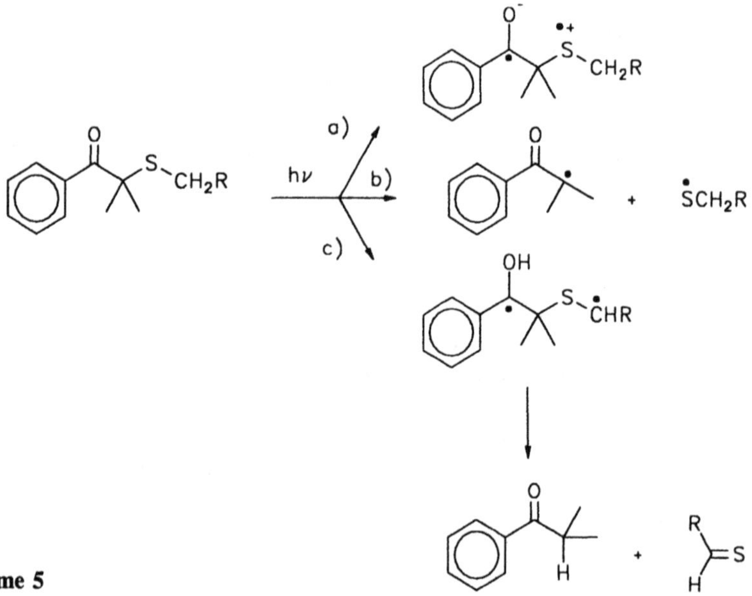

Scheme 5

From this context it is interesting to note that isothiochroman-4-ones—
which essentially represent (bi)cyclic derivatives of phenacylsulfides—behave
differently on irradiation as they photoisomerize to thiochroman-3-ones, albeit
in moderate yield [32]. As a reasonable proposition, 1-methylene-9-thia-
spiro[5,3]nona-2,4-dien-7-ones were assumed to be the intermediates, which
would then undergo a consecutive light induced α-cleavage to give the final
products (Scheme 6).

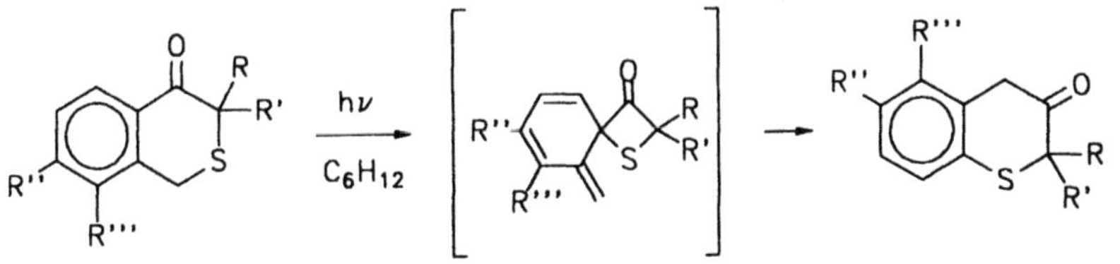

R	R'	R''	R'''	yield
H	H	H	H	21%
H	H	OCH$_3$	H	40%
H	H	H	CH$_3$	30%
H	CH$_3$	H	H	30%
CH$_3$	CH$_3$	H	H	37%

Scheme 6

Indeed, the formation of 2-alkenyl-thietan-3-ones represents the typical photochemical behavior for 2*H*,6*H*-thiin-3-ones and bicyclic compounds containing this ring system. Depending on the substitution pattern and on the wavelength chosen, these 2-alkenyl-3-thiacyclobutanones are stable [13, 33] or undergo further light induced conversions, e.g. a 1,3-acyl shift, to the final product(s) [34–36] as illustrated in Scheme 7.

A sulfuranyl-alkyl biradical has been proposed [13] as intermediate in this ring contraction (Scheme 8). Nevertheless, the question whether trialkylsulfuranyl radicals are to be considered stable intermediates [37] or just transition states [38] remains open.

It has been established [13] that this rearrangement of 2*H*,6*H*-thiin-3-ones to 2-(alk-1-enyl)-thietan-3-ones occurs from the excited singlet state with high quantum yields ($\Phi \approx 0.8$). This contrasts the very inefficient ($\Phi \approx 0.01$) triplet state rearrangement of 4,4-dialkylcyclohex-2-enones to bicyclo[3.1.0]hexan-2-ones [39]. In order to understand better the effect of this (formal) substitution of C(5) in the cyclohexenones by an S-atom, the corresponding 2*H*,6*H*-pyran-3-ones [15] were investigated. These compounds behave again in a different way on irradiation, as they undergo a *retro-Diels–Alder* reaction to a carbonyl compound and an alkenylketene, again from the triplet state. The behavior of these three types of compounds on irradiation is summarized in Scheme 9. The common trend for all three compounds seems to be a weakening of the bond between the allylic carbon atom (C(4) in the cyclohexenone,

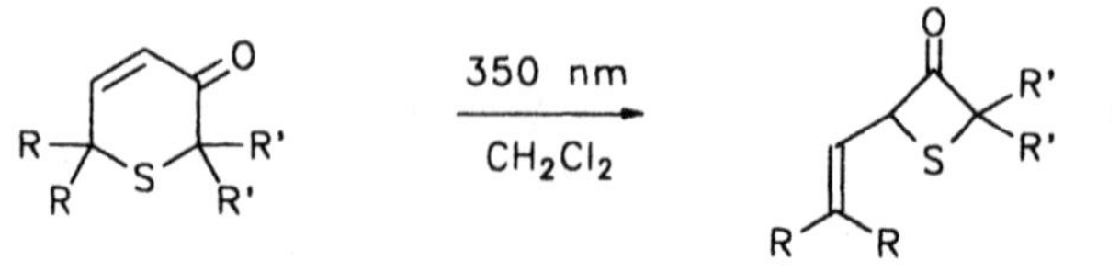

Scheme 7

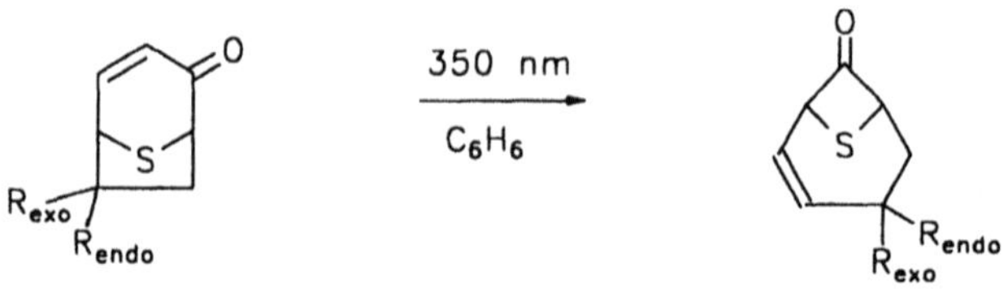

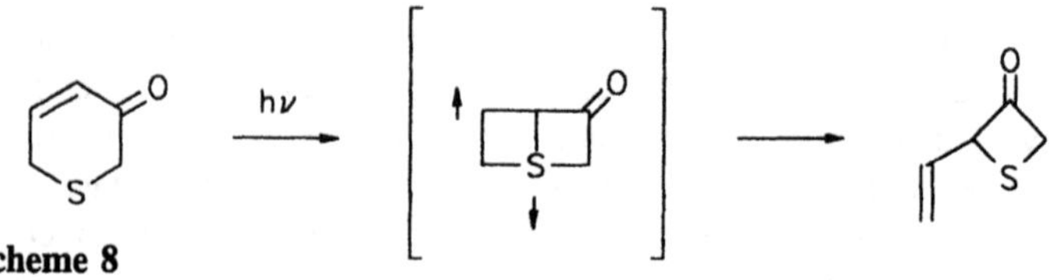

Scheme 8

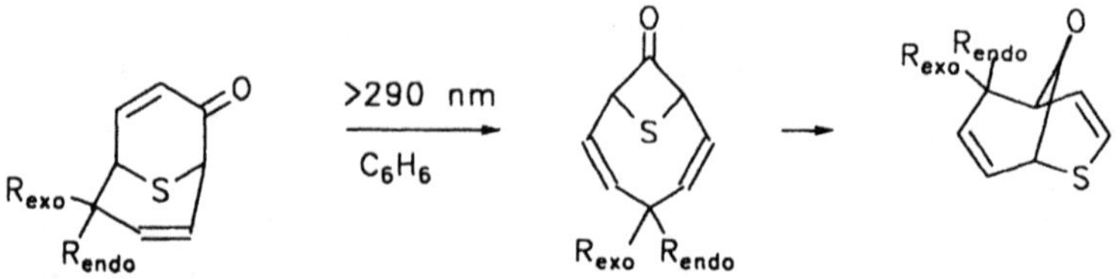

C(6) in the heterocycles) and the neighboring atom (C, S or O) concomitant with either, 1.) bridging of C_β to the CH_2 group to form a five-membered ring, 2.) bridging of the sulfur to form a four-membered ring, or 3.) cleavage of the C–C(O) bond to give a carbonyl compound.

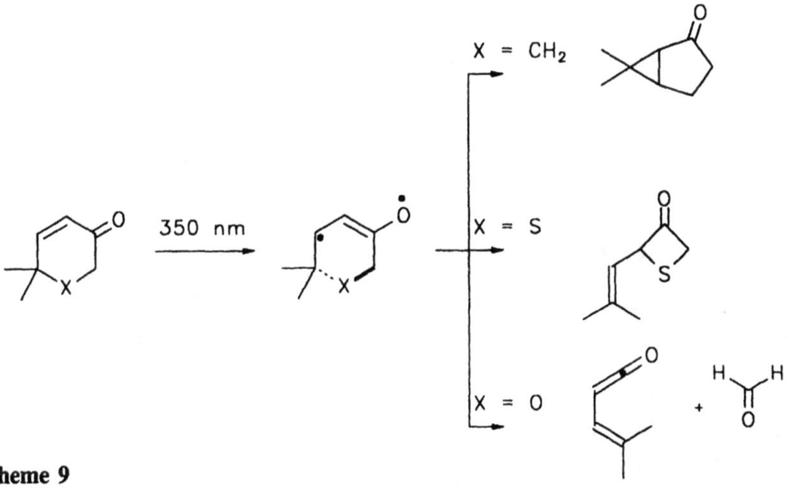

Scheme 9

This weakening of the bond between the allylic carbon and the neighboring atom can lead to bond rupture as illustrated in the photoisomerization of 2H,6H-thiin-3-one-1-oxides to 3H,7H-1,2-Oxathiepin-4-ones [14]. These seven-membered ring systems undergo further thermal ring contraction to thietanones via a 2,3-sigmatropic rearrangement followed by a *Pummerer* reaction (Scheme 10).

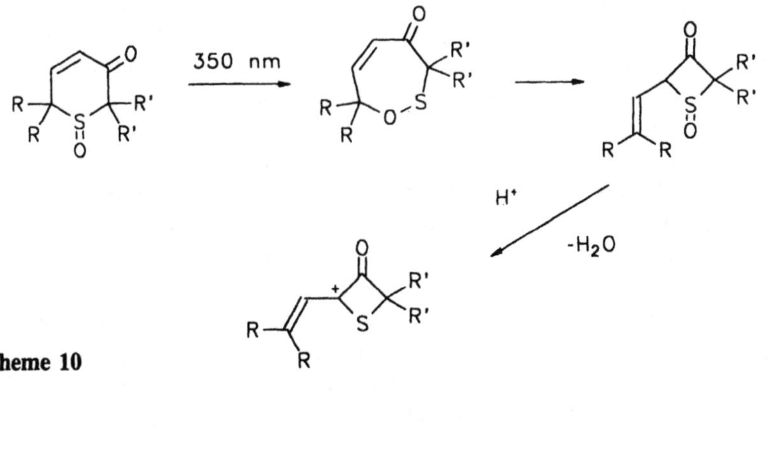

Scheme 10

Compounds where the sulfur atom is in through space bonding distance to the C=O bond, and the C=C double bond is distant from this same carbonyl group, undergo CT-interaction (S to CO) on irradiation, as shown in Scheme 11 for 9-thiabicyclo[3.3.1]non-6-en-2-one [40].

Scheme 11 Φ = 0.16   35%

IV. THE –S–C=C–C(O)– CHROMOPHORE

Both intermolecular [41–44] and intramolecular [45–48] enone/olefin photocycloadditions have gained increasing importance in organic syntheses. Heteropenta- and hexacyclic enones having an oxygen atom instead of C(4) have been extensively investigated in such reactions, as on the one side these oxaenones represent valuable model compounds for the elucidation of mechanistic aspects of such reactions, e.g. the regioselectivity in these [2 + 2]-photocycloadditions [49–52], and on the other side the photocycloadducts resulting from these reactions have found interesting synthetic applications [53–55].

    4-Thiacyclopent-2-enones behave very much alike the corresponding 3(2H)-furanones in undergoing photocyclodimerization in the absence, and mixed [2 + 2] photocylcoadditions in the presence, of alkenes [21]. Similarly, 2-(prop-2-enyl)thiophen-3(2H)-ones on irradiation isomerize to 7-thiatricyclo[3.2.1.0^{3,6}]octan-2-ones [56] (Scheme 12).

    The enhanced regioselectivity observed in photocycloadditions of both 3(2H)-thiophenones and 3(2H)-furanones as compared to the corresponding carbocyclic enones, e.g. cyclopent-2-enones has been ascribed to the possibility of CT-interaction in the excited heteracyclopentenones thus increasing the charge separation on the olefinic carbon atoms [49] (Scheme 13).

    The photochemical behavior of 2,3-dihydro-4H-thiin-4-ones (= 4-thiacyclohex-2-enones) resembles strongly that of cyclohept-2-enones; due to the longer bond length of the C–S bond, these rings become less rigid, and therefore the twisting around the C–C double bond in the excited state is facilitated. It is therefore not surprising that the parent compound only photodimerizes with very low efficiency and that even photocycloaddition to 2-methylpropene is accompanied by polymerization of starting material to the extent of 60–70% [57]. On the other hand, 2,2-dimethyl-2,3-dihydro-4H-thiin-4-one in irradiation in furan affords two [4 + 2] cycloadducts and on irradiation in methanol gives

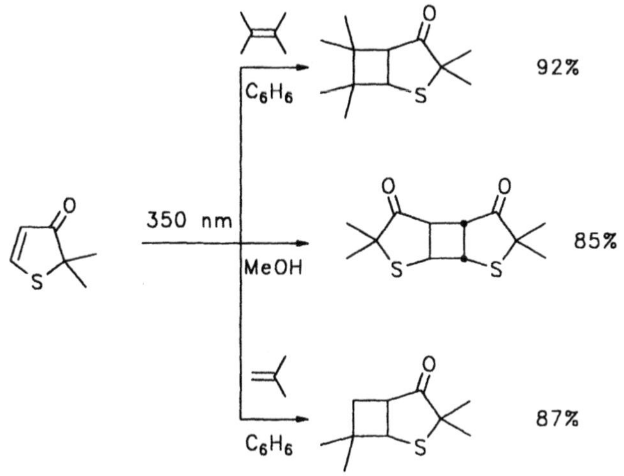
350 nm
C_6H_6
MeOH
C_6H_6
92%
85%
87%

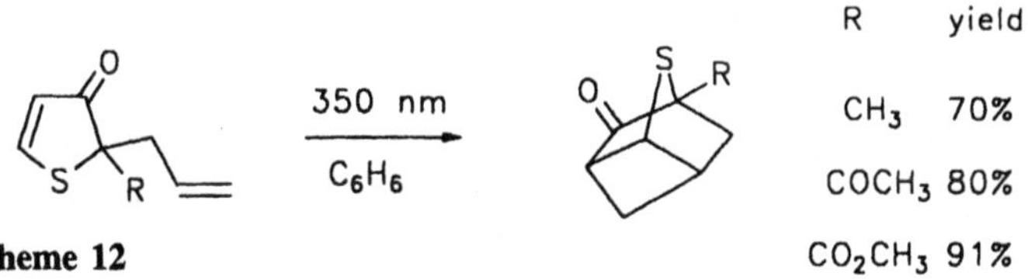
350 nm
C_6H_6
R yield
CH_3 70%
$COCH_3$ 80%
CO_2CH_3 91%

Scheme 12

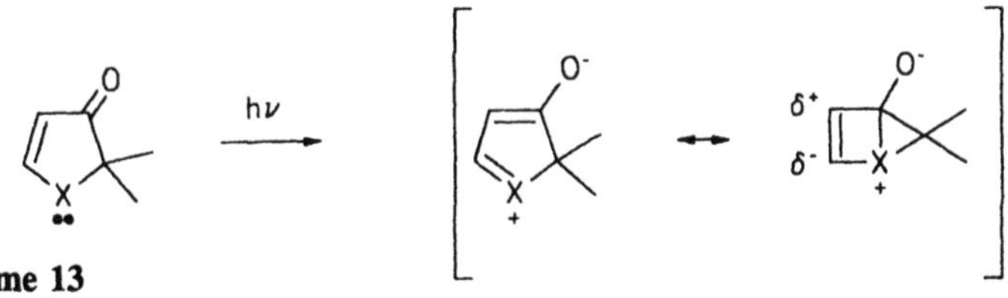
hν

Scheme 13

a 3:2 mixture of 5- and 6-methoxy-2,2-dimethylthian-4-one [58]. Irradiation in CD_3OD affords the same (deuterated) adducts with the CD_3O and D groups trans to each other, results compatible with cis addition of methanol to a E-configured ground state enone (Scheme 14).

From the facts that on irradiation of the same thiinone in furan/methanol *only* the furan cycloadducts are formed, and that the formation of these cycloadducts is quenched on addition of *E*-stilbene, it was stated [58] that the thiinone represents the first monocyclic cyclohexenone analogue undergoing

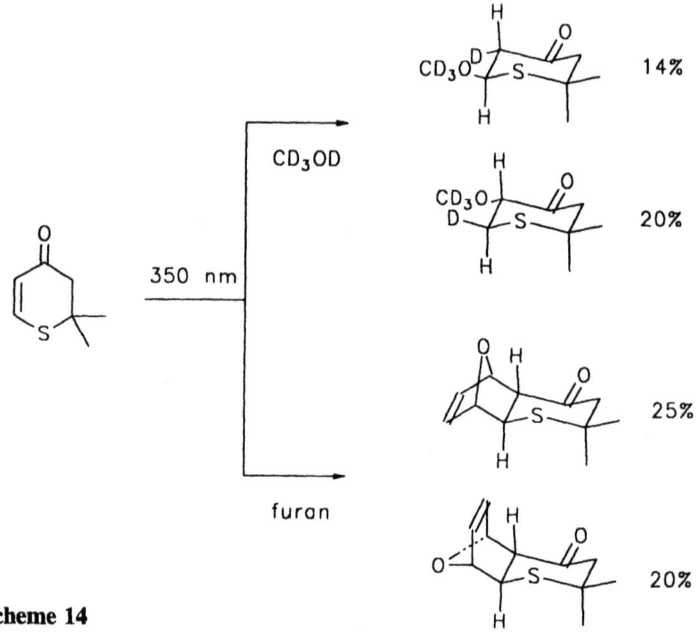

Scheme 14

efficient methanol photoaddition, and it was assumed that the enone triplet, i.e. the precursor to the *E*-cycloalkenone, is trapped by furan to give biradicals with an allylic moiety in the furan ring (Scheme 15).

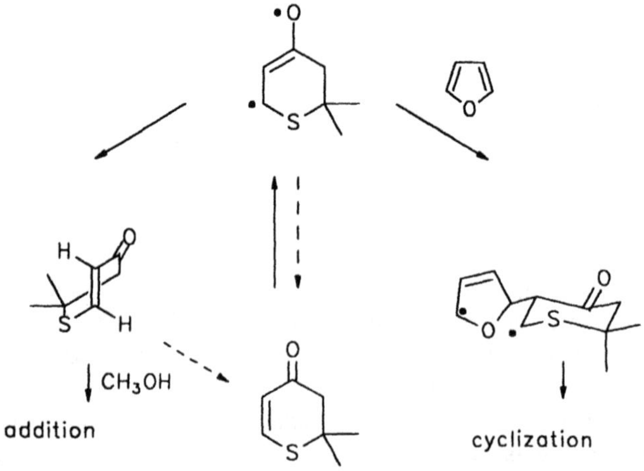

Scheme 15

The corresponding S-oxides show a totally different behavior on irradiation as they do not undergo any cycloadditions to alkenes. 2,2-Dimethyl-3(2*H*)-

thiophenon-1-oxide isomerizes quantitatively on iradiation (350 nm) to 5-isopropylidene-1,2-oxathiol-1-oxide [22]. This rearrangement is initiated by cleavage of the S–C(CH₃)₂ bond (α-cleavage of a sulfoxide) from the excited singlet state, the quantum yield for this conversion (Φ = 0.65) indicating that the resulting biradical closes preferentially to the sulfinate (Scheme 16).

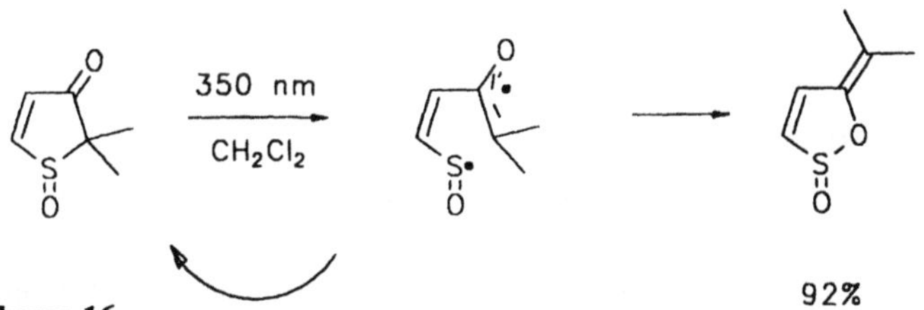

Scheme 16

The above-mentioned ring size effect, i.e. that excited thiin-4-ones deactivate efficiently via twisting of the C–C double bond, is reflected in the almost total unreactivity of 2,3-dihydrothiin-4-one-1-oxides on irradiation [59]. Both the parent and the 2,2-dimethyl compound exhibit the same behavior on irradiation in undergoing very slow decomposition without noticeable new product formation, except for traces of the corresponding (deoxygenated) thiins (Scheme 17).

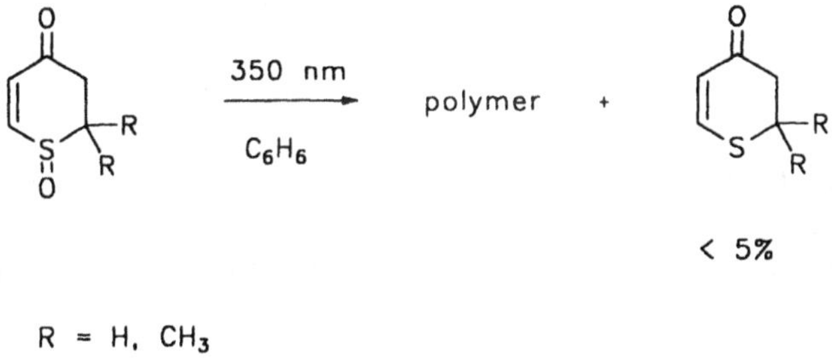

R = H, CH₃

Scheme 17

V. THE –C=C–C(O)–S– CHROMOPHORE

The photochemistry of monocyclic α,β-unsaturated thiolactones has only been investigated recently. One reason for this is that the first syntheses of 2(5H)-thiophenones from thiophenes were only developed in the late 1960s and early 1970s and that the first report on a 5,6-dihydrothiin-2-one was published even later [60].

A. 2(5H)-Thiophenones

Irradiation (305 nm) of 2(5H)-thiophenone in water or in alcohols affords 4-mercaptocrotonic acid or the corresponding esters, respectively [18,61,62]. The reaction (Scheme 18) has to be stopped at about 50% conversion, as the products themselves undergo consecutive light induced reactions (*vide infra*, Sec. B).

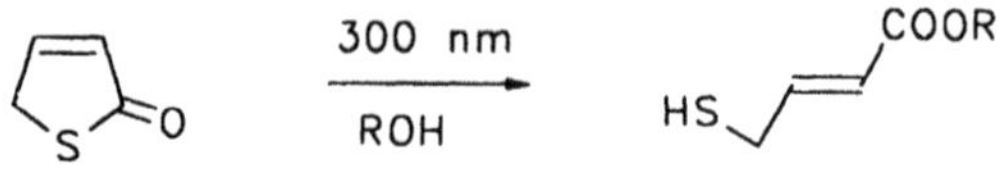

Scheme 18   R = H, CH$_3$, C$_2$H$_5$, i-C$_3$H$_7$, t-C$_4$H$_9$

This reaction is not quenched by 2,5-dimethyl-2,4-hexadiene or naphthalene. C(3)- and C(5)-alkylsubstituted 2(5H)-thiophenones behave similarly, but alkyl groups on C(4) prevent this photosolvolysis [18]. The rates for these reactions are similar in water, methanol, ethanol, 2-propanol or *t*. butanol but about ten times slower in 2,2,2-trifluoroethanol. All this suggests that the (singlet) excited thiolactone is attacked at C(4) by nucleophilic reagents to give an intermediate ketene which then affords the *E*-configured ester (or acid) selectively (Scheme 19).

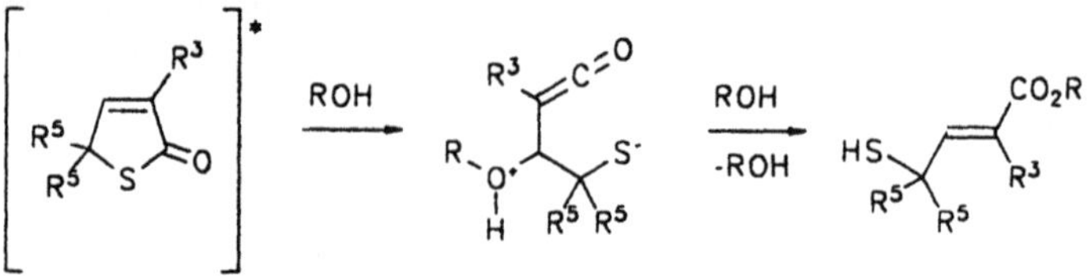

Scheme 19

This same preference for nucleophilic attack is reflected in the reaction of 2(5H)-thiophenones with alkenes (Scheme 20). This is again a singlet state reaction, wherein moderate yields of [2 + 2]-photocycloadducts are obtained with 2,3-dimethylbut-2-ene but none with 2-methylpropene [18].

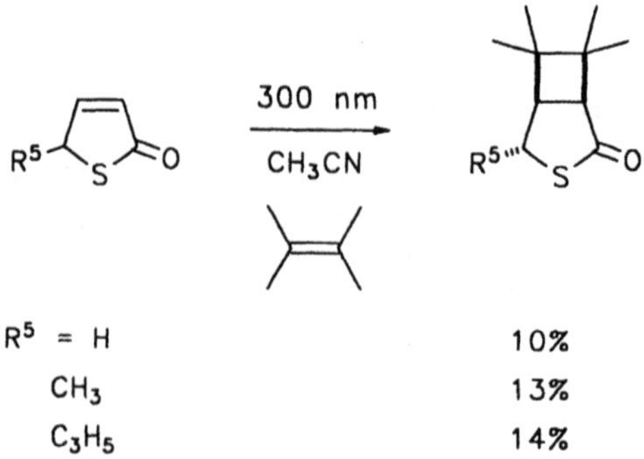

R^5 = H	10%
CH$_3$	13%
C$_3$H$_5$	14%

Scheme 20

On irradiation in alcohols in the presence of a triplet sensitizer, e.g. xanthone, 2(5*H*)-thiophenone affords *RH*-addition products [62] (Scheme 21). This behavior is identical to that of the corresponding furanones, which also undergo H-atom abstraction from the excited triplet state [63].

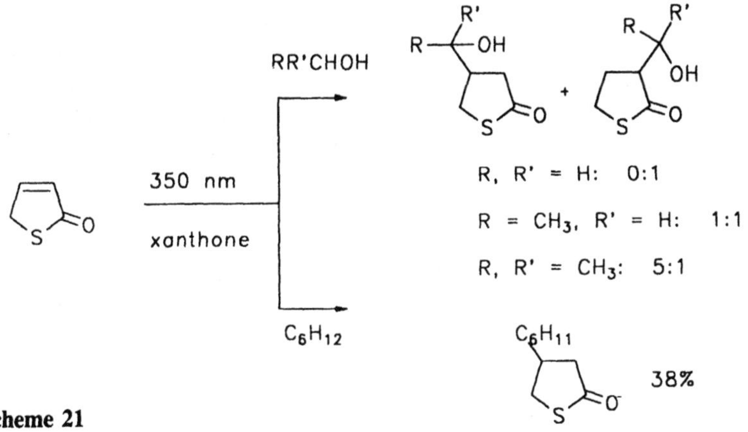

$$R, R' = H: \quad 0:1$$
$$R = CH_3, R' = H: \quad 1:1$$
$$R, R' = CH_3: \quad 5:1$$

38%

Scheme 21

B. 4-Mercapto-(*E*)-But-2-Enoic Acid Esters

As mentioned above, the title compounds are formed by photosolvolysis of 2(5*H*)-thiophenones in alcohols. As the extinction coefficient ε of these bifunctional (allylic thiols, α,β-unsaturated esters) molecules at the irradiation wavelength (300 nm) is about the same as that of the thiolactone precursors, they undergo a consecutive light induced homolysis of the S–H bond to give the corresponding thiyl radicals [63]. These intermediates have become versatile synthons for the preparation of a multitude of five- and six membered S-heterocycles, either through direct rearrangement (path *a*) or via trapping with alkenes (path *b*) or alkynes (path *c*) and subsequent cyclization (Scheme 22).

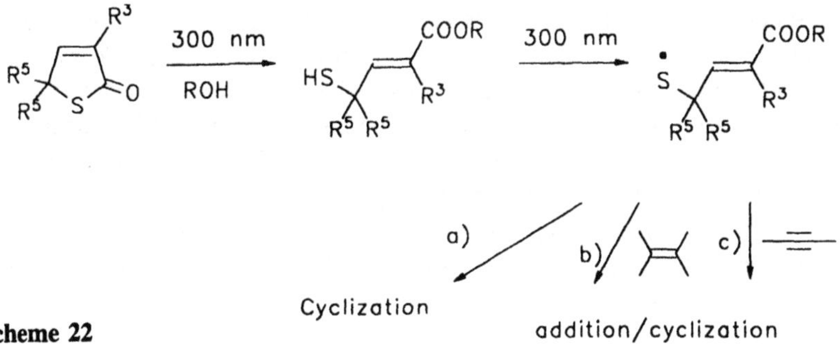

Scheme 22

While cyclization of a pent-4-enyl radical to a cyclopentyl radical—being inconsistent with the rules for ring closure—has not been observed, 1-thiapent-4-enyl radicals undergo ring closure to thiolanes, albeit in low yields [18,64]. The longer C–S bond(s) make this *endo-trig* cyclization less unfavorable than for the all-C species. A diastereoisomeric mixture of 5,6-dihydrothiin-4-carboxylates is formed via *exo-trig* cyclization of 1-thiahepta-3,6-dienyl radicals obtained from 3-(prop-2-enyl)-2(5H)-thiophenones. In the absence of a conveniently located C=C bond, the intermediate thiyl radical can eliminate sulfur, as illustrated in the (overall) conversion of 5,5-dimethyl-2(5H)-thiophenone to s-*trans*-1,1-dimethoxy-4-methyl-1,3-pentadiene [65] (Scheme 23).

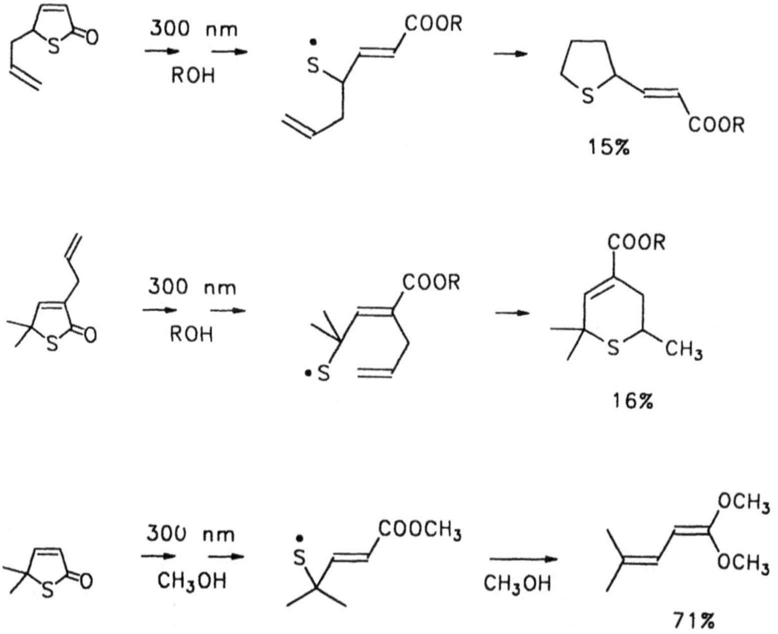

Scheme 23

When the alkenylthio radicals are generated in the presence of alkenes, a highly efficient addition/cyclization sequence leads to carboalkoxy-tetrahydrothien-3-yl alkyl radicals. In agreement with the stereochemical rules concerning the *exo-trig* cyclization of substituted hexenyl radicals [66], 3-thiahex-5-enyl radicals bearing a substituent on C(4) (originally on C(5) of the thiophenone) undergo stereoselective ring closure to 2,3-*trans*-disubstituted thiolanes, while those bearing a substituent on C(1)—stemming from the alkene—undergo stereoselective ring closure to 3,4-*cis*-disubstituted thiolanes [18,67] (Scheme 24).

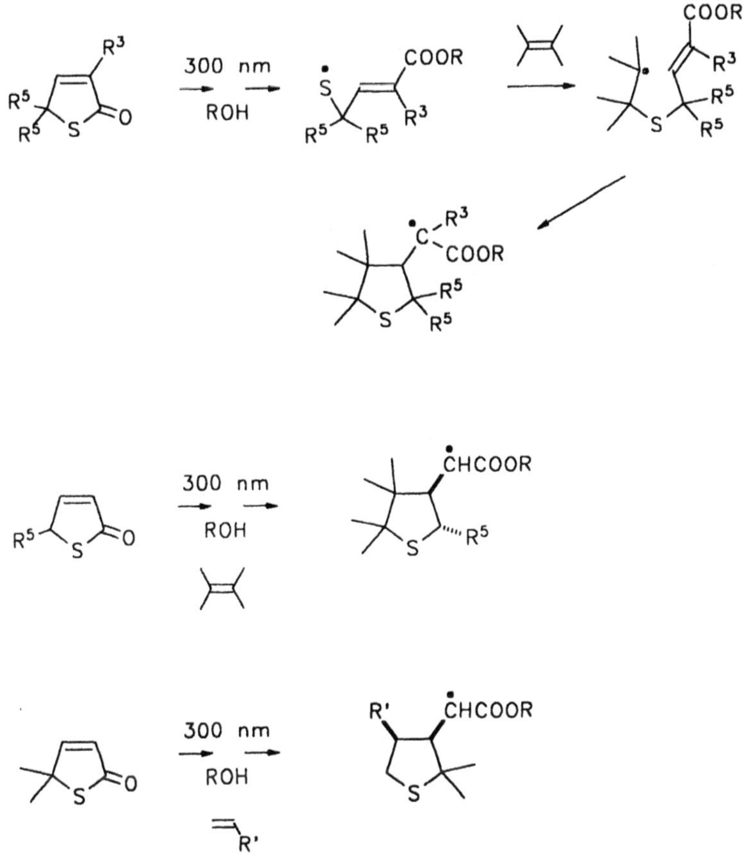

Scheme 24

These new radical intermediates either react with H-atom donors directly or interact—both inter- and intramolecularly—with a C=C double bond to give a new radical that then reacts with an H-atom donor [18,68] (Scheme 25).

Irradiation of 2(5H)-thiophenones in alcohols in the presence of alkynes gives carboalkoxy-dihydrothienylalkyl radicals by the same sequence as discussed above for alkenes. These intermediates can again react with an H-atom donor, be trapped by excess alkyne, or undergo a but-3-enyl ⇒ cyclopropanemethyl rearrangement to afford bicyclic thiolane derivatives [18,69] (Scheme 26).

VI. THIOCOUMARINS

While the photochemistry of coumarins was investigated in detail already at the beginning of this century [70], research on thiocoumarin has for the most part

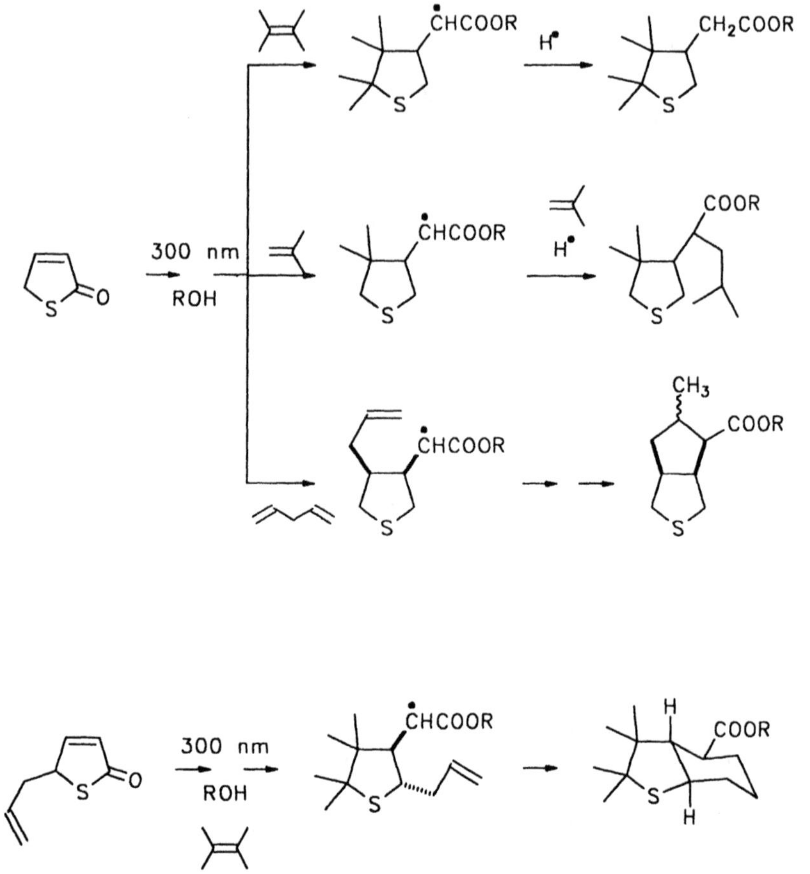

Scheme 25

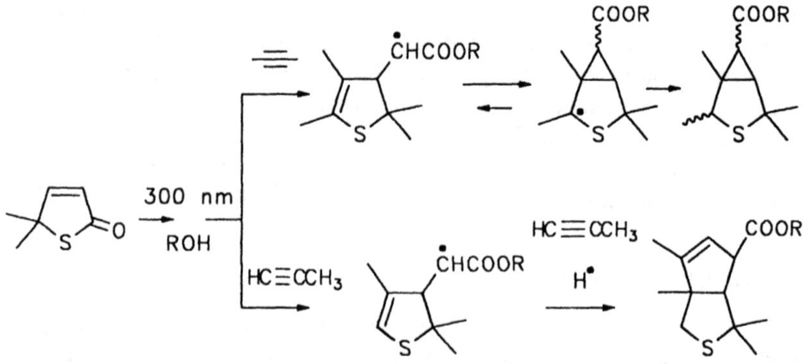

Scheme 26

been neglected [71]. It is therefore not surprising that the first brief report on a light induced reaction of a thiocoumarin derivative appeared in 1974 [72], followed some ten years later by one publication [73] on some more detailed investigations, again on 4-O-substituted thiocoumarins (Scheme 27).

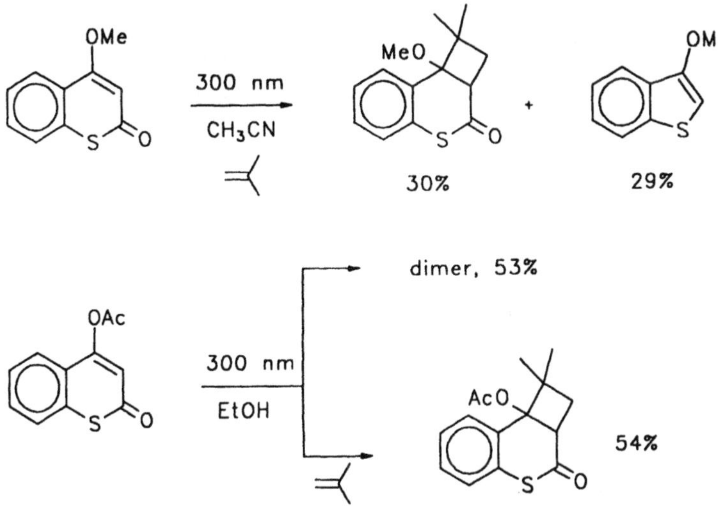

Scheme 27

Irradiation (350 nm) of the parent (unsubstituted) compound in benzene or acetonitrile produces one photodimer (head-to-head, *anti*) [74] selectively, while irradiation (390 nm) in the solid state affords the head-to-head, *syn* dimer exclusively [75]. On irradiation in diluted solutions ($< 10^{-2}$ M) minor amounts of benzo[*b*]thiophene are also formed. Irradiation in the solid state at shorter wavelengths (> 340 nm) affords a mixture of all four *cis*-fused tricyclic dimers (Scheme 28). This unprecedented wavelength dependence could be due to the population of two excited states through excitation with higher energy radiation. If one of them had a much longer lifetime, this would allow exciton migration to defect sites, leading to the—observed—nontopochemical behavior [75]. An alternative explanation would be that only irradiation with wavelengths corresponding to the chromophore's absorption tail induces a single-crystal to single-crystal transformation [76].

Irradiation of 1-thiocoumarin in benzene in the presence of an excess of 2,3-dimethylbut-2-ene affords a 4:1 mixture of the corresponding *cis*- and *trans*-fused[2 + 2]-cycloadducts [74]. Under similar conditions irradiation of coumarin yields only the *cis*-fused cycloadduct [77] (Scheme 29). This reflects the higher flexibility of the thiinone ring as compared to the pyranone ring, which is due to the differences in bond lengths (C–S–C > C–O–C).

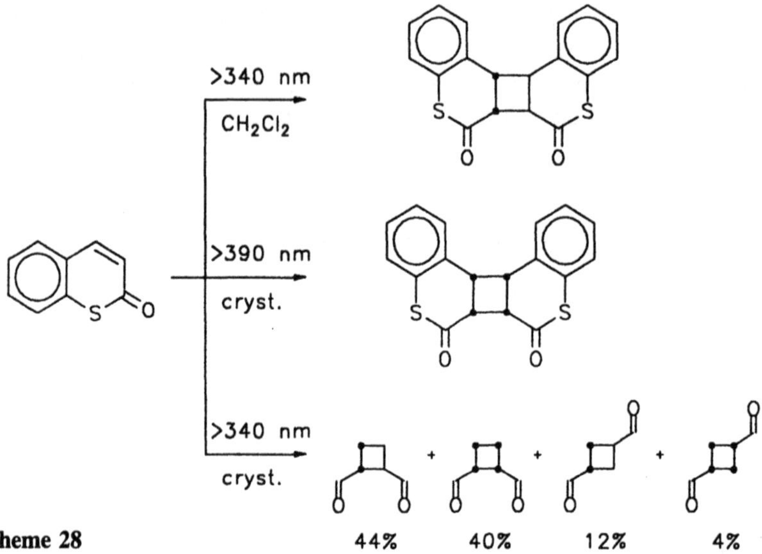

>340 nm

CH$_2$Cl$_2$

>390 nm

cryst.

>340 nm

cryst.

Scheme 28   44%   40%   12%   4%

X = O, S

hν

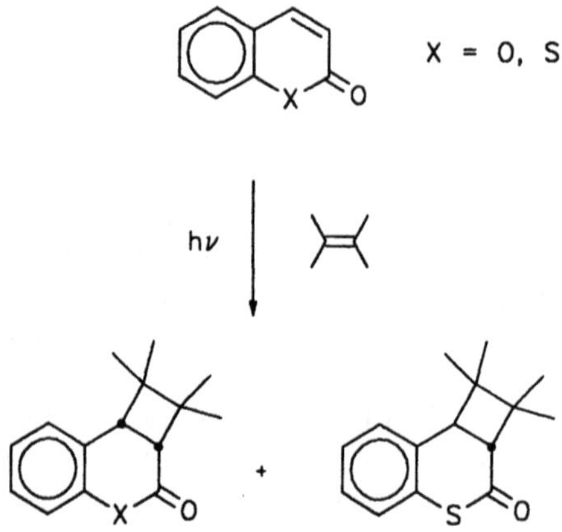

Scheme 29

A similar behavior is observed for tricyclic [g]-fused-1-thiocoumarins on irradiation in solution in the presence of the same alkene [24]. On long-wavelength (>395 nm) irradiation, both 2H,8H-benzo[1,2-b:5,4-b']bisthiopyran-2,8-dione and 2H,8H-benzo[1,2-b:4,5-b']bisthiopyran-2,7-dione afford 4:1 mixtures of cis- and trans-fused monocycloadducts (Scheme 30).

Interestingly, no regioselectivity is observed in the irradiation of 2H,8H-thiopyrano[3,2-g]benzo-pyran-2,8-dione in the presence of 2,3-dimethylbut-2-ene, as 50% of a 4:1 mixture of cis- and trans-fused monocycloadducts at the

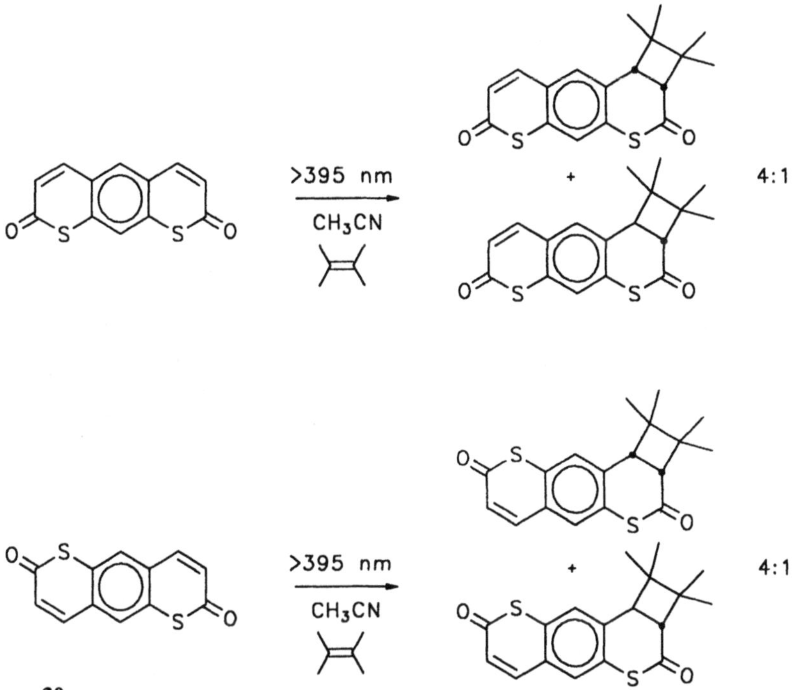

>395 nm

CH₃CN

+   4:1

>395 nm

CH₃CN

+   4:1

Scheme 30

thiinone moiety on the one side and in addition 50% of the *cis*-fused mono-cycloadduct at the pyranone ring on the other side are formed [24] (Scheme 31).

VII. THE –C(O)–S–C=C– CHROMOPHORE

Both enethiol- and aromatic thiol esters of aliphatic carboxyxlic acids undergo a common reaction on excitation with light of 254 nm, namely cleavage of the S–acyl bond (Scheme 32). The fate of the thus formed thiyl radicals varies: enthiyl radicals dimerize to afford thiophenes in moderate yields [78], while arenethiyl radicals recombine efficiently, giving disulfanes in very good yields [79].

2(3*H*)-Thiophenones, i.e. S-ethenyl thiollactones, behave alike. Irradiation (300 nm) of 3,3,5-trimethyl-2(3*H*)-thiophenone in methanol, acetonitrile, or dichloromethane affords 2,2,4-trimethyl-2*H*-thiete via S–acyl bond cleavage [80]. The subsequent CO-elimination was proposed to occur via an intramolecular radical displacement rather than by simple cleavage followed by cyclization of the 1,4-thiyl-alkyl biradical (Scheme 33).

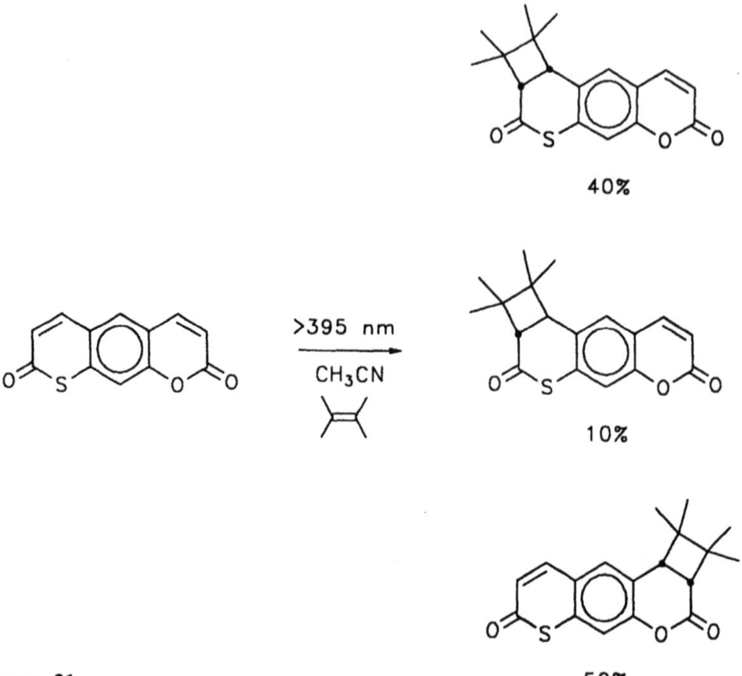

40%

>395 nm
————————
CH$_3$CN

10%

Scheme 31   50%

hν
————————→   CH$_3$ĊO +   →   PhSSPh

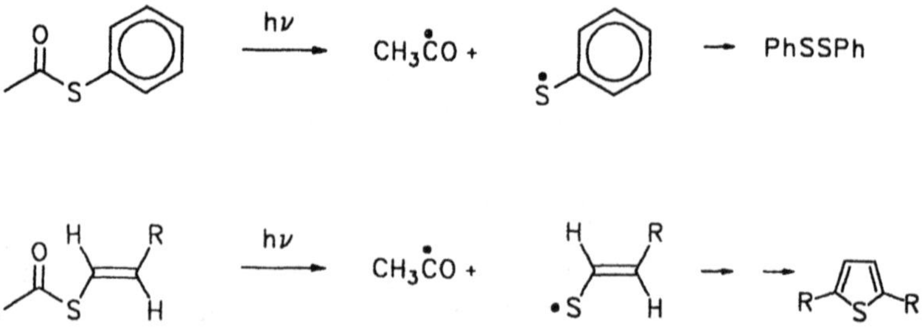

Scheme 32

hν
————————→   CH$_3$ĊO +   — —→

305 nm
————————→   -CO
C$_5$H$_{12}$ ————→

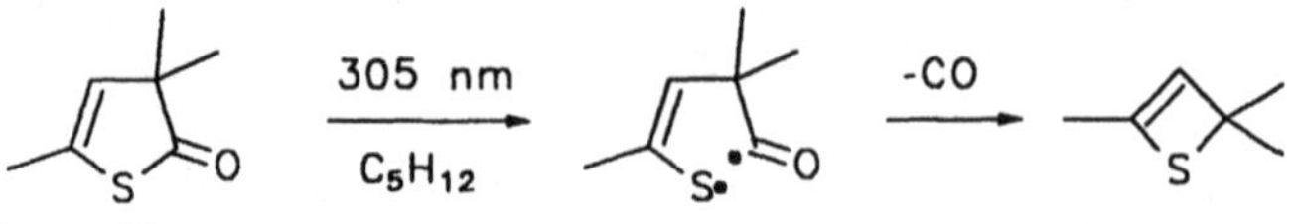

Scheme 33

In a more detailed investigation it was found [17] that independent of the substitution pattern 2(3H)-thiophenones do indeed undergo S–acyl bond cleavage on irradiation, but that the fate of the so-formed thiyl-acyl 1,5-biradical varies with the alkyl substituents, as hydrogen atom transfer with concommitant CO elimination affording dienethiols (Scheme 34) becomes competitive. These dienethiols undergo a consecutive light induced S–H bond homolysis and the so-formed thiyl radicals finally cyclize to thiophenes in an *exo-trig* addition step.

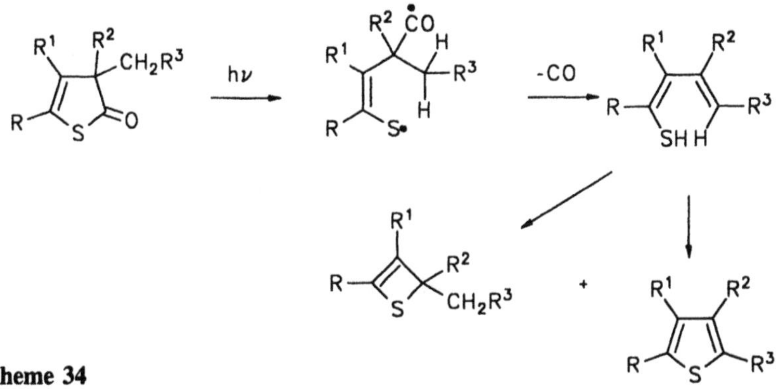

Scheme 34

On irradiation with light of shorter wavelengths (254 nm), the 2,2,4-trialkyl-2H-thietes themselves undergo efficient ring opening to (acyclic) α,β-unsaturated thiones [81]. By such a reaction 4-*tert*-butyl-2,2-dimethyl-2H-thiete is converted quantitatively to 2,2,5-trimethylhex-4-en-3-thione, which recloses to the thiete on irradiation with light of 300 nm or greater than 450 nm. The combination [2H-thiete-enthione] thus represents an interesting photochromic system (Scheme 35).

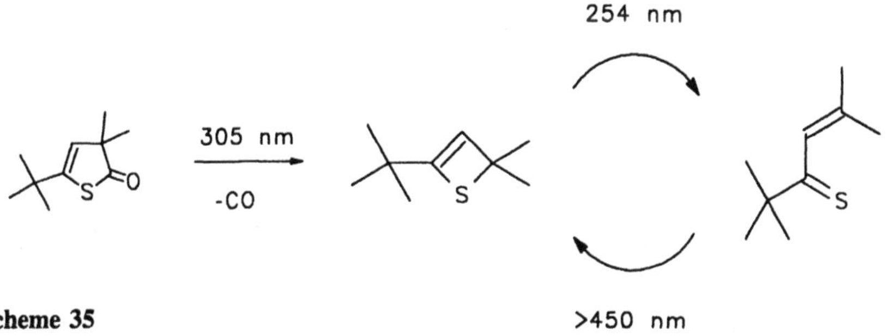

Scheme 35

VIII. CYCLIC UNSATURATED THIOLCARBONATES AND THIOANHYDRIDES

Aromatic thiolcarbonates, e.g. benzo- or naphtho[d]oxathiol-2-ones, behave like aromatic thiol esters on irradiation in undergoing S–acyl bond cleavage followed by CO elimination. If a phenolic OH group is located in *para* position to the C–O bond, a tautomeric equilibrium hydroxy-keto-thione / 2-mercapto-*p*-quinone is established [82]. This mercaptoquinone can be trapped with alkenes

in a consecutive [3 + 2] cycloaddition to give [b]-fused thiophene derivatives
[83] (Scheme 36).

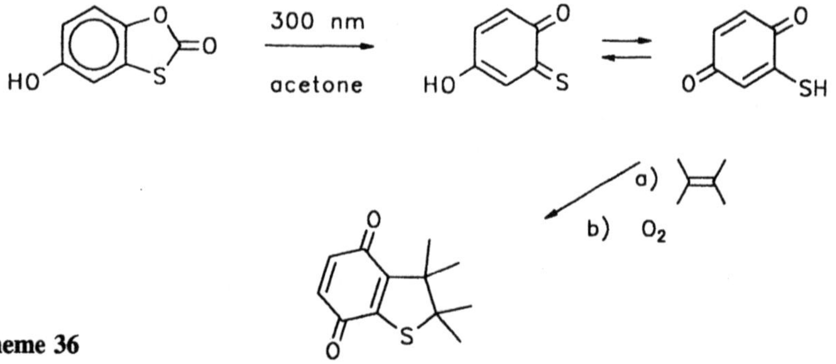

Scheme 36

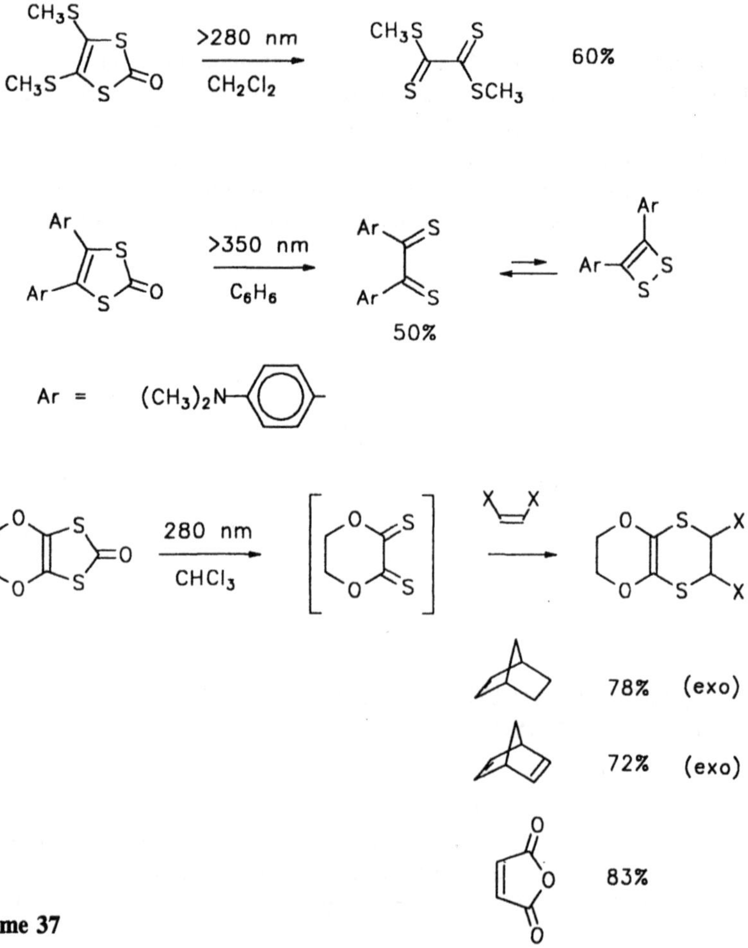

Ar = (CH₃)₂N—⟨⟩—

Scheme 37

Similarly, 1,3-dithiol-2-ones undergo S–acyl bond cleavage followed by CO elimination to afford α-dithiones, dithio- or perthiooxalates, respectively. These compounds can be either isolated directly [84,85] or trapped with dienophiles to afford 1,4-dithiins [86]. An equilibrium between an α-dithione and a 1,2-dithiete has also been observed [84] (Scheme 37).

via		R	R'	yield
		CH_3	CH_3	15%
		C_6H_5	CH_3	54%
		C_6H_5	C_6H_5	53%
Scheme 38		C_6H_5	H	25%

The first and only report on a photoreaction of a cyclic thioanhydride describes the insertion of alkenes into the C(O)–S bond of naphthalene-dicarboxylic thioanhydrides [87]. Thus irradiation of 1,2-naphthalenedicarboxylic thioanhydride or its 2,3-isomer with alkenes in acetonitrile gives naphthothiepindiones. The singlet excited molecules undergo S–acyl cleavage, either directly or via a zwitterionic intermediate. Reactions with *E*- or *Z*-but-2-ene proceed in a stereospecific manner (Scheme 38).

REFERENCES

1. Reactions of Sulphur-containing Compounds. Photochemistry, Specialist Periodical Reports, The Royal Society of Chemistry.

2. Fragmentation of Organosulfur Compounds. Photochemistry, Specialist Periodical Reports, The Royal Society of Chemistry.

3. Photochemical Synthesis and Transformations of Organosulfur Compounds. Vlasova, N. N. In *Chemistry of Organosulfur Compounds*; Belen'kii, L. I., Ed.; E. Horwood: Chichester, 1990, p. 69.

4. Photochemistry of Organic Sulfur Compounds. Still, I. W. J. In *Organic Sulfur Chemistry*; Bernardi, F., Csizmadia, I. G., Mangini, A., Eds.; Elsevier: Amsterdam, 1985, p. 596.

5. Photochemistry of Organic Sulphur Compounds. Coyle, J. D. *Chem. Soc. Rev.* 1975, *4*, 523.

6. Photochemistry of Sulphoxides and Sulphones. Still, I. W. J. In *The Chemistry of Sulphoxides and Sulphones;* Patai, S.; Rappoport, Z.; Stirling, C. M. J., Eds.; John Wiley: New York, 1988, p. 873.

7. Thiocarbonyl Photochemistry. Ramamurthy, V. Org. Photochem. 1985, 7, 231.
8. Photochemistry of Thiols. Knight, A. R. In *The Chemistry of Thiols;* Patai, S., Ed.; John Wiley: Chichester, 1974, p. 455.
9. Photochemical Reactions of Thiophenes. Lablache-Combier, A. In *Thiophene and its Derivatives;* Part 1, Gronowitz, S., Ed. (The Chemistry of Heterocyclic Compounds, Vol. 44); John Wiley: New York, 1985, p. 745.
10. The Photochemistry of Oxygen- and Sulfur-containing Heterocycles. Reid, S. T. *Adv. Heteroc. Chem.* 1983, *33*, 1.
11. UV-Atlas of Organic Compounds. H. H. Perkampus. VCh.
12. Fehnel, E. A.; Carmack, M. *J. Am. Chem. Soc.* 1949, *71*, 84.
13. Er, E.; Margaretha, P. *Helv. Chim. Acta* 1992, *75*, 2265.
14. Kowalewski, R.; Margaretha, P. *Helv. Chim. Acta* 1993, *76*, 1251.
15. Er, E.; Margaretha, P. *Helv. Chim. Acta* 1994, *77*, 904.
16. Nagata, S.; Yamabe, T.; Fukui, M. *J. Phys. Chem.* 1975, *79*, 2335.
17. Hinrichs, H.; Margaretha, P. *Chem. Ber.* 1992, *125*, 2311.
18. Kiesewetter, R.; Margaretha, P. *Helv. Chim. Acta* 1989, *72*, 83.
19. Trofimov, B. A.; Shainyan, B. A. In *Supplement S, The Chemistry of Sulphur-Containing Functional Groups*; Patai, S.; Rappoport, Z., Eds.; John Wiley: Chichester, 1993, p. 659.
20. Jaffé, H. H.; Orchin, M. *Theory and Application of Ultraviolet Spectroscopy,* John Wiley: New York, 1964, p. 175.
21. Anklam, E.; Ghaffari-Tabrizi, R.; Hombrecher, H.; Lau, S.; Margaretha, P. *Helv. Chim. Acta* 1984, *67*, 1401.
22. Kowalewski, R.; Margaretha, P. *Angew. Chem.* 1988, *100*, 1437.
23. Margaretha, P. *Tetrahedron Lett.* 1971, 4891.
24. Klaus, C. P.; Margaretha, P. *Liebigs Ann.* 1996, 291.
25. Ganguly, B. K.; Bagchi, P. *J. Org. Chem.* 1956, *21*, 1415.
26. Yamada, M.; Kimura, M.; Nishizawa, M.; Kuroda, S.; Simao, I. *Bull. Chem. Soc. Jp.* 1991, *64*, 1821.
27. Wagner, P. J.; Lindstrom, M. J. *J. Am. Chem. Soc.* 1987, *109*, 3057.
28. Wagner, P. J.; Lindstrom, M. J. *J. Am. Chem. Soc.* 1987, *109*, 3062.
29. Sumathi, R.; Chandra, A. K. *JCS Perkin Trans. 2,* 1992, 291.
30. Vedejs, E.; Eberlein, T. H.; Wilde, R. G. *J. Org. Chem.* 1988, *53*, 2220.
31. Vedejs, E.; Stults, J. S. *J. Org. Chem.* 1988, *53*, 2226.
32. Lumma, W. C.; Berchtold, G. A. *J. Org. Chem.* 1969, *34*, 1566.
33. Tsuruta, H.; Ogasawara, M.; Mukai, T. *Chem. Letters* 1974, 887.
34. Mellor, J. M.; Webb, C. F. *JCS Perkin Trans. 1,* 1972, 211.
35. Miyashi, T.; Suto, N.; Yamaki, T.; Mukai, T. *Tetrahedron Lett.* 1981, 22, 4421.
36. Miyashi, T.; Suto, N.; Mukai, T. *Chem. Letters* 1978, 157.
37. Anklam, E.; Margaretha, P. *Res. Chem. Intermed.* 1985, *11*, 127.
38. Franz, J. A.; Roberts, D. H.; Ferris, K. F. *J. Org. Chem.* 1992, *52*, 777.
39. Cruciani, G.; Margaretha, P. *Helv. Chim. Acta* 1990, *73*, 890.
40. Padwa, A.; Battisti, A. *J. Am. Chem. Soc.* 1972, *94*, 521.
41. Crimmins, M. T.; Reinhold, T. L. *Org. Reactions* 1993, *44*, 297.
42. de Mayo, P. *Acc. Chem. Res.* 1971, *4*, 41.

43. Cruciani, G.; Rathjen, H. J.; Margaretha, P. *Helv. Chim. Acta* 1990, *73*, 856.
44. Schuster, D. I.; Lem, G.; Kaprinidis, N. A. *Chem. Rev.* 1993, *93*, 3.
45. Cossy, J.; Carrupt, P. A.; Vogel, P. In *The Chemistry of Double-Bonded Functional Groups*; Patai, S., Ed.; John Wiley: New York, 1989, p. 1368.
46. Crimmins, M. T. *Chem. Rev.* 1988, *88*, 1453.
47. Becker, D.; Haddad, N. *Org. Photochem.* 1989, *10*, 1.
48. Gruciani, G.; Margaretha, P. *Helv. Chim. Acta* 1990, *73*, 288.
49. Margaretha, P. *Chimia* 1975, *29*, 203.
50. Gebel, R. C.; Margaretha, P. *Chem. Ber.* 1990, *123*, 855.
51. Margaretha, P. *Helv. Chim. Acta* 1974, *57*, 1866.
52. Margaretha, P. *Liebigs Ann. Chem.* 1973, 727.
53. Baldwin, S. W. *Org. Photochem.* 1981, *5*, 123.
54. De Keukeleire, D.; He, S. L. *Chem. Rev.* 1993, *93*, 359.
55. Winkler, J. D.; Bowen, C. M.; Liotta, F. *Chem. Rev.* 1995, *95*, 2003.
56. Gebel, R. C.; Margaretha, P. *Helv. Chim. Acta* 1992, *75*, 1633.
57. Anklam, E.; Lau, S.; Margaretha, P. *Helv. Chim. Acta* 1985, *68*, 1129.
58. Kowalewski, R.; Margaretha, P. *Helv. Chim. Acta* 1992, *75*, 1925.
59. Kowalewski, R. *Ph.D. thesis, University of Hamburg*; 1993, p. 55.
60. Vedejs, E.; Eberlein, T. H.; Mazur, D. J.; McClure, C. K.; Perry, D. A.; Ruggeri, R.; Schwartz, E.; Stults, J. S.; Varie, D. L.; Wilde, R. G.; Wittenberger, S. *J. Org. Chem.* 1986, *51*, 1556.
61. Anklam, E.; Margaretha, P. *Angew. Chem.* 1984, *96*, 360.
62. Anklam, E.; Margaretha, P. *Helv. Chim. Acta* 1984, *67*, 2198.
63. Chatgilialoglu, C.; Guerra, M. "Thiyl Radicals" In *Supplement S, The Chemistry of Sulfur Containing Functional Groups*; Patai, S.; Rappoport, Z., Eds.; John Wiley: Chichester, 1993, p. 363.
64. Kiesewetter, R.; Margaretha, P. *Helv. Chim. Acta* 1987, *70*, 121.
65. Kiesewetter, R.; Margaretha, P. *Helv. Chim. Acta* 1985, *68*, 2350.
66. Beckwith, A. L. J.; Schiesser, C. H. *Tetrahedron* 1985, *41*, 3925.
67. Kiesewetter, R.; Margaretha, P. *Phosphorus and Sulfur* 1988, *39*, 263.
68. Kiesewetter, R.; Graf, A.; Margaretha, P. *Helv. Chim. Acta* 1988, *71*, 502.
69. Kiesewetter, R.; Margaretha, P. *Helv. Chim. Acta* 1987, *70*, 125.
70. Ciamician, G.; Silber. P. *Ber.dtsch.chem.Ges.* 1902, *35*, 4128.
71. Meth-Cohn, O.; Tarnowski, B. *Adv. Heteroc. Chem.* 1980, *26*, 115.
72. Lehmann, J.; Wamhoff, H. *Liebigs Ann. Chem.* 1974, 1287.
73. Kaneko, C.; Naito, T.; Ohashi, T. *Heterocycles* 1983, *20*, 1275.
74. Karbe, C; Margaretha, P. *J. Photochem. Photobiol. A: Chem.,* 1991, *57*, 231.
75. Klaus, C. P.; Thiemann, C.; Kopf, J.; Margaretha, P. *Helv. Chim. Acta* 1995, *78*, 1079.
76. Enkelmann, V.; Wegner, G.; Novak, K.; Wagener, K. B. *J. Am. Chem. Soc.* 1993, *115*, 10390.
77. Hanifin, J. W.; Cohen, E. *Tetrahedron Lett.* 1966, 1419.
78. Grunwell, J. R.; Foerst, D. L.; Sanders, M. J. *J. Org. Chem.* 1977, *42*, 1142.
79. Grunwell, J. R.; Marron, N. A.; Hanhan, S. I. *J. Org. Chem.* 1973, *38*, 1555.
80. Hinrichs, H.; Margaretha, P. *Angew. Chem.* 1989, *101*, 1546.

81. Hinrichs, H.; Margaretha, P. *J. Photochem. Photobiol. A: Chem.* 1993, *71*, 103.
82. Chapman, O. L.; McIntosh, C. L. *JCS Chem. Commun.* 1971, 383.
83. Suginome, H.; Kobayashi, K.; Konishi, A.; Minakawa, H.; Sakurai, H. *JCS Chem. Commun.* 1993, 807.
84. Kusters, W.; de Mayo, P. *J. Am. Chem. Soc.* 1974, *96*, 3502.
85. Hartke, K.; Kissel, T.; Quante, J.; Matusch, R. *Chem. Ber.* 1980, *113*, 1898.
86. Hartke, K.; Lindenblatt, T. *Synthesis* 1990, 281.
87. Kubo, Y.; Okusako, K.; Araki, T. *Chem. Letters* 1987, 811.

4

Photochemistry of Conjugated Polyalkynes

Sang Chul Shim
The Korea Advanced Institute of Science and Technology,
Taejon, Korea

Since the first discovery of naturally occurring acetylenes in 1892 by Arnaud [1], many acetylenes were observed in tribus, umbelliferae, and other families of plants. Many natural acetylenes are linked with special hormonal functions in the plant metabolism and show powerful biological activity and toxicity to various kinds of microorganisms [2]. Polyalkynes have been studied in two categories with great efforts: polymerization and photobiological effects. Photochemical and topochemical solid state polymerization in some regulated linear polyalkynes have become important because of possessing potentially useful physical properties such as nonlinear optical properties [3]. Furthermore, certain naturally occurring and synthetic linear polyalkynes such as 1-phenyl-1,3,5-heptatriyne (PHT) and 1,4-diphenylbutadiyne (DPB) have been reported to be phototoxic to a variety of microorganisms through photodynamic effect and nonoxidative membrane damage, even though the exact mechanism has not yet been clarified [4–9]. Recently, it is reported that thiarubrine isolated from Asteraceae has antiviral activity against human immunodeficiency virus dependent upon UVA irradiation [10]. In spite of these important properties, the photophysical and photochemical behavior of conjugated linear polyalkynes have received relatively little attention, in contrast to those of the corresponding conjugated polyenes. Geometrical photoisomerization, electrocyclic reactions, [2+2] photocycloaddition reactions, sigmatropic rearrangements, and photo-ene reactions of conjugated polyenes have been extensively studied, and the reaction mechanisms are well understood on the molecular level. The photochemistry of simple acetylenes, especially monoalkynes, is, however, well established [11,12], and it has certain advantages of its own, especially in synthetic applications. For example, olefins undergo [2+2] photocycloaddition to give cyclobutanes, but acetylenes give cyclobutenes, which are potentially easier to functionalize or to modify in subsequent stages of a synthesis. The utility of cyclobutenes is increased, and a regiocontrolled ene-yne photochemical [2+2] cycloaddition reaction is developed by using silicon as a tether [13].

Simple acetylenes are known to undergo various photoreactions such as intermolecular and intramolecular addition reactions, cycloaddition reactions, and rearrangement reactions. The photocycloaddition of acetylenes to olefins or unsaturated carbonyl compounds has provided a fertile field of study, and many of the reactions are useful in synthesis. The common reaction between an acetylene and an olefin leads to a cyclobutene. The cyclobutene product may itself be photolabile, and if radiation is absorbed more strongly by the cyclobutene, the product isolated may be the 1,3-dienes derived by electrocyclic ring opening of cyclobutenes [14]. Though general photocycloaddition of acetylenes seems to be [2+2] photocycloaddition to produce a cyclobutene ring, a cyclopropane ring is formed via carbene intermediate, and/or bicyclo[2.2.0]hexane is formed by second photocycloaddition between the cyclobutene and the olefin in certain cases. In the photocycloaddition of dimethyl acetylene dicarboxylate with ethylene, a mechanism involving a carbene intermediate and consecutive [2+2] photocycloaddition reaction is operative. It is difficult to isolate the cyclobutene in such a reaction, and the bis-cyclobutane products undergo a ring-opening reaction on distillation to yield a 1,5-diene [15]. Some examples of photocycloaddition reaction of acetylenes to α,β-unsaturated carbonyl compounds, carbonyl compounds, or aromatic compounds have also been reported. However, these reactions are not acetylene photochemistry because α,β-unsaturated carbonyl group or carbonyl group absorb the light and the excited ketone attacks the ground state acetylene [16].

I. PHOTOCHEMISTRY OF ALKYNES

The electronic structure of acetylenes is related to that of olefins, and the photochemistry of the two classes of compounds reflects this similarity. Therefore the photochemistry of acetylenes has been studied and reported in the same range of olefin photochemistry [14,17,18].

From the spectroscopic data it seems likely that the excited states involved in alkyne photochemistry are (π,π^*) states in which an electron from a bonding π molecular orbital is promoted to an antibonding π^* molecular orbital, or Rydberg states, in which a π electron is promoted to an extended σ type orbital covering one nucleus. The UV absorption spectra of simple acetylenes such as propyne and 1-butyne show sharp Rydberg bands in the region of 110–160 nm, a weaker band with a maximum around 170–180 nm, and a weak tailing to 230 nm or 240 nm [19]. The longer wavelength region probably consists of two or three overlapping bands, of which the band at the longest wavelength can be attributed to a $\pi \rightarrow \pi^*$ transition. The lowest energy singlet state is (π, π^*) in nature; note that for some simple alkenes the Rydberg singlet state seems to be the lowest in energy. The energy of the lowest state is not easily assigned from the absorption spectrum, since there is a long, weak absorption tail (210–240 nm) associated with the changed geometry of the (π, π^*) singlet state: a value of 505 kJmol^{-1} (121 kcalmol^{-1}, equivalent to a wavelength of 237 nm) has been given [20] for the singlet state energy of acetylene. The Rydberg excited states of acetylene have the same linear geometry as the ground state [21], but the preferred geometry of the lowest (π, π^*) singlet state is nonlinear and *transoid*, with HCC angles of about 120° as shown in Scheme 1 [20].

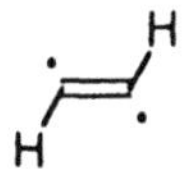

Scheme 1

The fluorescence of nonconjugated acetylenes is weak ($\Phi = 0.009$–0.014) [22], and the lifetime is about a few ns. Nonconjugated acetylenes have no appreciable phosphorescence, and their singlet–triplet absorption has not been characterized, while singlet–triple absorption spectra have been recorded for diynes, poliynes, and phenyl conjugated acetylenes, and triplet energies can be obtained from phosphorescence spectra [23].

Simple acetylenes are known to undergo various photoreactions such as fragmentation reactions, intermolecular and intramolecular addition reactions, cycloaddition reactions, and rearrangement reactions [12]. Ultraviolet irradiation of acetylenes leads to initial bond cleavage of either one or two bonds. The major products from the vapor phase photolysis of acetylene are butadiyne, ethylene, and hydrogen, and butenyne, benzene, and solid polymer are also formed [24] (Scheme 2). In this reaction, the primary process is the formation of diatomic carbon (C$_2$) and hydrogen (H$_2$), and radicals HCC· and H· [25]. 1- Haloacetylenes such as 1-bromo- or 1-iodohex-1-yne undergo homolytic carbon–halogen bond cleavage on irradiation [26].

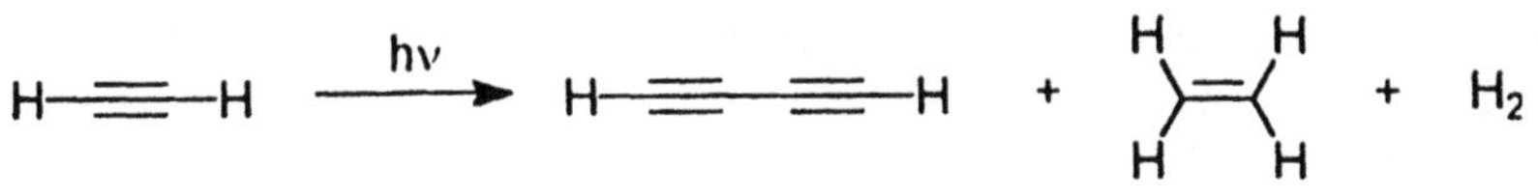

Scheme 2

When acetylenes are irradiated in aqueous solution [27], in acetic acid [28], or in alcohols [29,30], photoaddition reactions take place to give a ketone, an enol acetate, or an enol ether, respectively. In the photohydration reaction, a hydrated proton attacks the singlet excited state of the acetylene directly [31]. On the other hand, alcohols give addition products by attack on the excited states of acetylenes in a radical-like mechanism. Radical photoaddition to acetylenes occurs also with saturated hydrocarbons such as cyclohexane [29], and with cyclic ethers such as tetrahydrofuran [32]. Simple acetylenes are photoreduced on irradiation in hydrocarbon solvents; for example photolysis of dec-1-yne or dec-5-yne in pentane gives the corresponding alkene (dec-1-ene or *trans*- and *cis*-dec-5-enes) [32].

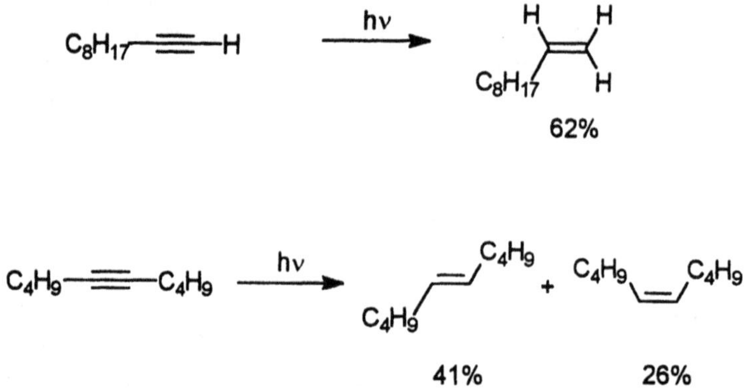

62%

41%   **26%**

Scheme 3

Irradiation of acetylenes with aromatic aldehydes or ketones gives α,β-unsaturated carbonyl compounds; a typical example is the reaction of benzophenone with 1-phenylpropyne [33].

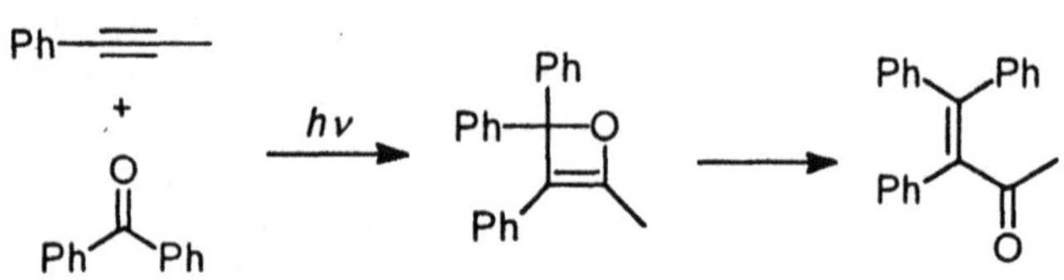

Scheme 4

The reaction mechanism involves an attack of the excited carbonyl compound on the ground state alkyne to give a biradical that ring closes and then undergoes ring opening in the alternating sense, which accounts for the product and the orientation of the addition reaction [33,34]. These reactions are not photochemistry of alkynes since carbonyl compounds absorb all the light.

Photocycloaddition of acetylenes to carbon–carbon double bonds in olefins or unsaturated carbonyl compounds has provided a fertile field of study, and many of the reactions are useful for synthesis of organic compounds. The basic reaction between an acetylene and an olefin leads to a cyclobutene, such as those formed from diphenylacetylene and 2,3-dimethyl-2-butene [16], cyclohexene, or dihydropyran [35]. In these examples it is clear that acetylene absorbs the light, and its excited state must be involved in the reaction with olefins. A mechanistic study on the reactions with 5-methyl-2,3-dihydrofuran showed that added pyrene quenches the reaction, whereas triphenylene sensitizes the formation of cyclobutene [36], suggesting that the reactive state is the lowest triplet state of acetylene. Although general photocycloaddition reactions of acetylenes seem to be [2+2] photocycloaddition reactions to produce a cyclobutene ring, it was reported in some cases that a cyclopropane ring is formed via carbene intermediate, and/or that bicyclo[2.2.0]hexane is formed by photocycloaddition between the primary photoadduct, cyclobutene, and the olefin [15]. The bicyclohexyl product is obtained by competing light absorption between acetylene and first-formed cyclobutene. Dimethyl acetylenedicarboxylate is less strongly absorbing than diphenylacetylene, and the first-formed cyclobutene competes effectively for light absorption. Because of the subsequent photochemical reactions, such as formation of a bicyclo[2.2.0]-hexane, it is difficult to isolate the cyclobutene in such reactions, and the bicyclohexyl product ring-opens on distillation to yield a 1,5-diene [15].

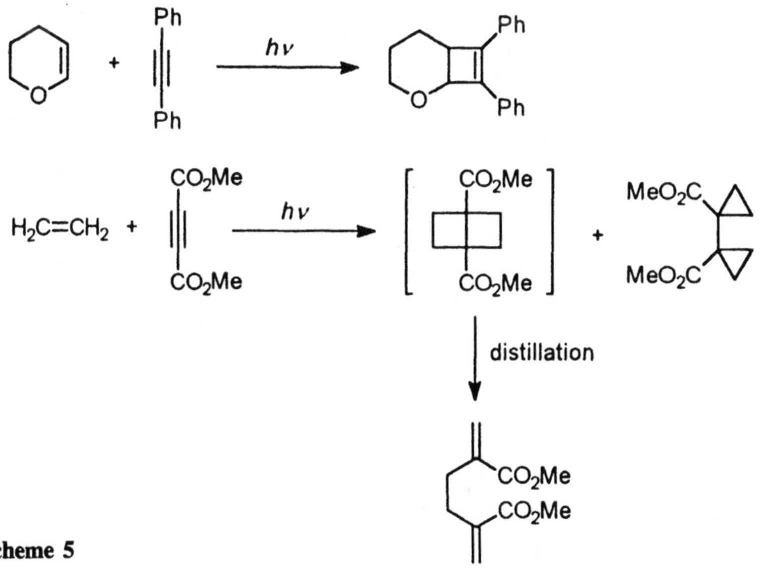

Scheme 5

II. PHOTOPHYSICAL PROPERTIES OF CONJUGATED POLYALKYNES

A series of conjugated polyalkynes with two or three conjugated C≡C triple bonds and various terminal groups such as phenyl, naphthyl, *tert*-butyl, methyl, and silyl groups have been synthesized in moderate to good yields by various methods as shown in Table 1 and Scheme 6 [37].

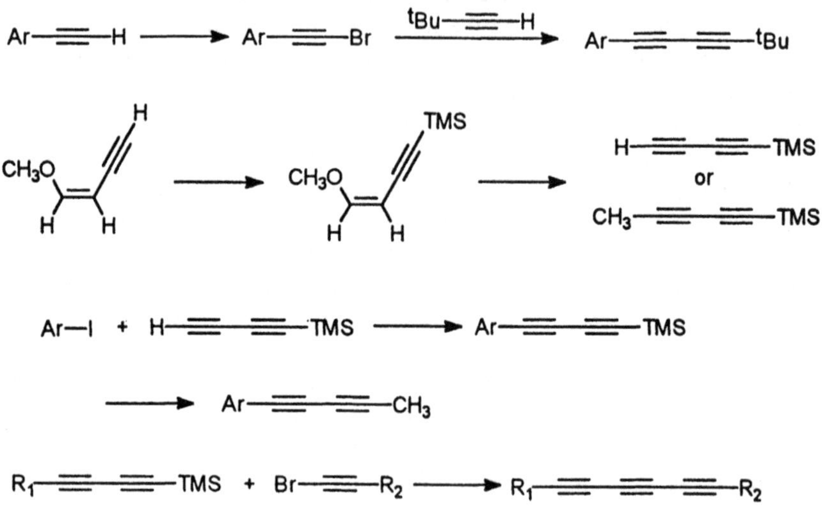

Scheme 6

All the polyalkynes, natural and synthetic polyalkynes, show characteristic spiky UV-VIS absorption bands separated by ~2300 cm^{-1} as C≡C stretching bands. Aromatic groups at both ends are conjugated through the triple bonds between them. No fluorescence was observed from poliynes except naphthyl derivatives. According to Beer [38], none of the polyalkynes with less than four triple bonds shows fluorescence except diphenylacetylene. Most of the polyalkynes show phosphorescence with the quantum yields shown in Table 2. In particular, these compounds show triplet–triplet absorption bands in the fluid solution at room temperature, and the triplet states have lifetimes in the range of a few tenths of a microsecond in methanol at room temperature and a few tenths of a second at liquid nitrogen temperature, unlike corresponding polyenes. For naphthyl conjugated polyalkynes, fluorescence is observed and quenched by olefins, particularly by electron deficient olefins such as dimethyl fumarate and fumaronitrile, and sometimes exciplex emissions are observed as shown in Table 3 [39]. Conjugated diynes can be either electron donors or acceptors in the excited state, but they are more likely to be electron donors because no fluorescence quenching is observed with electron-rich olefins such as 2,3-dimethyl-2-butene. The results indicate the conjugated polyalkynes to have poor electron-accepting ability [39].

Table 1 Synthesis of Some Conjugated Polyalkynes

R-(C≡C)$_n$-R'					
R	R'	n	T (°C)	Solvent	Yield (%)
Me	TMS	2	–78	THF	80
TMS	TMS	2	–78	THF	80
H	TMS	2	–78	THF	80
H	t-BDMS	2	–78	THF	50
Naph	TMS	1	30	Et$_3$N	94
Naph	H	1	25	MeOH	100
Ph	TMS	2	30	Et$_3$N	72
Naph	TMS	2	30	Et$_3$N	85
t-Bu	Ph	2	25	MeOH	45
t-Bu	Naph	2	25	MeOH	40
Ph	Me	2	–78	THF	89
Naph	Me	2	–78	THF	85
Naph	Naph	2	25	Acetone	95
Naph	Ph	2	25	MeOH	62
Ph	Ph	3	25	MeOH	55
Naph	Naph	3	25	MeOH	76
Me	Ph	3	25	MeOH	65
Me	Naph	3	25	MeOH	79
t-BDMS	Ph	3	25	MeOH	90
H	Ph	3	10	MeOH	100
Naph	Ph	3	25	MeOH	55

t-BDMS: t-butyldimethylsilyl; t-Bu: t-butyl; Me:methyl; Naph: α-naphthyl; Ph: phenyl; TMS: trimethylsilyl.

Table 2 Phosphoresence Quantum Yields of Some Conjugated Polyalkynes at 77 K in Methylcyclohexane

Conjugated polyalkynes	λ_{ex} (nm)	Φ_{ph}
1-Phenyl-1,3-pentadiyne	280	0.026
1-Phenyl-5,5-dimethyl-1,3-hexadiyne	280	0.090
1,4-Diphenylbutadiyne	320	0.012
1-(1-Naphthyl)-1,3-pentadiyne	330	0.002
1-(1-Naphthyl)-5,5-dimethyl-1,3-hexadiyne	330	0.002
1-(1-Naphthyl)-4-phenyl-1,3-butadiyne	350	0.001
1,4-Di(1-naphthyl)-1,3-butadiyne	370	0.000
1-Phenyl-1,3,5-heptatriyne	310	0.004
1,6-Diphenyl-1,3,5-hexatriyne	360	0.004
1-(1-Naphthyl)-6-phenyl-1,3,5-hexatriyne	377	0.001
1,6-Di(1-naphthyl)-1,3,5-hexatriyne	390	0.000

Relative to anthracene (Φ_f = 0.36 at 20–25°C in cyclohexane).

Table 3 Quenching Constants $(k_q\tau)$ of Some Olefins for 1-(1-Naphthyl)-1,3-Butadiynes of Fluorescence

Conjugated polyalkynes	In acetonitrile			In diethyl ether		
	DMB	DMFu	FN	DMB	DMFu	FN[a]
1-(1-Naphthyl)-1,3-pentadiyne	0	17	37	0	20	3.6
1-(1-Naphthyl)-5,5-dimethyl-1,3-hexadiyne	20.14[a]	156	130	0	129	157
1-(1-Naphthyl)-4-phenyl-1,3-butadiyne	0	9	6.9	0	9.8	2.3
1,4-Di(1-naphthyl)-1,3-butadiyne	0	14	5.4	0	14	1.7

[a]Exciplex emissions were observed.
DMB: 2,3-dimethyl-2-butene; DMFu: dimethyl fumarate; FN: fumaronitrile.

III. PHOTOREACTIONS WITH OLEFINS

A. Photoreaction of Conjugated Diynes with Some Olefins

Photolysis of 1,4-diphenyl-1,3-butadiynes(DPB) in 2,3-dimethyl-2-butene(DMB) gives three photoadducts 1–3 as shown in Scheme 7 [39–41]. Photoadducts 1 and 2 are formed initially, and prolonged irradiation results in the decrease of 1 and the formation of 3, indicating that 1 and 2 are primary photoproducts and 3 is a secondary photoproduct. Irradiation of pure 1 in DMB yields 3 strongly supporting that 1 is the primary photoadduct and 3 is a secondary photoadduct. The azulene (E_T = 129.9 kcal/mol) [42] quenching studies on the formation of 1 and 2 show a linear relationship with Stern-Volmer constants $(k_q\tau)$ of 4400 and 4800 M^{-1}, suggesting that both 1 and 2 originate from the triplet excited state of DPB.

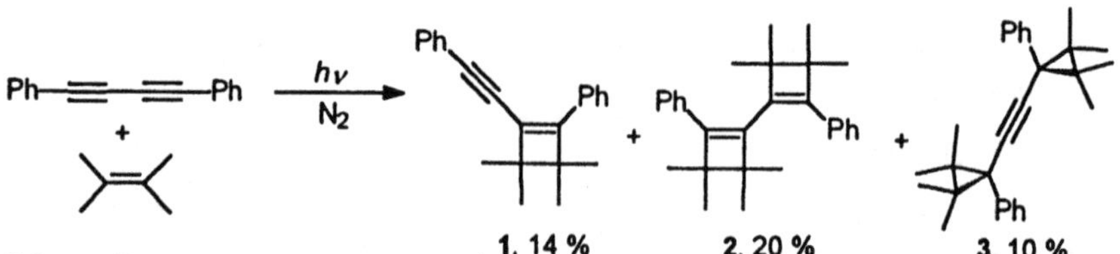

Scheme 7

1, 14 %   2, 20 %   3, 10 %

Photolysis of DPB with unsymmetrical olefins, acrylonitrile (AN), and ethyl vinyl ether (EVE) initially yields regioselective [2+2] type photoadducts 4–5 as primary photoproducts, and extended irradiation results in the formation of secondary photoadducts, 6–9 [43] (Scheme 8). The secondary photoproduct(s) of 5 are very unstable at room temperature, giving rearranged products at this temperature. The regiochemistry of these adducts is determined by analyzing coupling patterns in the ^{1}H-NMR spectra of hydrogenation products. The photoreaction of DPB with symmetrical olefins to give 1:1 photoadducts is efficiently quenched by azulene and 9,10-diphenylanthracene (E_T = 174.9

kcal/mol), yielding large $k_q \tau$ values (of 6300 M^{-1} for 4 and 17000 M^{-1} for 5, respectively), and the photoreaction proceeds from the triplet excited state of DPB. The photoreactions of 1:1 adducts with olefins to give 1:2 adducts are also efficiently quenched by oxygen, suggesting the reaction to proceed via triplet excited states. The reactive center for 1:1 photoadducts (4 and 5) to give secondary photoadducts is dependent on the electron density of olefins, i.e., the triple bond for AN but the double bond for EVE. Irradiation of 1:1 adduct 4 (A = 8) with EVE (400 mM) in tetrahydrofuran (THF) yields 1,5-diene products 10 and 11, which are formed from the bicyclohexyl intermediate(s). The photoreaction of 5 (A = 8) with AN (400 mM) in THF also results in the formation of biscyclobutene photoadducts (12 and 13), indicating the triplet excited state of 1:1 photoadducts to have a polar character, as shown in Scheme 9. A plausible reaction mechanism involving cumulene type triplet excited state and a ploar triplet structure of 1:1 photoadduct is proposed in Scheme 9. The regioselectivity of these photocycloaddition reactions is attributed to stabilization of α-position by substituents (CN, OEt) on olefins in diradical type intermediate.

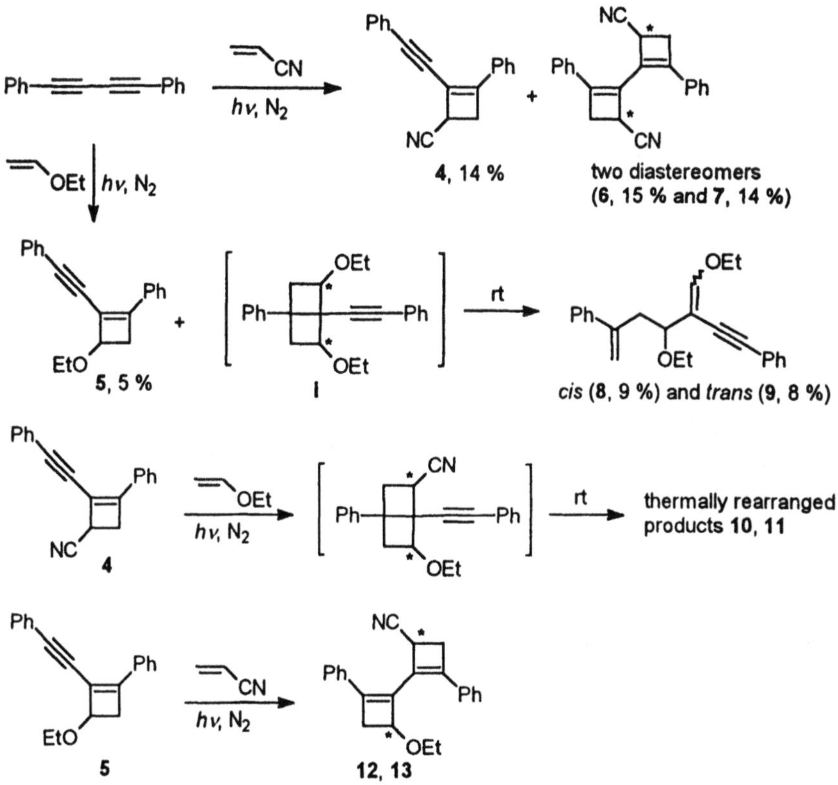

Scheme 8

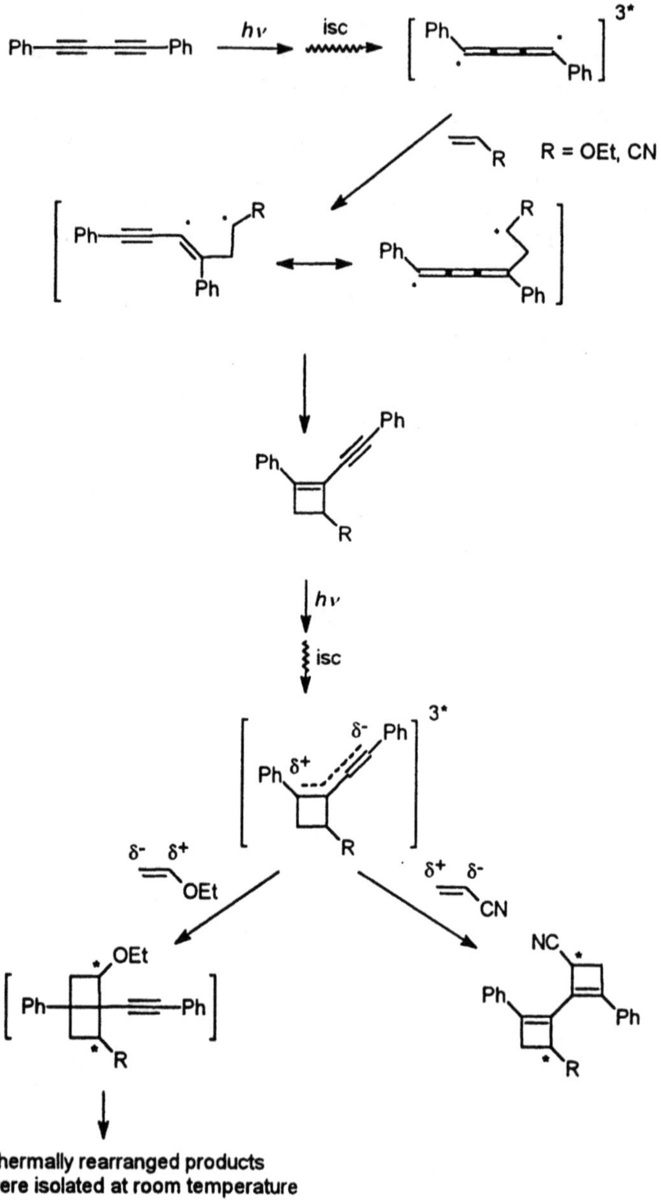

Scheme 9

Irradiation of DPB with dimethylfumarate (DMF) in deaerated THF yields a 1:1 photoadduct 14 and 1:2 photoadducts 15 and 16 [44,45]. The 1:1 photoadduct is initially formed, and extended irradiation results in decrease of 14 and formation of secondary 1:2 photoadducts 15 and 16. The photocycloaddition reaction of DPB with DMF to give 14 is efficiently quenched by tetracene (E_T = 122.8 kJ/mol) showing a large $k_q\tau$ value of 2400 M^{-1}, indicating that the reaction proceeds from the triplet excited state. In the oxygen quenching studies, no significant effect on the quantum yield of 15 is observed, but that of 16 decreased to 25% in aerated solution, indicating that 15 and 16 are produced through singlet and triplet excited states of 14, respectively. The fluorescence of 14 is efficiently quenched by DMF ($k_q\tau$ = 5 M^{-1}), strongly supporting that the secondary photoproduct 15 is produced via the singlet excited state of 14.

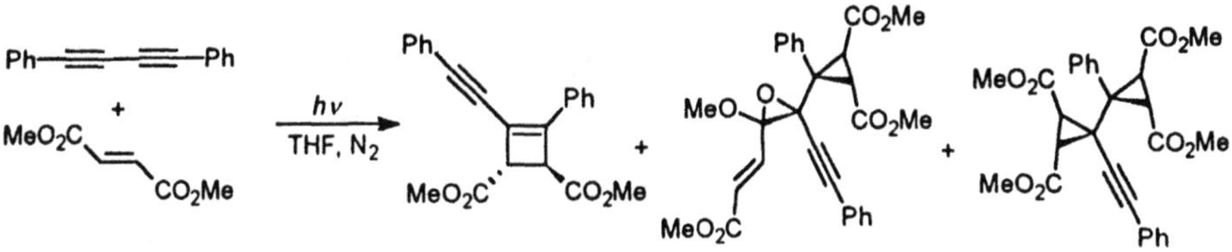

Scheme 10

14 and 15 are initially suggested to be formed via carbene intermediates [44,45], but every attempt to detect or trap these carbene intermediates in cryogenic conditions was not successful. Later Kim et al. [46] proposed that the photoaddition proceeds through a geminate radical ion pair based on the photolysis of 14 in methanol.

$$14 \xrightarrow{h\nu} 14^* \xrightarrow{DMFu} 14^{+\cdot}\text{---}DMFu^{-\cdot} \longrightarrow 15 + 16$$

Scheme 11

Contrary to DPB, photolysis of 1-phenyl-1,3-pentadiyne (PPD) [47] and 1-phenyl-5,5-dimethyl-1,3-butadiyne(PDB) [39] with DMB yields only 1:1

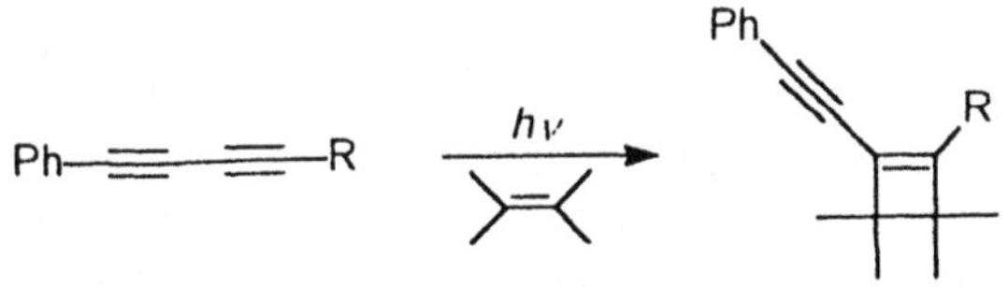

R = CH$_3$ 17, 10 %
tert-butyl   18 , 22%

Scheme 12

cyclobutene adduct, and the reaction site is C3 and C4 triple bond.

Irradiation of PPD (4 mM) with unsymmetrical olefins (EVE and AN) in deaerated solutions yields site selective and regioselective [2+2] type 1:1 photoadducts (19 and 20) and 1:2 photoadducts (21–26) [47]. The structure of these adducts was determined by various physical methods, including ^{13}C-NMR spectroscopy, which is vital for the determination of the reaction site. The ^{13}C-NMR spectra for all of these adducts show two sp hybridized carbon peaks indicating that one of the carbon–carbon triple bonds remains intact. The existence of ethynyl benzene moiety (120–125 ppm) in all the adducts indicates that C3–C4 triple bond is the reactive site. The 1:2 adducts (21–23) rearranged to give ring opened 1,5-diene products (27 and 28) on warming to 101°C (in methylcyclohexane).

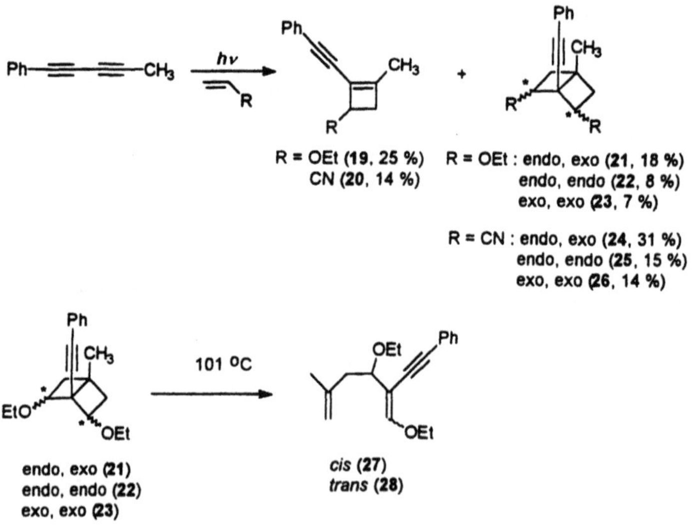
Ph
CH3
hν
R
Ph
CH3
R
+
Ph
CH3
R
R
R = OEt (19, 25 %)
CN (20, 14 %)
R = OEt : endo, exo (21, 18 %)
endo, endo (22, 8 %)
exo, exo (23, 7 %)
R = CN : endo, exo (24, 31 %)
endo, endo (25, 15 %)
exo, exo (26, 14 %)
Ph
CH3
EtO
OEt
101 °C
OEt
Ph
OEt
endo, exo (21)
endo, endo (22)
exo, exo (23)
cis (27)
trans (28)

Scheme 13

In the photolysis of PPD with EVE and AN, the 1:1 photoadducts 19 (20) are initially formed, and prolonged irradiation gives secondary 1:2 photoadducts 21–23 (24–26). The regiochemistry of 1:1 photoadducts (19 and 20) is deduced from the regiochemistry of 1:2 photoadducts (21–26). The Stern-Volmer plots on the photocycloaddition of PPD with unsymmetrical olefins to give 19 and 20 show a linear relationship with various concentrations of 9,10-diphenylanthracene with quenching constants ($k_q\tau$) of 11,000 and 1,600 M^{-1}, suggesting that 19 and 20 are originated from the triplet excited state of PPD. The secondary photoreactions show a similar tendency, suggesting that the secondary photoreaction also proceeds from the triplet excited state of 1:1 photoadduct.

1-(1-Naphthyl)-1,3-butadiynes were photolyzed in DMB to obtain [2+2] photoadducts (29–32) as shown in Scheme 14 [39]. The reaction site is identified as an acetylene moiety attached to a naphthyl group, and reactions proceed from both singlet and triplet excited states of diynes. Introduction of a 1-naphthyl group on the C1 carbon causes the participation of the singlet pathway and the localization of electrons (or diradical character) on the C1 and C2 carbons in the excited state, while the phenyl group on the C1 carbon causes the delocalization of electrons (or diradical character) throughout the conjugated system. The results are attributed to the difference of the inductive and resonance effects of the naphthyl and phenyl groups.

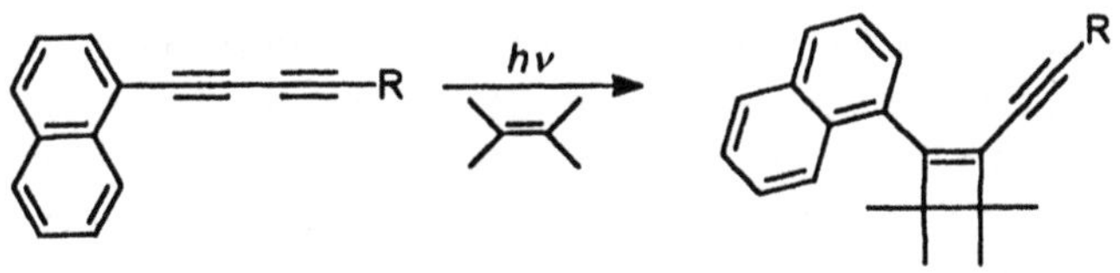

R = 1-Naphthyl   **29**, 5 %
    Phenyl   **30**, 10 %
    t-Butyl   **31**, 43 %
    Methyl   **32**, 77 %

Scheme 14

B. Photoreaction of Conjugated Triynes with Olefins

1-Phenyl-1,3,5-hetatriyne (PHT) was the one of the most thoroughly studied polyalkyne compounds. PHT is a naturally occurring compound in the plants of genera Dahlia, Bidens, Coreopsis, Zoegea, and Ferreyanthus [48]. The striking property of PHT is photocytotoxicity to mammalian systems, bacteria, fungi, nematodes, viruses, insects, plants (cercatiae) and fish. Two mechanistic models for the phototoxicity of PHT have been suggested: biological damage from the conversion of excited PHT to radical under anaerobic conditions [6,7,9,49,50] and the photodynamic effect from quenching of excited PHT by oxygen under aerobic conditions [48,51–53].

Many works have been performed to verify the exact mechanism of the phototoxicity of PHT. *E. coli* and *S. cerevisae* are actively studied organisms that can live even in the absence of oxygen. Many disputes have been raised on whether *E. coli* shows a photodynamic effect or not, while *S. ceresiae* clearly shows a photodynamic effect. The survival of *E. coli* was not significantly modified by the use of standard diagnostic reagents, such as sodium azide, D_2O, and superoxide dismutase [49]. Free radicals were detected by ESR when PHT in liposome was irradiated [50]. From these results, the semioxidized PHT radical is suggested to be the likely toxic species created in the nonoxidative sensitization. Kagan et al. proposed a different mechanism, raising an objection to other groups' work in anaerobic conditions. They carried out the experiment carefully in an anaerobic environment and concluded that the phototoxicity of PHT is only based upon a photodynamic effect, although they do not provide the exact extent of type I and type II photodynamic pathways [48,51–53].

The mechanism for photochemical membrane damage in microorganisms by PHT is not firmly established. Covalent bonding of PHT to membrane other than by the photodynamic effect is a strong possibility for the phototoxicity of PHT, and the photocycloaddition of polyalkynes to olefins has been investigated.

Deaerated solution of 1,6-diphenyl-1,3,5-hexatriyne(DPH) with dihydrofuran, AN, methyl acrylate(MA), and styrene(ST) yields regioselective [2+2] type 1:1 photoadducts (33–36) [54]. In the ^{13}C-NMR spectra for all of these adducts, a phenyl substituted sp^2 carbon peak of the cyclobutene ring appears at 150–155 ppm, indicating that the C1–C2 triple bond is the reactive site.

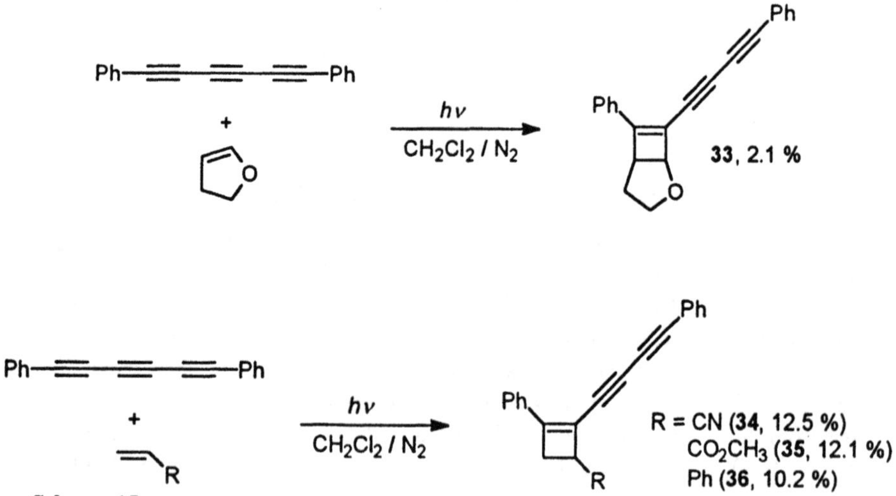

Scheme 15

Irradiation of DPH with electron-deficient olefins such as fumaronitrile (FN) and dimethyl fumarate (DMFu) in deaerated solutions also yield site selective [2+2] type 1:1 photoadducts (37–39).

Photolysis of deaerated 1-phenyl-1,3,5-hexatriyne with unsymmetrical olefins such as AN, MA, and ST in methylene chloride gives three 1:1 photoadducts (40–48) [55].

The photoreaction of 1-phenyl-1,3,5-hexatriynes with various olefins proceeds through a triplet excited state and shows site selectivity and regio-

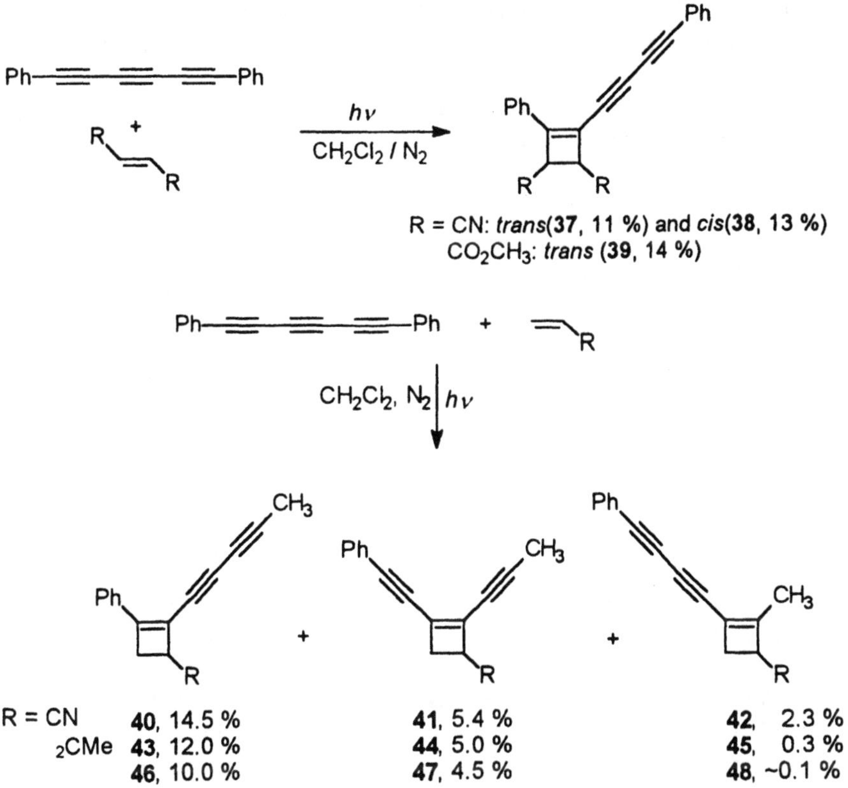

Ph———≡———≡———≡———Ph

R⤷ + ⤷R

$\xrightarrow[\text{CH}_2\text{Cl}_2 \ / \ \text{N}_2]{h\nu}$

R = CN: *trans*(**37**, 11 %) and *cis*(**38**, 13 %)
CO$_2$CH$_3$: *trans* (**39**, 14 %)

Ph———≡———≡———≡———Ph   +   ⤷R

$\text{CH}_2\text{Cl}_2, \ \text{N}_2 \Big| h\nu$

R = CN	**40**, 14.5 %	**41**, 5.4 %	**42**, 2.3 %
2CMe	**43**, 12.0 %	**44**, 5.0 %	**45**, 0.3 %
	46, 10.0 %	**47**, 4.5 %	**48**, ~0.1 %

Scheme 16

selectivity. Olefins with electron-withdrawing substituents, such as DMFu, FN, AN, MA, and ST are more reactive than electron-rich olefins. The triplet excited state of DPH is quenched by olefins [54]. The quenching rate constants are determined by laser photolysis studies, and these are rates of intermolecular addition of the DPH triplet excited state to the olefins, because the quenching rates are very different from energy transfer quenching rates, which are usually in the range of diffusion control. The reaction rate constants for highly reactive olefins such as FN, DMFu, AN, MA, and ST are in the range of 10^6–10^4, and unreactive or less reactive olefins such as EVE, dihydrofuran, DMB, and *trans*-1,2-dichloroethylene show very small quenching rate constants, and tetrachloroethylene does not quench the triplet excited state of DPH at all, consistent with the lack of photochemistry. This can be explained on the basis of polar effects of electron-withdrawing substituents. The regioselectivity and reactivity of olefins in the photoaddition reaction are in accord with that of the addition reaction of nucleophilic alkyl radical to olefins [56], and the results suggest that the triplet excited states of 1-phenyl-1,3,5-hexatriynes have a nucleophilic radical character.

Photolysis of 1-phenyl-1,3,5-hexatriynes in DMB gives dicyclopropyl photoadducts (49–53) [57–59].

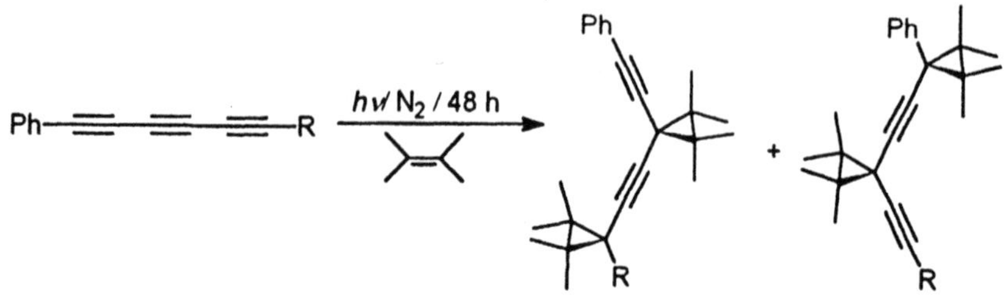

R = Ph   **49, 33 %**
CH_3 **50, 11 %**
H   **51, 7 %**
Scheme 17 $Si(CH_3)_2Bu^t$ **52, 11 %**   **53, 10 %**

The reaction seems to proceed from the cumulene type triplet excited states of triynes probably via carbene intermediates, even though all efforts to trap or detect carbenes at cryogenic conditions have failed. Radical-stabilizing groups such as the *tert*-butyldimethylsilyl group on terminal carbons of conjugated hexatriynes control the photoaddition reactions of the compounds. Hydrogen and methyl groups on C6 have no or little radical-stabilizing ability at C6, while phenyl ring and *tert*-butyldimethylsilyl groups can stabilize the radical at C6 by delocalization of the unpaired electron over the phenyl ring and empty d orbital of the silyl group, respectively. Consequently, the radical at C6 in 1-phenyl-1,3,5-hexatriyne and 1-phenyl-1,3,5-heptatriyne is more reactive than the C1 radical. However, when the R group is phenyl and *tert*-butyldimethylsilyl, the radical at C1 and C6 will have similar reactivity.

As a model reaction for the phototoxic membrane damage by PHT, PHT was photolyzed with unsaturated fatty acid, undecylenic acid methyl ester, and isolated four major photoadducts (54–57) as shown in Scheme 18 [60]. PHT was also irradiated with 1-hexene as a reference reaction [32]. Photoreactions occur through the [2+2] photocycloaddition reaction between double bonds of olefins and C1–C2, C3–C4, and C5–C6 triple bonds of PHT. These reactions most likely result from the cumulene type diradical species.

The results show that the photoaddition reaction of conjugated polyynes with olefins is dependent on the number of conjugated triple bonds, the nature of the terminal groups, and the structural properties of olefins.

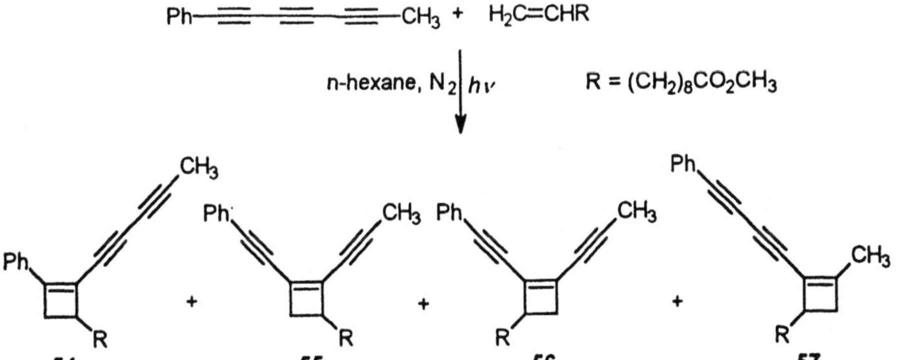

Ph—≡—≡—≡—CH₃ + H₂C=CHR

n-hexane, N₂ | hν   R = (CH₂)₈CO₂CH₃

54 + **55** + **56** + **57**

Scheme 18

IV. PHOTOHYDRATION REACTION OF POLYALKYNES

A. Photohydration Reaction

Neutral carbon of organic compounds is inherently a very weak base due to the lack of nonbonding electrons. However, the π-system of alkenes, alkynes, and allenes are potential basic sites due to the availability of π-electrons. Wooldridge and Roberts [61] first reported the facile photohydration of arylacetylenes to the corresponding ketones (in an overall Markownikoff sense) in dilute aqueous acid solutions. The facile protonation of the excited state was attributed to the enhanced basicity of the alkyne moiety in the excited state, more specifically as a result of extensive charge migration from the aryl group to the acetylene moiety. The photoaddition of methanol [62] and acetic acid to π-electron systems such as alkenes, alkynes, and allenes may be rationalized in the same way.

Yates and coworkers [63–74] have carried out systematic mechanistic investigations on the photohydration of aryl alkynes, alkenes, and allenes. In general, these photohydration reactions are acid-catalyzed and proceed via the S_1 state to give regiospecific hydration products in the Markownikoff sense. Exceptions are the *m*- and *p*-nitro derivatives, which add water via the T_1 state in the anti-Markownikoff directions as shown in Scheme 19 [64]. A mechanism involving a synchronous addition of H_3O^+ to the T_1 state has been proposed for the photohydration of (nitrophenyl)acetylenes. The driving force of the reaction was proposed to be the enhanced electron-withdrawing character of nitro groups in the T_1 state.

Irradiation of 1-(1-naphthyl)-1,3-butadiynes in aqueous sulfuric acid yields both type A and type B photohydration products [75]. Fluorescence of starting material is quenched by sulfuric acid, and azulene quenching experiments showed that the singlet excited state yielded type A and type B photoadducts,

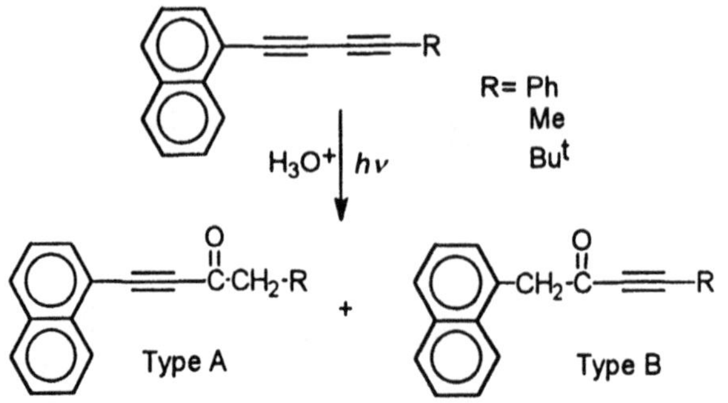

Scheme 19

supporting the dichotomy of charge separation in the singlet excited state. On the other hand, the triplet excited state gives only the type B photoadduct.

When R is the methyl group, the quantum yield for type A product is $\Phi_{pdt} = 4.2 \times 10^{-3}$. The quantum yields for the formation of type B photoproduct from the singlet and triplet excited states are $\Phi_{SB} = 3.5 \times 10^{-3}$ and $\Phi_{TB} = 1.6 \times 10^{-3}$. The S_1 state of 1-(1-naphthyl)-1,3-pentadiyne is probably highly charge separated, while the T_1 state is weakly dipolar, since T_1 state of acetylene is normally considered to be similar to that of diradicals and the T_1 state yields the type B photohydration products in low yield due to its very low reactivity to water. The charge in the singlet excited state seems to be delocalized throughout the conjugated cumulene system. This is supported by the structure of type A and type B photoadducts which are formed by protonation of C1 or C4 followed by the attack of water on C2 or C3.

The product quantum yields are increased with increasing acidity of the medium, indicating that the protonation step is the rate limiting step. A possible charge distribution of excited states is shown in the following scheme. It is clearly shown that the type A product is produced from both singlet and triplet excited states and the type B only from the triplet excited state. The singlet excited state has the dominant contribution of positively charged structure on C1, and the triplet excited state has the partial negative charge on C1.

Naphthyl substituent shows a different character in the excited state in comparison with phenyl substituent. The photohydration of 1-phenyl-5,5-dimethyl-1,3-hexadiyne gives three photohydration products (58–60) (two acetylenyl ketone and one allenyl ketone) through both S_1 and T_1 excited states (Scheme 20). An allenyl ketone product is obtained as a minor product. A proposed mechanism involves the protonation step as the rate limiting step in the formation of acetylenyl ketone products. On the other hand, the allenyl ketone product is formed by the synchronous addition mechanism of H_3O^+.

The aromatic nitro group has the electron withdrawing effect and enhanced intersystem crossing efficiency in the excited state. The enhancement

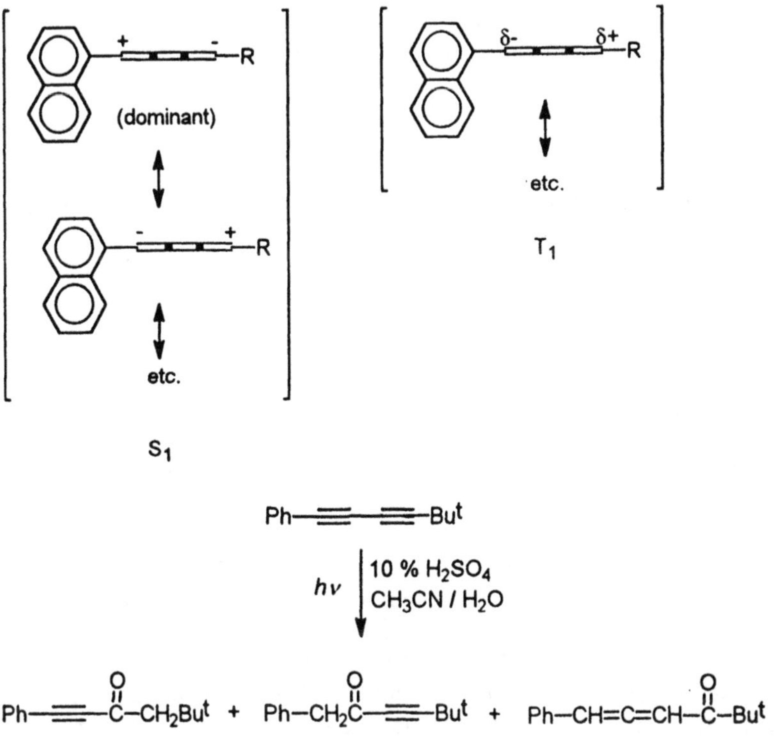

$$Ph \equiv \equiv But$$

$$hv \downarrow \begin{array}{l} 10\% \ H_2SO_4 \\ CH_3CN \ / \ H_2O \end{array}$$

$$Ph \equiv \overset{O}{\underset{\|}{C}} - CH_2But \ + \ Ph-CH_2\overset{O}{\underset{\|}{C}} \equiv But \ + \ Ph-CH=C=CH-\overset{O}{\underset{\|}{C}}-But$$

58, 7 %   **59, 11 %**   **60, 3 %**

Scheme 20

of mesomeric and inductive electron withdrawing character is greater in the *m*-isomer than in the *p*-isomer [76,77] in the photoreactions, and this is called the *meta* effect. Investigation of the photohydration of nitro substituted aryl diacetylenes is therefore warranted.

Irradiation of 1-(*p*-nitrophenyl)-5,5-dimethyl-1,3-hexadiyne (*p*-NDHD) in aqueous sulfuric acid gives four photoproducts (61–64) as shown in Scheme 21 [76]. Kinetic studies on the products' formation are carried out monitoring the disappearance rate of the starting material. The data clearly show that *p*-NDHD is initially converted photochemically into allenyl ketone compounds which undergo thermal hydration to give dicarbonyl compounds.

The oxygen and 9-fluorenone-1-carboxylic acid quenching experiments on the photohydration of *p*-NDHD show that the triplet excited states are involved to give the primary photoproducts. The photohydration of *p*-NDHD exhibits a maximum efficiency at $H_0 = -1.0$ for both primary photoproducts, as shown by the measurements of quantum yields as a function of acidity. The suggested mechanism involves a nucleophilic attack of water, synchronous with protona-

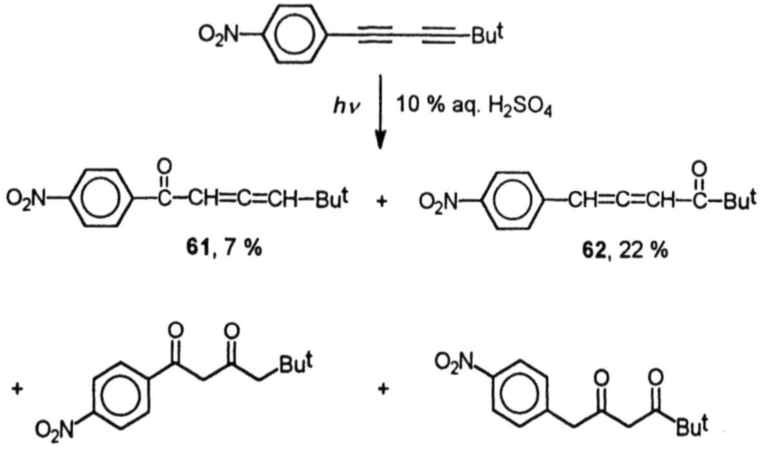

O_2N—⬡—≡—≡—Bu^t

hv | 10 % aq. H_2SO_4

O_2N—⬡—$\overset{O}{\overset{\|}{C}}$—CH=C=CH—$Bu^t$ + O_2N—⬡—CH=C=CH—$\overset{O}{\overset{\|}{C}}$—$Bu^t$

61, 7 %   **62, 22 %**

+   (diketone with O_2N-phenyl, Bu^t)   +   (diketone with O_2N-phenyl, Bu^t)

63, 11 %   **64, ~1 %**

Scheme 21

tion. The decrease or leveling off in photohydration efficiency is most likely due to the depletion of water on going to strongly acidic media.

Two photohydration products (65 and 66) are obtained on irradiation of 1-(*m*-nitrophenyl)-5,5-dimethyl-1,3hexadiyne in aqueous sulfuric acid [77]. The protonation products at C1 position only are obtained in contrast to *p*-NDHD. The photohydration quantum yields for the primary photoproduct show maximum efficiency at $H_0 = -0.6$, which is lower than that of *p*-NDHD, and the photohydration of *m*-NDHD proceeds in weak acid solutions, even at pH 7, in contrast to *p*-NDHD, which is not photohydrated in neutral water. These results indicate that the enhancement of the electron-withdrawing effect and basicity in the excited states of *m*-NDHD are greater than those of *p*-NDHD.

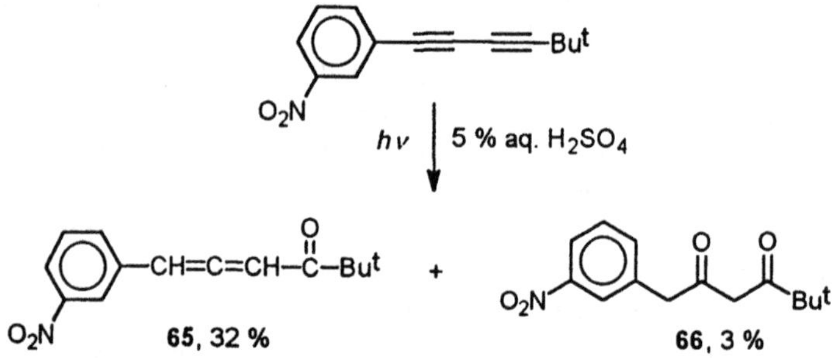

(structure: O_2N-phenyl)—≡—≡—Bu^t

hv | 5 % aq. H_2SO_4

(structure: O_2N-phenyl)—CH=C=CH—$\overset{O}{\overset{\|}{C}}$—$Bu^t$ +   (diketone structure with O_2N-phenyl, Bu^t)

O_2N **65, 32 %**   **66, 3 %**

Scheme 22

In summary, the photohydration of *p*-NDHD yields two allenyl ketones via both type A and type B excited states (type A has partial positive charge and type B partial negative charge on C1 position). On the other hand, only the type B allenyl ketone is obtained from the photohydration of *m*-NDHD, indicating that the *m*-nitro group has the stronger electron withdrawing effect in the excited state, which is consistent with the *meta* effect, originally noted by Zimmerman [78,79].

B. Photoaddition of Methanol to Polyalkynes

When methanol solutions of naphthyl substituted diacetylenes are irradiated, polar addition photoproducts are obtained [80]. The protonated photoproducts on C1 carbon only are obtained and the protonated product at C4 position is formed when a phenyl group is substituted on C4 carbon.

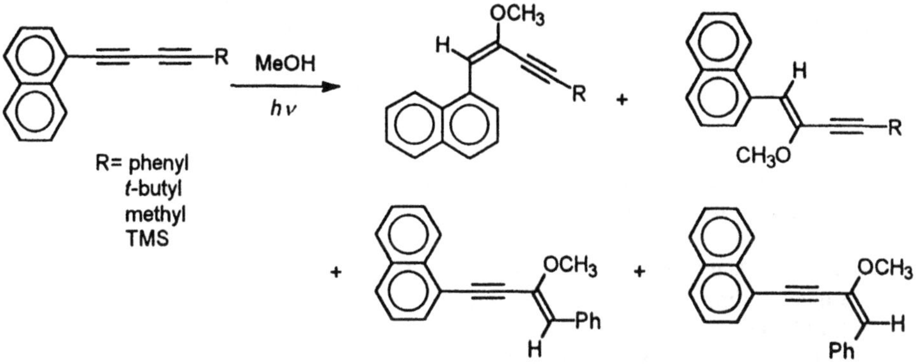

Scheme 23

Irradiation of phenyl substituted diacetylenes yields both protonated products and reduction products. The photoreduction products seem to originate from the triplet excited state, since oxygen quenches the photoreduction almost completely. However, oxygen does not quench the polar addition reaction, supporting a singlet excited state mechanism for the formation of type B photoproducts (C1 protonation product).

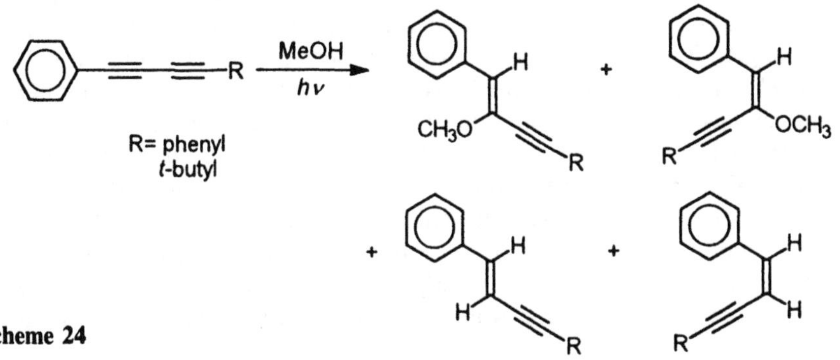

Scheme 24

The geometry of the lowest singlet excited state of acetylene is the *transoid* zwiterionic structure with an HCC angle of about 120°C. The basicity of the α,β-unsaturated group of the α,β-unsaturated compounds increases markedly in excitation. In more extensively conjugated systems, the charge distribution in the excited state can be determined by evaluating the site and type of the photoaddition reaction. If the charge separation of the excited state is localized on the C1-C2 triple bond, only type A photoadducts (C2 or C4 protonation products) should be obtained. In fact, type B photoadducts are obtained, suggesting that the C1 carbon is negatively charged and the excited state of diacetylenes has the cumulene type structure. The singlet excited state has highly dipolar character, and the triplet excited state shows a neutral or slightly charge separated state. Among the resonance structures of the singlet excited state, the structure with negatively charged C1 carbon seems to contribute more than that containing a positive charged C1 carbon due to the inductive effect of terminal groups attached to C1 and C4 of the conjugated 1,3-butadiyne backbone.

V. PHOTOCHEMISTRY OF 1-ARYL-4-(PENTAMETHYLDISILANYL)-1,3-BUTADIYNES

The excited state chemistry of aryldisilanes has been extensively studied because these compounds show interesting photophysical and photochemical properties [81–86]. Aryldisilanes show dual fluorescence, local and intramolecular charge transfer (CT) fluorescence. The CT emission has a broad and structureless band with a large Stokes shift. This intramolecular charge transfer state has a strongly polar structure with complete charge separation, which plays an important role in the photochemical reaction [81]. Shizuka et al. suggested that the CT emission originates from the $^{1}(2p\pi,3d\pi)$ state produced by the $2p\pi^{*}$ (aromatic ring) → vacant $3d\pi$ (Si-Si bond) intramolecular charge transfer transition on the basis of the steric twisting effect on the emission [81]. On the other hand, Sakurai et al. reported that the CT bands of aryldisilanes and related compounds can be explained satisfactorily by the emission from the intramolecular charge transfer states where σ(Si-Si) and π systems act as donors and acceptors, respectively [84]. They proposed the orthogonal intramolecular charge transfer (OICT) model as a general term on the basis of both substituent and geometric effects on the dual fluorescence of aryldisilanes. Interestingly, photolysis of alkynyl substituted disilane derivatives give highly reactive silacyclopropene intermediates [82,85,86]. In general, most of the silacyclopropenes are thermally stable, but they are extremely reactive toward atmospheric oxygen and moisture. These silacyclopropenes formed from the photolysis of alkynyl substituted disilanes in methanol or acetone react readily with methanol or acetone [82,85,86]. Silacyclopropenes also react with unsaturated functional groups such as aldehydes, ketones, styrenes, conjugated terminal acetylenes, benzyne, terminal 1,3-dienes, and conjugated imines to give five-membered cyclic organosilicon products in which $C=O$, $C=C$, $C\equiv C$, or $C=N$ bonds are inserted into the Si–C bond of the silacyclopropene ring [87]. Considerable attention has been devoted to the investigation of transition metal catalyzed reactions of silacyclopropenes and it has been known that the nickel(0)- and palladium(0)- catalyzed reactions of silacyclopropenes produce reactive intermediates such as metalasilacyclobutenes arising from the insertion of metals into the Si–C bond in a silacyclopropene ring [86,88–93]. In the absence of acetylenes or even in the presence of an unreactive acetylene such as diphenylacetylene, palladium catalyzed photolysis of (phenylethynyl)pentamethyldisilane gives 1,4-disilacyclohexa-2,5-diene type dimerization, while with acetylene dicarboxylic ester the acetylene is incorporated in the product [86]. Seyferth *et al.* have suggested that the silylene formed from silacyclopropene reacts with an alkyne to give another silacyclopropene in the presence of $PdCl_2(PPh_3)_2$, and that $(PPh_3)_2Pd^0$ is the active catalyst [88].

A. Photophysical Properties of 1-Aryl-4-(pentamethyldisilanyl)-1,3-butadiynes

No fluorescence was observed from 1-phenyl-4-(pentamethyldisilanyl)-1,3-butadiyne (PDSB), 1-(*p*-methoxyphenyl)-4-(pentamethyldisilanyl)-1,3-butadiyne (MDSB), or 1-(*p*-nitrophenyl)-4-(pentamethyldisilanyl)-1,3-butadiyne (NDSB) even at 77K, but 1-(1-naphthyl)-4-(pentamethyldisilanyl)-1,3-butadiyne (NaDSB) shows fluorescence with a maximum at 353 nm in methanol [94]. No charge transfer emission is observed in any of these compounds. All the 1-aryl-4-(pentamethyldisilanyl)-1,3-butadiyne derivatives show strong phosphorescence with strong 0-0 bands around 460–550 nm at 77K in organic glass, indicating that triplet energies of these compounds lie around 218–260 kJ/mol. Laser photolysis of NDSB in ethanol results in the formation of an interesting transient (zwitterion species) observed 500 ns after laser pulsing and quenched by oxygen (electron acceptor) and N,N-dimethylaniline (electron donor) [94]. The transient is probably formed by an intramolecular electron transfer from disilanyl to the nitro group in the triplet excited state in polar solvents such as ethanol and acetonitrile.

B. Photochemical Reactions of 1-Aryl-4-(pentamethyldisilanyl)-1,3-butadiynes

Upon irradiation 1-aryl-4-(pentamethyldisilanyl)-1,3-butadiynes give silacyclopropene intermediates from the singlet excited state, except for NDSB, which gives silacyclopropenes from the triplet excited state as shown in Scheme 25

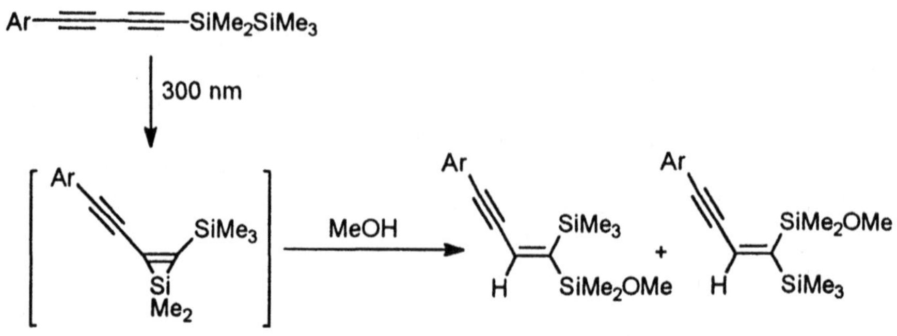
Ar—SiMe2SiMe3
300 nm
Ar
SiMe3
Si
Me2
MeOH
Ar
SiMe3
H
SiMe2OMe
+
Ar
SiMe2OMe
H
SiMe3

Ar = Phenyl (PDSB) — **67**, 13 % — **68**, 17 %
1-Naphthyl (NaDSB) — **69**, 25 % — **70**, 25 %
p-Methoxyphenyl (MDSB) — **71**, 13 % — **72**, 12 %

Scheme 25

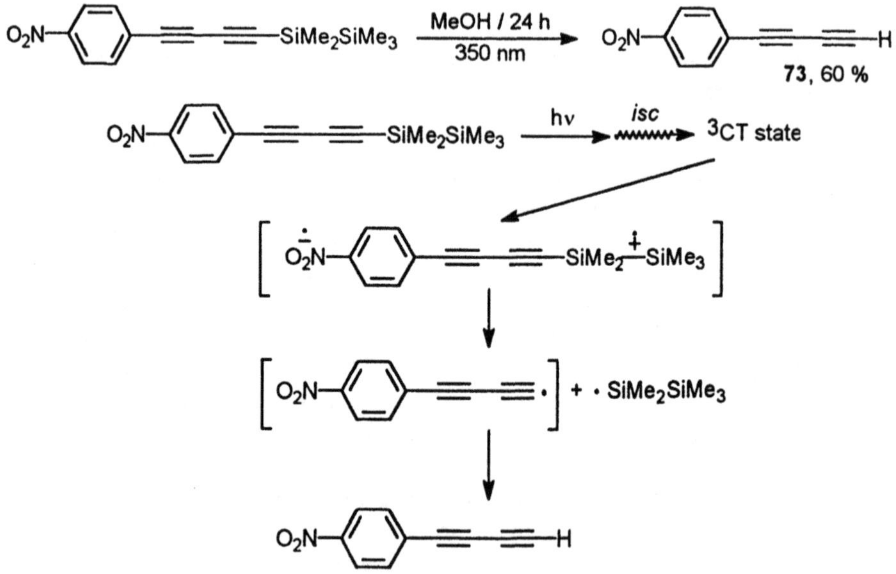
O2N
SiMe2SiMe3
MeOH / 24 h
350 nm
O2N
H
73, 60 %
O2N
SiMe2SiMe3
hν
isc
3CT state
O2N
SiMe2 SiMe3
O2N
+ · SiMe2SiMe3
O2N
H

Scheme 26

[94–96]. Irradiation of 1-aryl-4-(pentamethyldisilanyl)-1,3-butadiynes (PDSB, NaDSB, and MDSB) in methanol gives 1:1 photoadducts (67–72) via silacyclopropene intermediates [93].

Photochemical reactions of phenylethynyldisilanes in the presence of methanol were reported to give mainly photoaddition products through the silacyclopropene intermediates in the singlet excited states [82,86]. These photoreactions of 1-aryl-4-(pentamethyldisilanyl)-1,3-butadiynes in methanol are not quenched by oxygen, suggesting the photoreactions to proceed via silacyclopropene intermediates in the singlet excited state. On the other hand, photolysis of NDSB in methanol and ethanol results in the formation of 1-(p-nitrophenyl)-1,3-butadiyne (73) through C–Si bond cleavage as shown in Scheme 26 [94]. This photoproduct is formed from the triplet excited state of NDSB, and the plausible reaction mechanism involves an intramolecular electron transfer to give a long-lived transient (zwitterion species) in polar solvents. This transient is efficiently quenched by oxygen (electron acceptor) and N,N-dimethylaniline (electron donor). The C–Si bond is cleaved homolytically from the transient, and the radical abstracts a hydrogen from the methyl group of methanol. Irradiation of NDSB in CH$_3$OD does not give any deuterium incorporation in the acetylenic hydrogen, while photoreaction in CD$_3$OD yields deuterium incorporated product, indicating a radical mechanism.

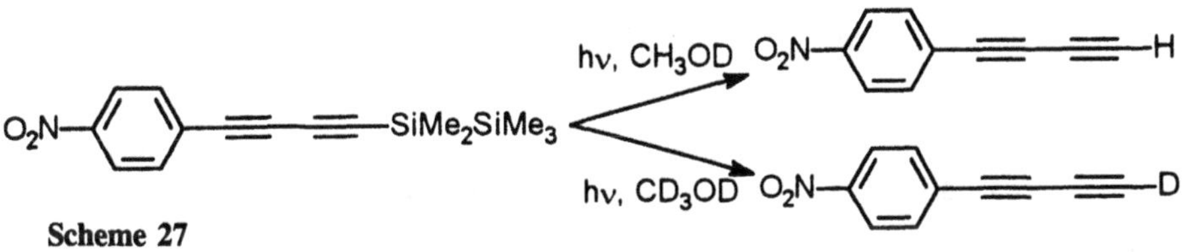

Scheme 27

Irradiation of 1-aryl-4-(pentamethyldisilanyl)-1,3-butadiynes (PDSB, MDSB, NDSB) with acetone yields site specific and regioselective 1:1 adducts (74–78) having 1-oxa-2-silacyclopent-3-ene structure through two-atom insertion of acetone to silacyclopropene intermediates [96].

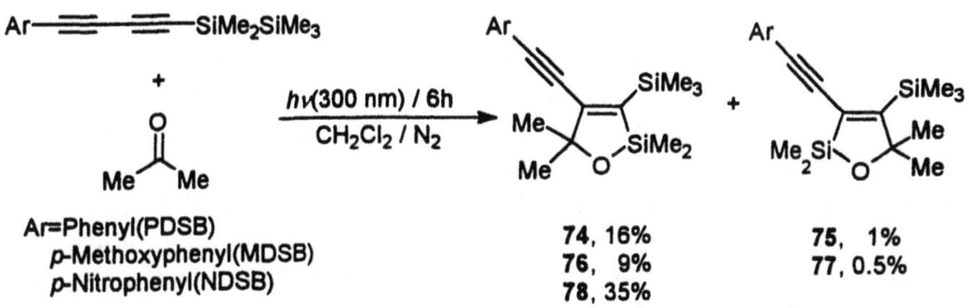

Ar=Phenyl(PDSB)
p-Methoxyphenyl(MDSB)
p-Nitrophenyl(NDSB)

74, 16%   75, 1%
76, 9%   77, 0.5%
78, 35%

Scheme 28

Experiments with various quenchers were carried out to quench the triplet excited state of acetone selectively [95]. From these experiments it is confirmed that the silacyclopropene formed from the photolysis of 1-aryl-4-(pentamethyldisilanyl)-1,3-butadiynes reacts with acetone via the triplet excited state of the silacyclopropene, and the triplet energy of the silacyclopropene lies around 62–68 kcal/mol.

Photoreaction of 1-aryl-4-(pentamethyldisilanyl)-1,3-butadiynes (PDSB, MDSB, NDSB) with dimethyl fumarate gives two-atom insertion products via silacyclopropene intermediates and/or [2+2] photocycloaddition products (79–84) [96].

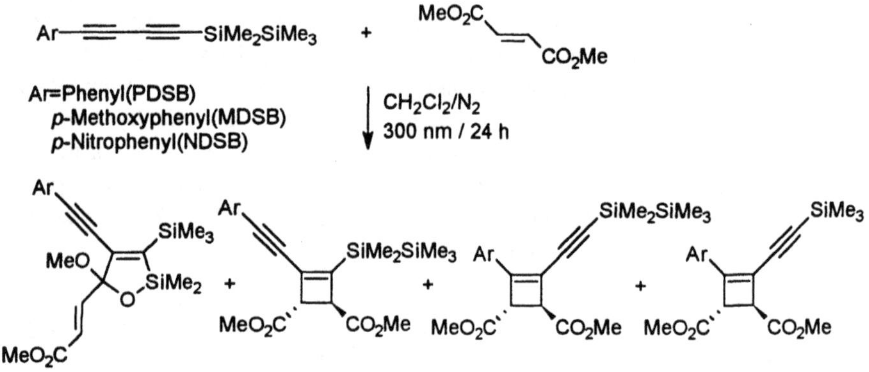

Ar=Phenyl(PDSB)
p-Methoxyphenyl(MDSB)
p-Nitrophenyl(NDSB)

CH$_2$Cl$_2$/N$_2$
300 nm / 24 h

79, 5 %
-
83, 20 %

80, 8 %
-

81, 10 %
84, 2 %

82, 8 %
-

Scheme 29

MDSB, which has an electron-donating group, yields only [2+2] photocycloaddition products (80–82), while NDSB, which has a nitro group, gives both a two-atom insertion product (83) via silacyclopropene intermediates and a [2+2] photocycloaddition product (84), the former in higher yields. These results can be attributed to the stabilization of silacyclopropene intermediates by substituents and also to the nucleophilic character of the 1-aryl-4-(pentamethyldisilanyl)-1,3-butadiyne radical in the [2+2] photocycloaddition reaction. The silacyclopropene formed from MDSB is rather unstable, and the triplet radical has an enhanced nucleophilic character due to the electron donating methoxy group, and only [2+2] photocycloaddition products are formed. On the other hand, for NDSB, due to the electron-withdrawing nitro group in NDSB, silacyclopropene is rather stable, and the triplet radical has a sustained nucleophilic character to give a two-atom insertion product in high yield via silacyclopropene intermediate; and the high efficiency of intersystem crossing may play another important role.

The formation of metallasilacyclobutene has been proposed in the transition metal catalyzed reaction of silacyclopropenes with an acetylene [86,88–93]. Ishikawa et al. [97] suggested the presence of nickelasilacyclobutenes from ^{13}C-, ^{29}Si-, and ^{31}P-nmr data. Irradiation of a solution of 2-mesityl-2-(phenylethynyl)-1,1,1,3,3,3-haxamethyltrisilane in hexane at room temperature gave silacyclopropene and then reacted with tetrakis(triethylphosphine)nickel(0) to afford the nickelasilacyclobutenes. The reaction of silylyene complex with phenyl(trimethylsilyl)acetylene gives 85 and 86, supporting the existence of nickelasilacyclobutene.

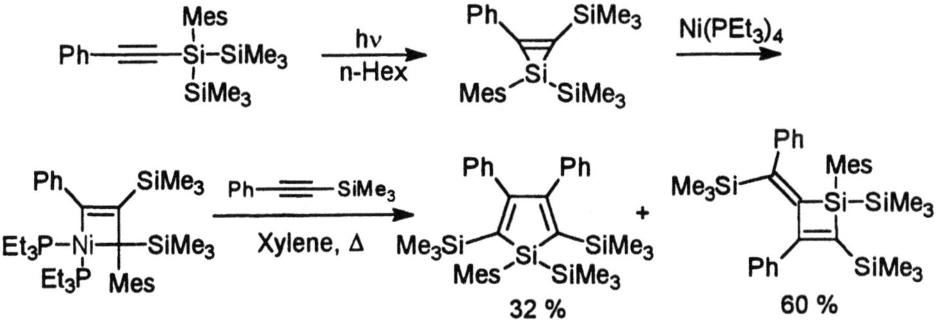

Scheme 30

Tungsten (0) catalyzed reaction of silacyclopropenes proceeds in a quite different fashion from the nickel and palladium catalyzed reactions [89]. Tungsten catalyzed abstraction of mesitylmethylsilylene from the silacyclopropene was observed, which is followed by the reaction with diphenylacetylene to give 87.

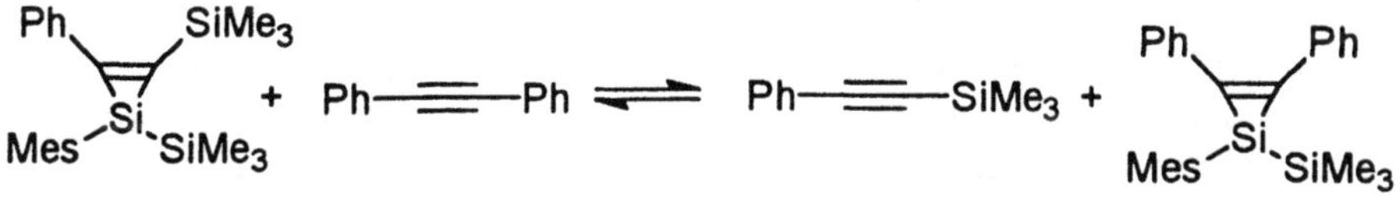

Scheme 31

PdCl$_2$(PPh$_3$)$_2$ catalyzed photolysis of 1-phenyl-4-(pentamethyldisilanyl)-1,3-butadiyne in dry benzene gives 1,4-disilacyclohexa-2,5-diene type dimerization products (89–92) and 1-phenyl-4-(trimethylsilyl)-1,3-butadiyne(88) [98].

A plausible reaction mechanism similar to the one proposed by Seyferth et al. [88] can explain the results as shown in Scheme 33. Silacyclopropene A is formed from the singlet excited state of 1-phenyl-4-(pentamethyldisilanyl)-1,3-butadiyne, and this silacyclopropene A reacts with (PPh$_3$)$_2$Pd^0 to form palladasilacyclobutene. The silylene complex formed from palladasilacyclobutene reacts with 1-phenyl-4-(pentamethyldisilanyl)-1,3-butadiyne and 1-phenyl-4-(trimethylsilyl)-1,3-butadiyne (88) to give silacyclopropene B and C. From these silacyclopropenes A, B, and C, dimerization products are obtained.

The results of PdCl$_2$(PPh$_3$)$_2$ catalyzed photolysis of 1-phenyl-4-(pentamethyldisilanyl)-1,3-butadiyne with other alkynes support the involvement of this silylene complex [98].

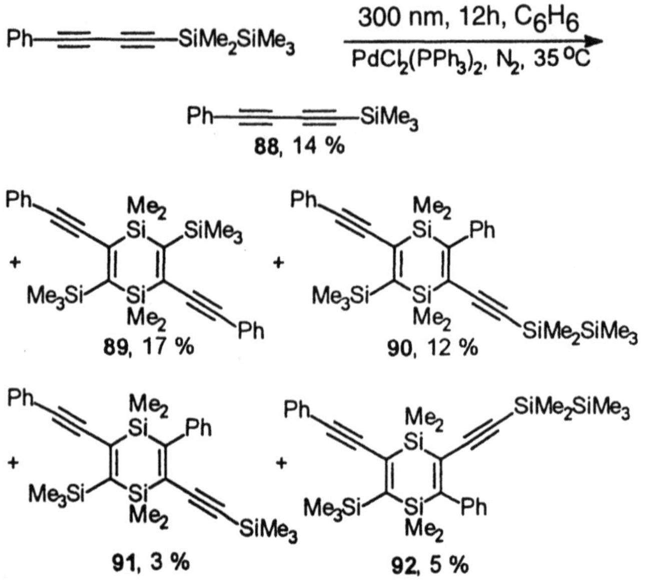

Ph———≡———≡———SiMe₂SiMe₃ $\xrightarrow[\text{PdCl}_2(\text{PPh}_3)_2, \text{ N}_2, \text{ 35 °C}]{\text{300 nm, 12h, C}_6\text{H}_6}$

Ph———≡———≡———SiMe₃

88, 14 %

89, 17 %   **90, 12 %**

91, 3 %   **92, 5 %**

Scheme 32

VI. SOLID-STATE POLYMERIZATION OF DIACETYLENES

Polydiacetylenes (PDAs) have been attracting much interest from basic and applicational viewpoints. They have an ordered array of fully extended polymer chains with conjugated backbone showing various forms such as high quality single crystals, films, and polymer solution or gels in organic solvents. Conjugated polymers like PDAs have many properties, such as the large quasi-one-dimensional structure of single crystal, high third-order optical susceptibility, large optical nonlinearity, and high photoconductivity [99,100]. Wegner first reported a class of monomers that can be polymerized in the solid state by high-energy irradiation (UV, γ-ray, etc.) or by annealing the monomer crystals below melting point to produce large, nearly defect-free polymer crystals [98,100]. There is no striking difference between the photopolymerization and thermal polymerization except for the polymers from bis-(p-toluene sulfonate) of 2,4-hexadiyn-1,6-diol. In this system, photopolymerization is faster than thermal conversion and requires no induction time. When the polymerization is carried out by thermal (spontaneous) polymerization, it shows the considerable induction period and very high viscosity values obtained for thermal polymers [99]. These polymer monocrystals were obtained by irradiation or thermally induced 1,4-addition reaction of diacetylene monomer crystals (Figure 1).

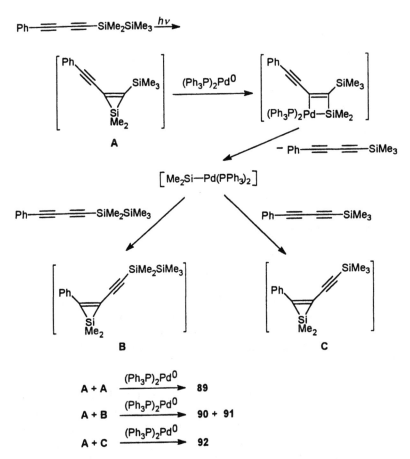

Scheme 33

Topochemical polymerization of diacetylenes proceeds in two pathways when the substituents have hydrogen bonding sites or show strong dipole–dipole interaction with each other. Diacetylenes having aromatic substituents that are not directly attached to the C(1) and C(4) polymerize through a specific rotation pathway, while diacetylenes with directly attached aryl groups polymerize by a shearing mechanism [101,102]. The systematic and detailed concept associated with the first polymerization pathway was elucidated by Baughman [103]. According to his report, the polymerization reactivity is directly correlated with the monomer's packing parameter.

In order to assess the relative reactivity of different diacetylene monomers, the packing parameters can be analyzed in terms of a reaction along a least mo-

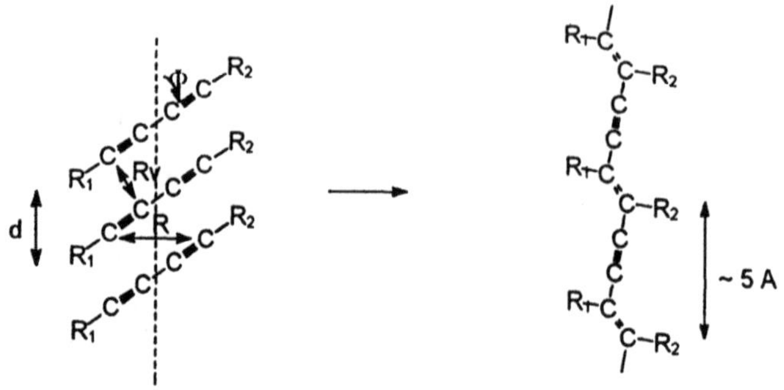

d: stacking distance
Φ: angle between the diacetylene rod and the stacking axis
R_v: van der Waals distance
R: separation between reacting atom C(1) and C(4)'

Figure 1 Packing model of monomer single crystal.

tion reaction path, i.e., the molecules simultaneously rotate and translate along the stacking direction. In this model, the maximal reactivity is expected for $d \approx 5$ Å and $\Phi \approx 45°$. The separation R should be less than 4 Å to be a more critical condition than the requirement of a least motion pathway as calculated by Baughman [103,104].

In the case of diphenyldiacetylene derivatives, only the ortho- and meta-disubstituted compounds are reactive, while the para-disubstituted isomer does not react at all. The ortho-disubstituted derivatives are somewhat less reactive than meta-substituted isomers, and the reactivity pattern of the photopolymerization and of the thermal polymerization is just the same. The plausible explanation has been suggested as shown in Figure 2 [101]. Wegner suggested that the simplest model of such a shearing reaction consists of a frame in which two rods "a" and two rods "b" are connected by four rods "c" which become axes of rotation. Such a frame is easily deformed by pulling in the direction of the arrows, thereby considerably shortening the distance between the two rods "a". When the diphenyldiacetylene derivatives have para-disubstituted rings, a totally stiff frame is obtained, and no shearing is possible at all, if the axes of rotation "c" are placed into the plane of the rods "a", which is the plane of shearing. For example, in the case of 93, the axes of rotation are provided by the C–N single bonds between phenyl rings and amide side groups, the two rods "b" are attributed by the –N–H···O=C–N bonds, and the remaining two rods "a" by the conjugated triple bonds with a phenyl ring at each side. Because the frame becomes too stiff to give a possible shearing pathway to obtain the 1,4-addition reaction, the compounds do not show any solid-state re-

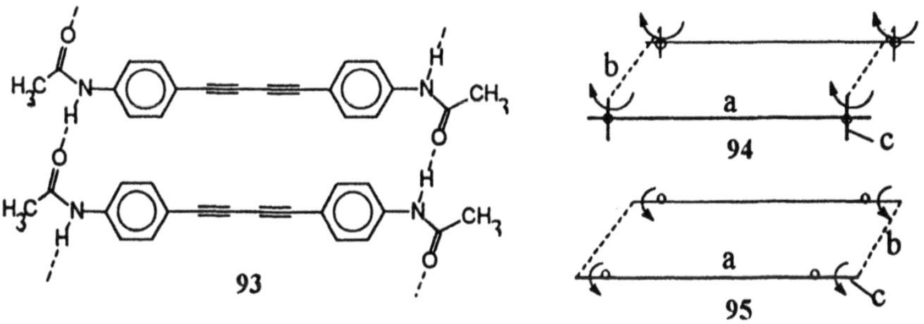

Figure 2 Plausible model for shearing mechanism.

activity in the case of para-disubstituted diphenyldiacetylene derivatives. However, if the polar substituents on the phenyl rings are placed into ortho and meta position relative to the conjugated triple bonds, the axes of rotation "c", that is, the C–N bonds between phenyl rings and amide groups, are placed at an angle of 60 or 120°, respectively, to the plane of the triple bonds. This corresponds to the model shown in 94. Therefore shearing is possible, and consequently ortho- or meta-substituted diphenyldiacetylenes with polar substituents exhibit solid-state reactivity. The only assumption is that the plane of the phenyl rings is perpendicular to the plane in which the triple bonds are arrayed. The reason why the ortho-disubstituted derivatives are less reactive than the meta-substituted isomers is presumably due to steric hindrance through the shearing action.

Suh et al. reported that the poly[6-(2-pyridyl)-3,5-hexadiyn-1-ol] [poly-(PyHxD)] [R_1 = 2-Pyridyl, R_2 = CH_2CH_2OH] cannot be obtained in bulk state because the monomers have the structure of dimeric forms having strong hydrogen bonding between –O–H$\cdots$N (96). PyHxD, however, could be polymerized by suspension polymerization in water although the polymerization yield is very low (<10%). They explained that this may be associated with the phase change of the monomer crystal to the polymerizable form during water insertion into the monomer matrix [105].

Tokura et al. [102] reported about PDAs substituted with fluorobenzenes, suggesting that aromatic rings directly attached to the diacetylene rods can have high conductivity when electrons (or holes) are produced in the aromatic rings by chemical doping or ion implantation, and these electrons or holes are very likely to be transferred to the backbone chain. This type of PDA crystal is also expected to be more conductive than other PDA crystals in which aromatic rings are separated from the backbone chain by nonconjugated methylene linkage. In addition, high conductivity might be expected from the PDA having heteroaromatic substituents (e.g., pyridine or pyrimidine) as a result of the proxim-

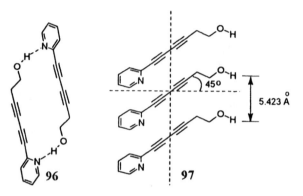

Figure 3 Stereo view of the PyPxD dimer having hydrogen bonding and unit cell packing diagram.

ity of the heteroaromatic ring to the polymer backbone [105]. It has been suggested that PDAs with aromatic substituents directly attached to the main backbone might be employed for enhancing and modulating the properties such as photonic and electronic applications because of the presence of extensive π-electron delocalization along the backbone. The number of π-electrons per repeating unit and the nature of π-delocalization might be increased through π-conjugation between main backbone and side groups. However, only a few such PDAs are polymerizabe and insoluble in common organic solvents. It is well known that longer, flexible side chains with the possibility of hydrogen bonding in PDAs promote the solubility of the polymers through the increased conformational entropy of the side groups such as poly[n,n'-dialkyn-m,m'-diol-bis(butoxy carbonylmethyl urethane)], poly(n-BCMU) [$R_1 = R_2 = (CH_2)_nOCONHCH_2COOC_4H_9$]. Kim et al. reported that poly[8-[[(butoxy-carbonyl)methyl]urethanyl]-1-(5-pyrimydyl)octa-1,3-diyne], poly(BPOD) [R_1 = 5-Pyrimidyl, $R_2 = (CH_2)_4OCONHCH_2COOC_4H_9$] obtained by γ-ray irradiation shows high solubility in chloroform and an absorption maximum of 510 nm in polymer solution. These are some of the longest wavelength absorptions for PDAs in solution [106].

Although there has been a lot of interest in the PDAs due to their unique chemical, optical, and electronic properties, no method is available yet that is generally applicable to the formation of polydiacetylene films possessing sufficiently high quality for technological applications, i.e., electronic and photonic devices, due to the insolubility of many polydiacetylenes. There is a continuous search for techniques to increase the processibility and quality of PDA film. Paley et al. reported a novel technique to obtain high-quality thin PDA films. They used photodeposition technique from monomer solutions onto UV transparent substrates. This process yields amorphous PDA films with thicknesses on the order of 1 μm that have optical quality superior to that of films grown by standard crystal growth techniques. These films give the high third-order nonlinearity ($\chi^{(3)} = 10^{-8} - 10^{-7}$ esu) [107].

ACKNOWLEDGMENTS

It is a great pleasure to acknowledge the contributions and hard work of my students whose names appear in the various references. This work is supported by the Organic Chemistry Research Centre, the Korea Science and Engineering Foundation, and the Korea Advanced Institute of Science and Technology.

REFERENCES

1. Arnaud, A. *Bull Soc. Chim. France* 1892, *3*, 233.
2. Bohlmann, F.; Burkhardt, T.; Zdero, C. *Naturally Occurring Acetylenes*, Academic: London and New York, 1973.
3. Chemla, D. S.; Zyss, J. *Nonlinear Optical Properties of Organic Molecules and Crystals*, Academic: London, 1987; pp. 1–119.
4. Kagan, J.; Gabriel, R.; Singh, S. P. *Photochem. Photobiol.* 1980, *32*, 607.
5. Hudson, J. B.; Graham, E. A.; Tower, G. H. N. *Photochem. Photobiol.* 1982, *36*, 181.
6. McLachlan, D.; Arnason, T.; Lam, J. *Photochem. Photobiol.* 1984, *39*, 177.
7. Weir, D.; Scaiano, J. C.; Arnason, T.; Avans, C. *Photochem. Photobiol.* 1985, *42*, 223.
8. Hudson, J. B.; Graham, E. A.; Tower, G. H. *N. Photochem. Photobiol.* 1986, *43*, 27.
9. McLachlan, D.; Arnason, T.; Lam, J. *Asteraceae. Biochem. Syst. Ecol.* 1986, *14*, 17.
10. Hudson, J. B,; Balza, F.; Harris, L.; Towers, G. H. N. *Photochem. Photobiol.* 1993, *57*, 675.
11. Coyle, J. D. *The Chemistry of the Carbon-Carbon Triple Bond*: Patai, S., ed., Wiley; Chichester, 1978; p. 528.
12. Coyle, J. D. *Organic Photochemistry*: Padwa, A., ed., Marcel Dekker: New York, 1985; p. 1.
13. Bradford, C. L.; Fleming, S. A.; Ward, S. C. *Tetrahedron Lett.* 1995, *36*, 4189.
14. Arnold, D. R.; Chan, Y. C. *J. Heterocyclic Chem.* 1971, *8*, 1097.
15. Owesley, D. C.; Bloomfield, J. J. *J. Am. Chem. Soc.* 1971, *93*, 782.
16. Chapman, O. L.; Adams, W. R. *J. Orgn. Chem.* 1976, *41*, 3931.
17. Serve, M. P.; Rosenberg, H. M. *J. Org. Chem.* 1970, *35*, 1237.
18. Owesley, D. C.; Bloomfield, J. J. *J. Am. Chem. Soc.* 1972, *93*, 782.
19. Nakayama, T.; Watanabe, K. *J. Chem. Phys.* 1964, *40*, 558.
20. Walsh, A. D. *J. Chem. Soc.* 1953, 2288.
21. Demoilin, D. *Chem. Phys.* 1975, *11*, 329.

22. Hamai, S.; Hirayama, F. *J. Chem. Phys.* 1979, *71*, 2934.
23. Evans, D. F. *J. Chem. Soc.* 1960, 1735.
24. Laufer, A. H.; Bass, A. M. *J. Phys. Chem.* 1979, *83*, 310.
25. Okabe, H. *Canad. J. Chem.* 1983, *61*, 850.
26. Inoue, Y.; Fukunaga, T.; Hakushi, T. *J. Org. Chem.* 1983, *48*, 1732.
27. Wooldridge, T.; Roberts, T. D. *Tetrahedron Lett.* 1973, 4007.
28. Fujita, K.; Yamamoto, K.; Shono, T. *Tetrahedron Lett.* 1973, 3865.
29. Büchi, G.; Feairheller, S. H. *J. Org. Chem.* 1969, *34*, 609.
30. Zimmerman, H. E.; Pincock, J. A. *J. Am. Chem. Soc.* 1973, *95*, 3246.
31. Martin, P. S.; Yates, K.; Csizmadia, I. G. *Theor. Chim. Acta* 1983, *64*, 117.
32. Ahlgren, G. *J. Org. Chem.* 1973, *38*, 1369.
33. Polman, H.; Mosterd, A.; Bos, H. J. T. *Recl. Trav. Chim.* 1973, *92*, 845.
34. Friedrich, L. E.; Bower, J. D. *J. Am. Chem. Soc.* 1973, *95*, 6869.
35. Kaup, G.; Stark, M. *Chem. Ber.* 1979, *111*, 3608.
36. Serve, M. P.; Rossenber, H. M. *J. Org. Chem.* 1970, *35*, 1237.
37. Shim, S. C.; Lee, T. S. *Bull. Korean Chem. Soc.* 1986, *7*, 357.
38. Beer, M. *Chem. Phys.* 1956, *25*, 745.
39. Shim, S. C.; Lee, T. S.; Lee, S. J. *J. Org. Chem.* 1990, *55*, 4544.
40. Shim, S. C.; Kim, S. S. *Bull. Korean Chem. Soc.* 1985, *3*, 153.
41. Chung, C. B.; Kim, G.-S.; Kwon, J. H.; Shim, S. C. *Bull, Korean Chem. Soc.* 1993, *14*, 506.
42. Murov, S. L. *Handbook of Photochemistry*, Marcel Dekker: New York, 1973.
43. Shim, S. C.; Lee, S. J.; Kwon, J. H. *Chem. Lett.* 1991, 1767.
44. Lee, S. J.; Shim, S. C. *Tetrahedron Lett.* 1990, *31*, 6197.
45. Shim, S. C.; Kim, S. S. *Tetrahedron Lett.* 1985, *26*, 765.
46. Kim, G.-S.; Hong, E.; Shim, S. C. *J. Photochem. Photobiol., (A) Chem.,* 1996, 101(1).
47. Kwon, J. H.; Lee, S. J.; Shim, S. C. *Tetrahedron Lett.* 1991, *32*, 6719.
48. Kagan, J. *Chemosphere* 1987, *16*, 2405.
49. Arnason, T.; Wat, C.-K.; Downum, K.; Yamamoto, E.; Graham, E.; Towers, G. H. N. Can. *J. Microbiol.* 1980, *26*, 698.
50. McRae, D.; Yamamoto, E.; Towers, G. H. N. *Photochem. Photobiol.* 1987, *47*, 353.
51. Kagan, J.; Tadema-Wielandt, K.; Chan, G.; Dhawan, S. N.; Jaworsky, J.; Prakash, I.; Aroro, S. K. *Photochem. Photobiol.* 1984, *39*, 4605.
52. Gong, H.-H.; Kagan, J.; Seitz, R.; Stokes, A. B.; Meyer, F. A.; Tuveson, R. W. *Photochem. Photobiol.* 1988, *47*, 55.
53. Kagan, J.; Tuveson, R. W. *Bioactive Molecules*, Vol. 7: Lam, J.; Breteler, H.; Arnason, T.; Hansen, L., eds., Elsevier: Amsterdam, 1988; pp. 71-84.
54. Chung C. B.; Kwon, J. H.; Shim, S. C. *Photochem. Photobiol.* 1993, *34*, 2143.
55. Chung, C. B.; Kwon, J. H.; Shim, S. C. *Tetrahedron Lett.* 1993, *58*, 159.
56. Giese, B. *Angew. Chem. Int. Ed. Engl.* 1983, *22*, 753.
57. Shim, S. C.; Lee, T. S. *Bull. Korean Chem. Soc.* 1986, *7*, 304.
58. Shim, S. C.; Lee, T. S. *J. Org. Chem.* 1988, *53*, 2410.
59. Shim, S. C.; Lee, T. S. *Chem. Lett.* 1986, *7*, 1075.

60. Lee, C. S., Shim, S. C. *Photochem. Photobiol.* 1992, *55*, 323.
61. Wooldrige, T.; Roberts, T. D. *Tetrahedron Lett.* 1973, 4007.
62. Roberts, T. D.; Ardemagri, L.; Sbechter, H. *J. Am. Chem. Soc.* 1969, *91*, 6185.
63. Wan, P.; Yates, K. *Rev. Chem. Intermed.* 1984, *5*, 157.
64. Wan, P.; Culshaw, S.; Yates, K. *J. Am. Chem. Soc.* 1982, *104*, 2509.
65. Wan, P.; Yates, K. *J. Org. Chem.* 1983, *48*, 869.
66. Rafizadech, K.; Yates, K. *J. Org. Chem.* 1984, *49*, 1500.
67. McEwen, J.; Yates, K. *J. Am. Chem. Soc.* 1987, *109*, 5800.
68. Anderson, S. W.; Yates, K. *Can J. Chem.* 1988, *66*, 2412.
69. Isaks, M.; Yates, K.; Kalanderopoulos, P. *J. Am. Chem. Soc.* 1984, *106*, 2728.
70. Kalanderopoulos, P.; Yates, K. *J. Am. Chem. Soc.* 1986, *108*, 6290.
71. Yates, K.; Martin, P.; Csizmadia, I. G. *Appl. Chem.* 1988, *60*, 205.
72. Martin P.; Yates, K.; Csizmadia, I. G. *Can. J. Chem.* 1989, *67*, 2178.
73. Sinha, H. K; Thompson, P. C. P.; Yates, K. *Can J. Chem.* 1990, *68*, 1507.
74. Wan, P.; Davis, M. J.; Teo, M. A. *J. Org. Chem.* 1989, *54*, 1354.
75. Shim, S. C.; Lee, T. S. *J. Chem. Soc. Perkin Trans 2* 1990, 1739.
76. Baek, E. K.; Shim, S. C. *J. Phys. Org. Chem.* 1995, *8*, 699.
77. Baek, E. K.; Lee, S. T.; Chae, Y. S.; Shim, S. C. *J. Photosci.* 1995, *2*, 73.
78. Zimmerman, H. E.; Somasekhara, S. *J. Am. Chem. Soc.* 1963, *85*, 922.
79. Zimmerman, H. E.; Sandel, V. R. *J. Am. Chem. Soc.* 1963, *85*, 915.
80. Shim, S. C.; Lee, T. S. *J. Photochem. Photobiol. A: Chem.* 1990, *53*, 323.
81. Shizuka, H.; Okazaki, K.; Tanaka, H.; Ishikawa, M. *J. Phys. Chem.* 1987, *91*, 2057.
82. Ishikawa, M.; Sugisawa, H.; Fuchikami, T.; Kumada, M.; Yamabe, T.; Kawakami, H.; Ueki, Y.; Shizuka, H. *J. Am. Chem. Soc.* 1982, *104*, 2872.
83. Shizuka, H.; Sato, Y.; Ueki, Y.; Ishikawa, M.; Kumada, M. *J. Chem. Soc. Faraday Trans. 1.* 1984, *80*, 341.
84. Sakurai, H.; Sugiyama, H.; Kira, M. *J. Phys. Chem.* 1990, *94*, 1837.
85. Ishikawa, M.; Fuchikami, Y.; Kumada, M. *J. Am. Chem. Soc.* Kumada,, *99*, 245.
86. Sakurai, H.; Kamiyama, Y.; Nakadaira, Y. *J. Am. Chem. Soc.* 1977, *99*, 3879.
87. Seyferth, D.; Vick, S. C.; Shannon, M. L. *Organometallics* 1984, *3*, 1897.
88. Seyferth, D.; Shannon, M. L.; Vick, S. C.; Lim, T. F. O. *Organometallics* 1985, *4*, 57.
89. Ohshita, J.; Ishikawa, M. *J. Organomet. Chem.* 1991, *407*, 157.
90. Ishikawa, M.; Matsuzawa, S.; Higuchi, T.; Kamitori, S.; Hirotsu, K. *Organometallics* 1985, *4*, 2040.
91. Ishikawa, M.; Fuchikami, T.; Kumada, M. *J. Chem. Soc. Chem. Comm.* 1977, 352.
92. Ishikawa, M.; Sugisawa, H.; Kumada, M.; Higuchi, T.; Matsui, K.; Hirotsu, K. *Organometallics* 1982, *1*, 1473.
93. Belzner, J.; Ihmels, H. *Tetrahedron Lett.* 1993, *34*, 6541.
94. Kwon, J. H.; Lee, S. T.; Shim, S. C.; Hoshino, M. *J. Org. Chem.* 1994, *59*, 1108.
95. Shim, S. C.; Lee, S. T. *Bull. Korean Chem. Soc.* 1995, *16*, 988.

96. Shim, S. C.; Lee, S. T. *J. Chem. Soc., Perkun Trans.* 2 1994, 1979.
97. Ishikawa, M.; Ohshita, J.; Ito, Y.; Iyodo, J. *J. Am. Chem. Soc.* 1986, *108*, 7417.
98. Lee, S. T.; Baek, E. K.; Shim, S. C. *J. Photosci.* 1994, *2*, 119.
99. Wegner, G. *Makromol. Chem.* 1971, *145*, 85.
100. Chu, B.; Xu, R. *Acc. Chem. Res.* 1991, *24*, 384.
101. Wegner, G. *J. Polym. Sci., Polym. Lett.* 1971, *9*, 133.
102. Tokura, Y.; Koda, T.; Itsudo, A.; Miyabayashi, M.; Okuhara, K.; Ueda, A. *J. Chem. Phys.* 1986, *85*, 99.
103. Baughman, R. H. *J. Polym. Sci., Polym. Phys. Ed.* 1974, *12*, 1511.
104. Enkelmann, V. *Adv. Polym. Sci.* 1984, *63*, 92.
105. Suh, M. C.; Kim, S. N.; Lee, H. J.; Shim, S. C.; Suh, I.-H.; Lee, J.-H.; Park, J.-R. *Synth. Met.* 1995, *72*, 51.
106. Km, W. H.; Kodali, N. B.; Kumar, J.; Tripathy, S. K. *Macromol.* 1994, *27*, 1819.
107. Paley, M. S.; Frazier, D. O.; Abdeldeyem, H.; Armstrong, S; McManus, S. P. *J. Am. Chem. Soc.* 1995, *117*, 4775.

5

Photochemistry and Photophysics of Carbocations

Mary K. Boyd
Loyola University of Chicago, Chicago, Illinois

I. INTRODUCTION

Carbocations are important intermediates in an array of organic chemical reactions, resulting in extensive investigations of their ground state chemistry, including reactivity, structure, and product studies [1]. Early studies in carbocation photochemistry focused primarily on photoproduct determination following irradiation of cations thermally generated in acidic media [2–4]. In a recent comprehensive review, Childs and Shaw noted two broad classes of carbocation photoreactions [2]. In very strongly acidic media such as FSO_3H, the typical reaction is photoisomerization. In less strongly acidic media, the predominant reaction is electron transfer to give radical intermediates which undergo coupling reactions to form dimeric materials or substitution products.

More recently, considerable attention has been centered on the characterization, properties, and mechanistic studies of excited state carbocations, and this attention is the main subject of this chapter. A separate, yet intimately related aspect of carbocation photochemistry involves the use of photochemical methods to generate the cations. The photogeneration of carbocations and the study of their subsequent (mainly thermal) chemistry have been the subject of two recent extensive reviews [5,6] and will be discussed only briefly here.

This chapter will describe methods used to generate cations for their photochemical studies, as well as the characterization of their singlet and triplet excited states. Mechanistic studies of excited state carbocations are described, including reactions with nucleophilic species, electron transfer reactions, oxygen quenching, and isomerization processes.

II. CARBOCATION GENERATION TECHNIQUES FOR PHOTOCHEMICAL STUDIES

A. Acidified Media

The predominant method for carbocation generation is treatment of the corresponding alcohol or olefin with acidic media [Eq. (1)]:

$$
-\overset{|}{\underset{|}{C}}-OH \xrightarrow{\text{H}^+} -\overset{|}{\underset{|}{C}}+ \xrightarrow{h\nu} -\overset{|}{\underset{|}{C}}+^* \tag{1a}
$$

$$
\overset{\diagdown}{\diagup}C=C\overset{\diagup}{\diagdown} \xrightarrow{\text{H}^+} -\overset{H}{\underset{|}{C}}-\overset{|}{\underset{|}{C}}+ \xrightarrow{h\nu} -\overset{H}{\underset{|}{C}}-\overset{|}{\underset{|}{C}}+^* \tag{1b}
$$

Protonation of the olefin, or protonation and subsequent dehydration of the parent alcohol, gives cations which are then subjected to laser excitation or steady-state irradiation. Cations generated in this way were identified by their characteristic absorption spectra, which also indicated cation stability over the time scale of the individual experiments by the lack of change in their absorption spectra. Among the numerous cations generated in acidified solution for photochemical studies are the xanthyl and thioxanthyl [7–15], dibenzosuberenyl [10], triphenylmethyl [10,15], α,ω-diphenylpolyenyl [16], and 1,1-diarylethyl [17] cations. Media included acetonitrile acidified with trifluoroacetic acid (TFA-ACN) or aqueous sulfuric acid [7–9,11,14,15], TFA in 2,2,2-trifluoroethanol (TFA–TFE) [10,12,13], n-heptane acidified with TFA [9], and BF$_3$-etherate in methylene chloride [16]. The absorption spectral data for several cations have been previously reviewed [6]. Characterization of the cation excited states will be discussed in Section III.

B. Two-Laser Pulse Techniques

Excited state carbocations have been generated via two-step laser photolysis experiments. The first laser pulse results in ionic photodissociation of a carbon–heteroatom bond to generate the cation. The second pulse, tuned to the cation absorption, electronically excites the cation. For example, the singlet-excited

9-phenylxanthyl cation 1* was generated by 355- or 337.1-nm laser pulse excitation, following initial cation generation from the corresponding alcohol by 248- or 308-nm laser flash photolysis [Eq. (2)] [8]:

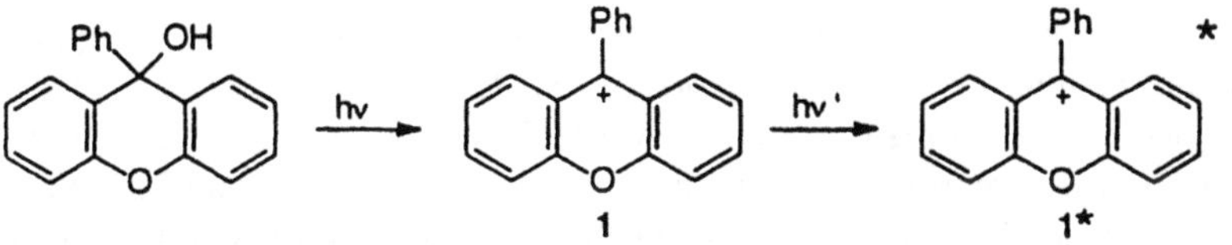

Polar solvents such as 1 : 1 MeOH : H$_2$O, neat MeOH, 1 : 1 ACN : H$_2$O, and TFE promoted the initial heterolytic bond cleavage. Nonpolar solvents such as n-heptane resulted in a preferential homolytic bond cleavage to give the corresponding 9-phenylxanthyl radical. Neat acetonitrile gave both the 9-phenylxanthyl radical and cation 1.

Photodehydroxylation of all trans-retinol 2 in nitrogen-saturated acetonitrile by 355-nm pulsed laser photolysis gave the all-trans-retinyl cation 3 as identified by an intense absorption at 580 nm [18]. The transient cation 3 was then photoexcited with a pulse from a flashlamp-pumped dye laser at 590 nm [Eq. (3)]. Laser excitation of alcohol 2 in cyclohexane yielded an oxygen-sensitive major band with λ_{max} = 380–400 nm, identified as the alcohol triplet–triplet absorption.

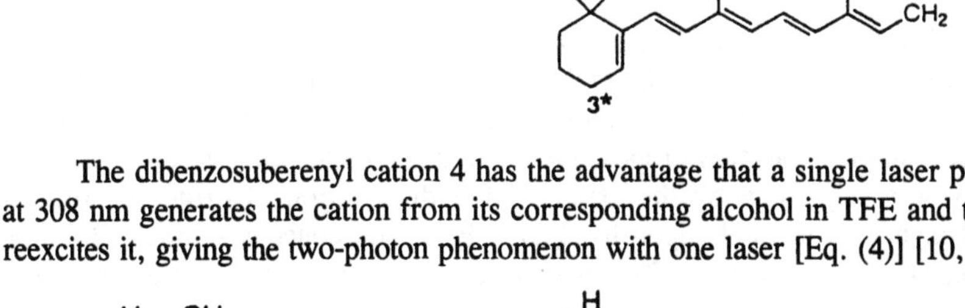

The dibenzosuberenyl cation 4 has the advantage that a single laser pulse at 308 nm generates the cation from its corresponding alcohol in TFE and then reexcites it, giving the two-photon phenomenon with one laser [Eq. (4)] [10,19].

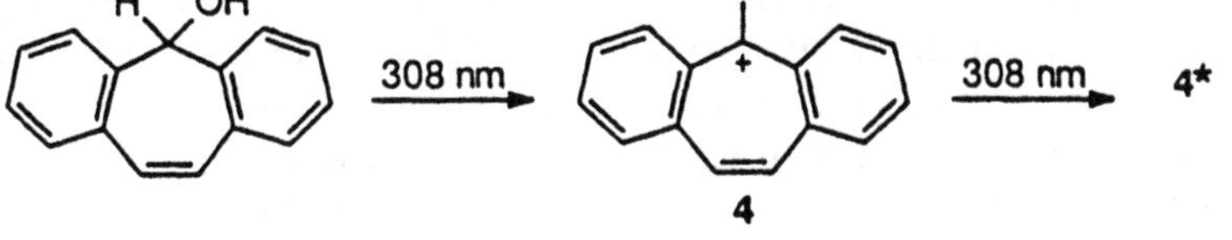

C. Adiabatic Dehydroxylation

Steady-state irradiation of 9-phenylxanthen-9-ol in neutral aqueous acetonitrile gave a weak fluorescence emission centered at 540 nm attributed to the 9-phenylxanthyl cation 1* [7]. The ground state cation does not form via thermal ionization in this solution; furthermore, irradiation at the cation absorption maximum resulted in no cation fluorescence. An excitation spectrum of the cation fluorescence emission corresponded to the alcohol absorption spectrum. The results were interpreted as an adiabatic photodehydroxylation of the alcohol on the S_1 surface to give the cation. Solvent isotope effects were consistent with a mechanism involving product-determining proton transfer from the solvent to the incipient hydroxide ion leaving group of the photoexcited alcohol to give the cation (Scheme 1). Hydronium ions were found not to catalyze the photodehydroxylation throughout the pH range 13–3.

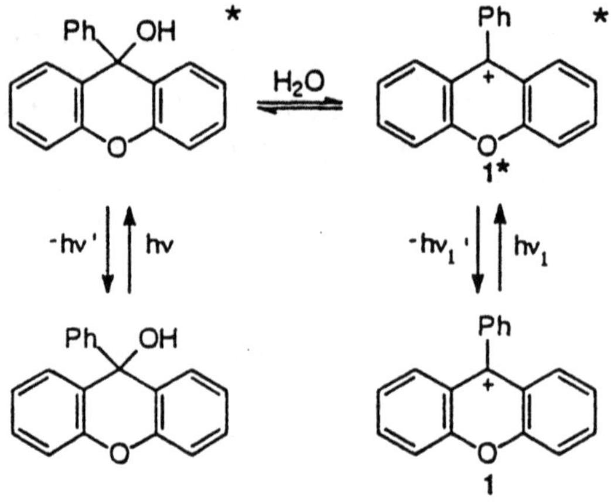

Scheme 1

The facile photodehydroxylation of 9-phenylxanthen-9-ol was attributed to the increased basicity of the molecule in S_1 compared to its ground state (S_0) [7]. The enhanced electron-donating effect of the central ring oxygen in the excited–singlet state was proposed to be responsible for the effect. This possibility was suggested by the efficient dehydroxylation of ortho-substituted methoxybenzyl alcohols upon irradiation in neutral aqueous solution [20] and the enhanced conjugative effect from ortho- and meta-substituted electron-donating groups in benzyl derivatives [21–23]. It was also noted that the rigid backbone of the xanthyl moiety permitted a sufficiently long excited singlet cation lifetime to allow for detection of adiabatic fluorescence.

Minto and Das observed that the amount of adiabatically produced 9-phenylxanthyl cation increased as the mole fraction of water increased in aqueous acetonitrile solutions [8]. However, the presence of water was not crucial for alcohol photodehydroxylation, as the cation could be generated (albeit not adiabatically) from laser excitation of 9-phenylxanthen-9-ol in 1,2-dichloro-ethane or acetonitrile. Das reported that only 1% of the 9-phenylxanthen-9-ol photoheterolysis in 1 : 1 H_2O : acetonitrile occurred via an adiabatic pathway [8]. McClelland and co-workers suggested that the quantum yield for adiabatic heterolysis of the xanthyl alcohol may be high, but that the majority of the excited cation decays to the ground state by nonradiative pathways [24].

Related 9-alkylxanthyl alcohols (R = Me, i-Pr), xanthenol and dibenzo-suberenol show a similar formation of adiabatic cation upon steady-state irradiation in neutral aqueous solution [11,25]. Fluorescence emission at 550 nm characteristic of the dibenzosuberenyl cation 4 was similarly observed following excitation of its parent alcohol in dilute acidic conditions [26]. Because the absorbing species is the parent alcohol in these conditions, the cation emission implies excited singlet state mediated ionization to give the singlet excited cation. Laser excitation at 308 nm of dibenzosuberenol in TFE resulted in a weak fluorescence emission attributable to cation formation [19]. It was suggested that some of the cation may be formed adiabatically at low light intensities, although the majority of the emission was attributed to formation and reexcitation of the cation within the laser pulse [19]. The 5-phenyldibenzosuberenyl cation 5 did not yield adiabatic fluorescence emission following irradiation of its corresponding alcohol in neutral aqueous acetonitrile [25].

4   **5**

Thioxanthyl cations 6–9 exhibit adiabatic cation formation upon irradiation of the corresponding thioxanthenols in neutral aqueous acetonitrile [25,27]. Shukla and Wan noted an enhanced adiabatic fluorescence intensity for the thioxanthyl cations compared to the xanthyl cations [27]. It was suggested that the thioxanthyl cations are less susceptible to nucleophilic attack by water and thus are not deactivated as readily as the xanthyl systems [27].

6 R = H
7 R = Me
8 R = Ph
9 R = CH_2Ph

D. Organic Salts

Stable organic salts have been used to eliminate the requirement for acidic media and to avoid the problems associated with competitive radical generation that occur via homolytic bond cleavage of carbon–heteroatom bonds in laser flash photolytic experiments [28–30]. Treatment of several 9-arylxanthenols and thioxanthenols with fluoroboric acid in propionic anhydride gave crystalline xanthyl and thioxanthyl cation tetrafluoroborate salts [Eq. (5)].

$$X = O, S \tag{5}$$

The salts are stable both in the solid phase and dissolved in dry acetonitrile. Absorption and fluorescence spectra of these cation salts in acetonitrile exhibit identical spectral properties as the same cations generated in acidified organic solvents or aqueous acidic solution [7–15]. The photooxidation of the triphenylcarbenium ion has also been studied via its tetrafluoroborate salt [31].

E. Heterogeneous Media

Berger and Weir reported that 9-phenylxanthen-9-ol adsorption on silica gel results in formation of the 9-phenylxanthyl cation 1, as indicated by a characteristic absorption band at approximately 380 nm using diffuse reflectance spectroscopy [32]. A vacuum-outgassed sample produced a strong luminescence characteristic of the cation. Excitation spectra indicated that the cation luminescence derives from two sources: from adiabatic photodehydroxylation of the alcohol and from direct excitation of cation thermally formed on the silica gel surface.

In contrast, deposition of the alcohol on acidic, neutral, or basic alumina gave no evidence for cation formation [32]. Vacuum-outgassed samples with similar substrate loadings exhibited only fluorescence from the alcohol. Similar alcohol fluorescence intensities and the lack of other luminescence bands strongly suggested that the cation was not thermally produced on the alumina surfaces. However, treatment of the alcohol-loaded alumina samples in a vacuum oven (130°C, 12 h) to activate the dehydroxylation process resulted in a weak cation fluorescence emission from the acidic alumina, with no signal from the neutral or basic samples. This suggested that the alumina surfaces have insufficient proton availability or cation-stabilizing ability to produce significant steady-state cation concentrations.

Cozens and co-workers found that the xanthyl cation was spontaneously generated by adsorption of 9-xanthenol within several acidic zeolites [33]. The cation was characterized spectroscopically and was found to be stable over a long period of time. Diffuse reflectance spectra of the zeolite composites exhibited absorption bands similar to those for the xanthyl cation in solution [10]. Fluorescence spectra also corresponded to cation solution spectra. The 9-phenylxanthyl, 9-phenylfluorenyl, and triphenylmethyl carbocations are similarly formed and readily detectable on dry montmorillonite clay minerals [34].

The triphenylmethyl, xanthyl and thioxanthyl, 9-phenylxanthyl, and 9-phenylthioxanthyl cations were also generated as stable species on Nafion films by Samanta and colleagues [15]. The carbocations were produced by dipping the Nafion films into *n*-heptane or acetonitrile solutions of the corresponding alcohols. The carbocation fluorescence spectra and lifetimes on the Nafion films were very similar to those in the ACN-TFA mixtures [15]. Fluorescence from the triphenylmethyl cation was easily observed under these conditions.

III. CHARACTERIZATION OF CARBOCATION EXCITED STATES

A. Singlet State

1. Xanthyl and Thioxanthyl Cations

Cations with the xanthyl backbone have been the subject of considerable photochemical study and have been generated by each of the techniques described in the previous section. The xanthyl cations studied include those with 9-alkyl (R = Me, c-Pr, *i*-Pr, and *t*-Bu) and 9-aryl (aryl substituent = H, *p*-F, *m*-F, *p*-Me, *m*-Me, *m*-OMe, and *p*-OMe) substituents, as well as the parent xanthyl cation. The xanthyl cation exhibits a characteristic absorption spectrum, with maxima at 260, 370, and 450 nm, invariant with the different alkyl or aryl substituents, and with the various techniques and media used for cation generation [6–8,10–15,24,28,32,33]. Only the *p*-OMe-substituted cation exhibits a different absorption spectrum, with the long-wavelength band less well resolved and shifted to 500 nm [13].

Excitation at 370 nm of the xanthyl cations generated in various media gives a broad unstructured fluorescence emission band centered at 540 nm, exhibiting a reasonable mirror image relationship to the long wavelength absorption band [6–11,13–15,28,32,33]. Figure 1 illustrates absorption and fluorescence spectra for the 9-phenylxanthyl cation. Fluorescence was not observed for the *p*-OMe cation in TFE at room temperature, although a structureless emission band at 560 nm was observed at 77 K [13]. Fluorescence spectra of the xanthyl cation generated in acid zeolites correspond to the reported solution spectra, although an enhancement of resolution was seen in the very tight

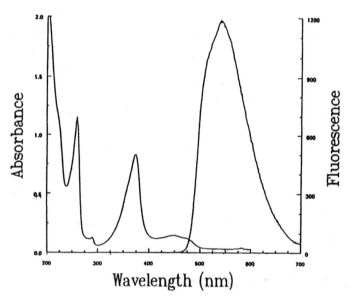

Figure 1 Absorption and fluorescence spectra for the 9-phenylxanthyl cation. The fluorescence intensity is in arbitrary units. (From Ref. 30.)

ZSM5 (Zeolite) environment, giving a true mirror image of the long-wavelength absorption band [33].

Fluorescence lifetimes and quantum yields measured for several 9-substituted xanthyl cations show some variation according to the media in which the cations were generated (Tables 1 and 2). Cation lifetimes measured in TFA-TFE exhibit significantly longer lifetimes than those in other solvent systems,

Table 1 Fluorescence Lifetimes and Quantum Yields for 9-Alkylxanthyl Cations

9-Substituent	τ_F (ns)	Φ_F	Media	Ref.
H	19.4		77% H_2SO_4/ACN[a]	11
	31	0.12	0.1 M TFA in TFE	10
	17.5	0.16	1 : 1 ACN : TFA	15
Me	35.0		77% H_2SO_4/ACN[a]	11
	41		0.1 M TFA in TFE	10
i-Pr	35.4		77% H_2SO_4/ACN[a]	11
c-Pr	32.6		77% H_2SO_4/ACN[a]	11
t-Bu	10.9[b]		77% H_2SO_4/ACN[a]	11

[a]2 : 1 Aqueous acid : acetonitrile. Acid concentration uncorrected for co-solvent.
[b]Cation not thermally stable under acidic conditions.

Table 2 Fluorescence Quantum Yields and Lifetimes for 9-Arylxanthyl Cations

9-Aryl substituent	τ_F (ns)	Φ_F	Media	Ref.
H	26.6		77% H_2SO_4/ACN[a]	11
	36	0.33	0.1 M TFA in TFE	10
	25	0.42	ACN[b]	8
	10		TFA-saturated n-heptane	8
	28.5	0.45	3M TFA in ACN	9
	28.5	0.48	1 : 1 ACN : TFA	15
	37		Silica gel[c]	32
	27.6	0.47	ACN[d]	28, 30
		0.80	0.01 M TFA in TFE	13
p-CF$_3$	24.1		77% H_2SO_4/ACN[a]	11
p-F	18.8	0.28	ACN[d]	28
	47	0.55	0.01 M TFA in TFE	13
m-F	14.0	0.18	ACN[d]	28, 30
p-Me	3.2[e]		77% H_2SO_4/ACN[a]	11
	2.1	0.014	ACN[d]	28, 30
m-Me	2.1[e]		77% H_2SO_4/ACN[a]	11
	2.0	0.084	ACN[d]	28, 30
p-OMe	0.14[e]		77% H_2SO_4/ACN[a]	11
m-OMe	0.18[f]		77% H_2SO_4/ACN[a]	11
	0.028[f]	6.6×10^{-4}	ACN[d]	28, 30

[a]2 : 1 Aqueous acid : acetonitrile. Acid concentration not corrected for co-solvent.
[b]Φ_F measured in ACN + 8% H_2SO_4.
[c]Cation formed through absorption of 9-phenylxanthenol on silica gel. Sample vacuum-outgassed prior to measurement.
[d]Measured as cation tetrafluoroborate salt in dry ACN.
[e]τ_F estimated from single photon counting.
[f]τ_F estimated from relative fluorescence quantum yield.

with a similar trend for the fluorescence quantum yields. Because the singlet-excited cations are quenched by nucleophilic species such as water (see Section V.A), the shorter lifetimes and lower quantum yields in other media may be due to trace amounts of water.

Fluorescence lifetimes for the 9-alkyl-substituted cations (R = Me, c-Pr, i-Pr, and t-Bu) were measured in strongly acidic aqueous acetonitrile in which the water activity is effectively zero (Table 1) [11]. With the exception of the t-Bu cation, the fluorescence lifetimes show little variation due to alkyl substitution, with each of the cations exhibiting lifetimes of approximately 35 ns. Absorption spectra for the t-Bu substituted cation indicated that it was not stable under the specific acid conditions, which may account for its shorter measured lifetime.

Numerous values for the fluorescence lifetime and quantum yield have been reported for the 9-phenylxanthyl cation (Table 2). Lifetime values of approximately 27–28 ns have been reported [8,9,11,15,28], whereas measurements in TFA-TFE and for the cation generated on silica gel gave a lifetime of 36–37 ns [10,32]. Similarly, the fluorescence quantum yield is reported as 0.42–0.48 [8,9,15,28], with outlying values of 0.80 and 0.33 in TFA-TFE [10,13]. Trace amounts of water may be responsible for the variation in lifetimes and quantum yields.

Fluorescence lifetimes have been determined for several 9-arylxanthyl cations generated from their corresponding alcohol in strongly acidic media, or as their tetrafluoroborate salt in acetonitrile (Table 2) [11,28,30]. Excellent agreement between the two methods was found for the parent 9-phenyl-, p-Me-, and m-Me-substituted cations, with poorer agreement in the case of the m-OMe cation. The error associated with the very weak fluorescence of the m-OMe-substituted cation may be responsible for the difference in the lifetime estimates. The two values reported for the fluorescence lifetime of the p-F xanthyl cation show the widest variation, with a report of 18.8 ns for the cation prepared as its tetrafluoroborate salt in acetonitrile versus 47 ns in TFA-TFE [13,28,30].

Aryl substitution has a dramatic effect on the lifetimes and quantum yields for the xanthyl cations. Lifetimes vary from the subnanosecond range for the m-OMe-substituted cation to the tens of nanoseconds for the 9-phenyl- and p-F-substituted cations. A similar trend is seen with the fluorescence quantum yields, those cations that exhibit short lifetimes are also more weakly fluorescent [11,30]. Implications of the substituent effects on cation photophysical parameters will be discussed in Section IV.

Thioxanthyl cations exhibit absorption and fluorescence maxima red-shifted in comparison to the xanthyl cations. The parent thioxanthyl cation and several 9-aryl-thioxanthyl cations (aryl substituents = H, p-F, m-F, and p-Me) each exhibit absorption maxima at 280, 384, and 494 nm. Excitation at 382 nm gave a broad unstructured fluorescence emission band centered at 575 nm [13,15,30]. Figure 2 illustrates absorption and fluorescence spectra for the 9-phenylthioxanthyl cation. Little or no substituent effect is seen on the fluorescence lifetimes for the 9-arylthioxanthyl cations, with each exhibiting a lifetime of approximately 1 ns (Table 3) [13,15,30]. Excellent agreement is found among the lifetime measurements for the 9-arylthioxanthyl cations generated in acidified acetonitrile or trifluoroethanol, or measured as their tetrafluoroborate salt in acetonitrile. Only a modest substituent effect is observed on the fluorescence quantum yields, ranging from 0.011 for the p-Me-substituted cation to 0.025–0.031 for the 9-phenyl cation. In comparison to the 9-arylthioxanthyl cations, the parent thioxanthyl cation exhibits a relatively long fluorescence

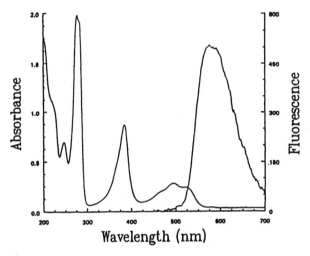

Figure 2 Absorption and fluorescence spectra for the 9-phenylthioxanthyl cation. The fluorescence intensity is in arbitrary units. (From Ref. 30.)

lifetime of 22.8 ns in acetonitrile acidified with trifluoroacetic acid and a large fluorescence quantum yield of 0.24 [15].

2. Di- and Triarylmethyl Cations

The triphenylmethyl cation exhibits absorption maxima at 248, 292, 405, and 430 nm in 1 : 1 TFA : ACN [15]. Excitation of the triphenylmethyl cation at 370–380 nm gave a weak fluorescence emission with maximum at 530 nm. The weak fluorescence gave an estimated lifetime of <1 ns in 1 : 1 TFA-ACN and TFA-TFE [10,15]. In contrast to the xanthyl and 9-phenylxanthyl cations, the

Table 3 Fluorescence Lifetimes and Quantum Yields for Thioxanthyl Cations

9-Substituent	τ_F (ns)	Φ_F	Media	Ref.
H	22.8	0.24	1 : 1 ACN : TFA	15
Ph	1.0	0.020	ACN[a]	30
	1.0	0.025	1 : 1 ACN : TFA	15
	<2.0	0.031	0.1 M TFA in TFE	13
p-F	1.0	0.018	ACN[a]	30
	<2.0	0.020	0.1 M TFA in TFE	13
m-F	1.3		ACN[a]	30
p-Me	1.0	0.011	ACN[a]	30

[a]Measured as cation tetrafluoroborate salt in dry ACN.

fluorescence lifetime of the triphenylmethyl cation is temperature dependent, increasing to 12 ns at $-40°C$ [10]. A fluorescence quantum yield of $< 10^4$ has been reported in TFA-acidified acetonitrile, in contrast to a value of 0.48 for the 9-phenylxanthyl cation in the same solvent [15]. The planar, rigid backbone of the xanthyl moiety is suggested to be responsible for its longer excited singlet lifetime and greater fluorescence efficiency in comparison to the triarylmethyl cation. Generation of the triphenylmethyl cation on Nafion films gave an enhanced fluorescence emission [15].

The p-anisylmethyl cation 10, generated by dehydration of its corresponding alcohol in TFA-TFE, exhibits absorption and fluorescence maxima of 500 nm and 600 nm, respectively, with a reported fluorescence lifetime of 3 ns [10]. The 1,1-di-p-anisylmethyl cation 11 was generated by protonation of 1,1-di-p-anisylethylene 12 in TFA-TFE. The cation exhibits absorption and fluorescence maxima of 475 nm and 535 nm, respectively, with a fluorescence lifetime of 6 ns [30]. Protonation of 12 in TFA-benzene gave a reported absorption maximum for cation 11 of 500 nm [17].

3. Dibenzosuberenyl Cations

Johnston and co-workers reported that 308-nm laser excitation of 5-dibenzosuberenol in TFE gave a transient absorption spectrum of the dibenzosuberenyl cation 4 with maxima at 390 and 525 nm (Fig. 3) [19]. The spectrum was identical to that obtained for the cation generated from its corresponding alcohol in acidified TFE. Fluorescence emission from the cation was readily detectable upon 308-nm excitation of the alcohol (Fig. 4). The fluorescence maximum at 550 nm was similarly identical to that of the cation generated from a solution of the alcohol in acidified TFE [19]. The dibenzosuberenyl cation has a relatively long lifetime of 40 ns in TFA-TFE or 98% H_2SO_4, with a fluorescence quantum yield of 0.087 in TFA-TFE [10,19,26]. The 5-phenyldibenzylsuberenyl cation exhibits a slight red shift in its absorption and fluorescence spectra and a dramatically shortened lifetime of < 1 ns [10]. Laser excitation of 5-phenyldibenzosuberen-5-ol in TFE did not produce transients corresponding to cation generation [19].

4. α,ω-Diphenylpolyenyl Cations

Young and co-workers reported cation generation from 1,3-diphenylpropen-1-ol (DP3+), 1,5-diphenylpenta-1,4-dien-3-ol (DP5+), 1,7-diphenylhepta-1,4,6-

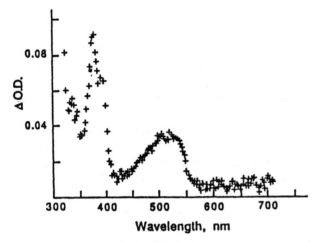

Figure 3 Transient absorption spectrum of dibenzosuberenyl cation measured 1 μs after 308-nm excitation of 5-dibenzosuberenol in TFE. (From Ref. 19.)

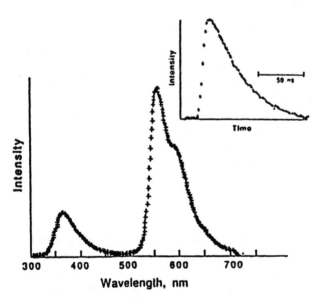

Figure 4 Fluorescence spectrum of dibenzosuberenyl cation obtained by 308-nm excitation of 5-dibenzosuberenol in TFE. Inset: Fluorescence decay for the dibenzosuberenyl cation monitored at 560 nm. (From Ref. 19.)

trien-3-ol (DP7+), and 1,9-diphenylnona-1,3,6,8-tetraen-5-ol (DP9+) in BF_3-etherate/methylene chloride [16]. Absorption spectra for DP3+, DP5+, DP7+, and DP9+ consist of a single band with absorption maxima at 495, 560, 615, and 675 nm, respectively. Excitation at their absorption maxima gave fluorescence emission spectra with maxima for the same ordering of cations at 530, 590, 660, and 725 nm [16,35]. Both the absorption and emission maxima red shift, as expected, with increasing chain length. The red shifts exhibited good regularity, with the position of the absorption maxima corresponding quite closely to the formula λ_{max} (nm) = 433 + 62n, where n is the number of olefinic double bonds in the parent polyene. The emission maxima give a good fit to the formula λ_{max} (nm) = 458 + 66n. The 1,5-di-p-tolylpentadienyl cation (DT5+) was also generated. Its absorption maximum of 595 nm is redshifted compared to that of DP5+, as is its emission spectrum which exhibits a maximum at 624 nm.

The absorption and emission maxima and band shape are independent of excitation wavelength and solution temperature over the range -20°C to -85°C. Fluorescence decays were well fitted by single exponentials, suggesting the presence of a single kind of ion pair. The insensitivity of the absorption spectra to temperature suggests that the type of ion pair does not change with the carbocations present as loose ion pairs (with the large counterion $HOBF_3^-$) in the dichloromethane solvent.

Fluorescence lifetimes for the diphenylpolyenyl carbocations were found to increase sigmoidally with decreasing temperature (Fig. 5). The measured lifetimes vary from 0.3–0.6 ns (0°C) to limiting values of 2.3, 2.0, and 1.7 ns at 77 K for DP5+, DP7+, and DP9+ respectively. The τ_0 values decrease with increasing chain length, whereas Arrhenius preexponential factors are typical for an intramolecular rotational process. Similar results were obtained for the related carbanions, strongly suggesting that the cation and anion excited states deactivate by a common mechanism involving competing pathways of fluorescence and skeletal twisting [16].

B. Triplet State

Johnston and Wong characterized the triplet excited state of the 9-phenylxanthyl cation 1 by phosphorescence and laser flash photolysis techniques [12,13]. The cation was generated by protonation and subsequent dehydration of its parent alcohol in 1–10 mM TFA in TFE. Laser excitation at 355 nm resulted in a transient absorption spectrum exhibiting bleaching of the ground state cation absorption in the 350–450-nm region and the appearance of a new transient with maximum absorption below 300 nm. This transient was assigned to the triplet excited state of the cation, which exhibits a lifetime of 5–10 μs at low alcohol

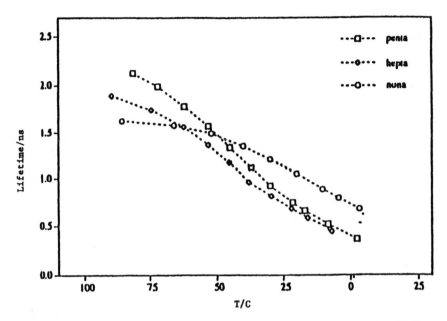

Figure 5 The temperature dependence of the fluorescence lifetimes of DP5+, DP7+, and DP9+. (From Ref. 16.)

concentrations. The lifetime of the cation decreased in the presence of air or oxygen, and with increasing 9-phenylxanthenol concentrations due to electron transfer processes. For example, triplet excited 1 reacts with excess alcohol to yield the 9-phenylxanthyl radical and the alcohol radical cation (Scheme 2).

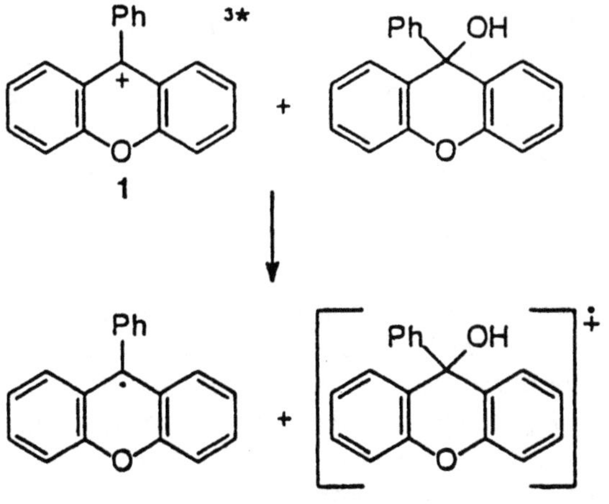

Scheme 2

Excitation of cation 1 in TFA-TFE or acidified ethanol glasses at 77 K gave a weak phosphorescence spectrum with maximum at 600 nm. A triplet energy of approximately 48 kcal was determined from the onset of phosphorescence. This triplet energy value is reasonable based on a singlet energy value of 57 kcal [8,11,28].

Cozens and colleagues also generated the triplet excited 9-phenylxanthyl cation via 355-nm laser excitation of the cation formed by adsorption of the corresponding alcohol within acid zeolites [33]. Transient spectra recorded 1 μs after laser excitation show reflectance near 300 nm previously assigned to the triplet cation [12,13], as well as a strong band at 520–570 nm (Fig. 6). Both bands were formed promptly after the laser pulse, decay at the same rate in nitrogen-purged samples, and are similarly quenched by oxygen, leading to the conclusion that both bands are due to the presence of the triplet transient cation [33]. The intense signal at 540 nm compared to the solution spectrum suggests that the electronic spectra of the xanthyl cation triplet is modified by inclusion within zeolites. In addition, the measured lifetime for the triplet xanthyl cation within the zeolite is approximately 50 μs, compared to 5–15 μs in TFA-TFE [12,13].

The triplet excited state of the p-F-substituted xanthyl cation was characterized and shown to behave similarly to its parent 9-phenylxanthyl cation [13]. The p-F triplet cation exhibits an absorption spectrum with maximum below 300 nm, and a lifetime of 4–5 μs in the absence of air or oxygen and at low concentrations of its precursor alcohol. In contrast, the p-OMe-substituted cation gave no detectable transient corresponding to a triplet cation following laser excitation of the ground state cation at 355 nm. A weak phosphorescence spectrum at 77 K was observed for both the p-F- and p-OMe-substituted cations with emission maxima at approximately 600 nm, giving estimates for their triplet energies of 47 and 48 kcal-mol^{-1}, respectively. Very weak phosphorescence spectra were reported for several 9-arylxanthyl cations in acetonitrile at 77 K following cation excitation at 370 nm [30]. No difference in phosphorescence intensity as a function of substituent was observed, in contrast to large substituent effects on the fluorescence quantum yields for these cations [30].

The triplet-excited 9-phenylthioxanthyl cation and its p-F-substituted analog have been characterized. The cations exhibit weak phosphorescence emission in TFA-TFE or acetonitrile at 77 K, with maxima near 700 nm [13,30]. Triplet energies of 40 and 41 kcal-mol^{-1} were estimated based on the wavelength for onset of phosphorescence. Laser excitation at 355 nm of these cations result in transients with λ_{max} at 300 nm and a very weak absorption in the 500–600 nm region attributed to the triplet cations. The triplet-excited 9-phenylthioxanthyl cation has a lifetime of approximately 15 μs at low alcohol concentrations [13].

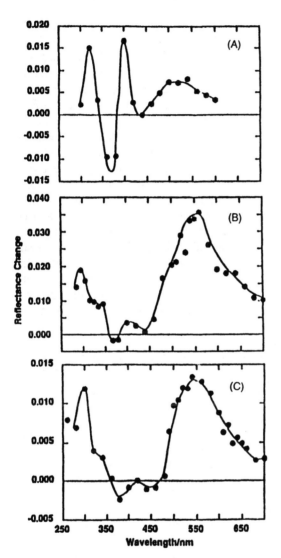

Figure 6 Transient reflectance spectra recorded 1 μs after 355-nm excitation of the xanthyl cation in three N_2-purged zeolites. (A) HY, (B) HZSM5, (C) HMor. (From Ref. 33.)

IV. PHOTOPHYSICAL PROPERTIES

Measurement of carbocation fluorescence quantum yields and lifetimes as described in the preceding section allowed a detailed analysis of cation photophysical properties. A study of the photophysical parameters for several 9-arylxanthyl and thioxanthyl cations has been recently reported, in which the cations were studied as their tetrafluoroborate salts in acetonitrile solution [30]. Fluorescence lifetimes and quantum yields are listed in Tables 2 and 3. Total decay rate constants k_{dt} were calculated from the measured fluorescence lifetimes and are given in Table 4. A dramatic substituent effect is observed for the xanthyl series as the k_{dt} values increase as the substituents become more strongly electron donating [30]. Decay rate constants range from 3.6×10^7 s^{-1} for the 9-phenyl cation to 3.6×10^9 s^{-1} for the m-OMe-substituted cation.

Fluorescence and nonradiative rate constants, k_f and k_{nr}, respectively (Table 4), were calculated from the fluorescence quantum yields and lifetimes. The k_f value for the 9-phenylxanthyl cation is in very good agreement with fluorescence rate constants calculated by Das and colleagues using the Strickler–Berg formula or from Φ_F/τ_F [15]. The fluorescence rate constants exhibit only a modest substituent effect for the 9-arylxanthyl cations. However, a large substituent effect is seen on the k_{nr} values for the xanthyl cations, ranging from 2.0×10^7 s^{-1} for the 9-phenyl cation to 3.6×10^9 s^{-1} for the m-OMe-substituted cation. As with the k_{dt} values, the nonradiative decay rate constants increase as the electron-donating ability of the substituent increases. The possibility of a substituent effect on the rate constant for intersystem crossing was suggested to account for the substituent dependence on the photophysical parameters [30].

Table 4 Photophysical Parameters for 9-Arylxanthyl and Thioxanthyl Cations

9-Aryl substituent	k_{dt} (s^{-1})	k_f (s^{-1})[a]	k_{nr} (s^{-1})	Ref.
Xanthyl cations				
H	3.6×10^7	1.7×10^7	1.9×10^7	30
		1.7×10^7		15
		2.5×10^{7b}		15
p-F	5.3×10^7	1.5×10^7	3.8×10^7	30
m-F	7.1×10^7	1.3×10^7	5.8×10^7	30
p-Me	4.8×10^8	6.7×10^6	4.7×10^8	30
m-Me	5.0×10^8	4.2×10^7	4.6×10^8	30
m-OMe	3.6×10^9	2.4×10^7	3.6×10^9	30
Thioxanthyl cations				
H	1.0×10^9	2.0×10^7	9.8×10^8	30
		2.5×10^7		15
		2.4×10^{7b}		15
p-F	1.0×10^9	1.8×10^7	9.8×10^8	30
m-F	7.8×10^8			30
p-Me	1.0×10^9	1.1×10^7	9.9×10^8	30

[a]Calculated from $k_f = \Phi_F/\tau_F$, except where indicated.
[b]Calculated from the Strickler–Berg formula.

In contrast to the behavior exhibited by the 9-arylxanthyl cations, no substituent effect is observed for the 9-arylthioxanthyl series on the total decay and nonradiative rate constants, with k_{dt} values of approximately 1.0×10^9 s^{-1} and k_{nr} values of $(9.8-9.9) \times 10^8$ s^{-1} for each of the cations studied (Table 4) [30]. A modest substituent effect is observed on the fluorescence rate constants for the 9-arylthioxanthyl cations, similar to that exhibited in the xanthyl series [30]. The k_f value of 2.0×10^7 s^{-1} for the 9-phenylthioxanthyl cation is in very good agreement with values reported by Samanta et al. [15]. The lack of a substituent effect on the photophysical parameters in the thioxanthyl series may be a result of an enhanced intersystem crossing due to the sulfur heteroatom, or a twisting of the 9-aryl ring away from planarity decreasing conjugative interactions [30]. An alternative possibility is the presence of a second excited state that is close in energy to the fluorescent state, as observed for the benzyl radical [36].

Samanta et al. calculated a fluorescence rate constant of 2.4×10^8 s^{-1} for the triphenylmethyl cation, an order of magnitude larger than for the xanthyl or thioxanthyl cations [15]. The fluorescence quantum yield for the triphenylmethyl cation is $< 10^4$, suggesting that its nonradiative rate constant is $> 10^{12}$ s^{-1}.

V. EXCITED STATE REACTIVITY

A. With Nucleophilic Species

Xanthyl cations have been substrates for numerous studies of bimolecular excited state cation reactivity due to their relatively long lifetimes and strong fluorescence emission, permitting use of Stern–Volmer fluorescence quenching techniques to measure rate constants. Minto and Das measured bimolecular rate constants for quenching of the 9-phenylxanthyl cation 1 in acetonitrile by several nucleophilic species [8]. The cation was generated by double laser photolysis from 9-phenylxanthen-9-ol, with the k_q values determined from the fluorescence decay measured as a function of the quencher concentration. The quenching rate constants (Table 5) are several orders of magnitude higher than the corresponding values for reaction of the ground state cation [24], suggesting an increase in cation electrophilicity in the excited state attributed to the enhanced exothermicity arising from electronic excitation. In addition to the k_q values in Table 5, Minto and Das measured a rate constant of 3.0×10^{10} M^{-1} s^{-1} for quenching of singlet-excited 1 by triethylamine or tetraethylammonium chloride [8]. Stern–Volmer quenching experiments were also conducted by monitoring the fluorescence emission from the adiabatically generated cation in 3 : 1 H$_2$O : acetonitrile in the presence of added quenchers. The Stern–

Table 5 Rate Constants (M^{-1} s^{-1}) for Quenching of Singlet-Excited 9-Arylxanthyl Cations by Water and Alcohols[a]

9-Aryl Substituent	Quencher				Ref.
	H_2O	MeOH	i-PrOH	t-BuOH	
H	3.3×10^7	9.7×10^7	1.2×10^8	6.9×10^7	8
	1.6×10^{7b}				8
	1.49×10^{7c}				11
	2.02×10^7	4.56×10^7	8.94×10^7	3.33×10^7	28
p-CF$_3$	6.42×10^{6c}				11
p-F	8.88×10^6	4.11×10^7	8.05×10^7	1.80×10^7	28
m-F	1.20×10^8	8.10×10^7	1.05×10^8	2.34×10^7	28
p-Me	2.64×10^{8c}				11
	6.23×10^8	2.06×10^9	3.28×10^9	1.02×10^9	28
m-Me	2.99×10^{8c}				11
	2.16×10^9	8.03×10^9	1.02×10^{10}	3.36×10^9	28
p-OMe	3.06×10^{8c}				11
m-OMe	2.51×10^{9c}				11
	1.76×10^{10}	4.97×10^{10}	6.14×10^{10}	3.12×10^{10}	28

[a]Solvent acetonitrile unless otherwise indicated.
[b]Solvent acetonitrile + 10% H_2SO_4.
[c]Solvent 2 : 1 77% H_2SO_4 : acetonitrile.

Volmer quenching constants show a similar trend for the quenching rate constants.

The bimolecular excited state quenching rate constants show some variation depending on the solvent media. Values for k_q of 3.3×10^7 M^{-1} s^{-1} and 1.6×10^7 M^{-1} s^{-1} were determined for water quenching of the cation 1 generated by double laser flash photolysis in acetonitrile or by steady-state irradiation of the cation generated in acidified acetonitrile, respectively [8]. The lower quenching rate constant may be explained by a decrease in the water activity in the acidified medium. It was alternatively suggested that the photoexcited cation in the ion-paired state may be less susceptible to electrophilic quenching.

Table 5 also lists rate constants determined using Stern–Volmer fluorescence quenching techniques for water quenching of the 9-phenylxanthyl cation generated from its corresponding alcohol in very strongly acidic aqueous acetonitrile and for the cation tetrafluoroborate salt in acetonitrile [11,28]. Both k_q values are in excellent agreement with the values reported by Minto and Das when the difference in solvent systems is taken into account [8].

Bimolecular excited state rate constants for water quenching of several 9-alkylxanthyl cations and the parent xanthyl cation were determined by Stern–

Volmer fluorescence quenching techniques in strongly acidic aqueous acetonitrile solution [11]. The k_q values are all in the range $(2-9) \times 10^6$ M^{-1} s^{-1} and follow the reactivity order H > Me > i-Pr > c-Pr. This trend suggested a small steric effect toward the attack of water at the 9-position, although the relative rate differences are not very large.

Rate constants were similarly determined for quenching of several singlet-excited 9-arylxanthyl cations by water, alcohols, and dialkyl ethers (Tables 5 and 6) [11,28,29]. The water quenching rate constants determined for the cations generated by acid-catalyzed dehydration of the corresponding alcohols in strongly acidic aqueous acetonitrile and for the cation tetrafluoroborate salts in acetonitrile exhibit excellent agreement [11,28]. Regardless of the solvent media or the quencher, the k_q values increase as the electron-donating ability of the aryl substituent increases. The rate constants also show that substituents in the meta position exhibit greater conjugative effects that those in the para position, a manifestation of the meta/para dichotomy of excited state substituent effects [21–23].

Excited state Hammett plots were constructed for each of the water, alcohol, and ether quenchers by plotting values of log k_q versus the σ^{hv} substituent parameter [37]. This σ scale is appropriate for correlation with these reactivities because it is based on the rate-determining formation of the structurally similar benzyl cation obtained from photoprotonation of singlet-excited aryl-substituted styrenes and phenylacetylenes. Attempted correlations versus σ and $\sigma+$ gave poor correlation, due mainly to scatter from points corresponding to the meta-substituted cations. Figure 7 shows the excited state Hammett plot for quenching of the 9-arylxanthyl cations by water.

Very good to excellent correlation was obtained with the σ^{hv} substituent parameter in every case, with negative ρ values of -1.5 to -2.0 [28,29]. The

Table 6 Rate Constants (M^{-1} s^{-1}) for Quenching of Singlet-Excited 9-Arylxanthyl Cations by Ethers

9-Aryl substituent	Quencher				
	i-Pr$_2$O	Et$_2$O	t-BuOMe	t-BuOEt	THF
H	2.28×10^7	2.35×10^7	6.94×10^6	5.71×10^7	6.2×10^{8a}
p-F	1.91×10^7	1.34×10^7	4.90×10^7	5.17×10^7	
m-F	2.80×10^7	2.24×10^7	6.22×10^6	6.72×10^7	
p-Me	1.70×10^8	1.88×10^8	7.42×10^7	7.76×10^8	
m-Me	4.40×10^8	3.77×10^8	2.01×10^8	1.22×10^9	
m-OMe	2.11×10^{10}	1.88×10^{10}	5.86×10^9	4.59×10^{10}	

Note: All rate constants taken from Ref. 29, unless otherwise indicated. Solvent acetonitrile.
[a]Taken from Ref. 8.

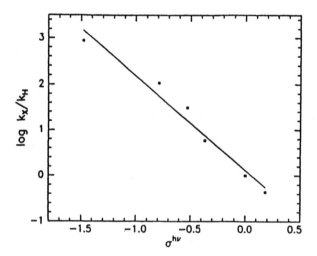

Figure 7 Excited state Hammett plot ($\log[k_q(X)/k_q(H)]$) versus σ^{hv} for quenching of 9-arylxanthyl cations by water. (From Ref. 28.)

ρ values are opposite in sign to those determined from Hammett plot analysis for the nucleophilic reaction of ground state 9-arylxanthyl cations with water and anionic species [24]. The ground state substituent dependence was attributed to weak electronic stabilization of the cation by the electron-donating 9-aryl substituents. The water and alcohol excited state quenching order of i-PrOH > MeOH > t-BuOH > H_2O also differs from the order determined for the quenching of ground state cations [38–42]. This excited state quenching order had been previously observed and was rationalized by the steric effect of increasing methyl substitution counterbalancing the increase in long pair availability of the nucleophile [8].

  With both the substituent effect and relative quenching order different from the ground state reactivities, the possibility that excited state cations may be quenched by water, alcohols, and ethers through an electron transfer process was considered [28,29]. Application of the Rehm–Weller equation [43,44] shows that for an electron transfer mechanism to be operating, cation reduction potentials must account for the observed substituent effect on reactivity [28,29]. In qualitative terms, an ease in reduction of the cations should parallel an increase in the quenching rate constants. Reduction potentials for the xanthyl cations [45] demonstrate that the trend for ease of reduction of the cations decreases with a concomitant increase in the magnitude of the quenching rate constants, ruling out the possibility of an electron transfer mechanism [29].

A possible explanation to account for the observed substituent dependence on the quenching rate constants was suggested, involving Hückel calculations that showed an increase in the lobe size at C_9 (i.e., the coefficient of the p orbital) for the singlet excited xanthyl cation as the substituents became more strongly electron donating [11]. Frontier orbital theory [46] suggests that the larger lobes would provide better overlap with the incoming nucleophile lone pair.

Shukla and Wan used Stern–Volmer fluorescence quenching techniques to obtain rate constants for quenching of the adiabatically generated thioxanthyl cations by several nucleophiles [27]. The k_q values for MeOH quenching was $4.1 \times 10^7 \, M^{-1} \, s^{-1}$, indicating that MeOH is more nucleophilic than H_2O, as previously observed for quenching of the 9-arylxanthyl cations [8,28]. Use of the more powerful nucleophiles N_3^- and NC^- (as their sodium salts) gave k_q values of 1.5×10^{10} and $1.3 \times 10^9 \, M^{-1} \, s^{-1}$, respectively. No electron transfer products were isolated following preparative photolyses, suggesting that if an electron transfer mechanism is operating, the subsequent geminate radical coupling is very fast or that the mechanism involves direct nucleophilic attack.

Johnston reported a rate constant of $1.1 \times 10^9 \, M^{-1} \, s^{-1}$ for quenching of the singlet-excited dibenzosuberenyl cation by azide in 4% aqueous TFE [19]. The cation was generated by laser flash photolysis of 5-dibenzosuberenol, with the rate constant obtained from a plot of the observed rate constant for decay of the transient cation at 525 nm as a function of added azide concentration. The water co-solvent was necessary for solubility of the sodium azide quencher.

B. Electron Transfer

Several early studies on the photochemistry of stabilized carbocations suggested the possibility of electron transfer processes to account for the observed photoproducts [2,47–49]. Al-Ekabi et al. obtained the first direct evidence for electron transfer processes through irradiation of the cation 11 obtained by protonation of 1,1-di-p-anisylethylene 12 in benzene-TFA in the presence of excess olefin [17]. Electron transfer to singlet-excited 11 from the ground state alkene gave the corresponding radical cation and radical. The radical ion intermediate was observed directly at 340–440 nm by transient absorption spectroscopy (Fig. 8). The assignment of this transient to the radical ion was confirmed by its generation when the alkene was irradiated in the presence of dicyanoanthracene. Further evidence for the intermediacy of the radical and the electron transfer process was obtained from the radical-derived photoproducts (Scheme 3).

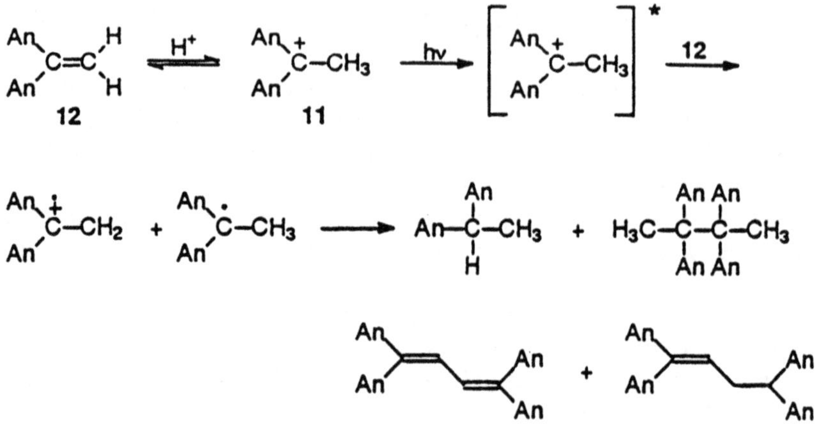

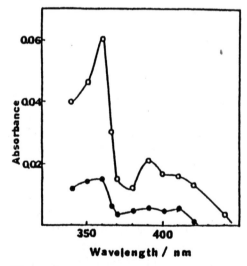

Scheme 3

Several singlet-excited cations have been shown to undergo photoinduced electron transfer from aromatic donors. Fluorescence from the 9-phenylxanthyl, xanthyl, thioxanthyl, and 9-phenylthioxanthyl cations is quenched in the presence of aromatic compounds [9,10,15]. Steady-state fluorescence quenching experiments gave bimolecular excited state rate constants for quenching of the cations by the aromatic donors (Table 7). The quenching rate constants increase

Figure 8 Transient absorption spectra produced in argon-saturated benzene solution containing 2.4×10^{-4} M of 1,1-di-p-anisylethylene and 0.52 M of TFA: 100 μs after flash (O); 200 μs after flash (●). (From Ref. 17.)

Table 7 Electron Transfer Quenching Rate Constants (M^{-1} s^{-1}) for Singlet-Excited Cations[a]

Cation	Quencher			
	Benzene ($E^{ox}_{1/2}$ = 2.08 V)	p-Xylene ($E^{ox}_{1/2}$ = 1.56 V)	Hexamethyl benzene ($E^{ox}_{1/2}$ = 1.16 V)	Anthracene ($E^{ox}_{1/2}$ = 0.84 V)[b]
9-Phenylxanthyl	4.5×10^9	1.4×10^{10}	1.9×10^{10}	2.6×10^{10}
Xanthyl	1.2×10^{10}	2.0×10^{10}	2.1×10^{10}	2.8×10^{10}
Thioxanthyl	9.7×10^7	5.3×10^9	2.2×10^{10}	2.9×10^{10}
9-Phenylthioxanthyl	8.5×10^7	5.2×10^9	3.1×10^{10}	4.0×10^{10}

[a]Solvent 1 : 1 acetonitrile : TFA.
[b]Versus Ag/0.1 N Ag$^+$ in acetonitrile.
Source: Ref. 15.

with decreasing oxidative potential of the aromatic quenchers, strongly suggestive of an electron transfer process. For example, k_q values for the 9-phenyl-
xanthyl cation range from $4.5 \times 10^9 \ M^{-1} \ s^{-1}$ for quenching by benzene ($E_{1/2}^{ox}$
$= 2.08$ V) to $2.6 \times 10^{10} \ M^{-1} \ s^{-1}$ for quenching by anthracene ($E_{1/2}^{ox} = 0.84$ V)
[9,15]. Laser flash photolysis experiments in the presence of 1-methylnaphtha-
lene, biphenyl, or anthracene gave transient spectra bands at long wavelengths
(500–700 nm) assignable to the aromatic radical cation (Fig. 9) [15]. Absorption peaks corresponding to the ground state cations were bleached, with the
appearance of a new peak at 345 nm corresponding to the xanthyl or thioxanthyl
radicals. The quantum efficiency of the net electron transfer is small (e.g.,
$\phi_{ET} = 0.02$ for biphenyl as the aromatic donor) and was rationalized as fast
back electron transfer between the radical/radical cation pair.

Electron transfer quenching rate constants for the thioxanthyl cations are
lower than those for the xanthyl cations, when the k_q values are below the
diffusion limit [15]. This was attributed to the lower singlet energies of the
thioxanthyl cations, by about 5 kcal mol^{-1}, compared to the corresponding
xanthyl cations. Values of k_q for the 9-phenylxanthyl cation are two to three
times lower than for the parent xanthyl cation [15]. Because the singlet energies and reduction potentials for the two cations are nearly equal, this suggests
a small steric effect in donor–acceptor interactions, with the phenyl moiety being
twisted out of plane with respect to the xanthyl backbone.

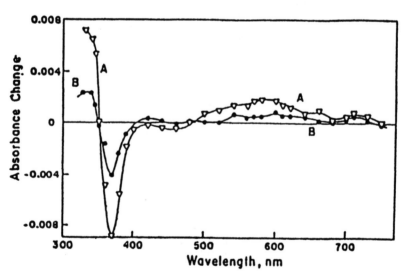

Figure 9 Transient absorption spectra at (A) 5 μs and (B) 45 μs following 355-nm
laser pulse excitation of the 9-phenylxanthyl cation in acetonitrile containing 0.40 M TFA
and 10 mM 1-methylnaphthalene. (From Ref. 15.)

Azarani et al. reported that the parent dibenzosuberenyl cation 4 also undergoes photoinduced electron transfer processes with aromatic compounds [10]. The singlet excited dibenzosuberenyl cation was generated by double laser excitation at 308 nm in TFE, and the fluorescence decay monitored as a function of the concentration of added quencher. Plots of the observed rate constant versus the quencher concentration gave fluorescence quenching rate constants usually greater than 10^9 M^{-1} s^{-1}. An increase in k_q values correlated with a decrease in the oxidation potential of the aromatic donor. Radical photoproducts were not isolated following preparative photolyses of the cations in aromatic concentrations sufficiently large to quench more than 90% of the excited cations, suggesting an efficient back electron transfer step. To resolve this problem, the dibenzosuberenyl cation 4 was irradiated in the presence of benzyltrimethylsilane which also quenches the excited cation but gives an unstable radical cation. Isolation of 5-benzyldibenzocycloheptene 13 confirmed an electron transfer process. It was proposed that 13 was generated through initial electron transfer to generate a dibenzosuberenyl radical and the silane radical cation. Cleavage of the radical cation produces a benzyl radical which couples with the dibenzosuberenyl radical to give 13 (Scheme 4).

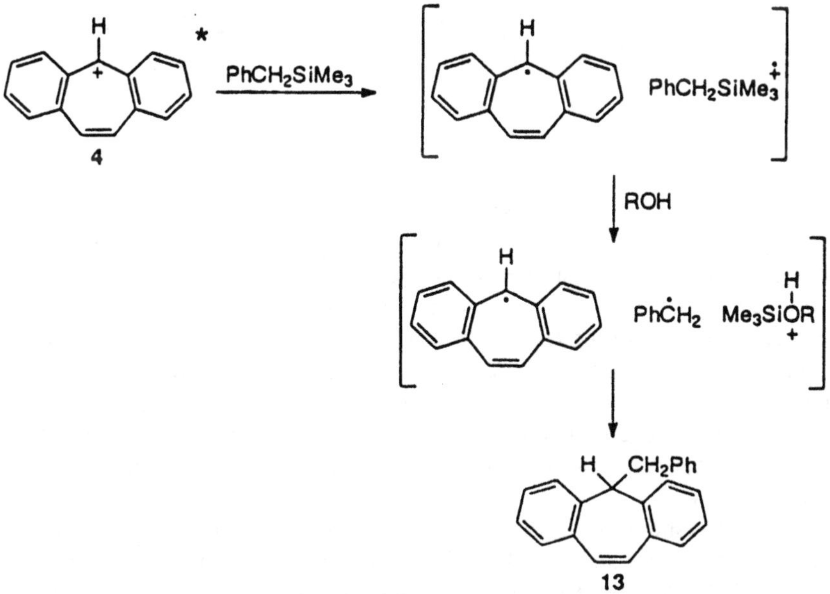

Scheme 4

Photoproduct studies by Shukla and Wan gave further evidence for the intermediacy of radicals in electron transfer reactions between photoexcited cations and aromatic donors [14]. Bis(9-phenylxanthen-9-yl) peroxide 14 was isolated and identified by x-ray crystallography following irradiation of the 9-phenylxanthyl cation in the presence of 1,3-dimethoxybenzene in aqueous acidic acetonitrile. Initial electron transfer from the aromatic donor was proposed to generate the 9-phenylxanthyl radical and the aromatic radical cation. Subsequent reaction of the radical or its dimer with oxygen gave the observed peroxide (Scheme 5).

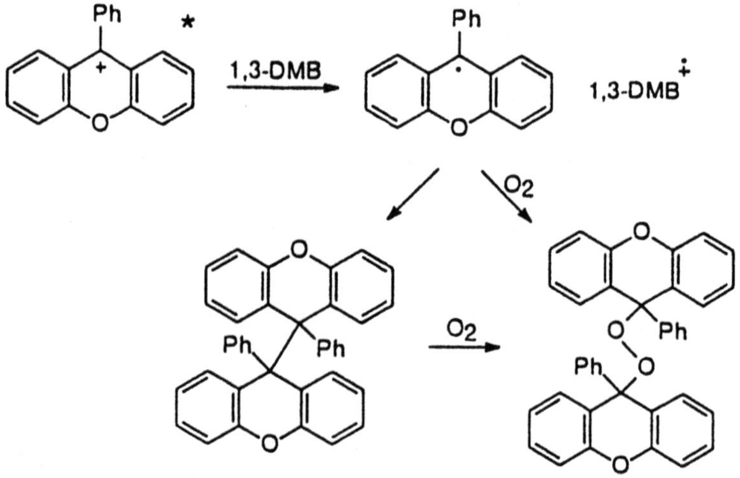

Scheme 5

Photooxidation of the triphenylmethyl cation (as its tetrafluoroborate salt) in the presence of an aromatic donor was similarly found to afford bis(triphenyl-methyl) peroxide [31]. The mechanism was proposed to proceed through initial electron transfer from the aromatic donor to the singlet-excited triphenyl-methyl cation to give the triphenylmethyl radical. Reaction of the radical with triplet oxygen, and subsequent coupling with another triphenylmethyl radical gave the observed peroxide. It was noted that electron transfer from the tetrafluoroborate counterion to the excited state cation could not be completely excluded, because the cation was slowly photooxygenated in the absence of an electron donor.

The 9-xanthylium radical was observed following irradiation of the xanthyl cation generated via adsorption of the corresponding alcohol in zeolite cavities [33]. This suggests that the singlet-excited cation has undergone electron transfer from a donor. It was speculated that zeolite environment may act as an electron donor in this case.

The singlet-excited *trans*-retinyl cation 3 also undergoes electron transfer quenching by aromatic donors [18]. Two laser photolysis of all-*trans*-retinol 2 to generate the retinyl cation in the presence of trimethoxybenzene results in enhanced bleaching of the cation absorption at 570 nm. Quenching of the excited state retinyl cation must occur competitively with trans-cis isomerization, as no transient signals due to production of the *cis*-retinyl cation were observed (see Section V.D). Bleaching of the trans-cation signal correlated with the oxidation potentials of the electron donors. Thus, bleaching was observed in the presence of tri- and dimethoxybenzene (E_{ox} = 1.42 V) and *trans*-stilbene (E_{ox} = 1.44 V) but not in the presence of anisole (E_{ox} = 1.72 V) or biphenyl (E_{ox} = 1.82 V). Direct evidence for an electron transfer process was obtained by irradiating the *trans*-retinyl cation in the presence of *trans*-stilbene. Transient absorption spectroscopy showed bleaching of the cation signal and enhanced absorption at 470 nm, corresponding to formation of the stilbene radical cation.

Triplet-excited xanthyl cations have also been shown to undergo electron transfer processes with aromatic donors [12,13]. The reactivity of the triplet state 9-phenylxanthyl cation, its *p*-fluoro analog, and the 9-phenylthioxanthyl cation was examined using transient absorption techniques. Irradiation of the 9-phenylxanthyl cation 1 in the presence of biphenyl resulted in an enhanced rate constant for decay of triplet 1, with concomitant production of transients corresponding to the 9-phenylxanthyl radical at 340 nm and the biphenyl radical cation at 670 nm (Fig. 10). Rate constants for triplet decay or radical growth were measured as a function of added quencher concentration for a variety of aromatic donors. Plots of the observed first-order rate constant versus the

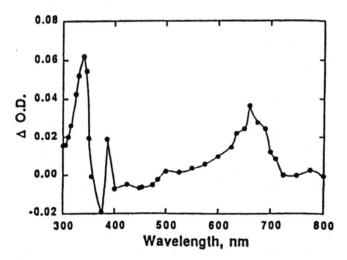

Figure 10 Transient absorption spectrum measured 7 μs after 355-nm laser excitation of the 9-phenylxanthyl cation in the presence of $2 \times 10^{-4} M$ biphenyl in TFA-TFE. (From Ref. 13.)

quencher concentrations gave the bimolecular excited state quenching rate constants (Table 8). An increase in the rate constants correlates with a decrease in the oxidation potential of the donor, consistent with reduction of the triplet cation via electron transfer quenching. Comparison of the singlet and triplet xanthyl cation quenching shows that the triplet leads to more efficient cage escape for the radical/radical cation pair [10,13].

C. Quenching by Oxygen

Minto and Das noted that the singlet-excited 9-phenylxanthyl cation 1 is resistant to quenching by oxygen [8]. Oxygen or air saturation of the cation in acetonitrile or acidified acetonitrile had a negligible effect on its fluorescence lifetime [8,28]. The fluorescence lifetime of the singlet-excited dibenzosuberenyl cation generated by laser flash photolysis of its corresponding alcohol was similarly not greatly affected by oxygen [10]. Oxygen quenching rate constants were measured directly by steady-state fluorescence quenching for several singlet-excited cations. Values of k_q of 1.1×10^9, 3.5×10^9, and 1.0×10^9 M^{-1} s^{-1} were obtained for oxygen quenching of the 9-phenylxanthyl, xanthyl, and thioxanthyl cations, respectively [15]. It was also noted that oxygen quenched the luminescence of cation 1 on silica gel surfaces. Fluorescence lifetimes for cation 1 of 37, 29, and 25 ns were measured under vacuum-outgassed, air-purged, and oxygen-purged conditions, respectively [32]. In contrast to singlet-excited cations in solution, the triplet-excited 9-phenylxanthyl cation and its p-F-substituted analog were reported to be efficiently quenched by oxygen [13].

D. *Trans-Cis* Isomerization

The all-*trans*-retinyl cation 3 was generated following 355-nm pulse laser photolysis of all-*trans*-retinol 2 in acetonitrile [18]. Subsequent photoexcitation of 3 by a pulse from a flashlamp-pumped dye laser (590 nm) resulted in bleaching of the absorption band corresponding to 3 at 570 nm, with the concurrent production of a new absorption with a maximum at 610 nm (Fig. 11). The red-shifted transient was identified as a *trans-cis* isomerized cation, although the number and positions of double bonds undergoing isomerization were not identified. Photoexcitation of the radical cations of 1,6-diphenyl-1,3,5-hexatriene and 1,8-diphenyl-1,3,5,7-octatetraene similarly results in *trans-cis* isomerization to give a red-shifted transient corresponding to formation of the *cis*-radical cation [50]. Photoexcitation of 3 at 590 nm in the presence of aromatic donors resulted in extensive bleaching at both 570 and 610 nm, indicating that the excited state cation 3 undergoes electron transfer instead of isomerization.

Table 8 Electron Transfer Quenching Rate Constants (M^{-1} s^{-1}) for Triplet-Excited Cations[a]

Cation	Quencher			
	o-Xylene ($E^{ox}_{1/2}$ = 1.89 V)	Biphenyl ($E^{ox}_{1/2}$ = 1.80 V)	Naphthalene ($E^{ox}_{1/2}$ = 1.62 V)	1-Methoxy naphthalene ($E^{ox}_{1/2}$ = 1.38 V)[b]
9-Phenylxanthyl	1.7×10^7	4.0×10^9	4.5×10^9	4.9×10^9
p-F xanthyl	2.3×10^7	3.8×10^9	4.7×10^9	4.3×10^9
9-Phenylthioxanthyl		1.2×10^7	7.1×10^7	4.1×10^9

[a]Solvent 0.01 M TFA in TFE.
[b]In acetonitrile versus SCE (Saturated Calomel Electrode).
Source: Ref. 13.

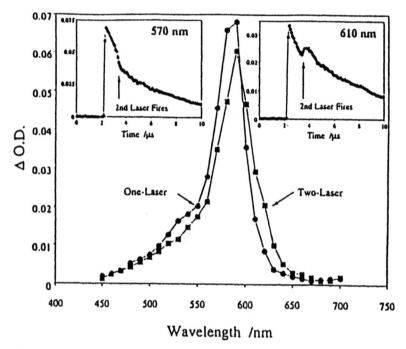

Figure 11 Transient absorption spectra obtained following one-laser photolysis (355 nm; ●) and two-laser photolysis (355 + 590 nm; ■) of all-*trans*-retinol in air-saturated acetonitrile. In the one-laser experiment, the spectrum was obtained 1.9 μs following the 355 nm pulse. In the two-laser experiment, the spectrum was obtained 1.9 μs after the 355-nm pulse and 400 ns following the 590-nm pulse. The kinetic behaviors observed at 570 and 610 nm are also shown. (From Ref. 18.)

VI. METHODS FOR PHOTOGENERATION OF CARBOCATIONS

A. Photoheterolysis of Carbon–Heteroatom Bonds

The major method for cation photogeneration involves photoheterolysis of carbon–heteroatom bonds leading to formation of carbocations and the heteroatom anionic leaving group [Eq. (6)] [5,6,51]. Leaving groups have included halide, acetate, hydroxide, tosylate, 4-cyanophenoxide, cyanide, and phosphonium chloride.

$$-\overset{|}{\underset{|}{C}}-X \xrightarrow{h\nu} -\overset{|}{\underset{|}{C}}{}^{+} + X^{-} \tag{6}$$

Numerous benzylic-type cations have been generated by this method, including di- and triarylmethyl [52–59], xanthyl [8,24], and fluorenyl cations

[60–69], as well as the retinyl cation [70]. Vinyl cations have similarly been generated from photolysis of vinyl halides [38–40,71–76]. Transient absorption spectroscopy and time-resolved conductivity have been used to detect the cations following laser flash photolysis of their precursors. The weakly nucleophilic solvents TFE and HFIP (hexafluoroisopropyl alcohol) have been used to increase the lifetime of reactive cations to allow for their observation and the study of their reactivity.

Heterolytic bond cleavage may be accompanied by homolytic bond cleavage to generate the corresponding radical pair, with nonpolar solvents increasing the extent of homolytic cleavage [8]. The nature of the leaving group also affects the partitioning between heterolytic and homolytic cleavage [52]. The possibility of initial homolytic cleavage and subsequent electron transfer to generate the ion pair, rather than direct photoheterolysis, has been suggested in the photochemistry of benzyl and naphthylmethyl esters [77–79]. Picosecond spectroscopy of diphenylmethyl chloride showed that both radicals and ions were formed concurrently following laser flash photolysis in acetonitrile, at least within the 20-ps temporal resolution of the instrument [59].

B. Carbon–Carbon Bond Cleavage

Pienta et al. prepared a series of molecules 15–17 containing covalently linked stable carbanions and carbocations [80]. Picosecond and nanosecond laser flash photolytic techniques showed that the molecules undergo a singlet state mediated intramolecular electron transfer to form linked radical ion pairs. These ion pairs undergo competitive back electron transfer to form the triplet-excited state of the triphenylcyclopropenyl moiety or homolytic cleavage of the carbon–carbon central bond to generate carbocations and carbanions (Scheme 6).

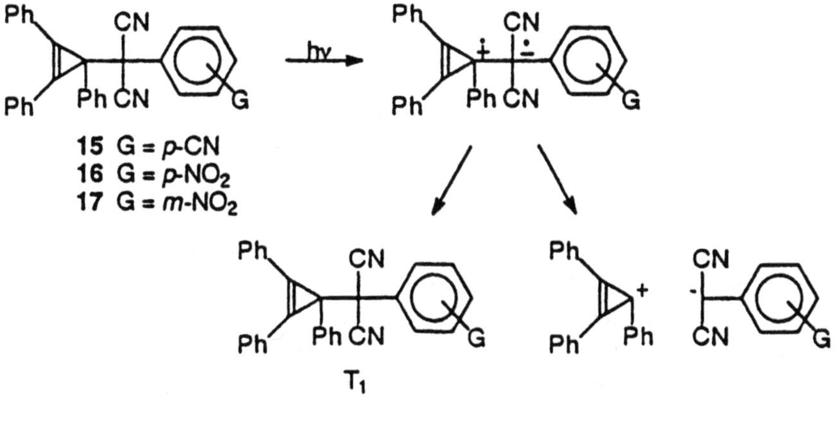

Scheme 6

Triplet-excited nitrobenzyl acetals 18-19 were proposed to undergo direct carbon–carbon bond photoheterolysis to generate the nitrobenzyl carbanion and the corresponding α-dialkoxy carbocation [Eq. (7)] [81,82]. Transient spectra obtained following laser flash photolysis of 19 showed the formation of the p-nitrobenzyl anion and the oxocarbocation [83]:

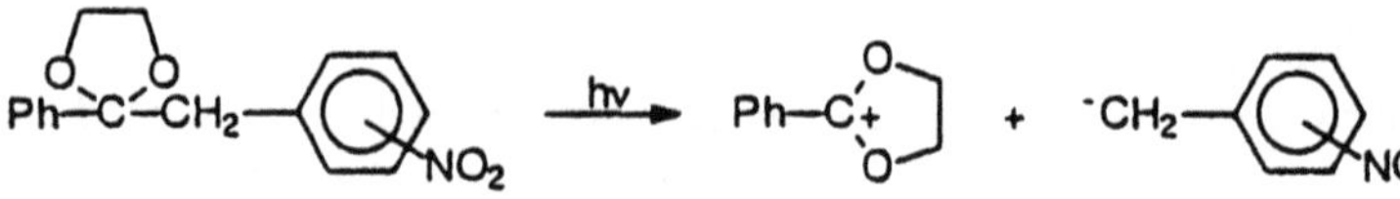

18 m-NO$_2$
19 p-NO$_2$

C. Heterolysis of Photogenerated Radical Cations

Radical cations may be photochemically generated through photoinduced electron transfer, photoionization, or electron transfer to a photolytically generated oxidant. These radical cations may undergo carbon–carbon bond cleavage to form a carbocation and a radical fragment [84]. For example, tetrakis(4-methylphenyl)ethanone 20 undergoes photoinduced electron transfer in the presence of 2,4,6-triphenylpyrylium tetrafluoroborate (TPP$^+$ BF$_4^-$) or dicyanoanthracene to form radical cation 21. Carbon–carbon bond cleavage gave the tris(4-methylphenyl)methyl cation, observed by transient absorption spectroscopy, and the 4-methylbenzoyl radical (Scheme 7) [85,86].

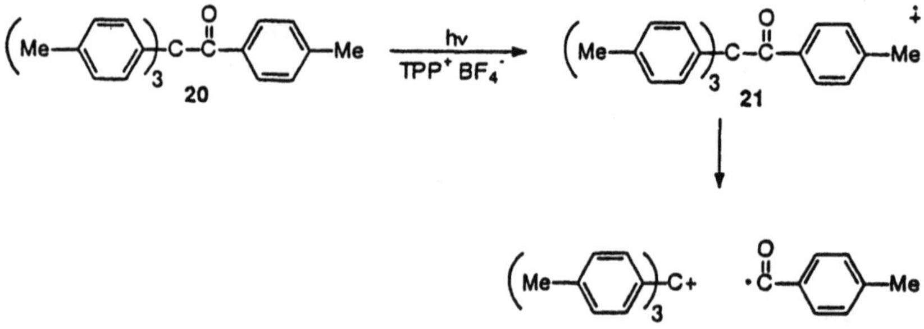

Scheme 7

In other examples, biphotonic ionization of 2,3-dimethyl-2,3-diphenylbutane (bicumene) in TFE gives the bicumene radical cation which undergoes carbon–carbon fragmentation to yield the cumyl cation and cumyl radical [87]. Photoinduced electron transfer from 2,2-dialkyldioxolanes to tetracyanoanthracene gives radical cations which fragment to yield dialkoxy carbocations and alkyl radicals [88]. Benzyl acetals were subjected to two-photon ionization or one-electron ionization from photolytically generated sulfate radical anions. The resulting acetal radical cation fragmented to the benzyl radical and the α-dialkoxymethyl cation [83].

Carbon–heteroatom bond cleavage from photogenerated radical cations also leads to carbocation formation. For example, photosensitized cleavage of the dithio-protecting group is proposed to proceed through electron transfer from the dithio derivative to the electronically excited methylene green sensitizer, with subsequent carbon–sulfur bond cleavage of the dithio radical cation [89]. Monothioacetal side chains have been used to initiate intramolecular electron transfer to a triplet-excited 1,8-naphthalimide chromophore, with subsequent fragmentation of the alkylthio group [90].

D. Oxidation of Photochemically Generated Radicals

Photochemically generated radicals undergo a subsequent photoinduced or thermal oxidation to the corresponding carbocation. Faria and Steenken produced the triphenylmethyl radical through two-photon ionization and subsequent decarboxylation of the triphenylacetate ion. Laser excitation of the triphenylmethyl radical resulted in photoinjection of an electron to give the triphenylmethyl cation and a solvated electron [Eq. (8)] [91].

$$\left(\!\left\langle\bigcirc\right\rangle\!\right)_{\!3}\!\!C\!-\!CO_2^{\cdot}\ \xrightarrow[-e,\,-CO_2]{h\nu}\ \left(\!\left\langle\bigcirc\right\rangle\!\right)_{\!3}\!\!C\!\cdot\ \xrightarrow[-e]{h\nu}\ \left(\!\left\langle\bigcirc\right\rangle\!\right)_{\!3}\!\!C^+ \qquad (8)$$

Scaiano and co-workers photogenerated the diphenylmethyl radical from laser flash photolysis of 1,1-diphenylacetone. Electron transfer from the excited state radical to aromatic electron acceptors such as p-dicyanobenzene gave the diphenylmethyl cation observed by transient absorption spectroscopy [Eq. (9)] [92]. The diphenylmethyl radical was also photogenerated by irradiation of 1,1,3,3-tetraphenylacetone in acetonitrile. Electrochemical oxidation gave the diphenylmethyl cation [93]:

$$\underset{Ph_2CHCCH_3}{\overset{O}{\underset{\|}{}}}\ \xrightarrow{2\,h\nu}\ (Ph_2\overset{\cdot}{C}H)^*\ \xrightarrow{DCB}\ Ph_2\overset{+}{C}H \qquad (9)$$

Irradiation of 9-xanthylacetone or 9-xanthylacetophenone results in homolytic bond cleavage to give the xanthyl radical. The transient spectrum of the radical decays in air-saturated solutions with the concomitant formation of a new transient spectrum corresponding to formation of the xanthyl cation, indicating that the radical is oxidized with oxygen acting as the oxidizing agent (Scheme 8) [94].

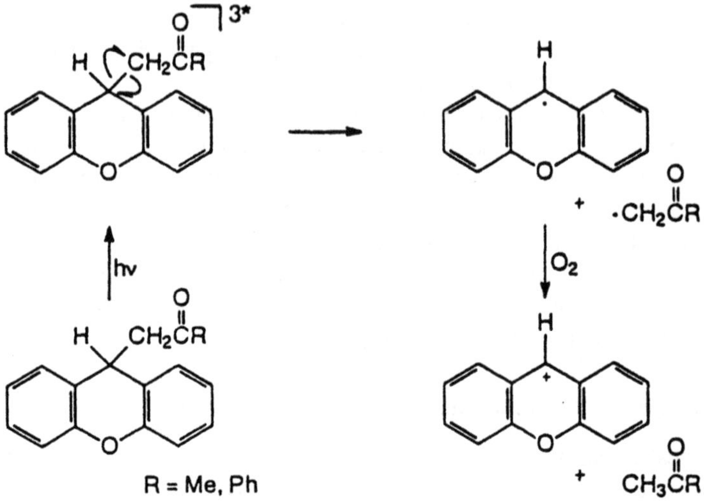

R = Me, Ph

Scheme 8

E. Photoprotonation of Aromatics, Styrenes, and Phenylacetylenes

Aromatic alkenes, alkynes, and allenes undergo intramolecular and intermolecular proton transfer in neutral or dilute aqueous acidic media to give Markovnikov-oriented hydration products [Eq. (10)] [95,96]:

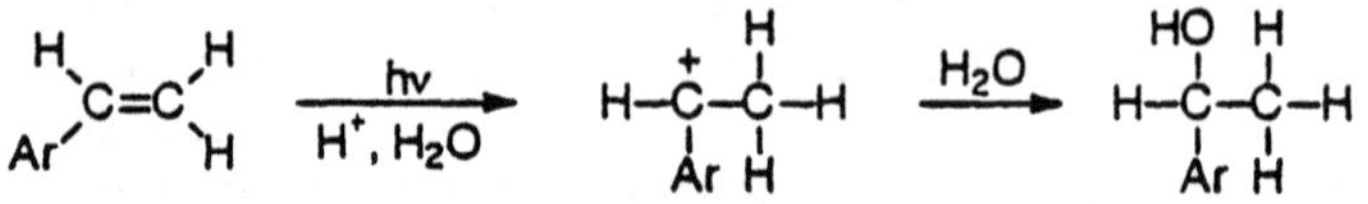

The benzylic cation obtained upon photoprotonation of aryl-substituted styrenes was observed directly by McClelland in weakly nucleophilic solvents of TFE or HFIP [56,97]. McClelland also used HFIP to observe the 1-phenylcyclohexyl cation upon laser flash photolysis of 1-phenylcyclohexene, and vinyl cations upon irradiation of phenylacetylenes [98,99].

Aromatic compounds similarly undergo photoprotonation, as evidenced by their fluorescence quenching in acidic media, acid-catalyzed ipso photosubstitution of alkoxy benzenes, and photochemical deuterium exchange of aromatic ring protons [100,101]. McClelland has summarized in a recent review the observation of transient spectra from cyclohexadienyl cations upon laser flash photolysis of several aromatic compounds in HFIP [5,98,102–105].

F. Protonation of Photochemically Generated Carbenes

Carbenes photogenerated from diazo precursors undergo proton transfer from alcohols to generate carbocations that are eventually trapped by nucleophiles [Eq. (11)] [106].

$$Ar_2CN_2 \xrightarrow{\text{hv}} Ar_2C: \xrightarrow{\text{ROH}} Ar_2\overset{+}{C}H \xrightarrow{\text{ROH}} Ar_2CHOR \qquad (11)$$

Conclusive evidence for the cation intermediate was obtained by detection of the absorption spectra of several diarylmethyl cations following nanosecond or picosecond laser flash photolysis of their respective diazo, diphenylaziridinylimine, or 3H-indazole precursors in acidic media [107–112]. Photochemically generated vinyl carbenes were recently shown to similarly protonate by deuterium labeling experiments to give allylic cations that were detected by transient absorption spectroscopy [113].

REFERENCES

1. Olah, G. A.; Schleyer, P. von R. *Carbonium Ions,* Wiley: New York, 1970; Vol. I–V.
2. Childs, R. F.; Shaw, G. B. *Org. Photochem.* 1991, *11*, 111.
3. Childs, R. F. *Rev. Chem. Intermed.* 1980, *3*, 285.
4. Cabell-Whiting, P. W.; Hogeveen, H. *Adv. Phys. Org. Chem.* 1973, *10*, 129.
5. McClelland, R. A. *Tetrahedron,* 1996, *52*, 6823.
6. Das, P. K. *Chem. Rev.* 1993, *93*, 119.
7. Wan, P.; Yates, K.; Boyd, M. K. *J. Org. Chem.* 1985, *50*, 2881.
8. Minto, R. E.; Das, P. K. *J. Am. Chem. Soc.* 1989, *111*, 8858.
9. Samanta, A.; Gopidas, K. R.; Das, P. K. *Chem. Phys. Lett.* 1990, *167*, 165.
10. Azarani, A.; Berinstain, A. B.; Johnston, L. J.; Kazanis, S. *J. Photochem. Photobiol. A: Chem.* 1991, *57*, 175.
11. Boyd, M. K.; Lai, H. Y.; Yates, K. *J. Am. Chem. Soc.* 1991, *113*, 7294.
12. Johnston, L. J.; Wong, D. F. *Can. J. Chem.* 1992, *70*, 280.
13. Johnston, L. J.; Wong, D. F. *J. Phys. Chem.* 1993, *97*, 1589.
14. Shukla, D.; Wan, P. *J. Photochem. Photobiol. A: Chem.* 1993, *76*, 47.
15. Samanta, A.; Gopidas, K. R.; Das, P. K. *J. Phys. Chem.* 1993, *97*, 1583.
16. Young, R. N.; Brocklehurst, B.; Booth, P. *J. Am. Chem. Soc.* 1994, *116*, 7885.
17. Al-Ekabi, H.; Kawata, H.; de Mayo, P. *J. Org. Chem.* 1988, *53*, 1471.
18. Wang, Z.; McGimpsey, W. G. *J. Photochem. Photobiol. A: Chem.* 1996, *93*, 151.
19. Johnston, L. J.; Lobaugh, J.; Wintgens, V. *J. Phys. Chem.* 1989, *93*, 7370.
20. Turro, N. J.; Wan, P. *J. Photochem.* 1985, *28*, 93.
21 Zimmerman, H. E.; Sandel, V. K. *J. Am. Chem. Soc.* 1963, *85*, 915.
22. Zimmerman, H. E.; Somasekhara, S. *J. Am. Chem. Soc.* 1963, *85*, 922.

23. Havinga, E.; de Jongh, R. O.; Dorst, W. *Recl. Trav. Chim.* 1956, *75*, 378.

24. McClelland, R. A.; Banait, N.; Steenken, S. *J. Am. Chem. Soc.* 1989, *111*, 2929.

25. Boyd, M. K. Ph. D. Thesis, University of Toronto, 1988.

26. Feldman, M. R.; Thame, N. G. *J. Org. Chem.* 1979, *44*, 1863.

27. Shukla, D.; Wan, P. *J. Photochem. Photobiol. A: Chem.* 1994, *79*, 55.

28. Valentino, M. R.; Boyd, M. K. *J. Org. Chem.* 1993, *58*, 5826.

29. Valentino, M. R.; Boyd, M. K. *J. Photochem. Photobiol. A: Chem.* 1995, *89*, 7.

30. Bedlek, J. M.; Valentino, M. R.; Boyd, M. K. *J. Photochem. Photobiol. A: Chem.* 1996, *94*, 7.

31. Futamura, S.; Kamiya, Y. *Chem. Lett.* 1989, 1703.

32. Berger, R. M.; Weir, D. *Chem. Phys. Lett.* 1990, *169*, 213.

33. Cozens, F. L.; Garcia, H.; Scaiano, J. C. *Langmuir* 1994, *10*, 2246.

34. Cozens, F. L.; Gessner, F.; Scaiano, J. C. *Langmuir*, 1993, *9*, 874.

35. Hafner, K.; Pelster, H. *Angew. Chem.* 1961, *73*, 342.

36. Meisel, D.; Das, P. K.; Hug, G. L.; Bhattacharyya, K.; Fessenden, R. W. *J. Am Chem. Soc.* 1986, *108*, 4706.

37. McEwen, J.; Yates, K. *J. Phys. Org. Chem.* 1991, *4*, 193.

38. Kobayashi, S.; Schnabel, W. *Z. Naturforsch. B* 1992, *47*, 1319.

39. Kobayashi, S.; Zhu, Q.; Schnabel, W. *Z. Naturforsch. B* 1987, *43*, 825.

40. Kobayashi, S.; Kitamura, T.; Taniguchi, H.; Schnabel, W. *Chem. Lett.* 1983, 1117.

41. Bartl, J.; Steenken, S.; Mayr, H. *J. Am. Chem. Soc.* 1991, *113*, 7710.

42. Dinnocenzo, J. P.; Todd, W. P.; Simpson, T. R.; Gould, I. R. *J. Am. Chem. Soc.* 1990, *112*, 2462.

43. Rehm, D.; Weller, A. *Isr. J. Chem.* 1970, *8*, 259.

44. Rhem, D.; Weller, A. *Ber. Bunsenges. Phys. Chem.* 1969, *73*, 834.

45. Arnett, E. M.; Flowers II, R. A.; Meekhof, A. E.; Miller, L. *J. Am. Chem.* 1993, *115*, 12603.

46. Fleming, I. *Frontier Orbitals and Organic Chemical Reactions;* Wiley: New York, 1976.

47. Van Tamelen, E. E.; Cole, T. M. *J. Am. Chem. Soc.* 1971, *93*, 6158.

48. Owen, E. D.; Allen, D. M. *J. Chem. Soc., Perkin Trans II* 1973, *95*.

49. Van Tamelen, E. E.; Greeley, R. H.; Schumacher, H. *J. Am. Chem. Soc.* 1971, *93*, 6151.

50. Wang, Z.; McGimpsey, W. G. *J. Phys. Chem.* 1993, *97*, 3324.

51. Cristol, S.J.; Bindel, T. H. *Org. Photochem.* 1983, *6*, 327.

52. Bartl, J.; Steenken, S.; Mayr, H.; McClelland, R. A. *J. Am. Chem. Soc.* 1990, *112*, 6918.

53. McClelland, R. A.; Kanagasabapathy, V. M.; Banait, N. S.; Steenken, S. *J. Am. Chem. Soc.* 1989, *111*, 3966.

54. McClelland, R. A.; Banait, N.; Steenken, S. *J. Am. Chem. Soc.* 1986, *108*, 7023.

55. Mathivanan, N.; McClelland, R. A.; Steenken, S. *J. Am. Chem. Soc.* 1990, *112*, 8454.

56. McClelland, R. A.; Kanagasabapathy, V. M.; Steenken, S. *J. Am. Chem. Soc.* 1988, *110*, 6913.

57. Chateauneuf, J. E. *J. Chem. Soc., Chem. Commun.* 1991, 1437.
58. Spears, K. G.; Gray, T. H.; Huang, D. *J. Phys. Chem.* 1986, *90*, 779.
59. Peters, K. S.; Li, B. *J. Phys. Chem.* 1994, *98*, 401.
60. Wan, P.; Krogh, E. *J. Am. Chem. Soc.*, 1989, *111*, 4887.
61. Krogh, E.; Wan, P. *Tetrahedron Lett.* 1986, *27*, 823.
62. Wan, P.; Krogh, E. *J. Chem. Soc., Chem. Commun.* 1985, 1207.
63. Krogh, E.; Wan, P. *Can. J. Chem.* 1990, *68*, 1725.
64. McClelland, R. A.; Mathivanan, N.; Steenken, S. *J. Am. Chem. Soc.* 1990, *112*, 4857.
65. Blazek, A.; Pungente, M.; Krogh, E.; Wan, P. *J. Photochem. Photobiol. A: Chem.* 1992, *64*. 315.
66. Johnston, L. J.; Kwong, P. Shelemay, A.; Lee-Ruff, E. *J. Am. Chem. Soc.* 1993, *115*, 1664.
67. Mecklenburg, S. L.; Hilinski, E. F. *J. Am. Chem. Soc.* 1989, *111*, 5471.
68. Cozens, F. L.; Mathivanan, N.; McClelland, R. A.; Steenken, S. *J. Chem. Soc., Perkin Trans. 2* 1992, 2083.
69. Lew, C. S. Q.; McClelland, R. A.; Johnston, L. J.; Schepp, N. P. *J. Chem. Soc., Perkin Trans. 2* 1994, 395.
70. Pienta, N. J.; Kessler, R. J. *J. Am. Chem. Soc.* 1992, *114*, 2419.
71. VanGinkel, F. I. M.; Visser, R. J.; Varma, C. A. G. O.; Lodder, G. *J. Photochem.* 1985, *30*, 453.
72. Johnen, N.; Schnabel, W.; Kobayashi, S.; Fouassier, J. -P. *J. Chem. Soc., Faraday Trans. 2*, 1992, *88*. 1385.
73. Kitamura, T.; Soda, S.; Nakamura, I. Fukuda, T.; Taniguchi, H. *Chem. Lett.* 1991, 2195.
74. Chiang, Y.; Eliason, R.; Jones, J.; Kresge, A. J.; Evans, K. L.; Gandour, R. D. *Can. J. Chem.* 1993, *71*, 1964.
75. Krijnen, E. S.; Zuilhof, H.; Lodder, G. *J. Org. Chem.* 1994, *59*, 8139.
76. Verbeek, J. -M.; Stapper, M.; Krijnen, E. S.; van Loon, J. -D.; Lodder, G.; Steenken, S. *J. Phys. Chem.* 1994, *98*, 9526.
77. DeCosta, D. P.; Pincock, J. A. *J. Am. Chem. Soc.* 1989, *111*, 8948.
78. Hilborn, J. W.; Pincock, J. A. *J. Am. Chem. Soc.* 1991, *113*, 2683.
79. Pincock, J. A..; Wedge, P. J. *J. Org. Chem.* 1994, *59*, 5587.
80. Pienta, N. J.; Kessler, R. J.; Peters, K. S.; O'Driscoll, E. D.; Arnett, E. M.; Molter, K. E. *J. Am. Chem. Soc.* 1991, *113*, 3773.
81. Wan, P.; Muralidharan, S. *J. Am. Chem. Soc.* 1988, *110*, 4336.
82. Wan, P.; Muralidharan, S. *Can. J. Chem.* 1986, *64*, 1949.
83. Steenken, S.; McClelland, R. A. *J. Am. Chem. Soc.* 1989, *111*, 4967.
84. Arnold, D. R.; Du, X.; Chen, J. *Can. J. Chem.* 1995, *73*, 307, and references therein.
85. Akaba, R.; Kamata, M.; Sakuragi, H.; Tokumaru, K. *Tetrahedron Lett.* 1992, *33*, 8105.
86. Akaba, R.; Niimura, Y.; Fukushima, T.; Kawai, Y.; Tajima, T.; Kuragami, T.; Negishi, A.; Kamata, M.; Sakuragi, H.; Tokumaru, K. *J. Am. Chem. Soc.* 1992, *114*, 4460.
87. Faria, J. L.; Steenken, S. *J. Phys. Chem.* 1992, *96*, 10869.

88. Mella, M.; Fasani, E.; Albini, A. *J. Org. Chem.* 1992, *57*, 3051.
89. Epling, G. A.; Wang, Q. *Tetrahedron Lett.* 1992, *33*, 5909.
90. Saito, I.; Takayama, M.; Sakurai, T. *J. Am. Chem. Soc.* 1994, *116*, 2653.
91. Faria, J. L.; Steenken, S. *J. Am. Chem. Soc.* 1990, *112*, 1277.
92. Arnold, B. R.; Scaiano, J. C.; McGimpsey, W. G. *J. Am. Chem. Soc.* 1992, *114*, 9978.
93. Nagaoka, T.; Griller, D.; Wayner, D. D. M. *J. Phys. Chem.* 1991, *95*, 6264.
94. Clifton, M. F.; Fenick, D. J.; Gasper, S. M.; Falvey, D. E.; Boyd, M. K. *J. Org. Chem.* 1994, *59*. 8023.
95. Wan P.; Yates, K. *Rev. Chem. Intermed.* 1984, *5*, 157.
96. Arnaut, L. G.; Formosinho, S. J. *J. Photochem. Photobiol. A: Chem.* 1992, *69*, 41.
97. McClelland, R. A.; Chan, C.; Cozens, F.; Modro, A.; Steenken, S. *Angew. Chem. Int. Ed. Engl.* 1991, *30*, 1337.
98. Cozens, F. L.; McClelland, R. A.; Steenken, S. *J. Am. Chem. Soc.* 1993, *115*, 5050.
99. McClelland, R. A.; Cozens, F.; Steenken, S. *Tetrahedron Lett.* 1990, *31*. 2821.
100. Shizuka, H. *Acc. Chem. Res.* 1985, *18*, 141.
101. Wan, P.; Wu, P. *J. Chem. Soc., Chem. Commun.* 1990, 822.
102. Mathivanan, N.; Cozens, F.; McClelland, R. A.; Steenken, S. *J. Am. Chem. Soc.* 1992, *114*, 2198.
103. Lew, C. S. Q.; McClelland, R. A. *J. Am Chem. Soc.* 1993, *115*, 11516.
104. Cozens, F.; McClelland, R. A.; Steenken, S. *Tetrahedron Lett.* 1992, *33*, 173.
105. Steenken, S.; McClelland, R. A. *J. Am. Chem. Soc.* 1990, *112*, 9648.
106. Kirmse, W.; Kilian, J.; Steenken, S. *J. Am. Chem. Soc.* 1990, *112*, 6399.
107. Belt, S. T.; Bohne, C.; Charette, G.; Sugamori, S. E.; Scaiano, J. C. *J. Am. Chem. Soc.* 1993, *115*, 2200.
108. Fehr, O. C.; Grapenthin, O.; Kilian, J.; Kirmse, W. *Tetrahedron Lett.* 1995, *36*, 5887.
109. Kirmse, W.; Meinert, T.; Modarelli, D. A.; Platz, M. S. *J. Am. Chem. Soc.* 1993, *115*, 8918.
110. Chateauneuf, J. E. *Res. Chem. Intermed.* 1994, *20*, 249.
111. Schepp, N. P.; Wirz, J. *J. Am. Chem. Soc.* 1994, *116*, 11749.
112. Kirmse, W.; Krzossa, B.; Steenken, S. *Tetrahedron Lett.* 1996, *37*. 1197.
113. Kirmse, W.; Strehlke, I. K.; Steenken, S. *J. Am. Chem. Soc.* 1995, *117*, 7007.

6

Regioselective and Stereoselective [2 + 2] Photocycloadditions

Steven A. Fleming, Cara L. Bradford, and J. Jerry Gao
Brigham Young University, Provo, Utah

I. INTRODUCTION

Several excellent reviews have been published concerning the synthetic utility of photochemical [2 + 2] cycloadditions [1]. Recent attention to the mechanistic details of photocycloaddition of enones and alkenes has resulted in additional reviews [2]. Particular examples of regiocontrol and/or stereocontrol in [2 + 2] photocycloaddition have been presented in many venues, but to our knowledge, no total review has been given for this topic. This chapter is an attempt to organize the data available on controlled [2 + 2] photocycloadditions.

The proposed mechanistic details for these cyclizations will be presented according to the type of reagent and excited state involved. This discussion will be followed by recent examples of regioselectivity and stereoselectivity that have been reported in the literature.

A. Methods for Control

There are a number of approaches that have been taken in an attempt to control photochemical [2 + 2] photocycloadditions. This chapter will deal primarily with tethered systems, intermediate stabilization, and steric template methodologies. Significant work has also been published in the area of regio-selectivity and/or stereoselectivity control via crystal lattice constraints. (For recent reviews of this work, see Ref. 3.) There are a number of useful reviews and recent publications concerning photoreactions in the solid state [4]. This technique is limited by the specific framework of the crystal structure (for excellent examples of regioselectivity and stereoselectivity of [2 + 2] photo-cycloadditions that are a function of the crystal structure, see Ref. 5.) Nevertheless, work in this area has been fruitful, especially for mechanistic understanding of certain photochemical processes [6]

Alternative approaches that have been explored for control of photochemical reactions include irradiation in liquid crystals [7], micelles [8], inclusion complexes [9], and zeolites [10]. In addition, photochemistry of monolayers [11] and on surfaces [12] (e.g., alumina, silica, clays, and semiconductors) has received considerable attention. Each of these methods has potential for regiocontrol and/or stereocontrol of certain photochemical processes. However, reactions between different groups (e.g., intermolecular photocycloadditions) are difficult to modulate with most of these approaches.

B. Types of Photoreactions

Control of solution photochemistry has been demonstrated with use of templates, tethering of reagents, and substituent stabilization [13] to affect the regio-selectivity or stereoselectivity. The types of photoreactions that have been investigated using these techniques are numerous. This chapter will focus on the [2 + 2] photocycloaddition (see Ref. 14 for reviews), but other reactions that have been studied include the [4 + 4], [2 + 3], [2 + 4], and meta-cycloadditions (see Refs. 15–18, respectively). Hydrogen abstractions [19], photo-rearrangements [20], and other photoreactions have been explored as well.

Stereocontrol of the [2 + 2] photocycloaddition has been an attractive goal for a number of reasons. First, the reaction offers an efficient pathway to small ring synthesis. The cyclobutane ring, or the oxetane ring in the case of the Paternò-Büchi reaction, can be expanded, contracted, or opened. Control of this carbon–carbon bond-forming process allows formation of four contiguous stereocenters in one reaction, not an easy process to duplicate with enolate chemistry or other typical carbon–carbon bond forming reactions.

Second, investigation of the regiochemical and stereochemical outcome of the cyclization process allows for a better understanding of the mechanistic pathway the reaction takes. The reaction is studied not only for synthetic exploitation but also for basic understanding of the photochemical process. Advances in the area of thymine–thymine dimerization, for example, can be traced to the increased comprehension of the stereoselectivity of the reaction. (For a recent example, see Ref. 21.)

A third reason for the interest in the [2 + 2] is that it allows for a logical analysis of orbital symmetry arguments. Few photochemical reactions are as predictable as the Diels–Alder reaction, but the [2 + 2] photocycloaddition is a good starting point for comparison.

II. MECHANISTIC DETAILS

Understanding and predicting the product selectivity for the [2 + 2] photocycloaddition has been a driving force for understanding the mechanism of cyclization. As will be noted in the later sections of this chapter, a majority of the regioselective and stereoselective [2 + 2] photocycloaddition examples can be justified by analysis of the mechanistic details.

Photocycloaddition between an excited state enone and a ground state enone can have different selectivity than the cycloaddition between the singlet excited state alkene and a ground state alkene. Thus, our survey of selective photocycloadditions will require a separate handling of the different types of reactions. In addition to the moiety that is involved (e.g., enone, carbonyl, aromatic ring, alkene), the nature of the excited state (i.e., singlet or triplet) plays an important role in the resultant selectivity and must be included in this summary.

A. Enone + Alkene

A common theme in enone cycloadditions is the involvement of a triplet enone excited state. The regioselectivity of triplet photocycloadditions is typically explained by formation of the most stable intermediate biradical species (see Scheme 1). Schuster et al. [22] and Weedon et al. [23] have made significant contributions to the understanding of enone photoreactivity. Corey et al. [24] originally suggested that the [2 + 2] reaction involved a polar π-complex, and this justification for the observed regioselectivity continues to appear in the literature as will be described below.

These two mechanisms, the reversible biradical intermediate and the intermediate exciplex [25] have both been useful for analysis of the regioselectivity and stereoselectivity observed in [2 + 2] photocycloaddition between enones and alkenes. Hoffman et al. [26], for example, describes his stereoselective photocycloadditions as arising from a biradical from the triplet excited enone which may or may not involve exciplex formation. Ground state *trans*-cycloalkenones have also been proposed as the reactive intermediates which lead to [2 + 2] cycloadducts [27]. The distance between the reacting partners is clearly an issue. If the π systems are not sufficiently close, then photocycloaddition will not occur [28].

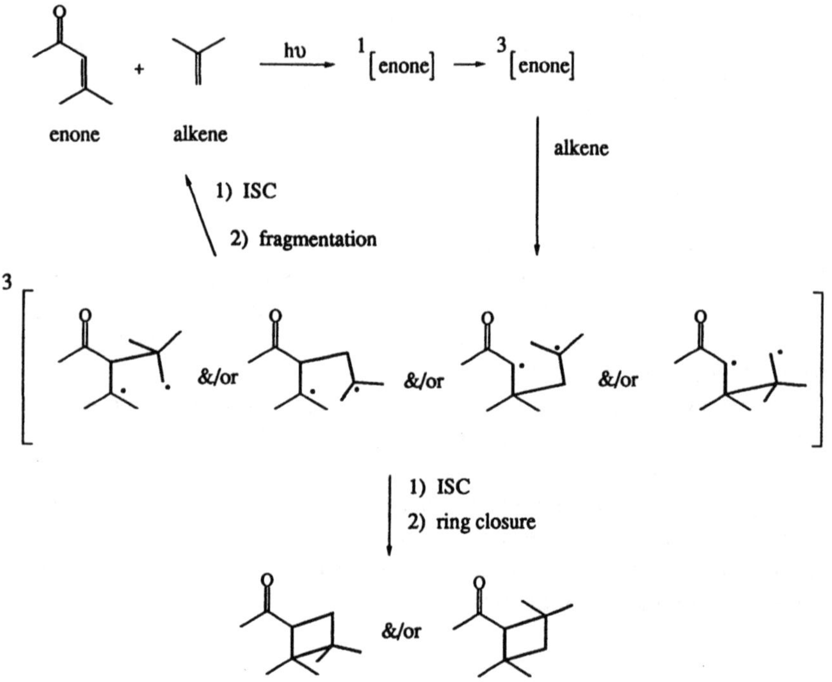

Scheme 1

There are a few reports of the enone singlet excited state adding to the partner π system [29]. These studies suggest that there is a triplet exciplex which intersystem crosses to a singlet exciplex prior to bond formation. It is not clear that this mechanistic detail would have any impact on the regiochemical or stereochemical outcome of the reaction.

B. Carbonyl + Alkene

The Paternò–Büchi reaction is one of the more predictable photocycloaddition reactions. Regiocontrol of the photoproduced oxetane is a function of the stepwise addition of the carbonyl chromophore to the alkene [30]. In the case of electron-rich alkenes, excitation of the carbonyl group produces a triplet species that adds to the alkene. The product regioselectivity is a result of addition that generates the most stable biradical, and the triplet lifetime of the intermediate biradical allows for substantial stereoselectivity prior to closing (see Scheme 2). Electron poor alkenes are more likely to undergo cycloaddition with carbonyl groups directly from an exciplex [31].

There are examples of alkyl aldehydes which undergo photoaddition from the singlet state [32]. This mode of addition has an impact on the stereochemical

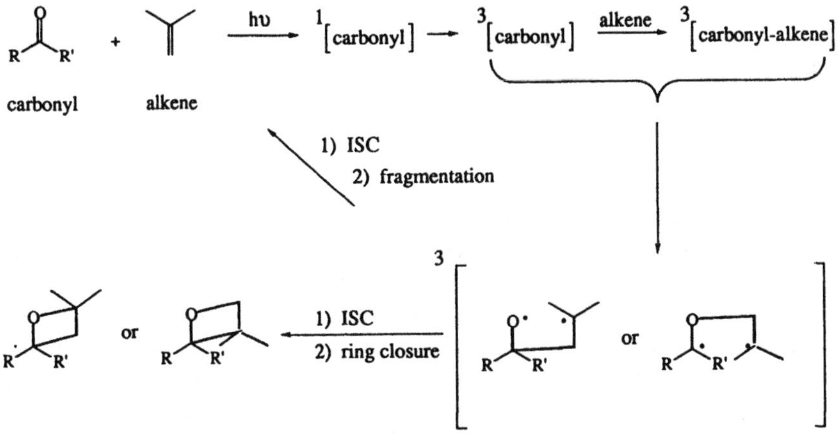

Scheme 2

outcome, due to the rapid formation of oxetane which does not involve a long-lived biradical. The singlet excited state pathway is typically regiospecific and stereospecific in its addition to electron-poor alkenes.

Cycloaddition with the carbonyl group in nonpolar solvents may involve initial formation of an exciplex, but recent evidence indicates that polar solvents preclude the exciplex to biradical pathway [33]. There are a number of carbonyl/alkene pairs that are capable of photoelectron transfer due to their redox potentials (e.g., quinones/tetracyanoethylene (TCNE)). The resultant radical ions can bond to give 1,4-biradicals that close to form oxetanes [34].

C. Arene + Alkene

Considerable attention has been given to the mechanistic details of 1,3-photo-cycloaddition of alkenes to aromatic systems [2d,35]. The mechanistic picture of 1,2-photocycloaddition of alkenes to aromatic rings (a [2 + 2] reaction) is less clear. Substituents on the aromatic ring play a significant role which implies that there is exciplex formation or precoordination of the addends. Singlet reactions have also been reported to involve biradical intermediates [36]. Nuss has studied the singlet photoreaction and has shown that excited state charge transfer is critical for [2 + 2] cycloaddition [35d]. Regioselectivity is apparently dictated by the polarity of the reactants and stereocontrol, of intramolecular systems at least, is a function of strain energy.

Triplet [2 + 2] photocycloadditions between arenes and alkenes have been reported (see below) and undoubtedly involve biradical intermediates. Wagner and Cheng have made significant contributions to the mechanistic understanding of this type of reaction [37]. Al-Qaradawi et al. recently reported the regioselective [2 + 2] photocycloaddition of ethyl vinyl ether to various cyanobenzenes and observed that the origin of substituent control "remains obscure" [38].

D. Alkene + Alkene

There are few experimental examples of singlet photoreaction between alkenes. Calculations have suggested the presence of exciplex [39] and/or biradical intermediates [40]. In general, regiocontrol and stereocontrol of the singlet alkene + alkene reaction have not been impressive.

There are a number of examples in the literature of triplet alkene excited states adding to alkenes in a [2 + 2] fashion. Complete stereoselectivity in this triplet reaction is not typical. The triplet biradical lifetime presumably allows for bond rotation prior to spin-flipping and cyclization. If this event is accurately understood, then the stereoselectivity should be significantly in favor of the least hindered cyclobutane product. If there is reversibility in the first bond forming step of this presumed stepwise reaction, then the regioselectivity can be driven by the prolonged lifetime of the more stable biradical. If there is no reversibility, then the regioselectivity may be explained by a Frontier Molecular Orbital argument based on a polarized excited state or perhaps a π-stacking phenomena which controls the preorientation.

Caldwell et al. [41] have recently published mechanistic details of the cycloaddition between an excited state alkene in the triplet manifold and a ground state alkene. A study of *p*-acylstyrene cycloaddition to styrene gave complete regioselectivity in cyclobutane formation. Only 1,2-diarylcyclobutane was observed upon irradiation at 355 nm. Similar results were reported from the triplet-sensitized irradiation of 1-phenylcyclohexene. Only head-to-head products 2 and 3 were observed (see Scheme 3).

The stereoselectivity is approximately 4 : 1 in favor of the trans isomer in the *p*-acylstyrene to styrene cycloaddition, and the sensitized cycloaddition of 1-phenylcyclohexene gave a 4 : 1 ratio of trans : cis-substituted cyclobutane.

The trans stereoselectivity may easily be rationalized as a result of sterics, although direct irradiation of the 1-phenylcyclohexene gives a similar ratio of trans : cis stereoselectivity. The singlet reaction presumably does not involve a biradical intermediate of significant lifetime. Thus, the trans selectivity may be enhanced by sterics in the case of the triplet reaction. The preference for trans isomer may be inherent to the approach of the alkenes or, perhaps , it is a function of selective reversibility of biradical intermediates (see Scheme 1). Caldwell has published a formula for predicting the potential for photocycloaddition of alkenes and arenes in the singlet excited state [42]. His analysis implicates an exciplex.

Another mechanism that has been useful for understanding the regioselectivity and stereoselectivity of alkene + alkene reactions is the electron

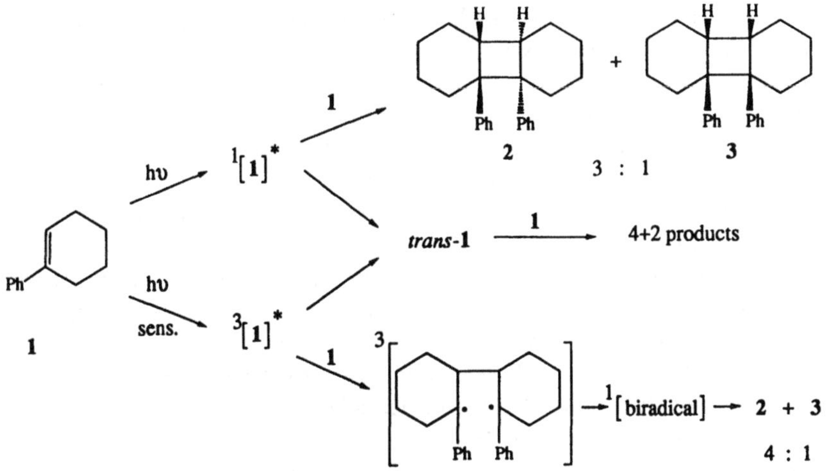

Scheme 3

transfer processes (see Scheme 4). (For reviews of PET reactions including [2 + 2], see Ref. 43.) This pathway is likely involved in photocycloadditions between electron-rich and electron-poor alkenes. However, Kim et al. recently reported that electron transfer is not occurring in the [2 + 2] reaction between the electron-rich p-methoxystyrene and the electron-poor tetracyanoethylene [44].

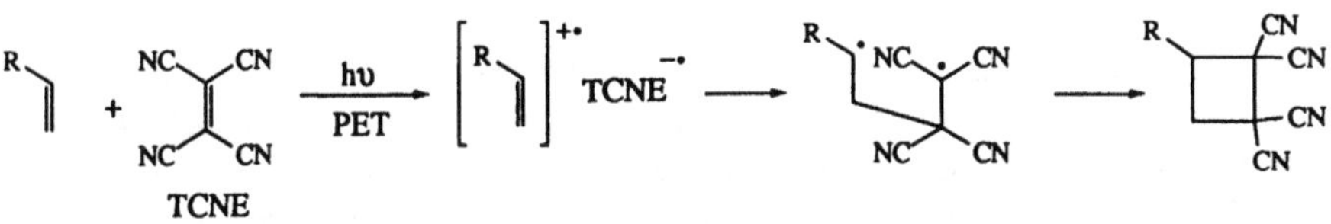

Scheme 4

It is obvious that further mechanistic studies are necessary to better understand [2 + 2] photocycloadditions. The following sections will describe useful examples of regiocontrolled and stereocontrolled photoreactions with an attempt to indicate the source of the observed selectivity.

III. REGIOSELECTIVITY OF PHOTOCYCLOADDITION

A. [2 + 2] Photocycloaddition of Enones to Alkenes

Mechanistic studies of [2 + 2] photocycloaddition of alkenes to α,β-unsaturated carbonyl compounds shows that alignment of reactants to form an exciplex plays an important role in the regiocontrol of cycloaddition. Orientation of the two reactants arises from the dipolar attraction between the ground state alk-ene and the triplet excited state enone or perhaps ground state orientation. Consequently, substituents on alkenes as well as enones become a major con-sideration for regioselectivity of reactions due to their effect on the charge distribution in reactants. Other factors influencing regioselectivity include the structure character of reagents and reaction conditions such as solvent, tempera-ture, and concentration. Several excellent reviews have been published on these related topics [45].

Table 1 [46] summarizes some important studies on [2 + 2] intermolecu-lar cycloaddition of enones to alkenes with different substituents. Groups at-tached to the alkene have been divided into electron-donating group (EDG) and electron-withdrawing group (EWG), with EDGs showing more control of the regioselectivity. Reactions of electron-rich alkenes provide head-to-tail (HT) cyclobutanes as major products. Generally, selectivity increases in the order of substituents on alkene as H < OAc < CH_3 < OMe = OEt = SEt. In addi-tion, two EDGs on one terminal of the alkene results in enhanced selectivity (Table 1, entry 1 versus 9 and entry 2 versus 7). In the case of electron-poor alkenes, regioselectivity of the reaction is reversed with head-to-head (HH) cyclobutane as the major product (entries 3 and 4). Entries 13–16 demonstrate two interesting studies. Alkenylboronate esters react with cyclopentenone to yield HH major products, which can then be converted to a wide variety of functional groups. Entries 14–15 show that the distance between the bromine atom and the double bond greatly influences the adduct distribution. This re-sult has been explained by heavy-atom effects as well as dipole–dipole inter-actions.

In the case of cycloalkenes, it is found that the size of the ring is an important factor in product distribution. Photo [2 + 2] cycloaddition of cyclohexenone derivatives (4) to carbomethoxy cyclobutene [47], cyclopentene [48], and cyclohexene [49] (see Scheme 5) demonstrates a gradual reversal of regioselectivity from head-to-head to head-to-tail adducts as indicated in Table 2 [50]. This result of head-to-tail products is not consistent with the dipole–dipole interaction theory. Stability of biradical intermediates is suggested to explain the reversed regioselectivity.

Cycloaddition with enone reactants has been studied using cyclopentenone or cyclohexenone derivatives in order to avoid side reactions resulting from cis-trans isomerization of the alkene. In the case of cyclopentenone or cyclo-hexenone, regioselectivity of cycloaddition is influenced by the substituents at the C-3 position as well as whether or not the 4-position is a heteroatom.

Photoaddition of cyclohexenones with moderate electron-donating groups such as methyl, phenyl, and acetoxy at the enone 3-position proceeds smoothly with the expected HH regioselectivity except for a few cases [24,51a]. 3-Me-thyl- (10) and 3-acetoxycyclohexenone (11) react with isobutene (12) or 1,1-diphenylethylene (13) to provide the opposite regioisomeric products as shown

Table 1 Examples of [2 + 2] Intermolecular Cycloaddition of Enones to Alkenes with Different Substituents

Entry	Alkene	Enone	HT : HH (%)	Ref.
1	OEt		80.2 : 19.8	58b
2	OAc		78 : 22	46a
3	CN		16 : 84	51a
4	ClF$_2$C CF$_2$Cl		0 : 100	46b
5			85 : 15	24
6	OMe		98.5 : 1.5	58b
7	OAc		82.1 : 17.9	58b
8	MeO OMe			58b
9	EtO SEt		100 : 0	58b
10	n-Bu OAc	Bu	100 : 0	8

(*continued*)

Table 1 Continued

Entry	Alkene	Enone	HT : HH (%)	Ref.
11	n-C₅H₁₁ / OAc	(cyclopentenone with Bu)	100 : 0	8
12	CO₂Et / EtO OEt	(cyclopentenone)	82.5 : 17.5	58b
13	H B(O—O) / Bu H	(cyclopentenone)	HH major 46c	
14	(CH₂)₄Br	(cyclopentenone with Bu)	84 : 16	46d
15	(CH₂)₃Br	(cyclopentenone with Bu)	61 : 39	46d
16	(CH₂)₂Br	(cyclopentenone with Bu)	29 : 71	46d

Table 2 Regioselectivity as a Function of Alkane Ring Size

Entry	Enone	Alkene	Adduct HH	Adduct HT	HH : HT
1	1a	Cyclobutene	5a	—	>95 : 5
2	1b	Cyclobutene	5b	—	> 95 : 5
3	1a	Cyclopentene	6a	7a	50 : 50
4	1b	Cyclopentene	6b	7b	60 : 40
5	1a	Cyclohexene	8a	9a	11 : 89
6	1b	Cyclohexene	8b	9b	< 5 : 95

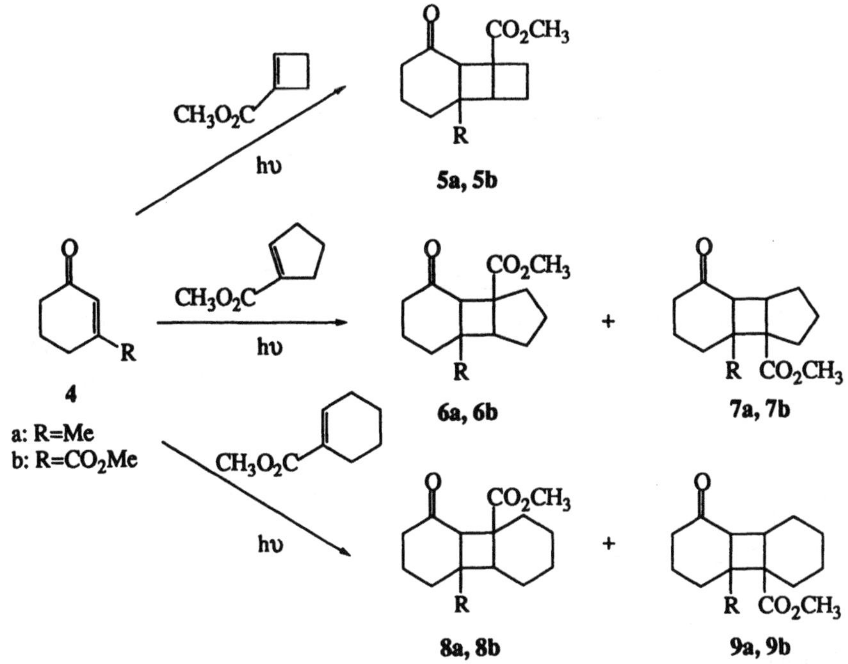

Scheme 5

in Scheme 6. The preference in the product distribution is rationalized by steric interference encountered by the two groups substituted on the alkene when approaching the substituted 3-position of the enone. This may outweigh the electronic effect and reverse the orientation of reactants. However, in the case of strong electron-donating groups on alkenes, head-to-tail adducts predominate regardless of sterics.

Cycloaddition studies have also been carried out on cyclohexenone with strong EDGs or EWGs at the 3-position [51b]. Reacitivity of this type of enone is greatly reduced (see Scheme 7). For example, the reaction of 15–18 is very sluggish or does not occur at all. 3-Cyanocyclohexenone 14 reacts similarly to the parent enone. The vinylogous imide 19, with the reduced electron-donating ability of the amide function, renders [2 + 2] adducts in good yield. Endocyclic oxa-, thia-, and aza-enones prove to be more reactive and provide better regioselectivity in the cycloaddition reaction. In some cases, these reactions can give high conversion in a regiospecific manner. A variety of oxa-enones (20–27) have been examined (see Scheme 8) for their photochemical behavior [52, 53].

Cycloadditions of oxa-enone 20, 21, and 22 to polarized alkenes proceed regiospecifically, whereas the nonpolarized isobutene yields only moderate regioselectivity with head-to-tail products dominating (Scheme 9) [52]. It has been suggested that the enhanced selectivity is due to larger charge polariza-

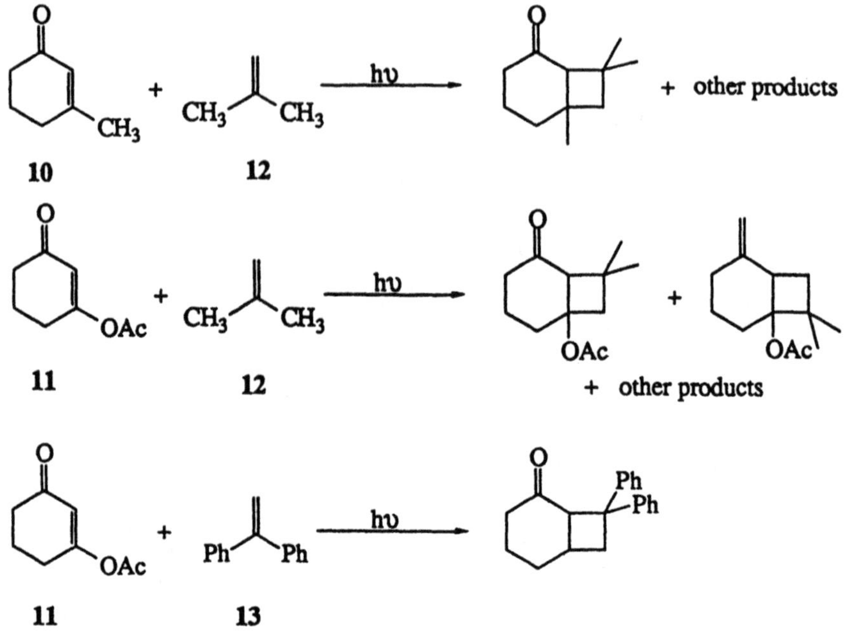

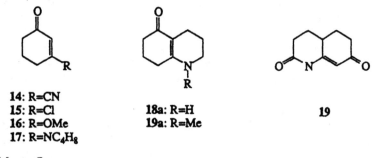

Scheme 6

tion in the C–C double bond of the oxa-enone excited state than for cyclo-hexenone. Substituents at C-3 position, however, substantially reduce the se-lectivity and in the case of 3-isopropyl oxa-enone 24, reversed regioselectivity is observed [52]. Comparison of endocyclic and exocyclic oxa-enones (26 ver-sus 27, and 28 versus 29) demonstrates a great difference in their reactivities, and smooth reaction with endo-enones is explained by the difference in the geometries of the corresponding enone–alkene exciplexes. Cycloadditions of 25 to moderately polarized alkenes, methyl acrylate, and vinyl acetate show less regioselectivity [54].

4-Thia-2-cyclopentenone 30 undergoes regiospecific cycloaddition with 2-methylpropene to provide the cyclobutane derivatives shown in Scheme 10 [55].

14: R=CN
15: R=Cl
16: R=OMe
17: R=NC₄H₈

18a: R=H
19a: R=Me

19

Scheme 7

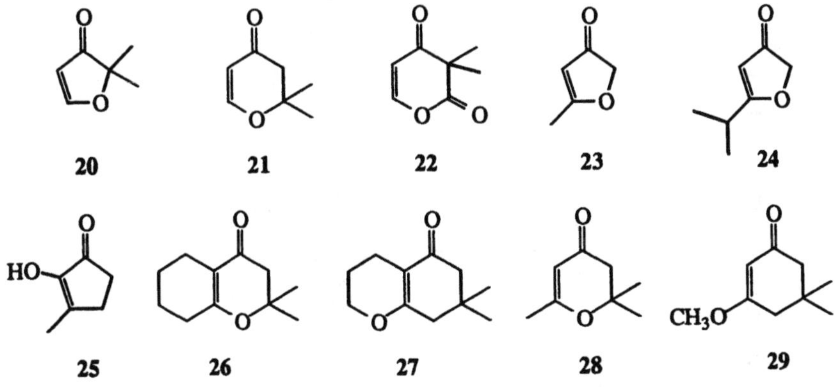

20 **21** **22** **23** **24**

25 **26** **27** **28** **29**

Scheme 8

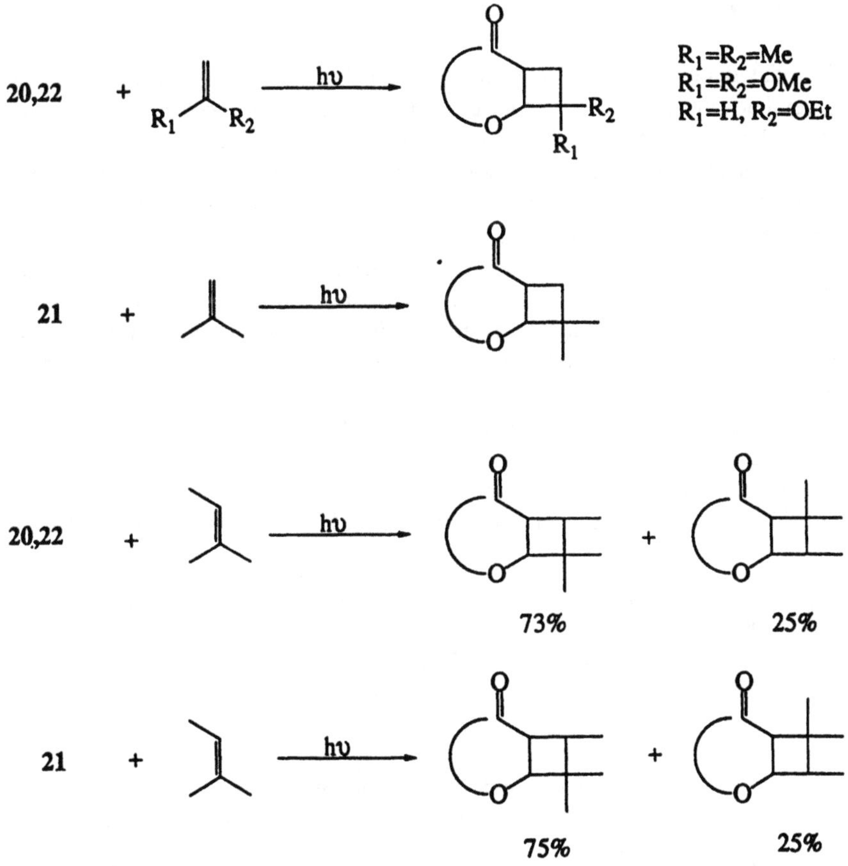

$R_1=R_2=Me$
$R_1=R_2=OMe$
$R_1=H, R_2=OEt$

20,22 + $R_1 \bigvee R_2$ $\xrightarrow{h\upsilon}$

21 + $\xrightarrow{h\upsilon}$

20,22 + $\xrightarrow{h\upsilon}$ 73% + 25%

21 + $\xrightarrow{h\upsilon}$ 75% + 25%

Scheme 9

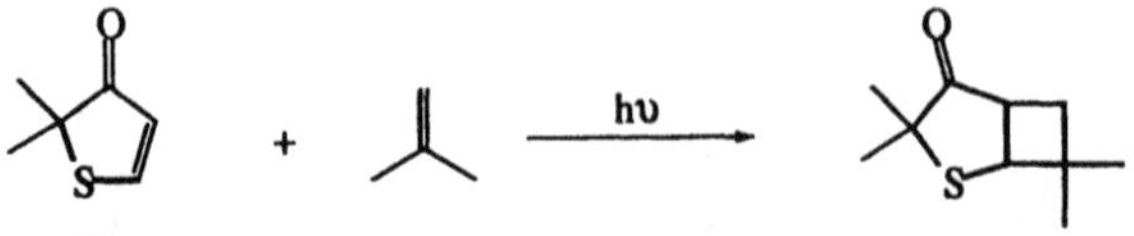

30

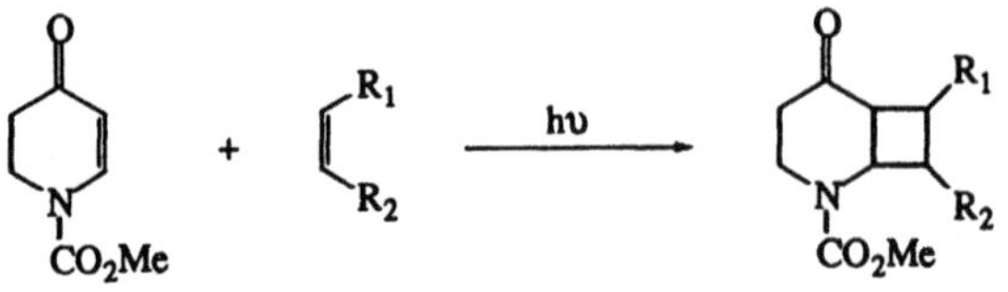

31   **32**

a: R_1=H, R_2=CO$_2$Me
b: R_1=H, R_2=O(CH$_2$)$_3$Me
c: R_1=H, R_2=CH$_2$Ph

Scheme 10

The vinylogous aza-enone, N-methoxycarbonyl-5,6-dihydro-4-pyridone **31**, has several advantages. It reacts with electron-rich as well as electron-poor alkenes to render cycloadducts in high yield and with extremely high regioselectivity [56]. It should be noted that photoaddition of the analogous cyclohexenone derivatives show moderate or no regioselectivity toward the electron-rich or electron-poor alkenes **32a** and **32c**.

Reaction conditions such as solvent, reaction temperature, and concentration of reagents may influence regioselectivity of the [2 + 2] enone-alkene cycloaddition. Due to the role of dipole–dipole interaction in determining the regioselectivity, polarity of solvents, which may influence the overall dipoles of reagents, logically effects the regioselective control of the reaction. Results from the cycloaddition of enone **33** and alkene **34** illustrate this solvent effect (see Scheme 11) [57]. The product ratio of **35** : **36** varies from 98 : 2 in dilute nonpolar hydrocarbon solvents to 45 : 55 in a polar solvent such as methanol. The results suggest that appropriate use of solvent may allow control of

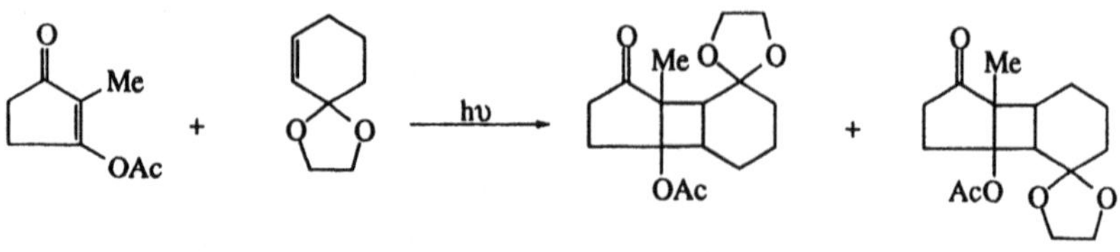

33		**34**		**35**		**36**
solvent	iso-octane	cyclohexane	diethyl ether	ethyl acetate	acetonitrile	methanol
35:36	98:2	97:3	85:15	71:29	48:52	45:55

Scheme 11

the reaction regioselectivity. Report on the use of potassium dodecanoate micelles (KDC) as a solvent serves as an example (see Scheme 12). The cycloaddition between enone 37 and alkene 38 proceeds with a totally reversed regioselectivity in KDC compared to that in organic solvents [8]. Some studies report that lowering the reaction temperature results in enhanced regioselectivity, even though a detailed explanation is not presented [51,58]. Presumably, it is a result of improved dipole–dipole interaction allowed at lower temperatures.

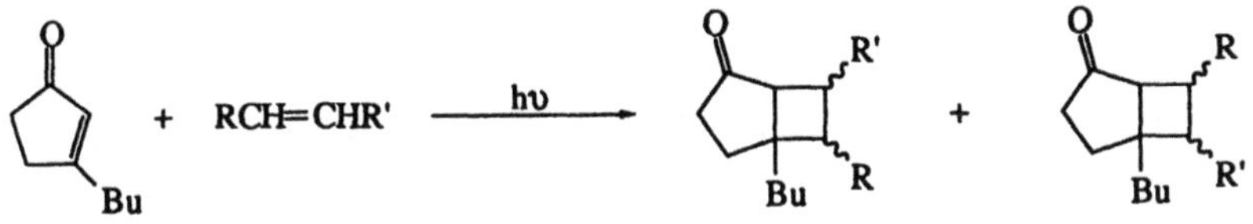

a: R=n-Bu, R'=OAc
b: R=n-C$_5$H$_{11}$, R'=OAc

entry	olefin	medium	39 (%)	40 (%)
1	38a	KDC	70	30
		Methanol	0	100
		Diethyl ether	0	100
2	38b	KDC	70	30
		Methanol	0	100
		Diethyl ether	0	100

Scheme 12

The regioselectivity of intramolecular [2 + 2] photocycloaddition is quite high, usually as a result of geometrical constraints within the reacting molecules. In general, a five-membered ring is preferentially formed in the initial bond forming step. In the case that a five-membered ring cannot be formed in the ring closure step, then a six-membered ring is favored. This trend is termed the "Rule of Five" by Srinivasan and Hammond [59], and is illustrated in Scheme 13. Two modes of addition to form the five-membered ring can be observed. They are described as cross addition and straight addition, respectively.

Numerous studies have been conducted on the intramolecular photoaddition of cyclic enones with tethered alkenes connected at different positions of the cyclic systems. Scheme 14a summarizes the different enone systems that have been studied. Compounds 37, 38, and 39 are examples of cyclohexenone derivatives with alkene chains tethered at C-2, C-3, and C-4 positions, respec-

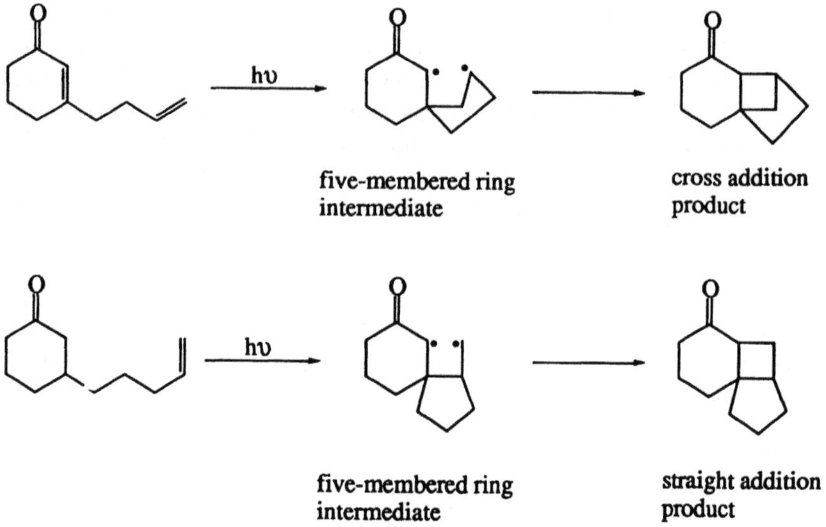
hυ
five-membered ring intermediate
cross addition product
hυ
five-membered ring intermediate
straight addition product

Scheme 13

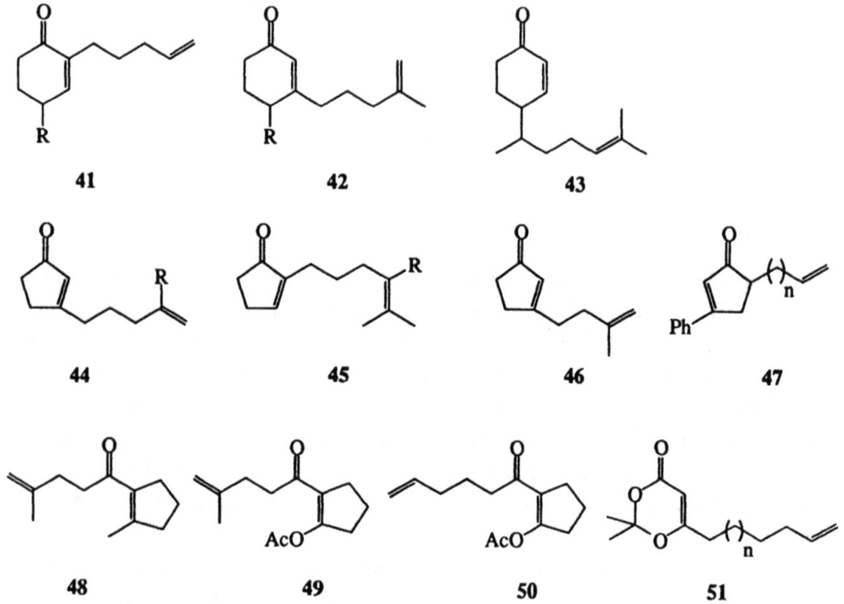
41
42
43
44
45
46
47
48
49
50
51

Scheme 14a

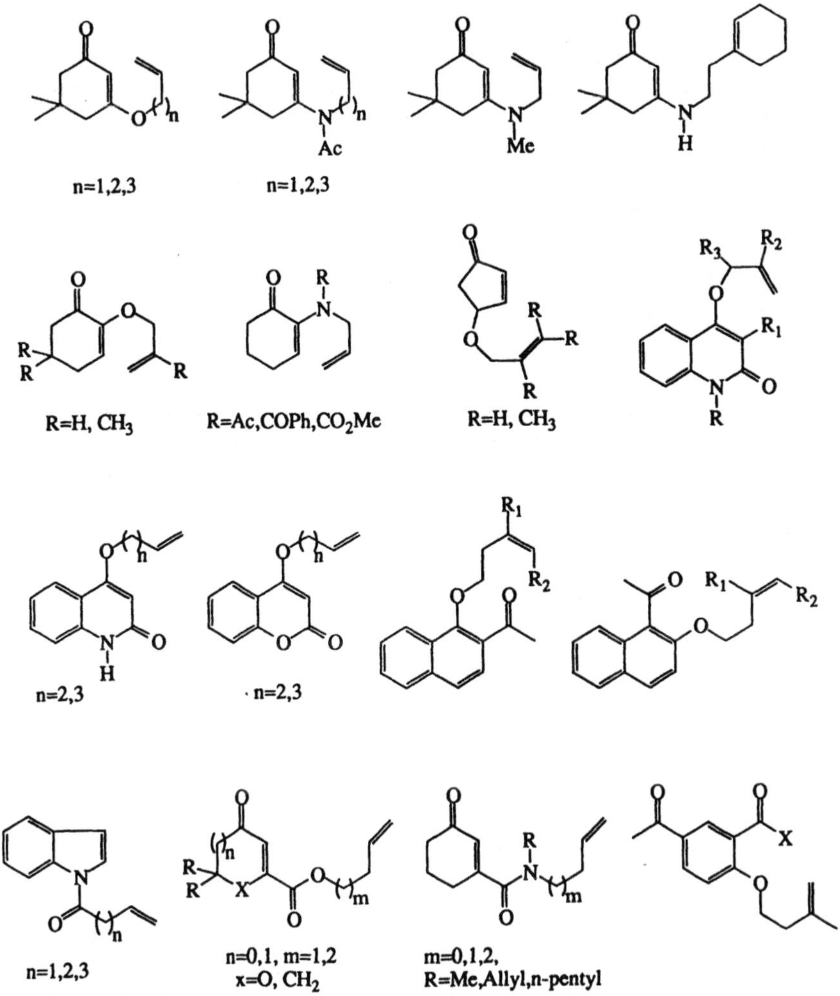

Scheme 14b

tively [60]. Compounds 44–47 are the analogous studies for substituted cyclo-pentenones [61]. Compounds 48–51 serve as examples for variant parent enones such as acyclic dienones and dioxolenones [61a,62]. In all cases except 45, the Rule of Five is preserved. Most of these examples give exclusive adducts in agreement with this rule. Irradiations of 42 and 50 produce exclusively the straight cycloadducts with six-membered rings, because no five-membered ring can be formed in the ring closure step. Photocycloaddition of 51 proceeds in a manner to give exclusive straight addition, which has been applied as a method to synthesize compounds with medium-sized rings [62c]. One phenomenon observed is that an alkyl group on the internal carbon of tethered alkenes on cyclopentenone (e.g., 46 and 48) reduces the regioselectivity of cycloaddition. In the extreme case of 45, photocycloaddition gives a straight addition product exclusively, whereas according to the Rule of Five, it should provide a cross adduct. This regiochemical reversal has been rationalized as a steric effect from the alkyl group precluding initial formation of the five-member ring.

Another approach to the intramolecular photocycloaddition is to use vinylogous esters and amides so that the tether connecting the enone and alkene functionalities includes an oxygen atom or a nitrogen atom. Many different systems have been examined, and the photoadducts, in most cases, are found to be consistent with the Rule of Five. Scheme 14b lists some examples, which range from normal vinylogous esters and amides [63] to some special cases such as alkenyloxyquinolones [64], alkenyloxybenzopyranones [65], enoxyacetonaphthones [66], N-alkenoylindoles [67], oxoesters or oxoamides [68], and enoxyacetobenzenes [69]. Scheme 15 illustrates one exception in which regioselectivity is reversed. The result is explained by the extra stability of the six-membered ring biradical intermediate 55.

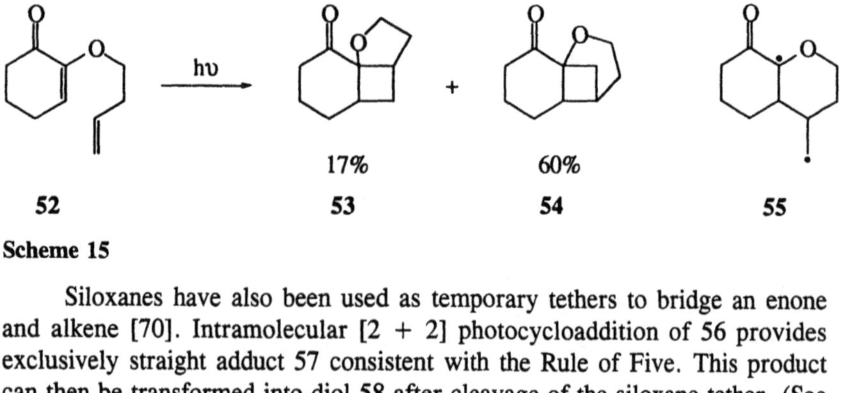

| 52 | 53 | 54 | 55 |
| | 17% | 60% | |

Scheme 15

Siloxanes have also been used as temporary tethers to bridge an enone and alkene [70]. Intramolecular [2 + 2] photocycloaddition of 56 provides exclusively straight adduct 57 consistent with the Rule of Five. This product can then be transformed into diol 58 after cleavage of the siloxane tether. (See Scheme 16.)

Investigation of [2 + 2] photocycloaddition of enones to monosubstituted alkynes indicates that the reaction proceeds with an opposite regioselectivity compared to that of alkenes [71]. Generally, head-to-head adducts are major products except when the group substituted on the alkyne is $-CO_2R$ or $-OR$ (see Scheme 17). Thus, methyl propynoate (60, R = CO_2CH_3) photoadds to cyclopentenone 59 to give 1 : 1 HH and HT products. Cycloaddition of cyclohexenone to 1-hexyne has also been studied and the HH : HT ratio varies from 2 : 1 to 7 : 1 as a function of substituents at the enone 3-position. It

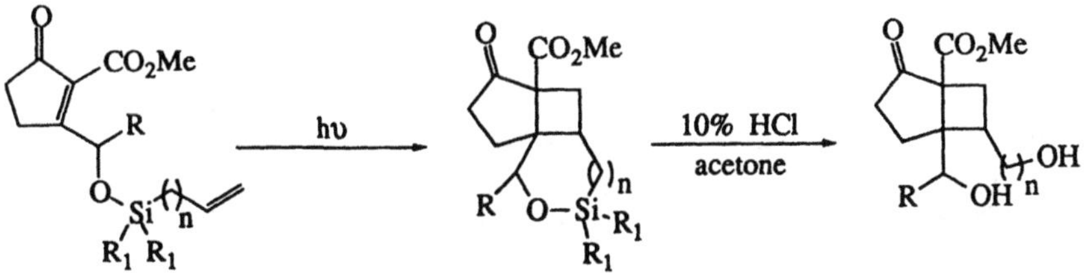

R_1=Ph, CH_3. n=0,1. R=CH_3, n-C_5H_{11}, C_6H_5.

56   **57**   **58**

Scheme 16

has been suggested that the opposite regioselectivity (HH major) for monosub-stituted alkynes is due to the dipole–dipole interactions between the triplet enone and ground state alkyne in addition to the weaker polarization of the π-bond in these alkynes [72].

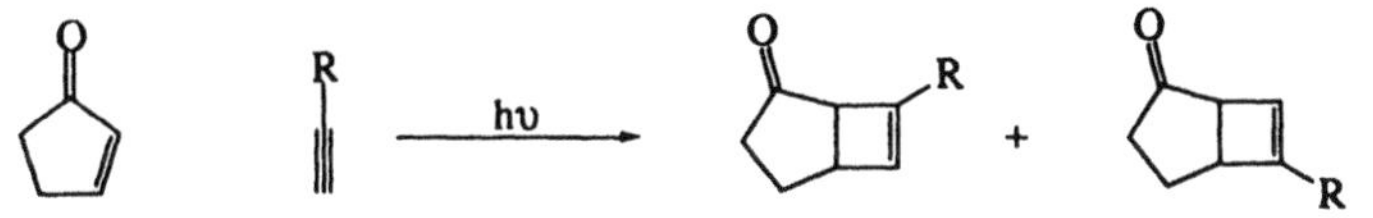

59   **60**   **HH**   **HT**

R =	n-C_5H_{11}	n-C_4H_9	n-C_3H_7	t-Bu	CH_2Cl	CH_2OH	CO_2CH_3	OC_2H_5
HH:HT	70:30	69:31	61:39	87:13	72:28	69:31	48.5:51.5	24:32

Scheme 17

Intramolecular [2 + 2] photocycloaddition of enones to alkynes has been utilized as a synthetic procedure [73]. The tether chain connecting the alkynes and enones at the C-3 position involves all carbon (61 or 62) or ether (63) link-ages. Straight adducts are obtained in all cases (consistent with the Rule of Five). Scheme 18 illustrates these photoreactants.

Photocycloaddition of enones 59, 64, 65, and 66 to allene produces head-to-head adducts as the major products (Scheme 19) [24,74]. Substituents on the allene make the reaction less regioselective, resulting in multiple regioisomeric products [74b,75]. Utilizing a hydrocarbon tether to transform the reaction into an intramolecular reaction improves the regioselectivity greatly. Cyclic enones with three-atom tethers at the 3- or 4-position (67 or 68) give one regioisomeric

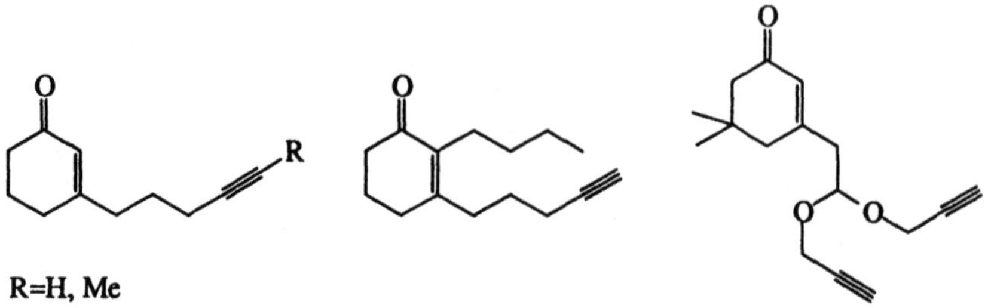

R=H, Me

61   62   63

Scheme 18

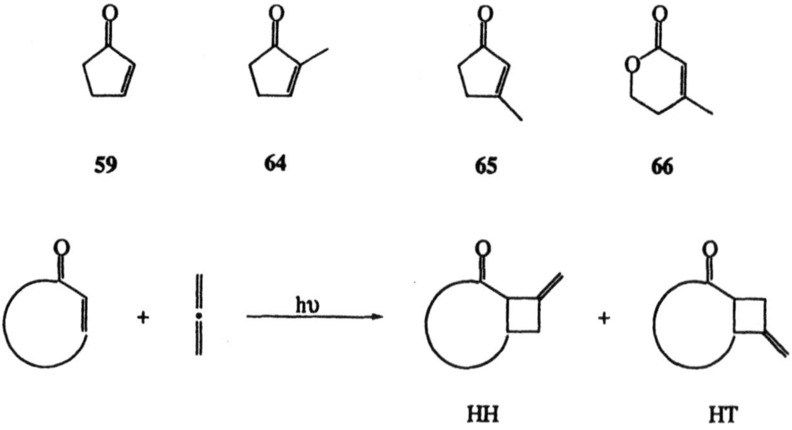

59   64   65   66

HH   HT

	59	64	65	66
HH:HT	90:10	92:8	80:20	60:40

Scheme 19

product exclusively, the straight head-to-head adduct as shown in Scheme 20 [73e,76]. The ratio of major product to other regioisomers can be influenced greatly by the ring size of the enone, the length of tether chain connecting enone and allene, the substituents on the tether chain, and the reaction temperature [77]. One interesting phenomenon that has been observed is intramolecular photocycloaddition of tethered ketones 69 and 70 (Scheme 20) [78]. The result of reversed regioselectivity has been rationalized by the opposite ground state charge distribution of ketones compared to allenes.

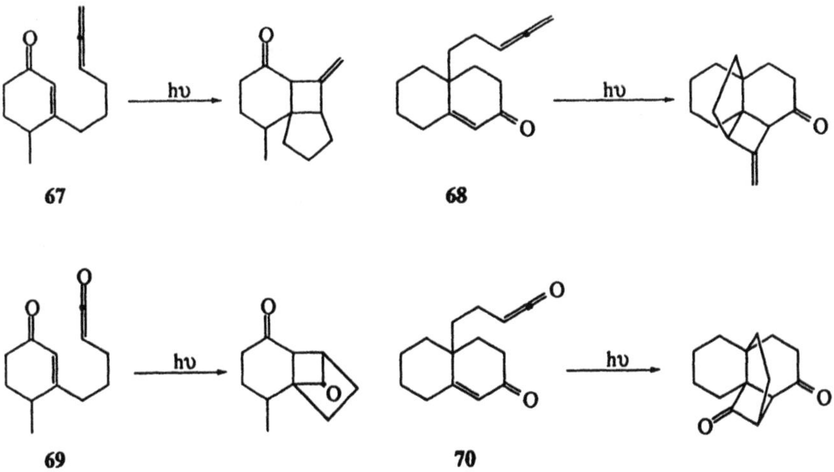

67   **68**

69   **70**

Scheme 20

B. Paternò–Büchi Regioselectivity

The Paternò–Büchi reaction is a powerful synthetic tool because it can be applied to a wide variety of alkenes and carbonyl compounds. However, the application of the reaction is greatly limited because the normal regioselectivity does not always satisfy the synthetic requirement. In general, the regioselectivity of the Paternò–Büchi reaction is controlled by the substituents on the alkenes and the type of carbonyl compound involved. Like the [2 + 2] photocycloaddition, studies on control of regioselectivity by electron-donating groups on alkenes outnumber those of alkenes with electron-withdrawing groups. A few examples that have been studied, such as reactions with 1,1-dimethylethene, 1,1-diphenylethene and α,β-unsaturated nitriles, yield good to excellent regioselectivity [79]. One interesting study is the investigation of the photochemical reaction of various benzophenones (71a–71c) with allylic silane 72 [80]. The reaction yields oxetanes with moderate yield but high regioselectivity (Scheme 21). The result of selectivity is explained by the β-silyl $\sigma_{Si-C}-P_\pi$ hyperconjugation effect stabilizing the biradical intermediate.

The addition of enol ethers to carbonyl compounds has been studied extensively with the expectation that the strong electron-donating group might enhance the possibility of regioselective oxetane formation. An early survey of this type of reaction suggests that satisfactory regioselectivity cannot be reached [81]. However, photoadditions of silyl enol ether 75 to benzophenone [82] and benzaldehyde [83] prove to be highly regioselective (Scheme 22). 3-Trimethylsilyloxy oxetanes 76 and 78 are favored by ~ 20 : 1 (Scheme 22). The model of the biradical intermediate has been cited to explain the observed regio-

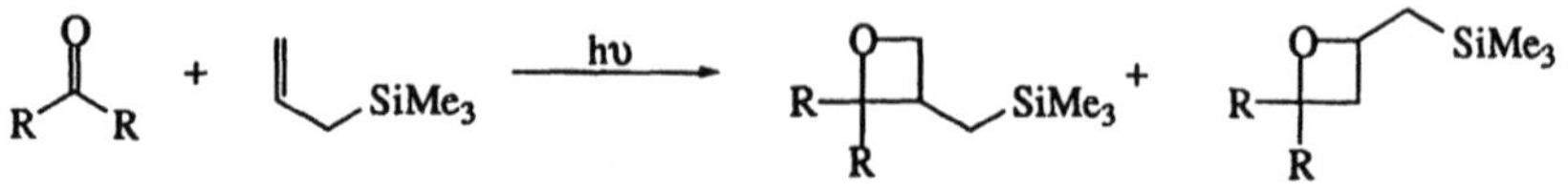

71 72 73 74

entry	benzophenone	yield of 73+74	73 : 74
71a	R=Ph	41	24:1
71b	R=p-ClC_6H_4	38	32:1
71c	R=p-$CH_3C_6H_4$	36	24:1

Scheme 21

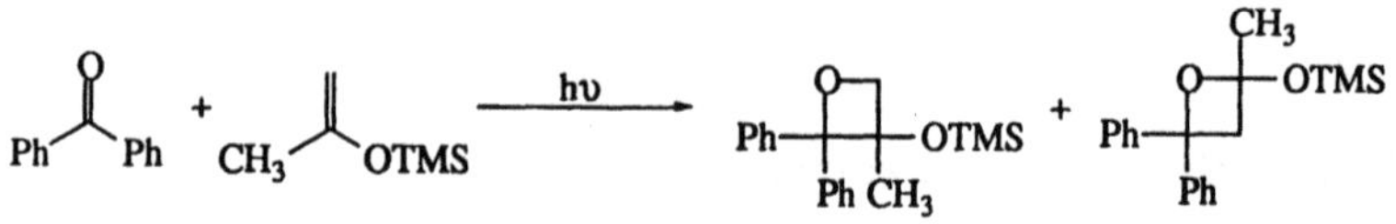

71a 75a 76 77

94 : 6

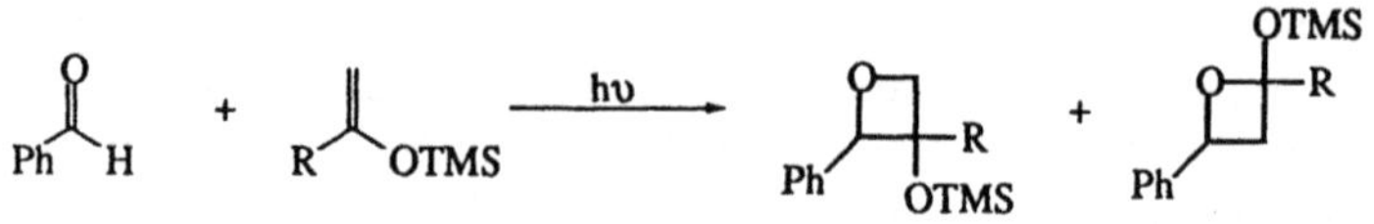

75a-h 78a-h 79a-h

entry	a	b	c	d	e	f	g	h
R =	Me	Et	iPr	t-Bu	Ph	$CH(OMe)_2$	$MeC(OCH_2)_2$	$HC(OCH2)_2$
78:79	90:10	95:5	95:5	95:5	95:5	95:5	95:5	95:5

Scheme 22

selectivity. Another approach to improve the regioselectivity has been to use alkenyl sulfides [84]. Thus, photolysis of benzophenone in the presence of alkenyl methyl sulfides 80a–80f provides 3-methylthio oxetane 81 as the major product with regioselectivity that varies from 9 : 1 to exclusive formation of the 3-thio oxetanes, depending on the conditions (Scheme 23). The stabilizing

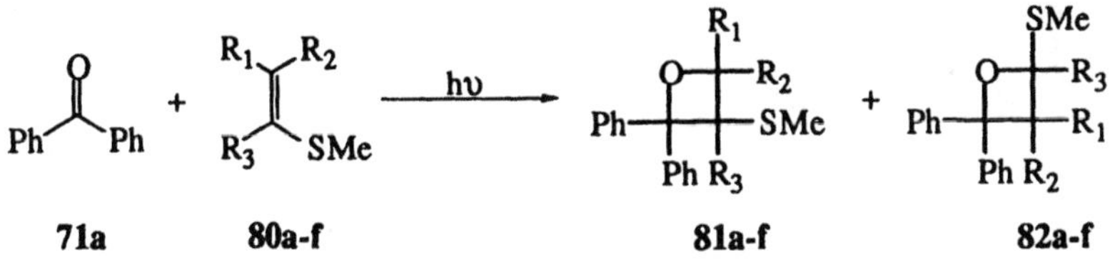

entry	R_1	R_2	R_3	81:82
a	Pr	H	H	13:1
b	i-Pr	H	H	100:0
c	t-Bu	H	H	9:1
d	PhCHMe	H	H	100:0
e	i-Pr	H	Me	100:0
f	THPO(CH$_2$)$_3$	H	H	100:0

Scheme 23

effect of the sulfur group on the putative biradical intermediate has been used to explain the high regioselectivity.

The excellent regiocontrol of photocycloaddition to furan is well known [1a,85]. Studies on a wide variety of aliphatic and aromatic aldehydes, ketones, and esters have shown that they add to furan with high regioselectivity [86]. In all cases, 2-alkoxy oxetanes are the major or only oxetane products. However, one limitation of this type of reaction is the lack of selectivity in the addition of aldehydes to unsymmetrically substituted furans. Addition of aldehydes 83 to silyl and stannyl-substituted furans 84 has been investigated, and in some cases, high regioselectivity has been attained (Table 3, entries 1–3) [87]. Recently, addition of 2-acetylfurans to aromatic aldehydes substituted with EWGs has been reported to proceed in a highly regioselective fashion (Table

Table 3 Examples of Regioselectivity in Photocycloadditions between Substituted Furans and Aldehydes

Entry	R_1	R_2	85 : 86
1	Si(i-Pr)$_3$	Et	4.0 : 1
2	Si(i-Pr)$_3$	Ph	>20 : 1
3	SnBu$_3$	CO$_2$Bu	>20 : 1
4	Ac	4-CNC$_6$H$_4$	1 : >20
5	Ac	4-MeO$_2$CC$_6$H$_4$	1 : >20
6	Ac	4-CHOC$_6$H$_4$	1 : >20

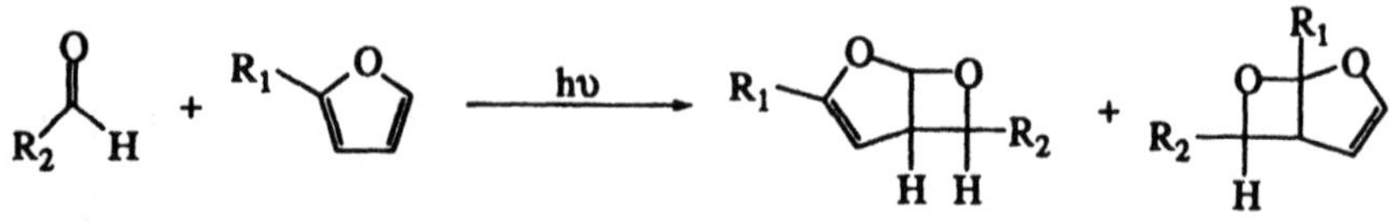

83   **84**   **85**   **86**

Scheme 24

3, entries 4–6) [88]. The favored regioisomers are oxetanes 86. (See Scheme 24.) Alternatively, the preference for 3-alkoxy oxetanes (88 and 89) has been found in the addition of acetone [89] and aromatic aldehyde [90] to 2,3-dihydrofurans (87a and 87b) as shown in Scheme 25. The reversed regioselectivity of addition to furan compared to 2,3-dihydrofuran has been explained by the relative stability of the biradical intermediates.

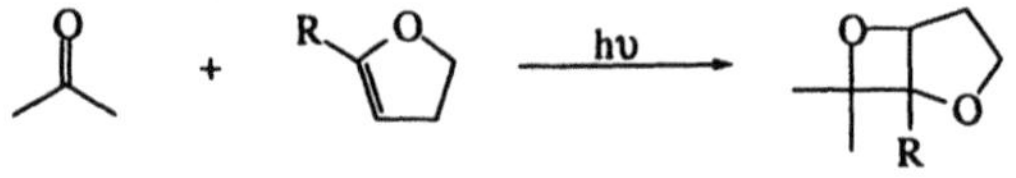

87a R=H
87b R=Me

88

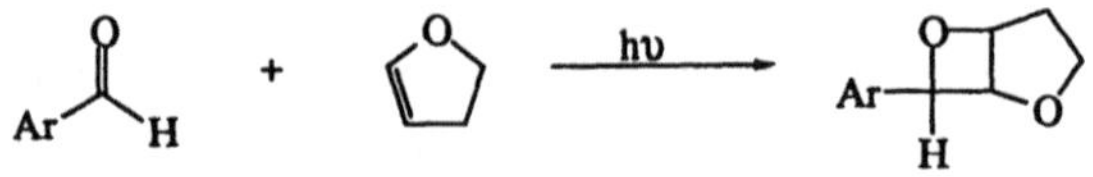

89a   **89**

Scheme 25

A wide variety of other carbonyl compounds can give Paternò–Büchi products when irradiated with alkenes. Studies have been reported on regioselectivity of photocycloadditions of carbonyl compounds such as thiocarbonyls [91], benzoquinones [92], diacetyls [93], and uracils [94]. A large number of intramolecular reactions have been reported, where the restraint of the reactant structure helps to give good regioselectivity.

C. Photocycloaddition of Arenes and Alkenes to Alkenes

Arene substituents are very important to the regiocontrol in the photocycloaddition of arenes to alkenes. Whereas electron-donating groups on the arene direct alkenes to meta-cycloaddition [95], olefins add to benzonitrile at the 1,2-positions [96]. Regioselective addition has been reported in some circumstances [97]. Intramolecular [2 + 2] cycloaddition of p-butenoxyacetophenones proceeds with high regioselectivity. In the case of substituents ortho to the tethered alkene, strong electron-donating and electron-withdrawing groups provide opposite regioselectivities (see Scheme 26) [98]. A methyl group on the double bond can also totally reverse the selectivity. The same high regioselectivity has been observed in the study of acetophenones with substituents meta to the tethered butenoxy group [99]. The strong inductive effect of substituents on the triplet state cycloaddition has been suggested to explain the high selectivity. A recent study on acid-catalyzed intramolecular photoadditions of 3-alkenyloxyphenols shows a similar reaction [100].

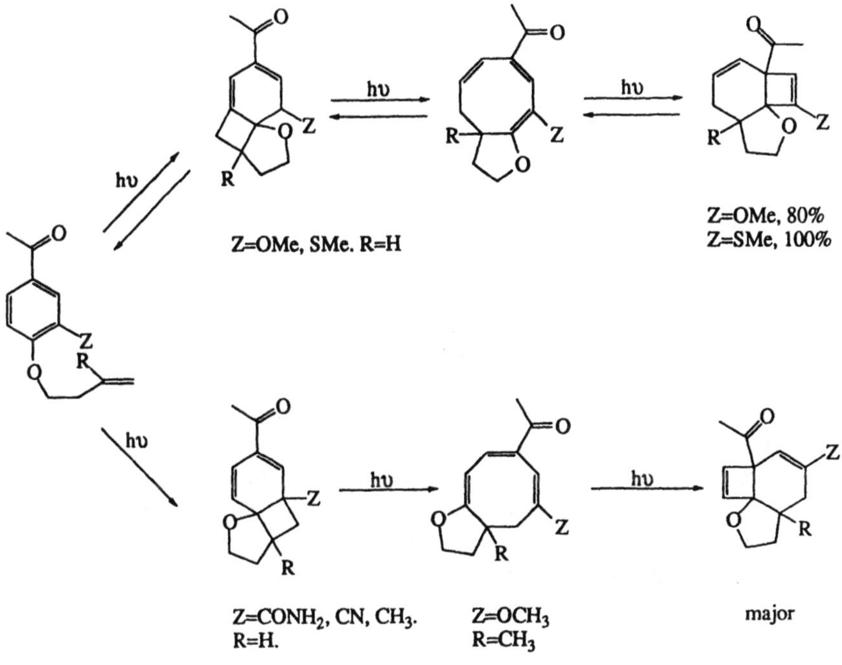

Z=OMe, SMe. R=H

Z=OMe, 80%
Z=SMe, 100%

Z=CONH₂, CN, CH₃.
R=H.

Z=OCH₃
R=CH₃

major

Scheme 26

In general, intermolecular [2 + 2] photocycloaddition of simple alkenes does not show satisfactory regioselective control. An approach to improve the reaction selectivity is use of a tether to constrain the reactive alkene functional groups proximate to each other and therefore allow the photocycloaddition to proceed. Tethers which have been employed include sugar alcohols such as D-mannitol and L-erythritol [101], diazacrown ethers [102], cyclophane [103], and silanes [104]. In all cases, the tether preorganized alkenes yield head-to-head adducts. The silane-tethered method can bind different alkene groups as well as alkynes, as shown in Scheme 27 [105]. Photoreactions of polyacetylenes and alkenes have also been investigated [106].

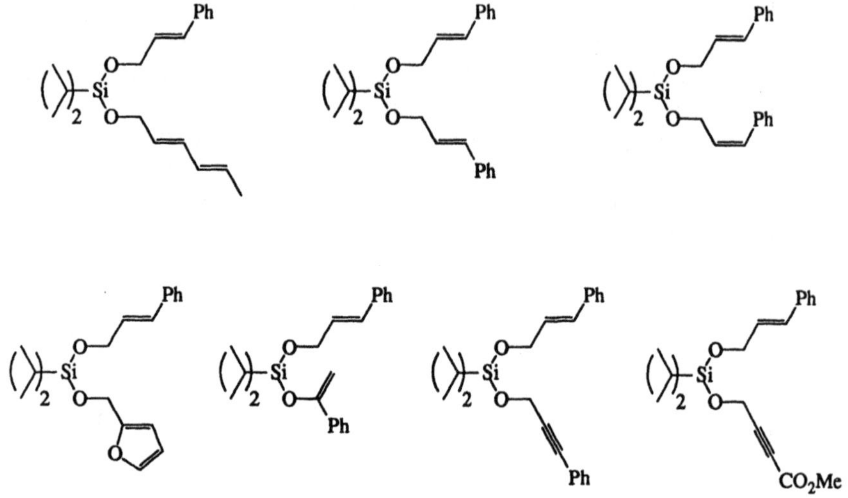

Scheme 27

D. Summary of Regioselectivity

Excellent regioselectivity has been achieved in many [2 + 2] photocycloaddition reactions. Selectivity can be controlled mainly by the groups substituted on the reactants. The approach to tether one reactant to the other shows some advantage. Photocycloaddition of enones to olefins is the reaction most studied and, therefore, more control of regioselectivity is available. Examples of regioselective addition of alkenes and arenes to olefins are limited and the reaction usually requires appropriate tethers to achieve the best regiocontrol. Paternò–Büchi reaction of some reactants yields regiospecific products. Further effort to study the regiocontrol in these photoreactions is still needed.

IV. STEREOSELECTIVITY

A. Stereoselective [2 + 2] Photocycloaddition of Alkenes to Enones

Stereocontrol of organic reactions has become an area of intense research in recent years, and the photochemical [2 + 2] reaction has not escaped inspection. The major emphasis in this area of research has been the reaction between enones and alkenes. Four main causes or reasons for stereoselectivity in these reactions come to view upon inspection of the recent literature: steric effects in the approaches of the reactants, thermodynamic and steric stability of the products, stereoelectronic efforts, and preorganization or tethering of the two reacting species.

Various groups have studied the photocyclizations of alkenes with 1,3-dioxinones. For example, Baldwin et al. [107] have found that substituted cycloalkenes add to the least hindered face of dioxinone 90 with excellent regioselectivity and moderate stereoselectivity, as in Scheme 28.

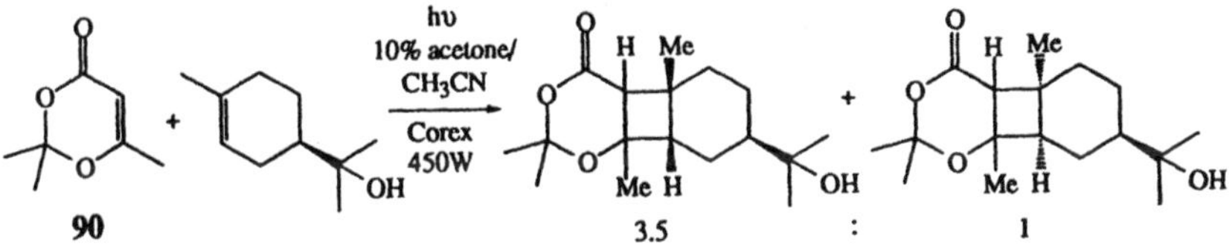

Scheme 28

Lange et al. [108], on the other hand, found that similar dioxinone 91 added to cycloalkenes in up to a 10 : 1 ratio of diastereomers. Interestingly, the addition in this case occurred predominantly from the face opposite that for ground state reactions, such as addition of cuprate reagents. Cuprates were found to add exclusively to the top face (opposite the *t*-butyl group), whereas the photocyclizations occurred mainly on the same side as the *t*-butyl group. One tentative explanation for this comes from Seebach et al., who suggest inverse pyramidalization of the dioxinone in the triplet excited state [109]. When dioxinone 92 is irradiated with cyclohexene (Scheme 29), there is no stereoselectivity. This suggests that the C-2 methyl group plays a large role in the selectivity of this reaction.

An interesting study by Little et al. [110] also shows the large impact of sterics on selectivity by showing a marked difference in reactivity in the [2 + 2] photocycloaddition between ethylene and the *trans*- and *cis*-bicyclononanone isomers 93 and 94. The irradiation of 93 in the presence of ethylene led to a 60% yield (42% isolated) of 95, whereas the cis isomer 94, when irradiated under the same conditions, did not react (see Scheme 30). The difference in reactivity comes from the large steric bulk over the reactive enone moiety in the cis isomer as opposed to the relative openness to attack of the enone moiety in the trans.

In a few cases, selectivity is a result of both steric and electronic factors. For example, Alibes and his co-workers [111] have studied the reactions of alkenes with chiral butenolides 96. They found that, in all cases, the predominant approach of the alkenes to the excited butenolide is from the side opposite the C-4 substituent to form the anti product (Scheme 31). When the alkene was tetramethylethylene, the size of the R group did not drastically affect the stereoselectivity, suggesting that an electronic effect was also taking part in the stereoselectivity of the reaction.

One possible explanation of this effect is the overlap of the nonbonding orbitals on oxygen with the *p*-orbitals of the carbonyl and α-carbons in the

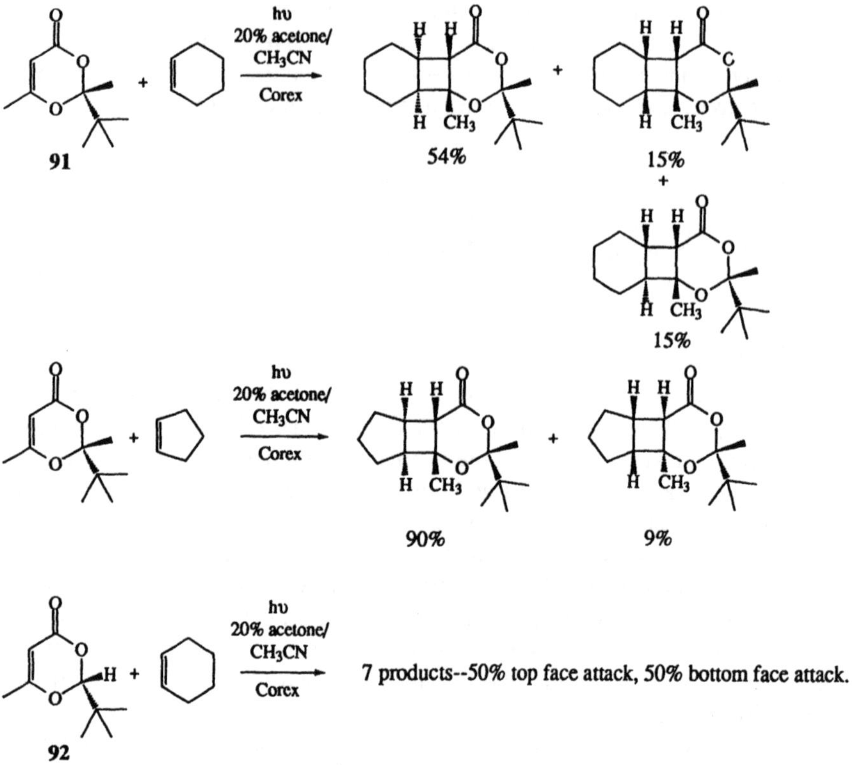

Scheme 29

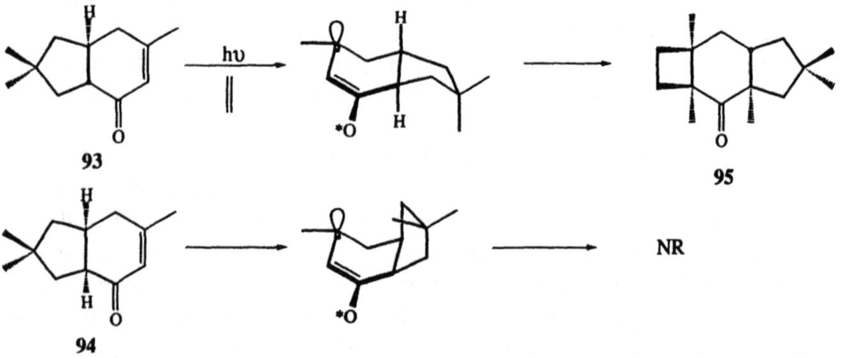

Scheme 30

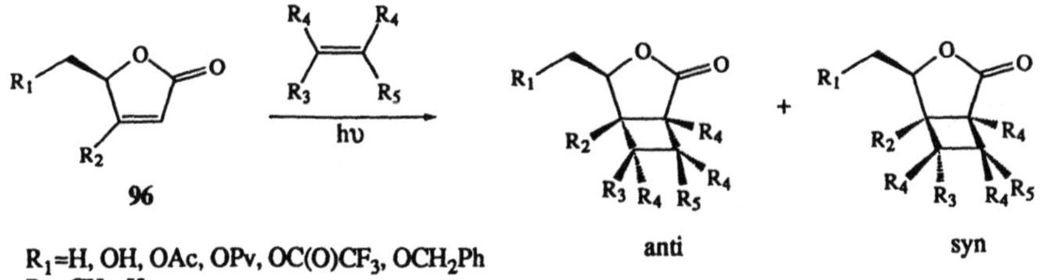

Scheme 31

R_1=H, OH, OAc, OPv, OC(O)CF$_3$, OCH$_2$Ph
R_2=CH$_3$, H
alkenes=ethylene, TME, vinylene carbonate

enone system (see Fig. 1). This effectively blocks the top face of the enone, regardless of what group is attached to the oxygen. Selectivity was not drastically different in the absence of oxygen (e.g., when the R group is methyl), suggesting that even a methyl group is large enough to block the top face from attack if the alkene is bulky. This is supported by the low selectivity observed when R = Me and the alkene used was ethylene.

The approach of orbitals prior to bonding is also an important factor in stereoselectivity of [2 + 2] photocycloadditions. For example, the alkene-tethered cyclopentenones 97 and 98 both gave 99 and 100 in a 1 : 2 ratio in low-conversion experiments (Scheme 32) [112].

The rationalization for this result is that both species involve a common biradical intermediate, and the more stable product is formed in greater abundance. In a similar experiment, 101 and 102, having a *t*-butyl-substituted olefin in the Z and E configurations, respectively, led exclusively to the adduct 104 (see Scheme 33). A similar biradical intermediate 103 was proposed, with the rationalization that the *t*-butyl group of this biradical orients itself in the least

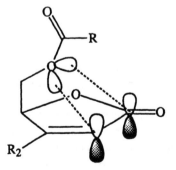

Figure 1 Transition state of alkene addition to chiral butenolides.

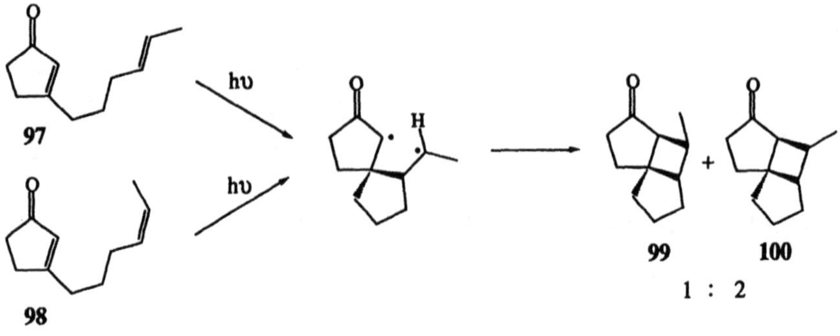

97

98

Scheme 32

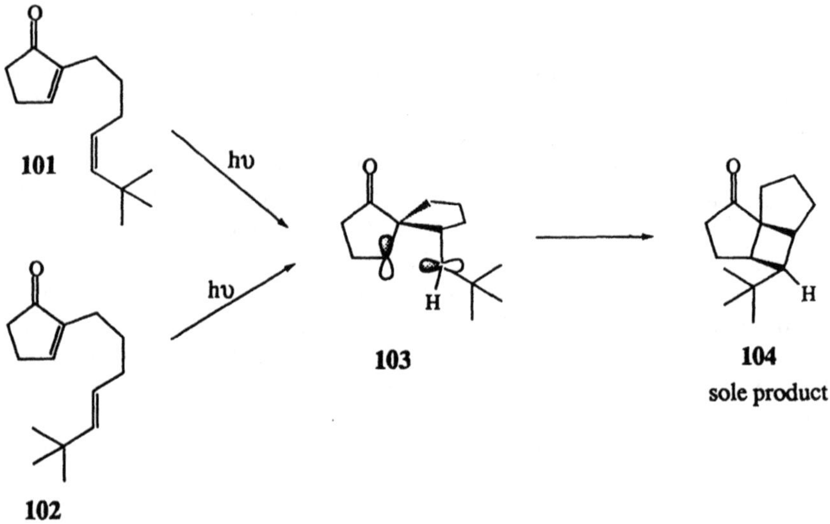

99   100

1 : 2

101

102

103

104
sole product

Scheme 33

sterically crowded position that allows the two radical centers to be perpendicular to each other for effective triplet to singlet spin inversion prior to final bond
formation.

An example of product-controlled [2 + 2] is in Schultz et al.'s work with
[2 + 2] photocycloaddition of dienones [113]. In this work, cyclohexadienone
105 was irradiated at –78°C in pentane to yield a single product 106, which
was found to have the isopropyl group in the more stable exo orientation as
shown in Scheme 34. An extension of this study showed that the isopropyl
group governed the regioselectivity as well as the stereoselectivity of the reac-

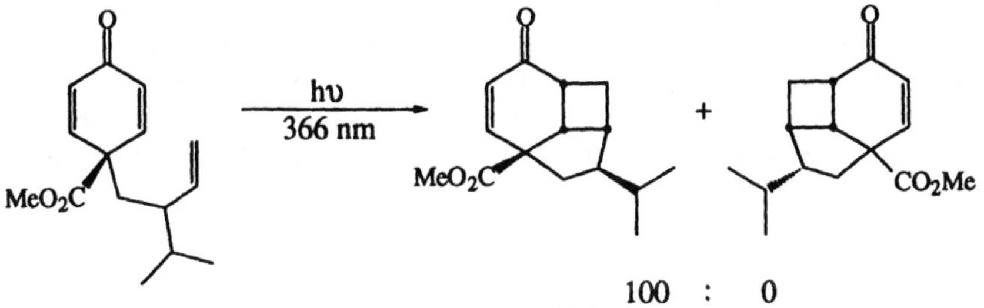

105 **106** 100 : 0

Scheme 34

tion. A mixture of dienones 107 and 108 in a 1 : 1.6 ratio was irradiated to give products 109 and 110 in an essentially unchanged 1 : 1.5 ratio, with no isomer 111 observed (see Scheme 35). These products were found to have the isopropyl group exclusively in the exo configuration, which is thermodynamically much more stable than the endo. Schultz et al. suggests that this reaction proceeds through a reversible 1,4-biradical intermediate, and that product development control may generally be operative in photocyclizations of the 4-(3'-butenyl)-2,5-cyclohexadien-1-ones.

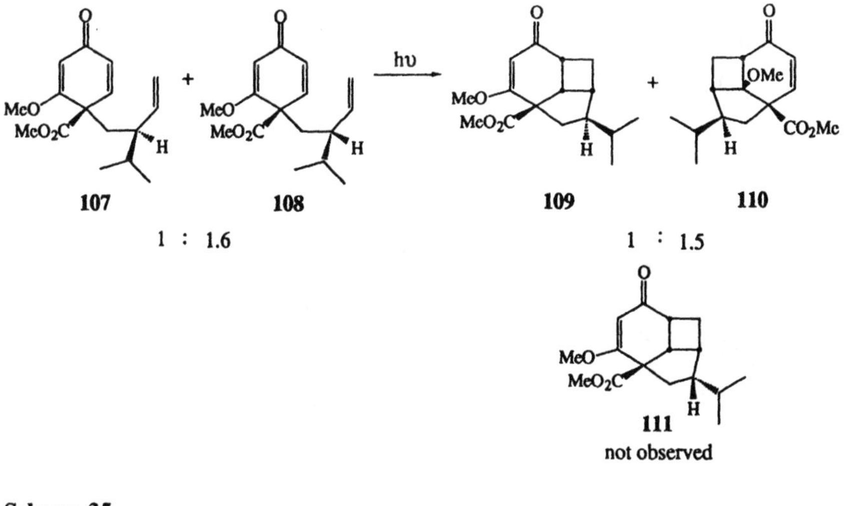

107   **108**   **109**   **110**

1 : 1.6   1 : 1.5

111
not observed

Scheme 35

A similar example is found in the irradiation of vinylogous amide-tethered tryptophan derivatives 112 [114]. The more bulky the substituent on the stereocenter of the tether, the greater the stereoselectivity, as shown in Scheme 36. This is rationalized by examining the steric effects in transition state structures 113 and 114 in Fig. 2. Conformer 113, which has the substituent in the pseudo-equatorial position, is clearly more favored because of fewer unfavorable steric interactions than is conformer 114, which has the substituent in the pseudo-axial position.

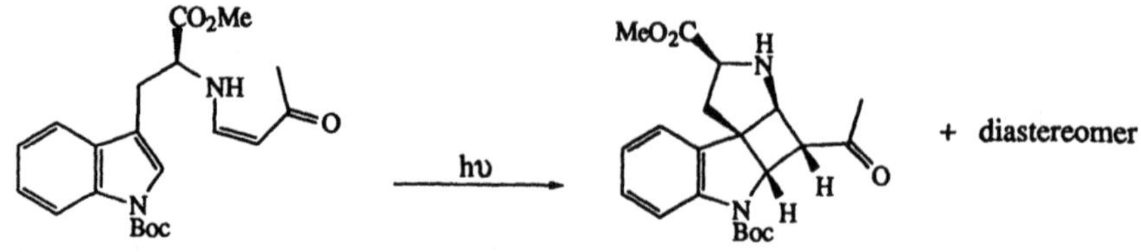

25°C	3	:	1
-40°C	5	:	1

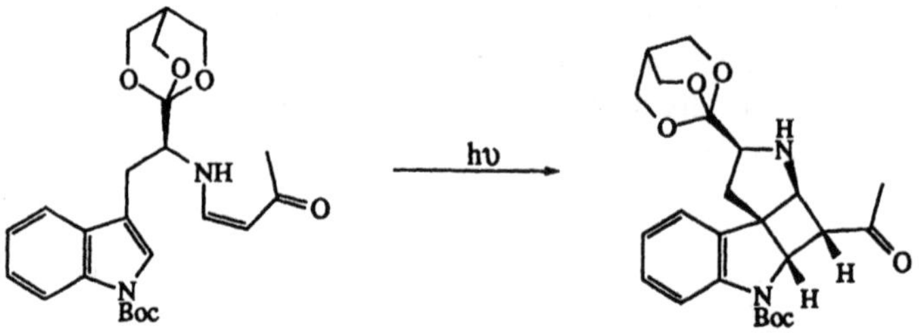

Scheme 36

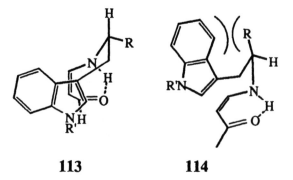

113   **114**

Figure 2   Steric effects in transition state structures 113 and 114.

Some examples of 1,3-dioxinones that employ a tether to control the regiochemistry and stereochemistry of the photochemical [2 + 2] include work by Sato [115] and Haddad [116]. Sato found that the length of the tether had a profound effect on the regioselectivity of the reaction (Scheme 37). The shorter the tether, the less selective the reaction, reportedly due to kinetic and torsional effects. When the compound 115c, where the alkene moiety is tethered to the enone by a 4-methylene chain, was irradiated, a 90% yield of product 116c was obtained, showing that the cycloaddition occurred preferentially from the less hindered side of the molecule. In the same vein, Haddad found that when $n = 1$ in compounds 117, the approach of the olefin to the enone

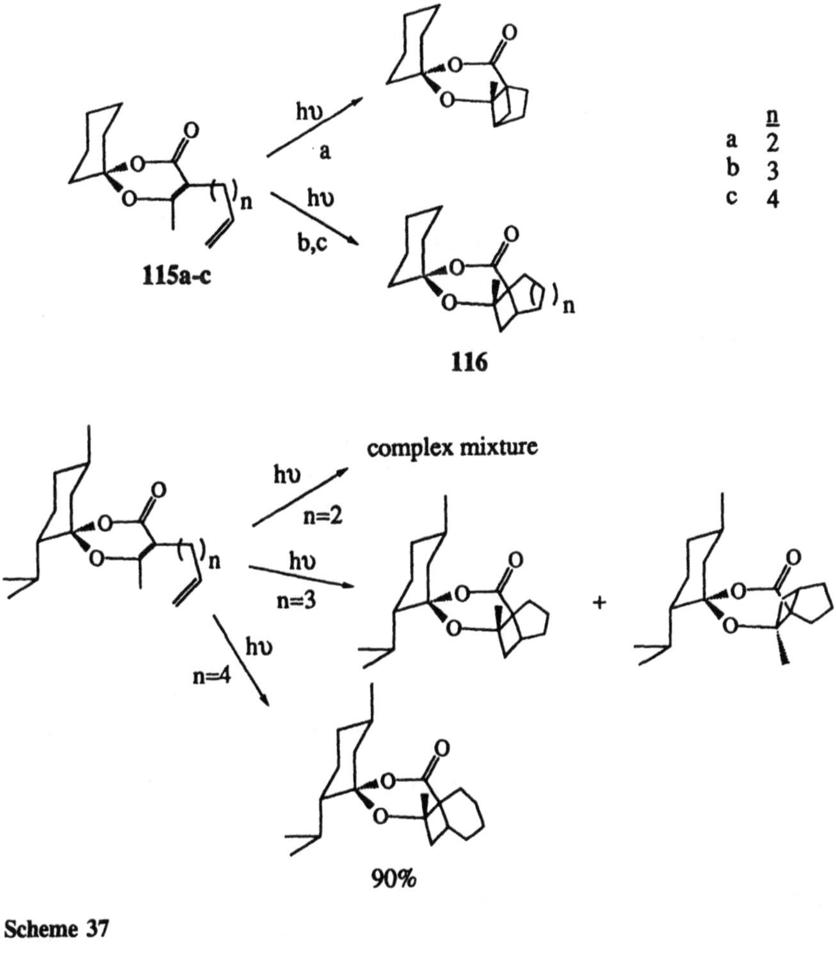

	n
a	2
b	3
c	4

Scheme 37

was from the less hindered face, whereas when $n = 2$, the approach was from the more hindered side, which supports the idea of pyramidalization of the β-carbon. (See Scheme 38.)

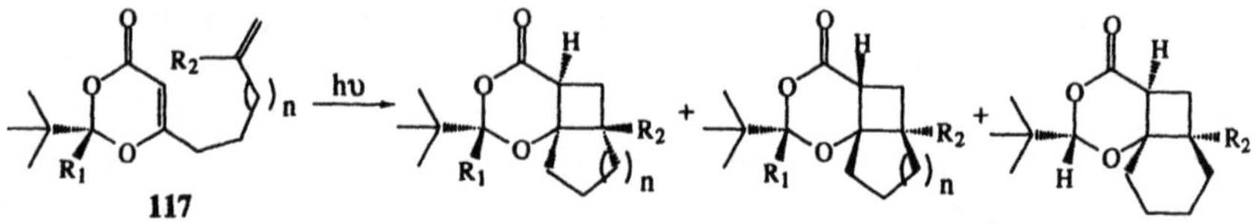

n	R_1	R_2						
1	Me	H	1	:	2			
1	Me	Me	1	:	2.4			
2	Me	H	1 (1.8 at -70)	:	1			
2	H	H	3.3	:	1.5	:	1	
2	H	H (at -70)	17.1	:	3.7	:	1	

Scheme 38

In some cases, tethers serve to preorganize the reactants, rather than simply to direct their approaches toward one another. For example, Greiving et al. [103] found that in the series of related cinnamates 118–121, only 118 underwent cycloaddition upon irradiation—the others simply showed isomerization of the double bonds (Scheme 39). This establishes that the reacting partners must be within a certain proximity in order to react, and if they are held by some framework or other method from coming within this range, they cannot react. This is also illustrated by a comparison of the photoreactions of tethered enones 122 [117] and 123 [118]. Enone 122 gave cycloadduct 124 as the predominant product at low (10^{-5} M) concentrations and a different [2 + 2] product at higher (10^{-3} M) concentrations, whereas enone 123 gave no cyclobutane adducts upon irradiation because the tethered enone and olefin partners could not get into the right orientation to form the [2 + 2] products (see Scheme 40).

These examples illustrate the factors that affect selectivity in the [2 + 2] photocycloaddition involving the enone moiety. It is evident that sterics play a large role in this selectivity and that employment of tethers to help control the regioselectivity can also enhance the stereoselectivity. Stereoelectronic effects also play a role, but it is not as large as that of sterics and preorganization.

B. Stereoselectivity of the Paternò–Büchi Reaction

The Paternò–Büchi reaction has also received considerable attention in the past several years. Stereoselectivity in this reaction seems to be a function of both steric and electronic effects. In some cases, one is dominant over the other, but both usually play a part. For example, in the addition of disubstituted alkenes to 5-substituted adamantanones, electronics plays a more major part than sterics

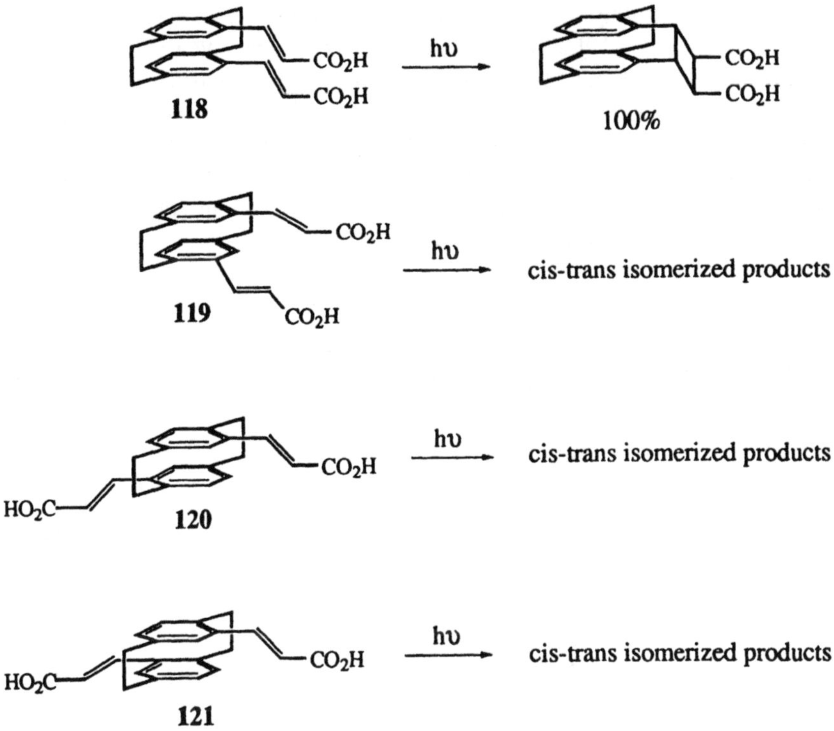

Scheme 39

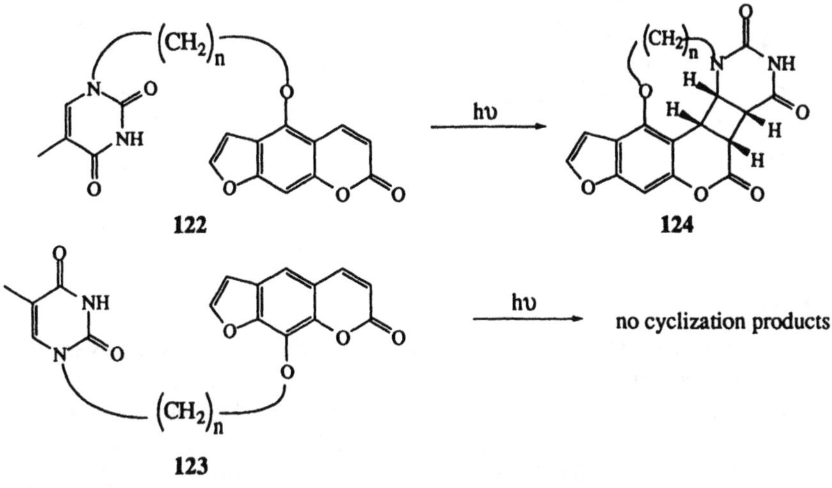

Scheme 40

because of the rigidity of the ring system. Addition of both electron-rich [31] and electron-poor [119] alkenes to adamantanones substituted with F, Cl, Br, or OH in the 5-position add predominantly syn to that substituent. This approach is antiperiplanar to the most electron-rich bonds in the system, allowing the most effective hyperconjugation, which stabilizes the transition state (Scheme 41). These additions were also found to occur with complete retention of double-bond stereochemistry at low conversions, indicating that any biradical formed would not have time to rotate or undergo inversion.

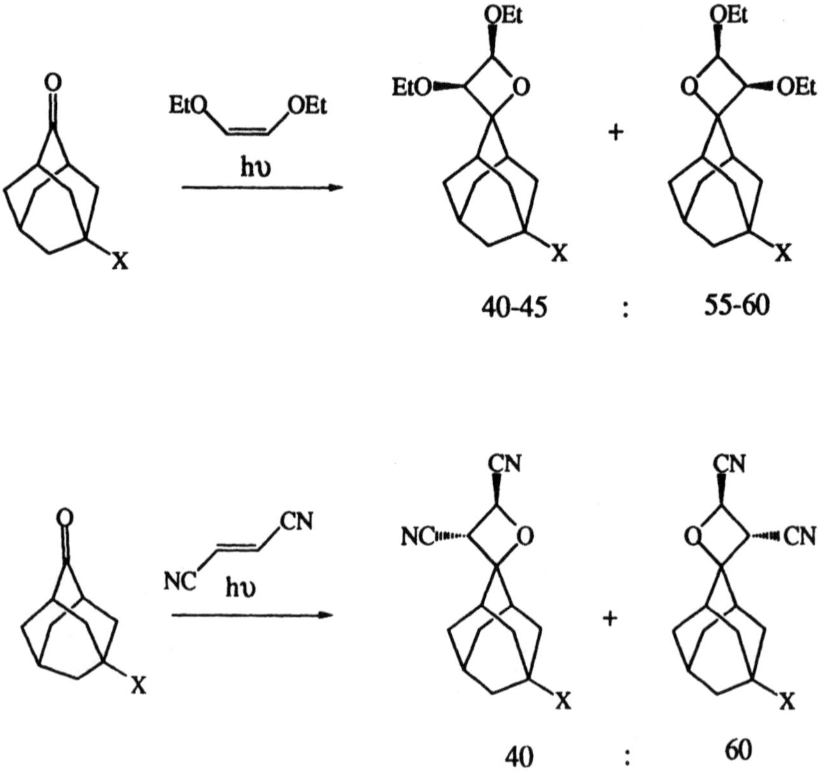

$$40\text{-}45 \quad : \quad 55\text{-}60$$

$$40 \quad : \quad 60$$

Scheme 41

A pair of examples where sterics are the predominant factor over electronics in selectivity are found in the work of Bach. He reported that the initial stereochemistry of a trisubstituted silyl enol ether [120] did not affect the final product ratio in the photoaddition to aromatic aldehydes and that the product found in greatest abundance was that with the fewest unfavorable steric interactions (see Scheme 42). This indicates that here the intermediate biradical

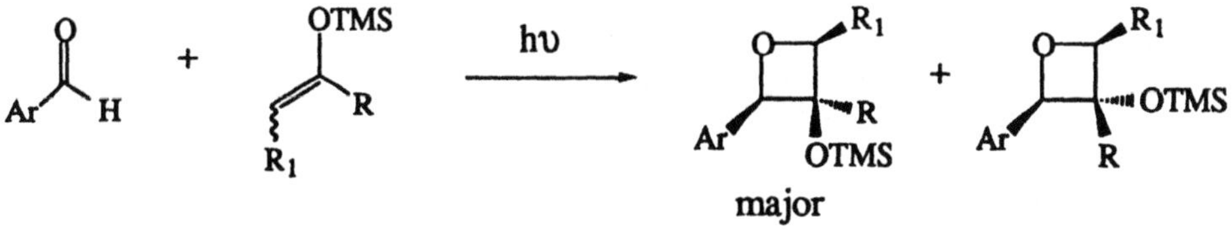

Scheme 42

does have time to rotate and equilibrate, so the thermodynamically most stable products predominate.

He found that this is also true in the [2 + 2] photocycloaddition between 1,1-disubstituted silyl enol ethers (e.g., 75a) [83a] and benzaldehyde, as in Scheme 43. The regiochemistry is largely determined on the basis of the most stable biradical, which, before closing, arranges itself so that the bulkiest substituents are trans to each other. Hence, in the cases of 1,1-disubstituted and trisubstituted silyl enol ethers, sterics is the determining factor in the stereoselectivity of the Paternò–Büchi reaction.

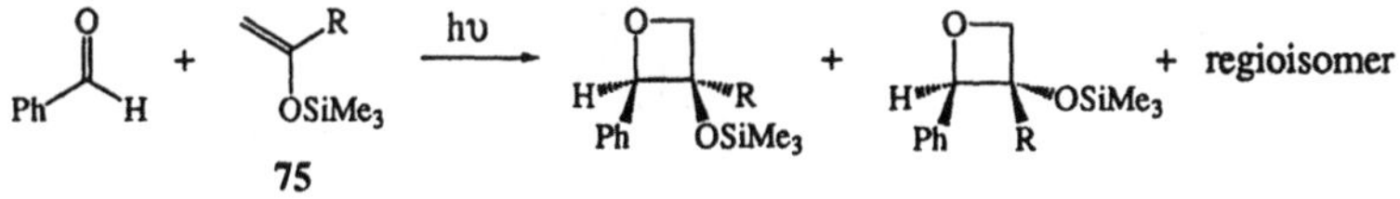

75	R	yield	ds
d	Ph	39	>95/5
e	t-Bu	59	93/7
i	t-BuS	27	>95/5

Scheme 43

There are numerous examples where sterics and electronics play equally important parts in stereoselectivity. Dopp et al. [121] have reported the stereoselective photocycloaddition of α-aminoacrylonitrile to α-diketones. His results are consistent with electronic control via an exciplex prior to bond formation.

Fleming and Jones [122] have found that the reaction between alkene-substituted pyridine 125 and acetophenone produces a 30 : 1 ratio of diastereomers 126 and 127 (see Scheme 44). This results was rationalized by arguing that the approach of the excited acetophenone to the alkene has to be perpendicular to allow for the necessary intersystem crossing prior to closure, and the sterics of this approach govern the diastereoselectivity.

This idea is supported by the work of Griesbeck, who has published several examples of stereoselective Paternò–Büchi reactions. The irradiation of

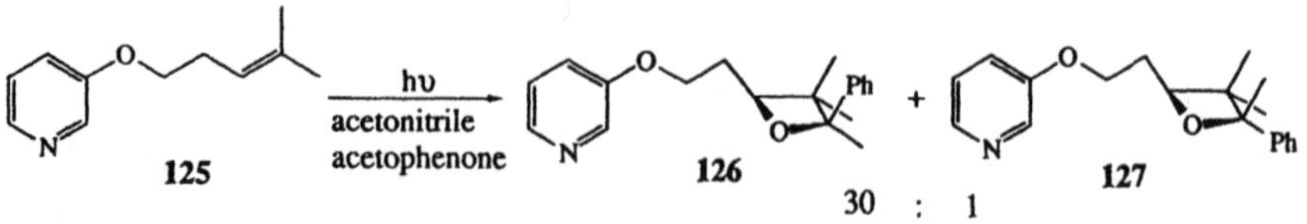

Scheme 44

substituted benzaldehydes (128) with furan [90] resulted in the formation of the exo product exclusively, whereas the irradiation of the same benzaldehydees with 2,3-dihydrofuran resulted in predominant endo-product formation. The rationalization for these apparently contradictory results lies in the approach of the orbitals in the 1,4-biradical intermediate. The p-orbitals of the triplet biradical resulting from the dihydrofuran reaction prefer to approach each other perpendicularly to facilitate the intersystem crossing (ISC) necessary for closure, whereas the first step of the reaction of benzaldehydes with furan results in a singlet biradical, which does not require perpendicular approach. The sterics of these different approaches result in the endo-product preference in one and exo-product preference in the other (see Scheme 45).

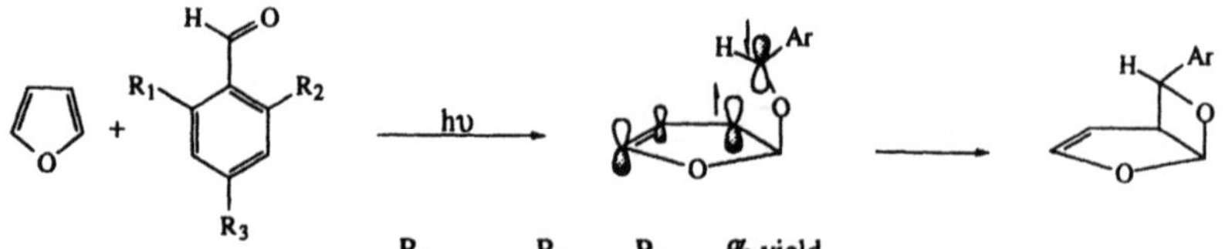

	R_1	R_2	R_3	% yield
a	H	H	H	78
b	Me	H	H	97
c	Me	Me	Me	90
d	t-Bu	Me	t-Bu	83

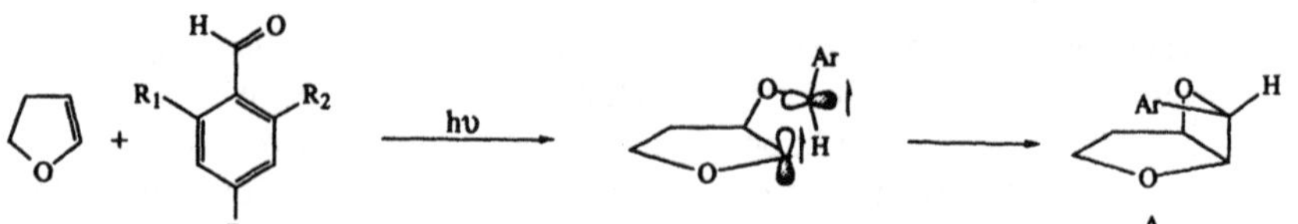

	A:B
a	7:1
b	15:1
c	>20:1
d	>20:1

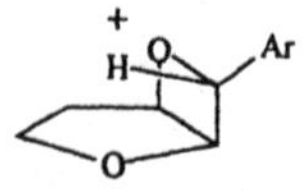

Scheme 45

In another study, Griesbeck examined the differences in selectivities between cyclic enol ethers, such as 2,3-dihydrofuran, and analogous all-carbon systems. The results (Scheme 46) showed that the "stereoselectivity of the triplet

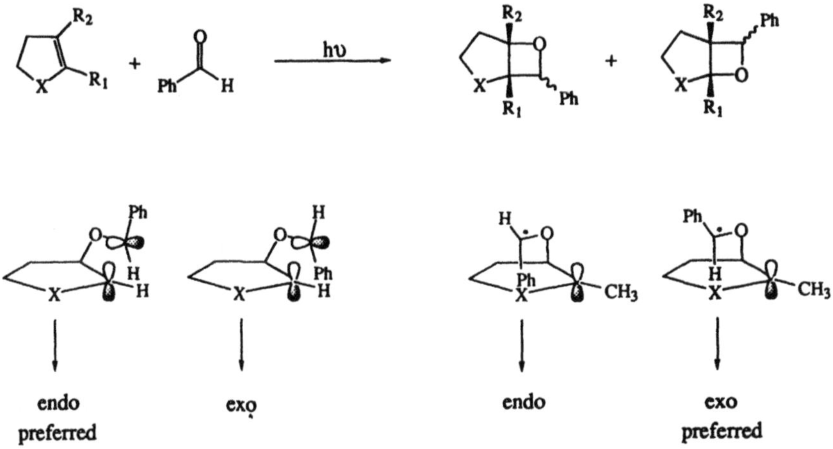

endo   exo   endo   exo
preferred   preferred

Scheme 46

[2 + 2] photocycloaddition reaction is controlled by reactive 1,4-biradical conformations of appropriate orbital alignment for rapid intersystem crossing" [123]. The regioselectivity is apparently governed by an exciplex like pre-orientation, which is more important, according to Griesbeck, than biradical stability, especially in the cases of cyclic enol ethers. Here, steric effects are also important because they help govern the preferred orientation of the biradical intermediate. Where there is no methyl substituent on the double bond, a pseudo-equatorial orientation is preferred for the appended radical, whereas when a methyl substituent is present on the double bond, a pseudo-axial conformation becomes more populated to avoid steric and gauche interactions between the appended radical and the methyl group. Thus, the preferred product, in which there is no substituent on the double bond, is the one with the phenyl group in the endo orientation, whereas when there is a substituent present, the exo orientation of the phenyl group is favored (see Scheme 46).

In yet another example, when alkene 129 is irradiated in the presence of various pyruvate ester derivatives 130, the resulting oxetanes show a marked preference for the R group of the ester to take the endo orientation, even though calculations show that the exo product is more stable (Scheme 47) [124]. Beck et al. concluded that this endo selectivity, which is independent of substituent steric demands, supports spin-orbit coupling (perpendicular approach for efficient ISC) control over the stereochemistry of this photocyclization.

An interesting extension of the Paternò–Büchi reaction is the photochemical formation of thietanes. A recent example [125] of stereoselective thietane formation shows that sterics play a large role in the selectivity. Silyl thioketone 131 (see Scheme 48) was irradiated in the presence of several electron-poor alkenes, which showed greater than a 12 : 1 propensity for the larger group on

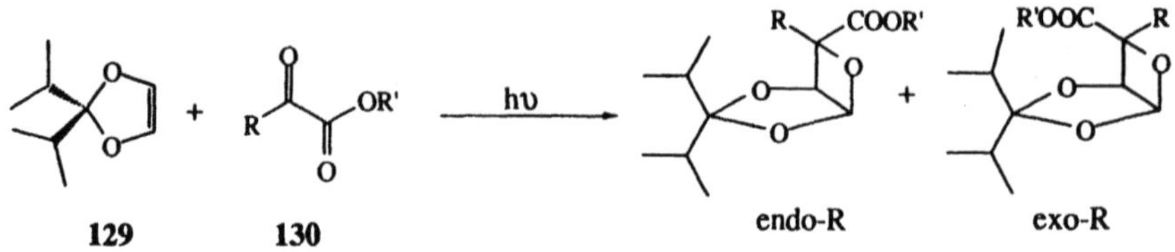

129   **130**   endo-R   exo-R

R	R'	endo-R/exo-R
Me	Et	80:20
t-Bu	Et	>98:2
t-Bu	Me	>98:2

Scheme 47

the alkene to be trans to the triphenylsilyl group. In this study, electron-rich alkenes were not regioselective nor stereoselective.

The stereoselectivity of the Paternò–Büchi reaction is more dependent on electronic factors than sterics. It is interesting to note that there are very few

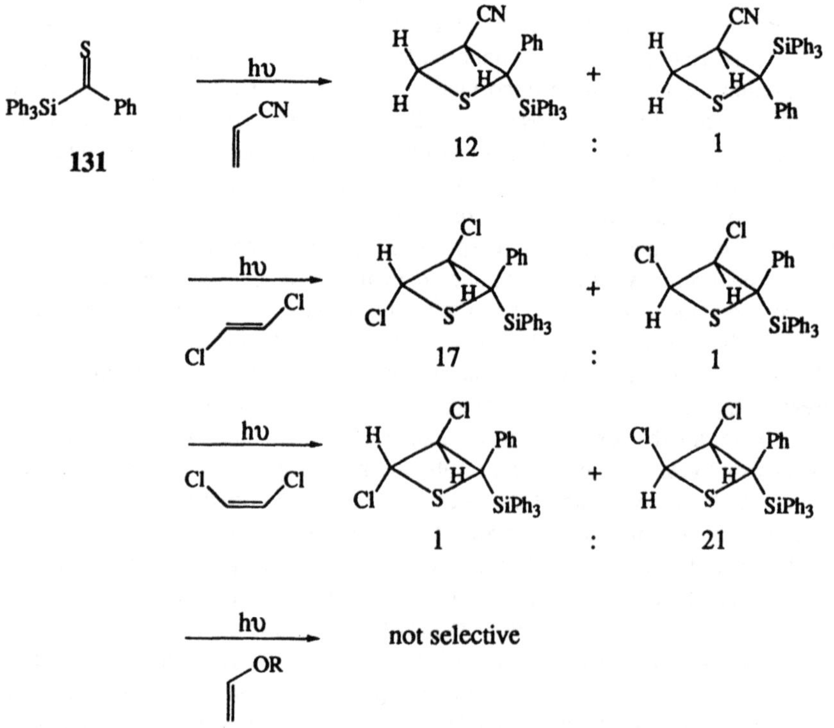

131   12   :   1

                          17   :   1

                          1   :   21

                    not selective

Scheme 48

examples of Paternò–Büchi reactions involving tethers to control the selectivity of the reaction. This is probably partly due to the high regioselectivity already exhibited by this reaction

C. Stereoselectivity of [2 + 2] Photocycloaddition Between Arenes and Alkenes

Aromatic systems have also been known to undergo [2 + 2] photocycloadditions to form cyclobutanes. Polycyclic aromatic systems are more reactive than benzene derivatives; therefore, they demonstrate a greater propensity to undergo this reaction. However, there are a few examples of benzene-derived systems that undergo fairly efficient cyclobutane formation.

Naphthalene rings are activated (i.e., their oxidation potentials are lowered) when they are substituted with silylmethyl and germanylmethyl substituents; therefore, they tend to undergo photocyclization reactions more readily than unsubstituted naphthalenes. Scheme 49 shows the reactions of several substituted naphthalenes with 1,1-disubstituted alkenes, where one of the substituents is an electron-withdrawing group. Exciplex formation is suspected although not observed, to account for the regioselectivity of the cycloaddition [126]. The stereoselectivity is a result of steric factors, although when the EWG was the fairly bulky carbomethoxy group, there was no selectivity when irradiated with 132 (X = SiMe$_3$), probably as a result of the interaction between the oxygens of the ester group and the silicon. When methyl crotonate was irradiated with the germanyl analog, there was no reaction. A decrease in selectivity was also observed when the R group of the alkene was larger than H, because of the decreased the difference in size between the two groups on the double bond, thus making the transition states of the two isomers more equal energetically. Naphthalenes substituted with tributylstannylmethyl groups did not undergo cycloaddition on irradiation, instead they decomposed, probably due to the very weak carbon–tin bond.

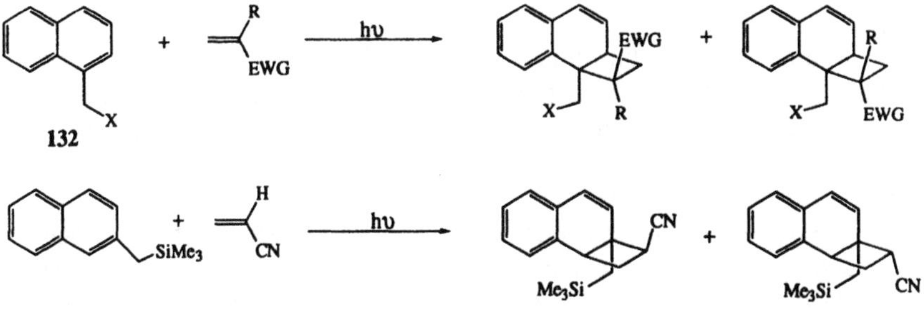

Scheme 49

The reaction between 9-cyanophenanthrene and electron-poor alkenes is stereospecific at low conversions, as shown in Scheme 50. The stereospecificity of this reaction is attributed to exciplex formation between the singlet-excited 9-cyanophenanthrene and the styryl group [127]. Evidence for this mechanism lies in the fact that methyl crotonate and cyanoethylene did not react with 9-cyanophenanthrene upon irradiation because they cannot form an exciplex as efficiently as a styryl group can. Dipole–dipole interactions contribute to the regioselectivity of the reaction.

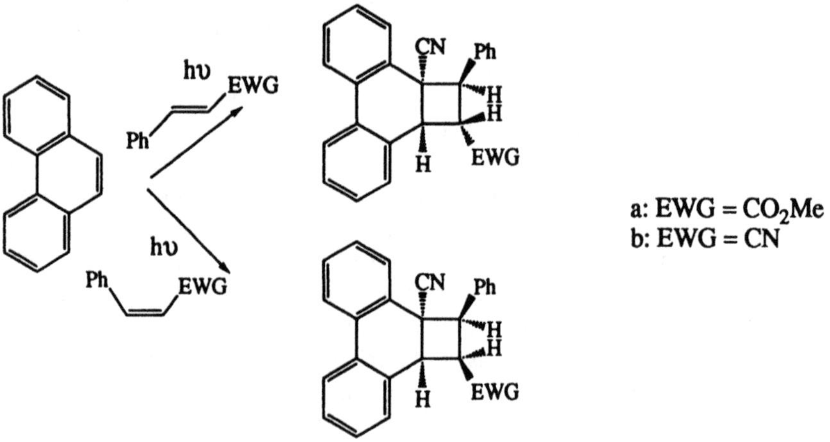

a: EWG = CO_2Me
b: EWG = CN

Scheme 50

Perfluorinated benzene and perfluorinated benzene derivatives can also undergo photochemical [2 + 2] photocycloaddition with cycloalkenes. Hexafluorobenzene reacts with cycloalkenes when irradiated to afford two products which react further to form three other products [128]. The anti addition products predominate, with the selectivity decreasing with increasing ring size of the cycloalkenes (see Scheme 51). When the alkene is indene or dihydronaphthalane, the syn product is the sole product, probably because of π-stacking.

Similarly, pentafluorophenyl esters 133 reacted regioselectivity and stereoselectivity [129] to form products similar to those of hexafluorobenzene, with the anti products again dominating (Scheme 52). In this case, selectivity also decreased with ring size, but the size of the R group of the ether did not affect the course of the reaction.

The [2 + 2] photocycloadditions involving aromatic systems as one of the reacting species are not common as a result of their relatively low reactivity. Selectivity in the examples discussed here depend predominantly on steric factors, but ring strain also begins to be an issue in some systems. Those ring

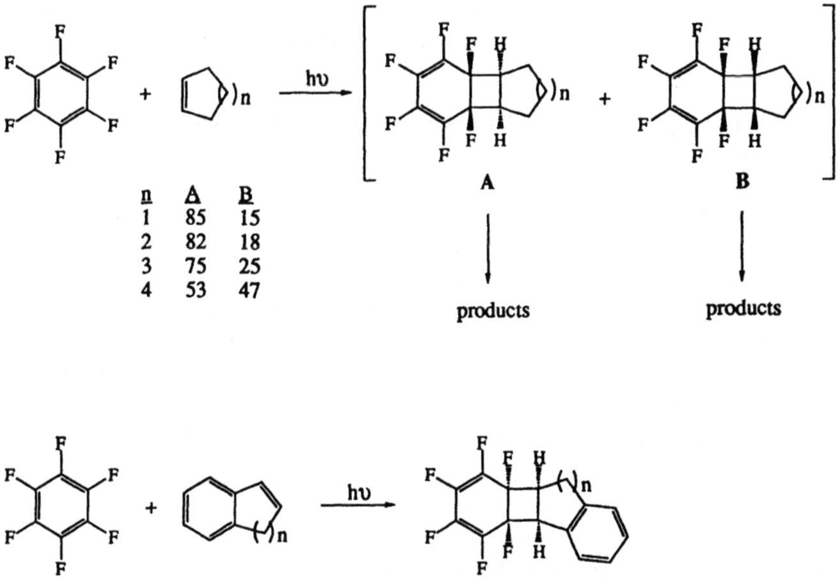

n	A	B
1	85	15
2	82	18
3	75	25
4	53	47

n = 1,2--exclusive product

Scheme 51

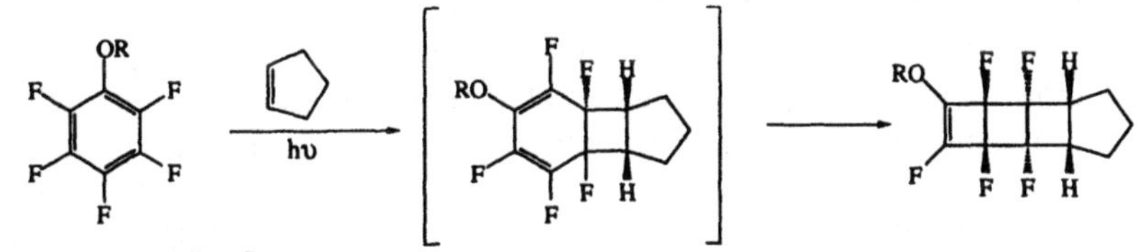

R = Me, Et, i-Pr, t-Bu

133

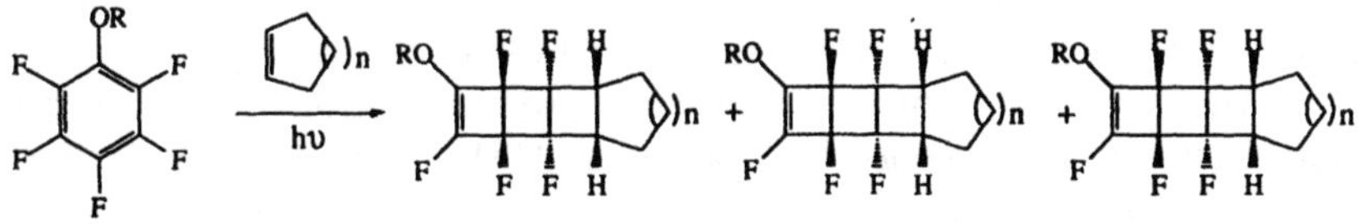

133 A B C

n	A : B : C
1	100:0:0
3	64:16:20
4	34:38:28

Scheme 52

systems with more flexibility exhibit lower selectivity than those that are relatively inflexible (such as 4- and 5-membered rings).

D. Stereoselective [2 + 2] Photocycloaddition Between Two Alkenes

The same basic principles of stereoselectivity apply in [2 + 2] photocycloaddition of two alkene moieties as in the enone, arene, and Paternò–Büchi cases. The sterics of approach and the thermodynamic stabilities of the products seem to have the most impact on stereoselectivity, but electronic and certainly intramolecular forces are very influential as well. An early example of this was published by Lewis and Hirsch in 1976 [130]. They reported that irradiation of diphenylvinylene carbonate (134) with various alkenes led to the formation of two or more products in unequal amounts. Furthermore, benzil triplet sensitization led to different product ratios than did direct irradiation, which presumably went through a singlet mechanism (see Scheme 53). The rationalization for the observed selectivity stems from steric and electronic considerations. The triplet reactions gave products with the highest thermodynamic stability, whereas the singlet reactions resembled concerted reactions, where the lowest-energy transition states had the greatest frontier orbital overlap. The product distribution indicates the π-stacking plays an important role in the reaction by stabilizing the transition states. The quantum yield data also suggest formation of a loose exciplex in both the singlet and triplet reactions.

Caldwell et al. have also reported a stereoselective [2 + 2] photocycloaddition where the major product was the most thermodynamically stable. Scheme 54 shows that the cyclobutane product with the two aryl groups trans to each other predominates, but it is not the exclusive product. The proposed intermediates include a 1,2-biradical, where the p-orbitals are perpendicular to each other, and a 1,4-biradical intermediate which has time to assume the most stable conformation before closing. The 1,2-biradical intermediate is supported by rate studies and quenching data, but these studies are not conclusive [41a]. In addition, the possibility of involvement of an exciplex prior to cycloaddition cannot be ruled out based on the studies Caldwell et al. have reported.

Another example of thermodynamic product control is found in the cyclophane synthesis reported by Okada et al. [131]. They found that irradiation of 135 yielded exclusively 136 (Scheme 55), which had both cyclobutane rings in the exo orientation. Calculations revealed that the bis-exo cycloadduct 137 had 12 kcal mol^{-1} less strain energy than the bis-endo adduct 139 and 6 kcal mol^{-1} less strain energy than the exo–endo compound 138 (see Fig. 3). With energy differences of this magnitude, the selectivity exhibited by this

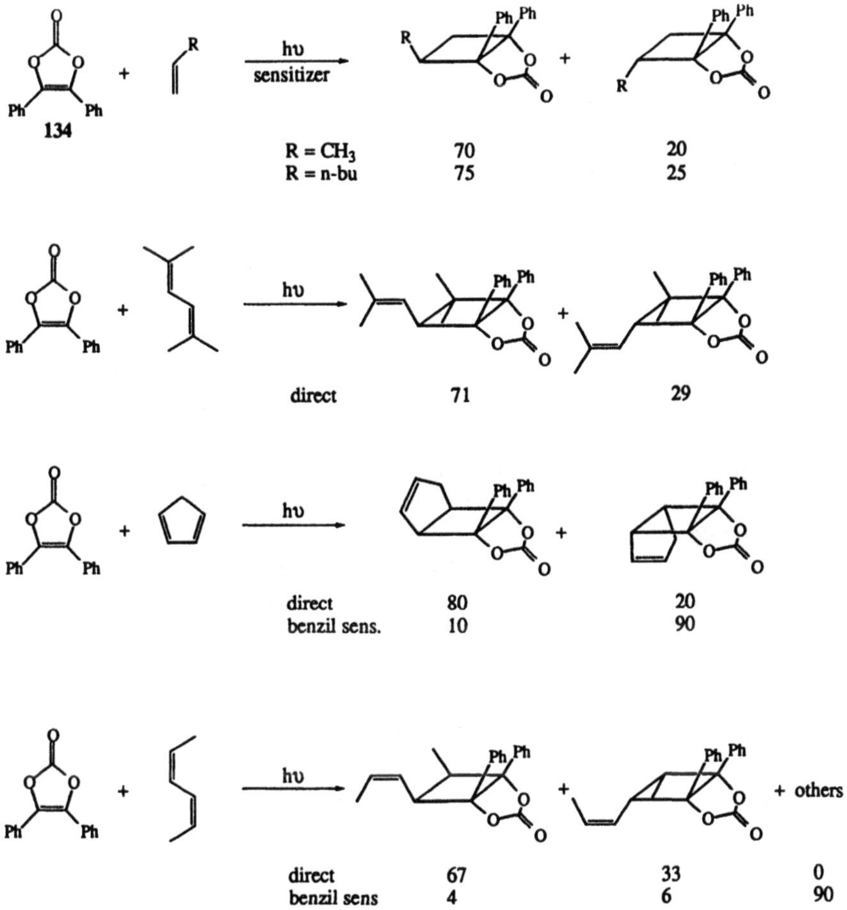
134
hυ
sensitizer
R = CH3 70 20
R = n-bu 75 25
hυ
direct 71 29
hυ
direct 80 20
benzil sens. 10 90
hυ
+ others
direct 67 33 0
benzil sens 4 6 90

Scheme 53

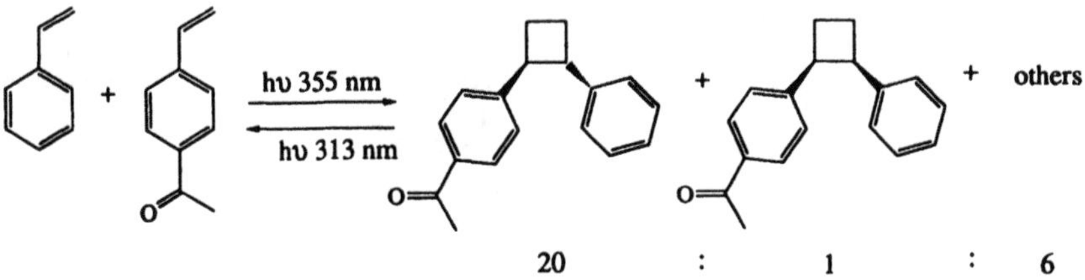
hυ 355 nm
hυ 313 nm
+ others
20 : 1 : 6

Scheme 54

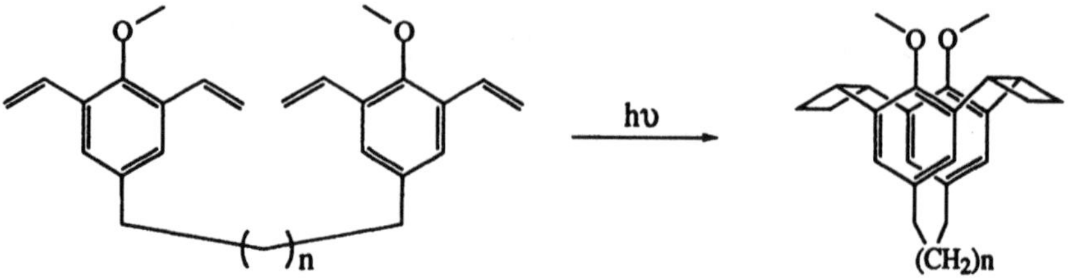

135

hυ

136

exclusive product

Scheme 55

reaction indicates reversibility, which leads to the most thermodynamically stable product; exciplex formation could also be involved.

Variation on the substitution pattern of this example yielded similar results. Compound 140 was irradiated to yield 141 exclusively in 5–45% yields [132] (Scheme 56). The endo product would be unfavored because of steric interactions between the tether and the cyclobutane ring.

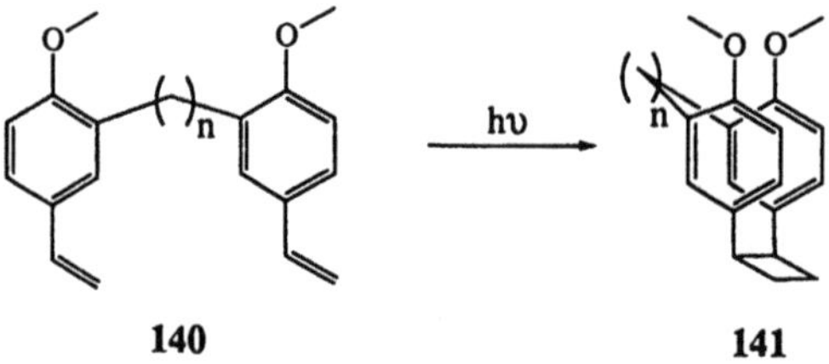

140   hυ   **141**

Scheme 56

Sterics play an important role in the [2 + 2] photocycloadditions of compounds 142, as well [133]. The tethered alkenylnaphthalene 142a, when irra-

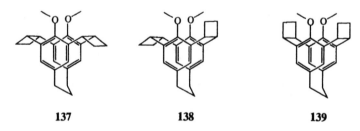

137   **138**   **139**

Strain Energy =   126.7 kcal/mol   132.7 kcal/mol   138.7 kcal/mol

Figure 3   Strain energies of structures 137, 138, and 139.

diated, yielded endo adduct 143, whereas 142b yielded the exo adduct 144. This reversal in selectivity is due to the decrease in steric interaction between the methyl hydrogens and the ring hydrogens with respect to the analogous exo product. Compound 144 predominates when the alkene moieties are substituted by R groups on the terminal end rather than 1,1-disubstituted because the exo conformation is less sterically demanding than the endo. It was also found that if the alkenes were trisubstituted, no cycloaddition occurred because the alkene simply became too crowded to react efficiently. Alkenylnaphthalene 145, in which the tether is linked in a different position from compound 142, gave similar results, producing 146 upon irradiation as the sole product (see Scheme 57), Pi-stacking as well as the tether help the two alkene substituents to come close enough together to cyclize upon irradiation.

A similar effect is seen in the reaction of tethered alkyl-substituted phenanthrenes [134] Irradiation of 147 yielded 148 with no detectable traces of 149,

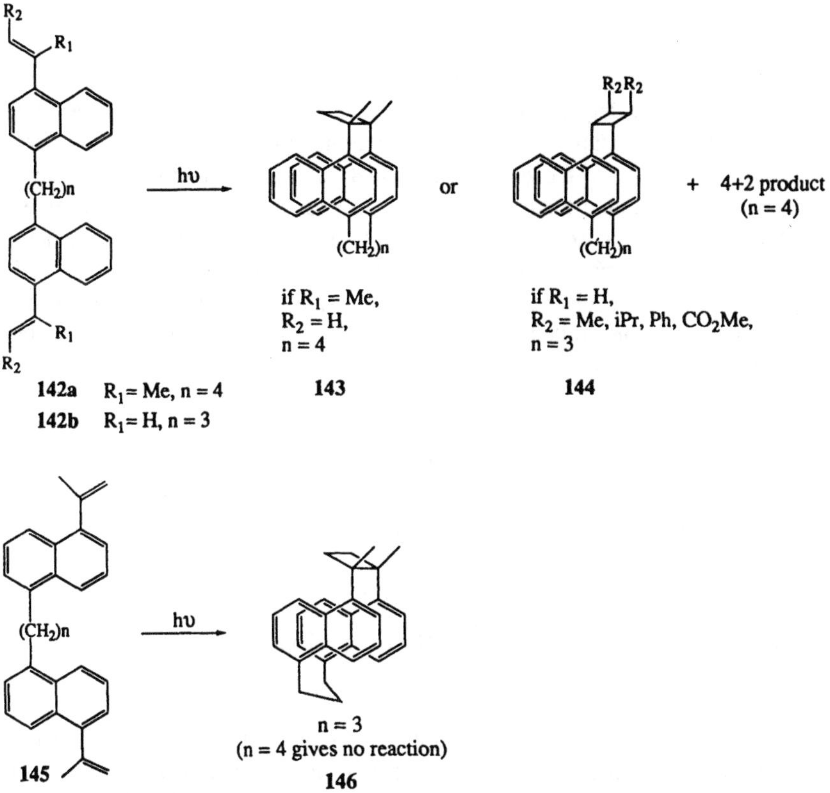

Scheme 57

as shown in Scheme 58. This result supports the idea that pi-stacking plays an important role in the preorganization of the reactants.

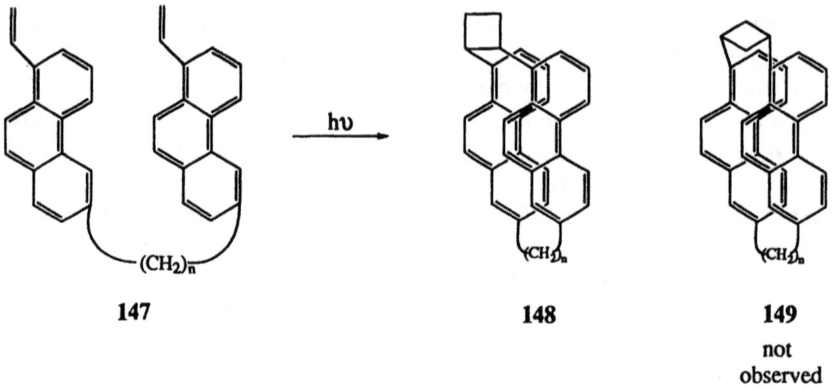

147   148   149

not
observed

Scheme 58

Another example of preorganization is found in our work with silicon tethers [104]. A variety of dialkenyloxysilanes 150 were prepared, where one of the alkenyloxy groups in each case was cinnamyloxy, to serve as the chromophore. Irradiation of those compounds that had radical-stabilizing π-systems resulted in cyclobutane formation in good to excellent yields, with the others yielding only cis-trans isomerized or hydrogen-abstraction products (see Scheme 59). (For a similar phenomenon in the aza-di-π photoarrangement, see Ref. 135.) The observed cyclobutane products all resulted from intermediates whose conformations in the excited state resulted in the maximum possible π-overlap between the two ligands.

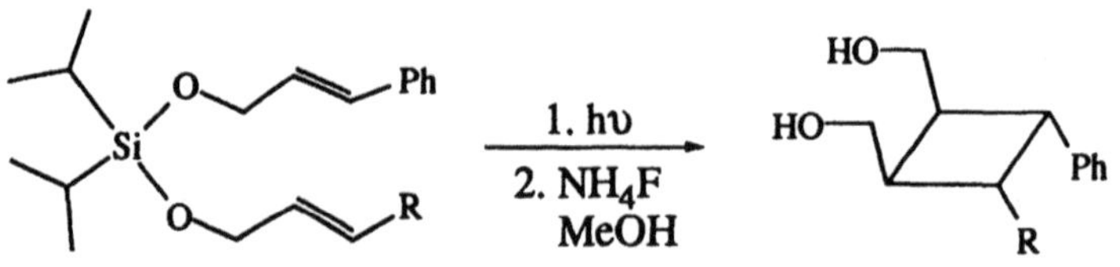

1. hυ
2. NH$_4$F
   MeOH

150

R= -H, -CH$_3$, -Ph, *trans* -CH=CHCH$_3$, -cyclopropyl

Scheme 59

Thus, the effects of a tether combined with electronic interactions between reactants results in highly diastereoselective singlet excited state reactions, as shown by our results and the results of the cyclophane studies. In a similar study, alkenes tethered by an amine group have been shown to undergo [2 + 2] cycloaddition with high stereoselectivity, which is attributed to steric interactions in the triplet biradical intermediate [136].

Cu(I) has also been used as a removable tether to preorganize alkene reactants such that they will undergo efficient [2 + 2] cyclization upon irradiation. Both diallyl ethers [137] and 1,6-heptadienes [138] will cyclize to from cyclobutane products when irradiated in the presence of copper triflate (CuOTf). Irradiation of diene 151 yields predominantly the exo product via transition state 152, which is favored over 153, which has unfavorable steric interactions between the methyl and R_2 groups (see Scheme 60).

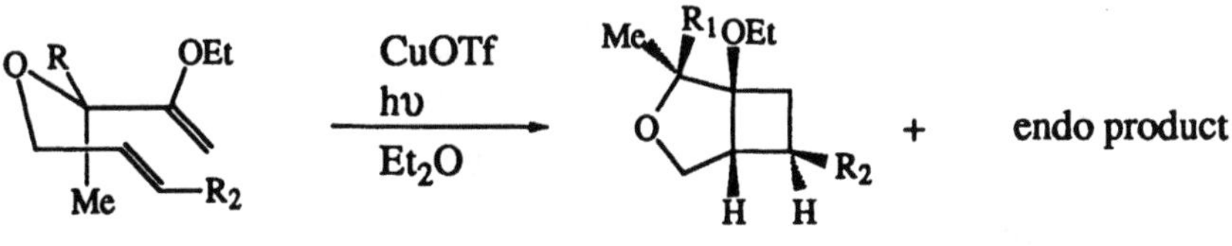

R_1	R_2	exo : endo
Me	Me	2.5 : 1
Me	Ph	100 : 0
Me	$CH=CHCH_3$	2.2 : 1
Bn	$CH=CHCH_3$	100 : 0

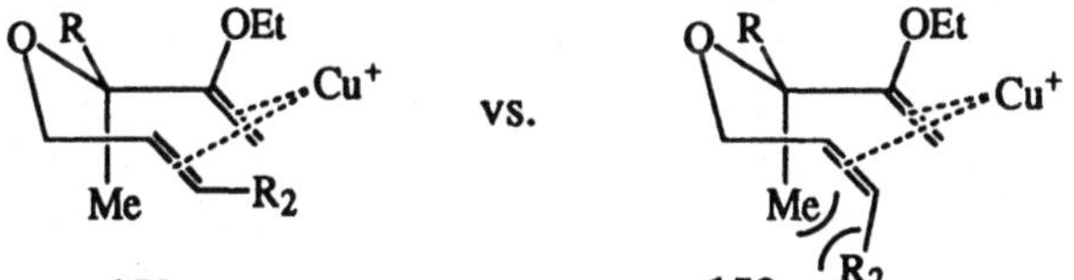

Scheme 60

Hydroxy-substituted 1,6-heptadienes, when irradiated in the presence of CuOTf, yield [3.2.0]bicycloheptanes diastereoselectively. For example, when (S)-154 was irradiated in the presence of CuOTf, 155 and 156 were formed, each in 98% enantiomeric excess (e.e.) (Scheme 61). The endoselectivity of this reaction increases in polar solvents [139] because it is claimed that 157 is favored over 158 in increasingly polar media.

The stereoselectivity of photochemical cyclobutane formation involving two alkenes has been shown to depend primarily on steric interactions especially for triplet reactions. Electronic effects such as π-stacking and the effects of tethers contribute particularly for stereocontrolled singlet photocycloadditions.

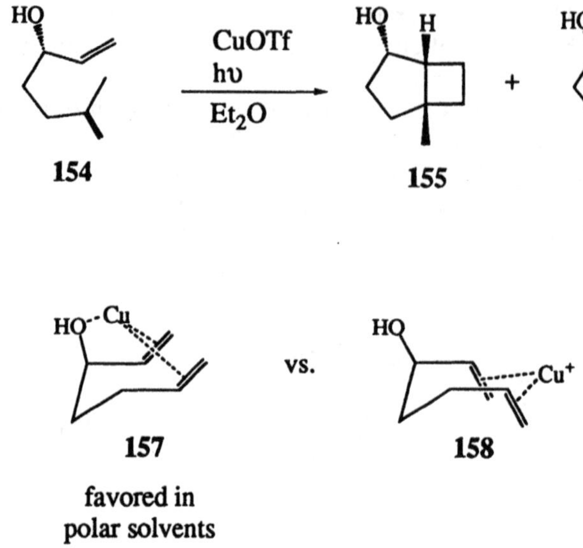

Scheme 61

E. Conclusion

The examples presented here indicate that the stereochemical outcome of a photochemical [2 + 2] cycloaddition is highly dependent on the sterics of approach of the two reactants as well as the sterics of the product. This is especially true when the reaction involves an enone moiety. However, in the case of the Paternò–Büchi reaction, the approach of the triplet radical orbitals prior to intersystem crossing and bonding seems to play a more important role than sterics.

Tethering the two reacting species (enone + alkene, carbonyl + alkene, arene + alkene, or alkene + alkene) often helps to facilitate the reaction and control the regiochemistry and the stereochemistry as well. There is still much room for the study of stereoselective cyclobutane and oxetane formation, especially in the area of asymmetric synthesis.

REFERENCES

1. (a) Schreiber, S. L. *Science* 1985, *227*, 857. (b) Oppolzer, W. *Acc. Chem. Res.* 1982, *15*, 135. For specific examples of recent applications, see (a) Bach, T.; Kather K. *J. Org. Chem.* 1996, *61*, 3900; (b) Somekawa, K.; Hara, R.; Kinnami, K.; Muraoka, F.; Suishu, T.; Shim, T. *Chem. Lett.* 1995, 407; (c) Meyers, A. I.; Fleming, S. A. *J. Am. Chem. Soc.* 1986, *108*, 306; (d) Crimmins, M. T.;

Gould. L. D. *J. Am. Chem. Soc.* 1987, *109*, 6199; (e) Pattenden, G.; Robertson, G. M. *Tetrahedron Lett.* 1986, *27*, 399; (f) Kitano, Y.; Fukuda, J.; Chiba, K.; Tada, M. *J. Chem. Soc., Perkin Trans. I* 1996, 829.

2. (a) Schuster, D. I. *Chem. Rev.* 1993, *93*, 3; (b) Weedon, A. C. In *Synthetic Organic Photochemistry*; W. M. Horspool, Ed.; Plenum: New York, 1984. (c) Becker, D.; Haddad, W. *Org. Photochem.* 1989, *10*, 1. (d) Müller, F.; Mattay, J. *Chem. Rev.* 1993, *93*, 99.

3. (a) Vaida, M.; Popvitz-Biro, R.; Leiserowitz, L.; Lahav, M. In *Photochemistry in Organized and Constrained Media*; V. Ramamurthy, Ed.; VCH Publishers Inc. (b) Schefer, J. R.; Pokkuluri, P. R. In *Photochemistry in Organized and Constrained Media*; V. Ramamurthy, Ed.; VCH Publishers Inc.; New York, 1991; p. 185. (c) Venkatesan, K.; Ramamurthy, V. In *Photochemistry in Organized and Constrained Media*; V. Ramamurthy, Ed.; VCH Publishers Inc.; New York, 1991; p. 133.

4. (a) Hasegawa, M.; Hashimoto, Y.; Chung, C.-M. In *Radiative Curing Polymer Sci. Tech.*, J.-P. Fouassier, J. F. Rabek, Eds.; Elsevier: London, 1993; Vol. 3; p. 341. (b) Sakamoto, M.; Takahashi, M. Fujita, T.; Watanabe, S.; Iida, I.; Nishio, T.; Aoyama, H. *J. Org. Chem.* 1993, *58*, 3476. (c) Schultz, A. G.; Taveras, A. G.; Taylor, R. E.; Tham. F. S.; Kullnig, R. K. *J. Am. Chem. Soc.* 1992, *114*, 8725. (d) Gudmundsdottir, A. D.; Lewis, T. J.; Randall, L. H.; Schetter, J. R.; Rettig, S. J.; Trotter, J.; Wu, C.-H. *J. Am. Chem. Soc.* 1996, *118*, 6167.

5. (a) Schultz, A. G.; Taveras, A. G.; Taylor, R. E.; Tham, F. S.; Kullnig, R. K. *J. Am. Chem. Soc.* 1992, *114*, 8725. (b) Ito, Y.; Borecka, B.; Olovsson, G.; Trotter, J.; Scheffer, J. R. *Tetrahedron Lett.* 1995, *36*, 6087.

6. (a) Pokkuluri, P. R.; Scheffer, J. R.; Trotter, J.; Yap, M. *J. Org. Chem.* 1992, *57*, 1486. (b) Garcia-Garibay, M. A.; Scheffer, J. R.; Watson, D. G. *J. Org. Chem.* 1992, *57*, 241. (c) Sakamoto, M.; Takahashi, M.; Hokari, N.; Fujita, T.; Watanabe, S. *J. Org. Chem.* 1994, *59*, 3131.

7. Nagamatsu, T.; Kawano, C.; Orita, Y.; Kunieda, T. *Tetrahedron Lett.* 1987, *28*, 3263.

8. DeMayo, P.; Sydnes, L. K. *J. Chem. Soc., Chem. Commun.* 1980, 994.

9. (a) Toda, F.; Miyamoto, H.; Kikuchi, S. *J. Chem. Soc., Chem. Commun.* 1995, 621. (b) Kaftory, M.; Tanaka, K.; Toda, F. *J. Org. Chem.* 1985, *50*, 2154.

10. Ramamurthy, V. In *Photochemistry in Organized and Constrained Media*, Ed. V. Ramamurthy, Ed.; VCH Publishers Inc.: New York, 1991; p. 429.

11. Spooner, S. P.; Whitten, D. G. In *Photochemistry in Organized and Constrained Media*; V. Ramamurthy, Ed.; VCH Publishers Inc.: New York, 1991; p. 691.

12. (a) Johnston, L. J. In *Photochemistry in Organized and Constrained Media*, V. Ramamurthy, Ed.; VCH Publishers Inc.: New York, 1991; p. 359. (b) Jones, W. In *Photochemistry in Organized and Constrained Media*;. V. Ramamurthy, Ed.; VCH Publishers Inc.: New York, 1991; p. 387. (c) Al-Ekabi, H. In *Photochemistry in Organized and Constrained Media*, Ed. V. Ramamurthy, Ed.; VCH Publishers Inc.: New York, 1991; p. 495.

13. Niwayama, S.; Kallel, E. A.; Spellmeyer, D. C. Sheu, C.; Houk, K. N. *J. Org. Chem.* 1996, *61*, 2813.

14. (a) Dilling, W. L. *Chem. Rev.* 1969, *69*, 845. (b) Baldwin, S. *Org. Photochem.* 1981, *5*, 123.

15. (a) Seibert, S. McN.; Hiel, G.; Lin, C.-H.; Kuan, D. P. *J. Org. Chem.* 1994, *59*, 80. (b) Seibert, S. McN.; Ravindran, K. *Tetrahedon Lett.* 1994, *35*, 3861.

16. (a) Suginome, H.; Sakurai, H.; Sasaki, A.; Takeuchi, H. Kobayashi, K. *Tetrahedron* 1994, *50*, 8293. (b) Kobayashi, K.; Takeuchi, H.; Seko, S.; Kanno, Y.; Kujime, S.; Suginome, H. *Helv. Chim. Acta.* 1993, *76*, 2942. (c) Kobayashi, K.; Kanno, Y.; Suginome, H. *J. Chem. Soc., Perkin Trans. I* 1993, 1449.

17. Takuwa, A. *Chem. Lett.* 1989, 5.

18. (a) Wender, P. A.; Siggel, L.; Nuss, J. M. *Org. Photochem.* 1989, *10*, 357. (b) Diether, D. A.; Fleming, S. A.; Turner, T. M.; 210th National ACS Meeting, 1995; (c) Van der Eycken, E.; De Keukeleire, C.; De Bruyn, A.; Van der Eycken, J.; Gilbert, A. *Recl. Trav. Chim. Pays-Bas* 1995, *114*, 480.

19. Wagner, P.; Park, B.-S. *Org. Photochem.* 1991, *11*, 227.

20. (a) Zimmerman, H. E. *Org. Photochem.* 1991, *11*, 1. (b) Demuth, M. *Organ. Photochem.* 1991, *11*, 37.

21. Suresh, C. G.; Gangamani, B. P.; Ganesh, K. N. *Acta Crystallogr., Sec. B* 1996, *52*, 376.

22. (a) Schuster, D. I.; Heibel, G. E.; Brown, P. B.; Turner, N. J.; Kumar, C. V. *J. Am. Chem. Soc.* 1988, *110*, 8261. (b) Schuster, D. I.; Heibel, G. E.; Woning, J. *Angew.* Chem. *Int. Ed. Engl.* 1991, *30*, 1345.

23. (a) Rudolph, A.; Weedon, A. C. *Can. J. Chem.* 1990, *68*, 1590. (b) Andrew, D.; Hastings, D. J.; Weedon, A. C. *J. Am. Chem. Soc.* 1994, *116*, 10870. Also; Audley, M.; Geraghty, N. W. A. *Tetrahedon Lett.* 1996, *37*, 1641.

24. Corey, E. J.; Bass, J. D.; LeMahieu, R.; Mitra, R. B. *J. Am. Chem. Soc.* 1964, *86*, 5570.

25. (a) Sakuragi, H.; Itoh, H. *Res. Chem. Interm.* 1995, *21*, 973. (b) Caldwell, R. A.; Hrneir, D. C.; Munoz, T.; Unett, D. J. *J. Am. Chem. Soc.* 1996, *118*, 8741.

26. Hoffmann, N.; Buschmann, H.; Raabe, G.; Scharf, H.-D. *Tetrahedon,* 1994, *50*, 11167.

27. Swapna, G. V. T.; Lakshmi, A. B.; Rao, J. M.; Kunwar, A. C. *Tetrahedron* 1989, *45*, 1777. For a thorough discussion, see: Schuster, C. I. In *The Chemistry of Enones, Part 2*; S. Patai, and Z. Rappaport, Ed.; Wiley: New York, 1989; p. 623.

28. Greiring, H.; Hopf, H.; Jones, P. G.; Bubenitschek, P.; Desvergne, J.-P.; Bouas-Laurent, H. *Liebigs Ann. Chem.* 1995, 1949.

29. (a) Chow, Y. L.; Cheng, X.-E. *Can. J. Chem.* 1991, *69*, 1331. (b) Chow, Y. L.; Wang, S.-S.; Cheng, X.-E. *Can. J. Chem.* 1993, *71*, 846.

30. (a) Turro, N. J. *Modern Molecular Photochemistry*; Benjamin-Cummings: Menlo Park, CA, 1978; p. 432. (b) Gilbert, A.; Baggott, J. *Essentials of Molecular Photochemistry*; Blackwell Scientific: Oxford, 1991; p. 340.

31. Chung, W. S.; Turro, N. J.; Srivastava, S.; LeNoble, W. J. *J. Org. Chem.* 1991, *56*, 5020.

32. Griesbeck, A. G.; Mauder, H.; Stadmüller, S. *Acc. Chem. Res.* 1994, *27*, 7, and references therein.

33. Eckert, G.; Goetz, M. *J. Am. Chem. Soc.* 1994, *116*, 11999.
34. Maruyama, K.; Imahori, H. *J. Org. Chem.* 1989, *54*, 2692.
35. (a) Mattay, J.; Rumbach, T.; Runsink, J. *J. Org. Chem.* 1990, *55*, 5691. (b) Cornelisse, J. *Chem. Rev.* 1993, *93*, 615. (c) Wender, P. A.; Siggel, L.; Nuss, J. M. *Org. Photochem.* 1989, *10*, 487. (d) Nuss, J. M.; Chinn, J. P.; Murphy, M. M. *J. Am. Chem. Soc.* 1995, *117*, 6801.
36. Kaupp, G. *Angew. Chem.* 1973, *85*, 766.
37. Wagner, P. J.; Cheng, K.-L. *Tetrahedon Lett.* 1993, *34*, 907.
38. Al-Qaradawi, S.; Gilbert, A.; Jones, D. T. *Rec. Trav. Chim. Pay. Bas.* 1995, *114*, 485.
39. Michl, J. *J. Photochem. Photobiol.* 1977, *25*, 141.
40. Bentzien, J.; Klessinger, M. *J. Org. Chem.* 1994, *59*, 4887.
41. (a) Caldwell, R. A.; Diaz, J. F.; Hrncir, D. G.; Unett, D. J. *J. Am. Chem. Soc.* 1994, *116*, 8138. (b) Unett, D. J.; Caldwell, R. A.; Hrncir, D. C. *J. Am. Chem. Soc.* 1996, *118*, 1682.
42. Caldwell, R. A. *J. Am. Chem. Soc.* 1980, *102*, 4004.
43. (a) Mattes, S. L.; Farid, S. *Org. Photochem.* 1983, *6*, 233. (b) Mariano, P. S.; Stavinoha, J. L. In *Synthetic Organic Photochemistry*; W. M. Horspool, Ed.; Plenum: New York, 1984; p. 185. (c) Mariano, P. S. *Org. Photochem.* 1987, *9*, 1.
44. Kim, T.; Mirafzal, G. A.; Bauld, N. L. *Tetrahedron Lett.* 1993, *34*, 7201. Also Sadlek, O.; Gollnick, K. Polborn, K.; Griesbeck, A. G. *Angew. Chem. Int. Ed. Engl.* 1994, *33*, 2300.
45. (a) Demuth, M.; Mikhail, G. *Synthesis* 1989, 145. (b) Crimmins, M. T. *Chem. Rev.* 1988, *88*, 1453. (c) Jones, G. II. Jones, G. II *Org. Photochem.* 1981, *5*, 1.
46. (a) Mori, K.; Sasaki, M. *Tetrahedron Lett.* 1979, *20*, 1329. (b) Margaretha, P. *Helv. Chim. Acta* 1974, *57*, 1866. (c) Hollis, W. G. Jr.; Lappenbusch, W. C.; Everberg, K. A.; Woleben, C. M. *Tetrahedron Lett.* 1993, *34*, 7517. (d) Sydnes, L. K.; Meling, H. L. *Acta Chem. Scand., Ser. B* 1987, *41*, 660.
47. Wender, P. A., Hubbs, J. C. *J. Org. Chem.* 1980, *45*, 365.
48. Lange, G. L.; Decicco, C.; Lee, M. *Tetrahedron Lett.* 1987, *28*, 2833.
49. Tada, M.; Nieda, Y. *Bull. Chem. Soc. Jpn.* 1988, *61*, 1416.
50. Lange, G. L.; Organ, M. G.; Lee, M. *Tetrahedon Lett.* 1990, *31*, 4689.
51. (a) Cantrell, T. S.; Haller, W. S.; Williams, J. C. *J. Org. Chem.* 1969, *34*, 509. (b) Cantrell, T. S. *Tetrahedron* 1971, *27*, 1227.
52. Margaretha, P. *Chimia* 1975, *29*, 203.
53. Margaretha, P. *Tetrahedron* 1973, *29* 1317.
54. (a) Ogino, T.; Kubota, T.; Manaka, K. *Chem. Lett.* 1976, 323. (b) Ogino, T. *Tetrahedon Lett.* 1979, *28*, 2445.
55. Anklam, E.; Ghaffari-Tabrizi, R.; Hombrecher, H.; Lau, S.; Margaretha, H. *Helv. Chim. Acta* 1984, *67*, 1402.
56. (a) Guerry, P.; Neier, R. *J. Chem. Soc., Chem. Commun.* 1989, 1727. (b) Guerry, P.; Blanco, P,; Brodbeck, H.; Pasteris, O.; Neier, R. *Helv. Chim. Acta* 1991, *74*, 163.

57. Challand, B. D.; De Mayo, P. *J. Chem. Soc., Chem. Commun.* 1986, 982.
58. (a) Van Audenhove, M.; Termont, D.; De Keukeleire, D.; Vandewalle, M. *Tetrahedron Lett.* 1978, *19*, 2057. (b) Termont, D.; De Keukeleire, D.; Vandewalle, M. *J. Chem. Soc. Perkin Trans. 1* 1977, 2349.
59. (a) Srinivasan, R.; Carlough, K. H. *J. Am. Chem. Soc.* 1967, *89*, 4932. (b) Liu, R. S.; Hammond, G. S. *J. Am. Chem. Soc.* 1967, *89*, 4930.
60. (a) Cargill, R. L.; Dalton, J. R.; O'Connor, S.; Michels, D. G. *Tetrahedron Lett.* 1978, *19*, 4465. (b) Becker, D.; Haddad, N. *Tetrahedron Lett.* 1986, *27*, 6393. (c) Pirrung, M. C. *J. Am. Chem. Soc.* 1979, *101*, 7130. (c) Pirrung, M. C. *J. Am. Chem. Soc.* 1981, *103*, 82. (d) Croft, K. D.; Ghisalberti, E. L.; Jeffries, P. R.; Stuart, A. D.; Raston, C. L.; White, A. R. *J. Chem. Soc. Perkin Trans. 1* 1981, 1473.
61. (a) Matlin, A. R.; George, C. F.; Wolff, S.; Agosta, W. C. *J. Am. Chem. Soc.* 1986, *108*, 3385. (b) Wolf, S.; Agosta, W. C. *J. Chem. Soc., Chem. Commun.* 1981, 118. (c) McMurry, T. B. H.; Work, A.; Mekenna, B. *J. Chem. Soc. Perkin Trans. 1* 1991, 811.
62. (a) Seto, H.; Fujimoto, Y.; Tatsuno, T.; Yoshioka, H. *Synth. Commun.* 1985, *15*, 1217. (b) Seto, H.; Tsunoda, S.; Ikeda, H.; Fujimoto, Y.; Tatsumo, T.; Yoshioka, H. *Chem. Pharm. Bull.* 1995, *33*, 2594. (c) Winkler, J. D.; Hey, J. P.; Hannon, F. J. *Heterocycles* 1987, *25*, 55.
63. (a) Tamura, Y.; Kita, Y.; Ishibashi, H.; Ikeda, M. *J. Chem. Soc., Chem. Commun.* 1971, 1167. (b) Tamura, Y.; Ishibashi, H.; Kita, Y.; Ikeda, M. *J. Chem. Soc., Chem. Commun.* 1973, 101. (c) Tamura, Y.; Ishibashi, H.; Jirai, M.; Kita, Y.; Iketa, M. *J. Org. Chem.* 1975, *40*, 2702. (d) Ikeda, M.; Ohno, K.; Homma, K.; Ishibashi, H.; Tamura, Y. *Chem. Pharm. Bull.* 1981, *29*, 2062. (e) Schell, F. M.; Cook, P. M. *J. Org. Chem.* 1984, *49*, 4067. (f) Ikeda, M.; Takahashi, M.; Uchino, T.; Ohno, K.; Tamura, Y.; Kido, M. *J. Org. Chem.* 1983, *48*, 4241. (g) Ikeda, M.; Uchino, T.; Takahashi, M.; Ishibashi, H.; Tamura, M.; Kido, M. *Chem. Pharm. Bull.* 1985, *33*, 3279. (h) Gariboldi, P.; Jommi, G.; Sisti, M. *Gazz. Chim. Ital.* 1986, *116*, 291.
64. Kaneko, C.; Suzuki, T.; Sato, M.; Naito, T. *Chem. Pharm. Bull.* 1987, *35*, 112.
65. Haywood, D. J.; Reid, S. T. *Tetrahedron Lett.* 1979, *20*, 2637.
66. Wagner, P. J.; Sakamoto, M. *J. Am. Chem. Soc.* 1989, *111*, 9454.
67. (a) Oldroyd, D. L.; Weedon, A. C. *J. Chem. Soc., Chem. Commun.* 1992, 1491. (b) Oldroyd, D. L.; Weedon, A. C. *J. Org. Chem.* 1994, *59*, 1333.
68. Blanc, S. L.; Piva, O. *Tetrahedron Lett.* 1994, *34*, 635.
69. Wagner, P. J.; McMahon, K. *J. Am. Chem. Soc.* 1994, *116*, 10827.
70. Crimmins, M. T.; Guise, L. E. *Tetrahedron Lett.* 1994, *35*, 1657.
71. Serebryakov, E. P.; Kulomzina-Pletneva, S. D.; Margaryan, A. K. H. *Tetrahedron* 1979, *35*, 77.
72. Burstrein, K. Y. A.; Serebryakov, E. P. *Tetrahedron* 1978, *34*, 3233.
73. (a) Koft, E. R.; Smith, A. B. *J. Am. Chem. Soc.* 1982, *104*, 5570. (b) Koft, E. R.; Smith, A. B. *J. Am. Chem. Soc.* 1984, *106*, 2115. (c) Koft, E. R.; Smith, A. B. *J. Org. Chem.* 1984, *49*, 832. (d) Wang, T. Z.; Paquette, L. A. *J. Org. Chem.* 1986, *51*, 5232. (d) Pirrung, M. C.; Thomson, S. A. *Tetrahedron Lett.* 1986, *27*, 2703.

74. (a) Eaton, P. E. *Tetrahedron Lett.* 1964, *5*, 3695. (b) McKay, W. R.; Ounsworth, J.; Sum, P. E.; Weiler, L. *Can. J. Chem.* 1982, *60*, 872. (c) Tobe, Y.; Kishida, T.; Yamashita, T.; Kakiuchi, K.; Odaira, Y. *Chem. Lett.* 1985, 1437.
75. (a) Sydnes, L. K.; Stensen, W. *Acta Chem. Scand. Ser. B* 1986, *40*, 657. (b) Becker, D.; Harel, Z.,; Nagler, M.; Gillon, A. *J. Org. Chem.* 1982, *47*, 3297. (c) Pasto, D. J.; Heid, P. F. *J. Org. Chem.* 1982, *47*, 2204.
76. (a) Becker, D.; Haddad, N. *Tetrahedron Lett.* 1986, *27*, 6393. (b) Becker, D.; Harel, Z.; Nagler, M.; Gillon, A. *J. Org. Chem.* 1982, *47*, 3297.
77. (a) Dauben, W. G.; Shapiro, G.; Luders, L. *Tetrahedron Lett.* 1985, *26*, 1429. (b) Dauben, W. G.; Shapiro, G. *Tetrahedron Lett.* 1985, *26*, 989. (c) Dauben, W. G.; Rocco, V.; Shapiro, G. *J. Org. Chem.* 1985, *50*, 3155.
78. (a) Becker, D.; Harel, Z.; Birnbaum, D. *J. Chem. Soc., Chem. Commun.* 1975, 377. (b) Becker, D.; Birnbaum, D. *J. Org. Chem.* 1980, *45*, 570.
79. (a) Arnold, D. R.; Hinman, R. L.; Glick, A. H. *Tetrahedron Lett.* 1964, *5*, 1425. (b) Tominaga, T.; Tsutsumi, S.; *Tetrahedron Lett.* 1969, *10*, 3175. (c) Barltrop, J. A.; Carless, H. A. J. *J. Am. Chem. Soc.* 1972, *94*, 1951.
80. Takuwa, A.; Fujii, N.; Tagawa, H.; Iwanoto, H. *Bull. Chem. Soc. Jpn.* 1989, *62*, 336.
81. (a) Schroeter, S. H.; Orlando, C. M. *J. Org. Chem.* 1969, *34*, 1181. (b) Turro, N. J.; Wriede, P. A. *J. Am. Chem. Soc.* 1970, *92*, 320.
82. Shimizu, N.; Yamaoka, S.; Tsuno, Y. *Bull. Chem. Soc. Jpn.* 1983, *56*, 3853.
83. (a) Bach, T. *Tetrahedron Lett.* 1991, *32*, 7037. (b) Bach, T.; Jödicke, K. *Chem. Ber.* 1993, *126*, 2457.
84. Kahn, N.; Morris, T. H.; Smith, E. H.; Walsh, R. *J. Chem. Soc. Perkin Trans. I* 1991, 865.
85. (a) Cantrell, T. S.; Allen, A. C. *J. Org. Chem.* 1989, *54*, 135. (b) Cantrell, T. S.; Allen, A. C.; Ziffer, H. *J. Org. Chem.* 1989, *54*, 140.
86. (a) Schenck, G. O.; Hartmann, W.; Steinmetz, R. *Chem. Ber.* 1963, *96*, 498. (b) Toki, S.; Shima, K.; Sakurai, H. *Bull. Chem. Soc. Jpn.* 1965, *38*, 760. (c) Shima, K.; Sakurai, H. *Bull. Chem. Soc. Jpn.* 1966, *39*, 1806.
87. Schreiber, S. L.; Desmaele, D.; Porco, J. A. Jr. *Tetrahedron Lett.* 1988, *29*, 6689.
88. Carless, H. A. J.; Halfhide, A. F. E. *J. Chem. Soc. Perkin Trans. I* 1992, 1081.
89. Carless, H. A. J.; Haywood, D. J. *J. Chem. Soc., Chem. Commun.* 1980, 1067.
90. Griesbeck, A. G.; Stadtmüller, S. *Chem. Ber.* 1990, *123*, 357.
91. (a) Ohno, A.; Ohnishi, Y.; Tsuchihashi, G. *J. Am. Chem. Soc.* 1969, *91*, 5038. (b) Takechi, H.; Machida, M.; Kanaoka, Y. *Synthesis* 1992, 778. (c) Nisho, T.; *Helv. Chim. Acta* 1992, *75*, 487. (d) Ooms, P.; Hartmann, W. *Tetrahedron Lett.* 1987, *28*, 2701.
92. Bryce-Smith, D.; Evans, E. H.; Gilbert, A.; McNeill, H. S. *J. Chem. Soc. Perkin Trans. II* 1991, 1587.
93. (a) Ryang, H.; Shima, K.; Sakurai, H. *J. Am. Chem. Soc.* 1971, *93*, 5270. (b) Ryang, H.; Shima, K.; Sakurai, H. *J. Org. Chem.* 1973, *38*, 2860.
94. Hyattt, J. A.; Swenton, J. S. *J. Chem. Soc., Chem. Commun.* 1972, 1144.
95. (a) Ors, J. A.; Srinivasan, R. *J. Org. Chem.* 1977, *42*, 1321. (b) Gilbert, A.; Taylor, G. N.; Collins, A. *J. Chem. Soc. Perkin Trans. I* 1980, 1218.

96. Cantrell, T. S. *J. Org. Chem.* 1977, *42*, 4238.

97. (a) Gilbert, A,; Heath, P. *Tetrahedron Lett.* 1987, *28*, 5909. (b) Al-Jalal, N. *J. Chem. Res.* (*S*) 1989, 110. (c) Hoffmann, N.; Pete, J.-P. *Tetrahedron Lett.* 1995, *36*, 2623.

98. Wagner, P. J., Sakamoto, M.; Madkour, A. E. *J. Am. Chem. Soc.* 1992, *114*, 7298.

99. Smart, R. P.; Wagner, P. J. *Tetrahedron Lett.* 1995, *36*, 5131.

100. Hoffmann, N.; Pete, J. *Tetrahedron Lett* 1996, *37*, 2027.

101. (a) Green, B. S.; Rabinsohn, Y.; Rejtö, M. *Carbohydrate Res.* 1975, *45*, 115. (b) Green, B. S.; Hagler, A. T.; Rabinsohn, Y.; Rejtö, M. *Isr. J. Chem.* 1976/ 1977, *15*, 124.

102. Akabori, S.; Kumagai, T.; Habata, Y.; Sato, S. *J. Chem. Soc. Perkin Trans. I* 1989, 1497.

103. Greiving, H.; Hopf, H.; Jones, P. G.; Bubenitschek, P.; Desvergne, P.; Bouas-Laurent, H. *J. Chem. Soc., Chem. Commun.* 1994, 1075.

104. (a) Fleming, S. A.; Ward, S. C. *Tetrahedron Lett.* 1992, *33*, 1013. (b) Ward, S. C.; Fleming, S. A. *J. Org. Chem.* 1994, *59*, 6476.

105. Bradford, C. L.; Fleming, S. A.; Ward, S. C. *Tetrahedron Lett.* 1995, *36*, 4189.

106. (a) Kwon, J. H.; Lee, S. J.; Shim, S. C. *Tetrahedron Lett.* 1991, *32*, 6719. (b) Shim, S. C.; Lee, S. J.; Kwon, J. H. *Chem. Lett.* 1991, 1767. (c) Lee, T. S.; Lee, S. J.; Shim, S. C. *J. Org. Chem.* 1990, *55*, 4544. (d) Chung, C. B., Kwon, J. H., Shim, S. C. *Tetrahedron Lett.* 1993, *34*, 2143.

107. Baldwin, S. W.; Martin, G. F. Jr.; Nunn, D. S. *J. Org. Chem.* 1985, *50*, 5720.

108. Lange, G. L.; Organ, M. G.; Froese, R. D. J.; Goddard, J. D.; Taylor, N. J. *J. Am. Chem. Soc.* 1994, *116*, 3312.

109. Seebach, D.; Zimmerman, J.; Gysel, U.; Ziegler, R.; Ha, T.-K. *J. Am. Chem. Soc.* 1988, *110*, 4763.

110. Little, R. D.; Moens, L.; Baizer, M. M. *J. Org. Chem.* 1986, *51*, 4497.

111. (a) Alibes, R.; Bourdelande, J. L,; Font, J.; Gregori, A.; Parella, T. *Tetrahedron* 1996, *52*, 1267. (b) Alibes, R.; Bourdelande, J. L.; Font, J. *Tetrahedron: Asymmetry* 1991, *2*, 1391.

112. Becker, D.; Klimovich, N. *Tetrahedron Lett.* 1994, *35*, 261.

113. Schultz, A. G.; Geiss, W.; Kullnig, R. K. *J. Org. Chem.* 1989, *54*, 3158.

114. Winkler, J. D.; Scott, R. D.; Williard, P. G. *J. Am. Chem. Soc.* 1990, *112*, 8791.

115. Sato, M.; Abe, Y.; Kaneko, C.; *Heterocycles* 1990, *30*, 217.

116. Haddad, N.; Abramovich, Z. *J. Org. Chem.* 1995, *60*, 6883.

117. Decout, J.-L,; Huart, G.; Lhomme, J. *Photochem. Photobiol.* 1988, *48*, 583.

118. Decout, J.-L.; Lhomme, J. *Photochem. Photobiol.* 1988, *48*, 597.

119. Chung, W.-S.; Turro, N. J.; Srivastava, S.; Li, Hi.; Le Noble, W. J. *J. Am. Chem. Soc.* 1989, *110*, 7882.

120. Bach, T. *Tetrahedron Lett.* 1994, *35*, 5845.

121. Dopp, D.; Fischer, M. A. *Rec. Trav. Chim. Pay. Bas.* 1995, *114*, 498.

122. Fleming, S. A.; Jones, R. W. *J. Het. Chem.* 1990, *27*, 1167.

123. Griesbeck, A. G.; Stadtmuller, S. *J. Am. Chem. Soc.* 1991, *113*, 6923.

124. Buhr, S.; Griesbeck, A. G.; Lex, J.; Mattay, J.; Schroer, J. *Tetrahedron Lett.* 1996, *37*, 1195.
125. Bonini, B. F.; Franchini, M. C.; Fochi, M.; Mazzanti, G.; Ricci, A.; Zani, P.; Zwanenburg, B. *J. Chem. Soc. Perkin Trans. I* 1995, 2039.
126. Mizuno, K.; Nakanishi, K.; Yasueda, M.; Miyata, H.; Otsuji, Y. *Chem. Lett.* 1991, 2001.
127. Mizuno, K.; Caldwell, R. A.; Tachibana, A.; Otsuji, Y. *Tetrahedron Lett.* 1992, *33*, 5779.
128. Sket, B.; Zupancic, N.; Zupan, M. *J. Chem. Soc. Perkin Trans. I* 1987, 981.
129. Sket, B.; Zupan, M. *Tetrahedron Lett.* 1989, *45*, 6741.
130. Lewis, F. D.; Hirsch, R. H. *J. Am. Chem. Soc.* 1976, *98*, 5914.
131. Okada, Y.; Ishii, F.; Kasai, Y.; Nishimura, J. *Tetrahedron Lett.* 1994, *50*, 12159.
132. Okada, Y.; Ishii, F.; Nishimura, J. *Bull. Chem. Soc. Jpn.* 1993, *66*, 3828.
133. Nishimura, J.; Takeuchi, M.; Takashashi, H.; Sato, M. *Tetrahedron Lett.* 1990, *31*, 2911.
134. Takeuchi, M.; Nishimura. J. *Tetrahedron Lett.* 1992, *33*, 5563.
135. Armesto, D.; Gallego, M. G.; Horspool, W. M.; Agarrabeitia, A. R. *Tetrahedron* 1995, *51*, 9223.
136. Steiner, G.; Munschauer, R.; Klebe, G.; Siggel, L. *Heterocycles* 1995, *40*, 319.
137. Ghosh, S.; Patra, D.; Samajdar, S. *Tetrahedron Lett.* 1996, *37*, 2073.
138. Langer, K.; Mattay, K. *J. Org. Chem.* 1995, *60*, 7256.
139. Langer, K.; Mattay, K.; Heidbreder, A.; Moller, M. *Leibigs Ann. Chem.* 1992, 257.

7

Photoinduced Redox Reactions in Organic Synthesis

Ganesh Pandey
National Chemical Laboratory, Pune, India

I. INTRODUCTION

Addition or removal (exchange) of an electron determines the chemical fate of the molecular entities to a large extent, although, at the primary stage, bonds are neither broken nor formed. Photoexcitation, which renders well-defined redox potential differences between two interacting species, has become an increasingly useful tool in initiating electron-exchange processes and to generate radical ions: a new type of reactive intermediates, from neutral substrates [1–4]. The importance of this concept in chemistry has grown rapidly during the last decade, and a subdiscipline known as photoinduced electron transfer (PET) appears to have emerged in the general arena of photochemistry. The knowledge gained over the years in the area of PET reactions concerning physical and mechanistic aspects has rapidly enhanced its scope in organic synthesis [5–7]. This may be imparted to the unique features of these transformations, as the key reactive intermediates are radical ions rather than photoexcited short-lived species. Sufficiently vast literature [5–7] has accumulated on the reactivity profiles of radical ions, which has led to the development of several new and novel synthetically useful chemical reactions. Product formation in these transformations is often governed by the secondary processes of initially formed radical ions because they serve as the precursors for the neutral radicals and ions by their mesolysis [8] (fragmentation of radical ions into radicals and ions).

The interaction of a donor (D) molecule with an acceptor (A) upon photoexcitation results either in partial charge transfer (exciplex formation) or electron transfer (radical ion formation) depending on the nature of the donor, acceptor, and solvent polarity [9]. Generally, the feasibility of producing radical ions via photoreactions is predicted by estimating the free energy change (ΔG_{et}) associated with their formation by using the Weller [10] equation [Eq. (1)], which employs experimentally derivable parameters such as oxidation potential of the donor [$E_{1/2}^{ox}(D)$], reduction potential of the acceptor [$E_{1/2}^{red}(A)$], energy of excitation ($E_{0,0}$), and Coulomb interaction (E_{coul}) in a given solvent:

$$\Delta G_{et} = E_{1/2}^{ox}(D) - E_{1/2}^{red}(A) - E_{0,0} + \Delta E_{coul} \qquad (1)$$

Under thermodynamically favorable PET reactions ($\Delta E_{et} < 0$), the radical ions are formed either as contact ion pair (CIP) or solvent-separated ion pair (SSIP). A closely related question is whether the primary intermediate is a SSIP or CIP. Gould et al. [11] have suggested in their recent study that in the polar solvents such as CH_3CN, electron transfer quenching results in the formation of SSIP directly, and in these solvents, the fully solvated ions (SSIP) can separate to form free radical ion pairs (FRIP) [12]. Therefore, under these reaction conditions, the anion radicals are potentially less reactive with the cation radicals than in nonpolar solvents in which CIP is more important. The use of polar solvents (e.g., CH_3CN and MeOH) thus facilitates ion radical chemistry [13].

Generally, the chemistry from the PET reactions is governed by the chemical properties and the reactivity profiles of the ion radicals formed. The most common pathways available to these reactive intermediates is the unimolecular dissociation to ions and radicals besides isomerization, cyclodimerization, nucleophilic/electrophilic addition, and substitution reactions.

This chapter is not intended to be the exhaustive survey of the literature, as many excellent reviews [1–7] have been written on the various aspects of this subject. The author has taken liberty to include only selected examples from the important contributions to highlight a general chemical reaction trend in order to encourage practicing organic chemists to use this emerging concept frequently for selective transformations during complex molecule synthesis.

II. CHEMISTRY FROM RADICAL CATIONS

The most common reaction observed from radical cation species is either the loss of proton or the dissociation of relatively weaker bond with consequent formation of radicals and cations with lifetime and chemical affinities different from those of parent species. However, in some instances, other types of reactions such as nucleophilic addition, cycloaddition, dimerization, and so on are also observed. This section has been devoted on the reactivity pattern of the radical cations.

A. Mesolytic Reactions

1. Carbon–Hydrogen Bond Dissociation (Deprotonation Reaction)

Usually, the formation of radical cation from a neutral substrate is associated with the increase in its acidity [14,15] and, therefore, facile deprotonation processes may be expected as the common step from some of these intermediates. In majority of instances, the proton transfer takes place between radical cation/ radical anion pairs, with the net result being the bimolecular coupling product. However, during sensitized PET reactions, deprotonation from radical cation is associated either with unilateral radical reactions or their further oxidation to produce carbocationic species [7]. Normally, the rate of proton transfer depends on the kinetic acidity of the cation radical and the basicity of the anion radical.

Efficient proton transfer from the benzylic position of alkylbenzene radical cations, formed by the electron transfer to excited 1,4-dicyanonaphthalene (DCN), to counter anion (DCN$^{-\cdot}$) is reported from Albini's group [16,17] to produce benzylic radical and DCN$^{\cdot}$, which upon mutual coupling yield photoaddition products (2, 3, and 4 in Scheme 1. Trace amount of bibenzyl is also formed by the dimerisation of benzylic radical. This reaction is shown to involve water-mediated proton transfer within the exciplexes and in cage coupling of the resultant radical pairs (Scheme 1) to form 2 and 3, whereas bibenzyl and 4 are suggested to arise from the escaped benzyl radical and coupling with DCN$^{-\cdot}$. The detailed study on the dependence of quantum yield on the solvent polarity and on the oxidation potentials of the alkylbenzenes, has led Lewis et al. [18] to suggest the involvement of FRIP in these reactions rather than initially produced CIP (nonemitting exciplex) or SSIP [17]. Detailed discussion on the deprotonation reaction from alkyl benzene radical cations may be found in the review article of Albini et al. [19]. Santamaria's [20] group has shown the selective reaction of the benzylic radical, formed by the deprotonation of alkyl arene radical cation generated by sensitized PET reaction using the singlet excited state of 9,10-dicyanoanthracene (^{1}DCA*) as electron acceptor and methyl viologen (MV^{++}) as electron relay, with the molecular oxygen to produce corresponding hydroperoxides in 57-100% yield as depicted through Eqs. (1)-(5) in Fig. 1. The photosensitized electron-transfer-(ET) generated arene radical cation from the excited p-methoxybenzyl protected ethers (5), in the presence of ground state DCN as an electron acceptor in wet acetonitrile, has been shown [21] to undergo efficient deprotonation reaction to produce benzyl radical, which gets oxidized further to a benzyl cation (7) by thermal ET

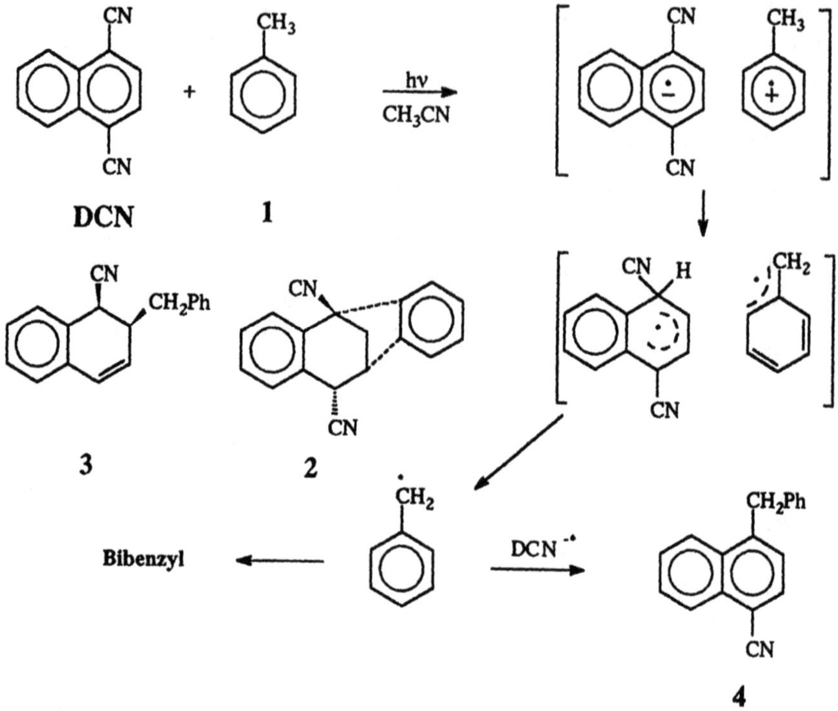

Scheme 1

to DCN. The hydroxylation of 7 and ensuing reaction pathways as shown in Scheme 2 leads to the net debenzylation product (9). An independent study by Nishida et al [22] has corroborated this photodebenzylation methodology. Albini's group in their recent publications [23] have demonstrated that PET between several alkanes and 1,2,4,5-tetracyanobenzene (TCB) is followed by efficient deprotonation of the radical cation with tertiary > secondary > primary

$$ArCH_2R + DCA^* \longrightarrow Ar\overset{+\bullet}{C}H_2R + DCA^{-\bullet} \quad \text{----} \quad (1)$$

$$DCA^{-\bullet} + MV^{++} \longrightarrow DCA + MV^{+} \quad \text{----} \quad (2)$$

$$MV^{+} + O_2 \longrightarrow MV^{++} + O_2^{-\bullet} \quad \text{----} \quad (3)$$

$$Ar\overset{+\bullet}{C}H_2R + O_2^{-\bullet} \longrightarrow Ar\overset{+}{C}HR + H\bar{O}_2 \quad \text{----} \quad (4)$$

$$Ar\overset{+}{C}HR + H\bar{O}_2 \longrightarrow ArCH(R)OOH \quad \text{----} \quad (5)$$

Figure 1 Equations depicting the selective reaction of the benzylic radical.

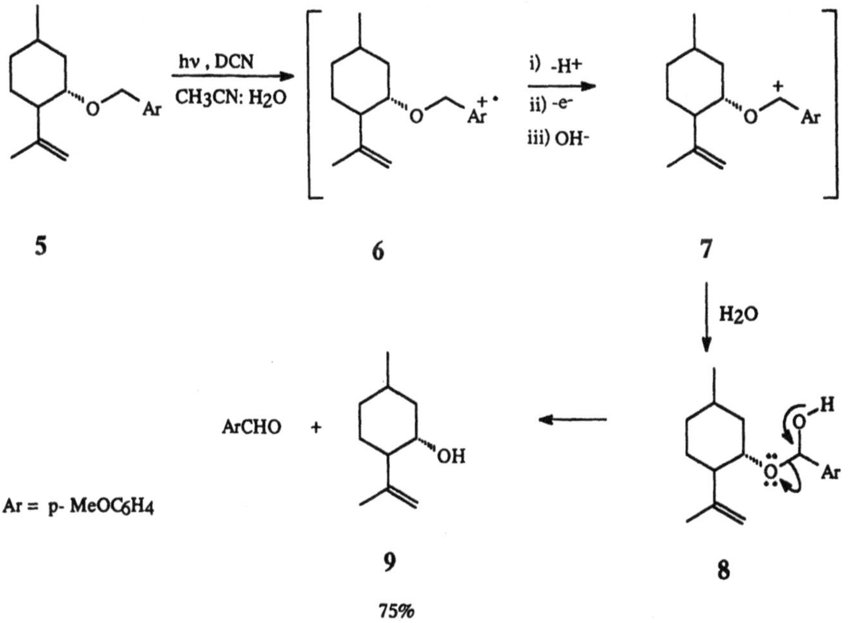

Scheme 2

selectivity. The alkyl radical is trapped by TCB⁻˙ or, when present, by oxygen. This reaction has been suggested to be useful for the functionalization of saturated alkanes.

Efficient proton transfer from tertiary amine radical cations from their α-C–H positions to counter the radical anion, formed by ET either between singlet excited arene–amine pairs [24–26] or triplet ketone–amine pairs, [27] has been known for a long time and cross-coupling of resultant radical pairs from donor–acceptor usually terminates the reaction. Incidentally, study of these reactions have contributed significantly to the present understanding of exciplexes and radical ion phenomena. Lewis et al. [28,29] have studied the ET-mediated photoaddition of tertiary amines to singlet excited state of *trans*-stilbene and has suggested that these reactions occur via CIP intermediates. Another report in this area has indicated [30] that the tertiary radical cation (10) formed under the sensitized PET reaction condition in the presence of ¹DCN*, a more potent electron acceptor than simple arenes, and in the solvent of high dielectric constant loses a proton to solvent medium rather than to the corresponding DCN radical anion. The α-amino radical (11) thus formed undergoes further one electron transfer, due to the reduced oxidation potential [31], thermally to DCN and generates an iminium cation intermediate (12) (Fig. 2) identical to electrooxidation [32] reactions. It appears that in sensitized PET reac-

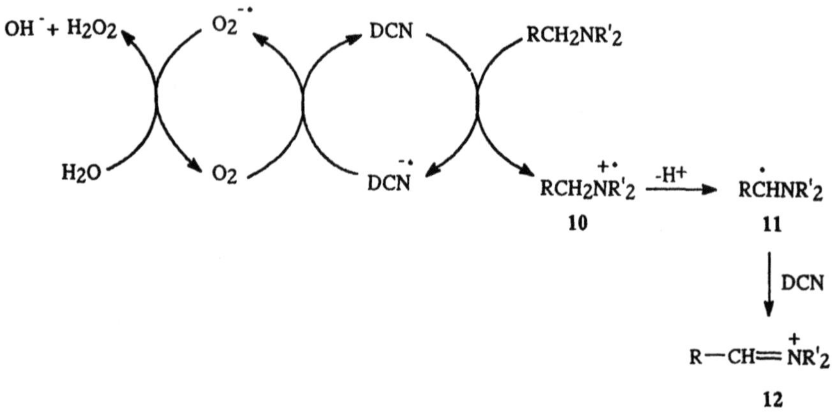

$OH^{-} + H_2O_2$ $O_2^{-\cdot}$ DCN $RCH_2NR'_2$

H_2O O_2 DCN $RCH_2\overset{+\cdot}{N}R'_2$ $\xrightarrow{-H^+}$ $RC\overset{\cdot}{H}NR'_2$

$\qquad\qquad\qquad\qquad\qquad\qquad$ 10 $\qquad\qquad\qquad\qquad$ 11

$\qquad\qquad\qquad\qquad\qquad\qquad\qquad\qquad\qquad$ DCN

$\qquad\qquad\qquad\qquad\qquad\qquad\qquad$ $R-CH\overset{+}{=\!\!=}NR'_2$

$\qquad\qquad\qquad\qquad\qquad\qquad\qquad\qquad\qquad$ 12

Figure 2 Generation of an iminium cation intermediate.

tions, the initially formed SSIP collapses to produce FRIP, and from there, the corresponding amine radical cation loses a proton to the aqueous solvent [33]. Several interesting applications of this electron–proton–electron (E-P-E) transfer sequence have been developed [30] for designing new synthetic reactions. For example, efficient transformation of N-hydroxyl amines (13) to corresponding nitrones (1,3-dipoles) (16) is achieved [34] by the PET reaction of (13), using DCN as a light-absorbing electron acceptor, in fairly good yields. The [3+2]-cycloaddition of 16 with a number of dipolarophiles give good yields of heterocyclic compounds (17) (Scheme 3). Subsequently, the sequential E-P-E transfer concept is extended to generate iminium cation intermediates from tertiary amines: an important reactive species utilized during the synthesis of biologically active heterocycles. Because, in Fig. 2, the iminium cation intermediate (12) is formed by further oxidation of α-amino radical (11), generated by the α-deprotonation of amine radical cation, the regiospecificity of iminium cation in unsymmetrically substituted tertiary amines would depend on the factors that influence the deprotonation site from the radical cations. Kinetic acidity is subject to stereoelectronic factors, solvent polarity, basicity of oxidizing agents, and the oxidation potentials of the amines, some of the important parameters that influence the site of deprotonation from the corresponding amine radical cation [33]. To probe this aspect, the PET reaction [35,36] of amine (18) is examined and has been found to produce imino ether (22) by the intramolecular trapping of the regioselectively formed iminium cation (21) by the tethered hydroxyl group. The regiospecificity of 21, as shown in Scheme 4, is likely to emerge due to the greater kinetic acidity of ring α-C–H protons. The quantum efficiency for the formation of imino ethers (22) in these cyclizations are in the range of $\varphi \cong 0.05$. The rate of these cyclizations is significantly

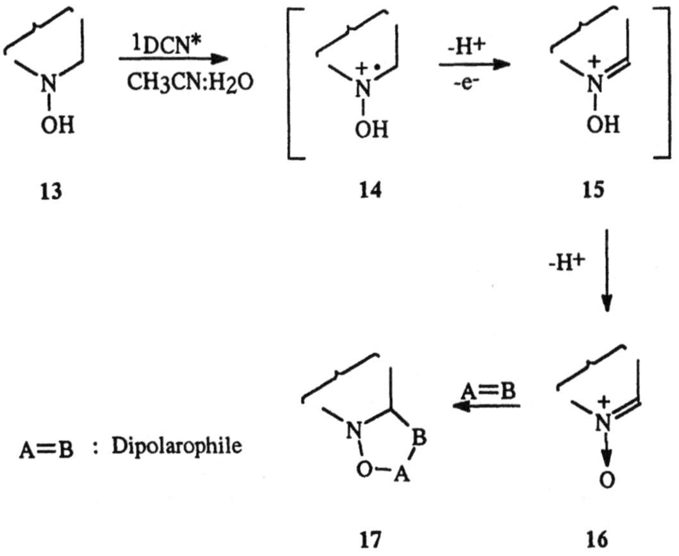

Scheme 3

enhanced [36] when methyl viologen (MV^{++}) is used as an electron relay. The electron transfer cycle in this case may be operating through the pathways as shown in Fig. 3. The lower reduction potential of MV^{++} (MV^{++}/MV^{+} = -0.45 eV) compared to oxygen ($O_2/O_2^{-\cdot}$ = -0.78 eV) possibly facilitates the reaction. Regioselectivity of iminium cation from 23, where two ring α-deprotonation sites are available, is further probed [36,37] by studying the sensitized PET reaction of 23 which produces regioselective and stereoselective tetrahydro-1,3-oxazine (24). The observed product indicates the complete regioselectivity for the ring closure toward the less substituted α-C–H moiety. As the deprotonation step from the amine radical cation requires the overlap of the half-vacant nitrogen p orbital with the incipient carbon radical p orbital, the steric factor in these cases forces the generation of the least substituted α-amino radical and thus the regioselective generation of the iminium cation. The

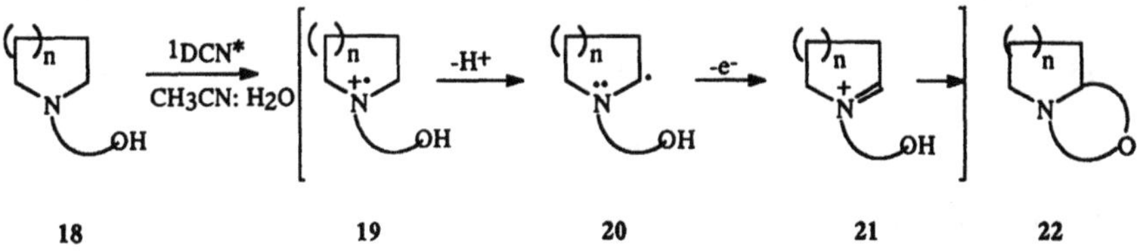

Scheme 4

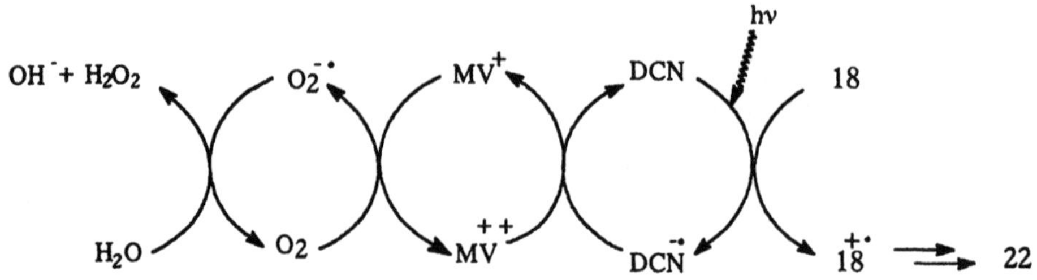

Figure 3 Pathways for the ET cycle for production of imino ether (22).

formation of 24 as a major diastereomer is explained on the basis of the preferential frontal attack of the hydroxyl group to the iminium cation, possibly to avoid steric crowding [37]. The stereoselectivity during the formation of 24 and its unique structural features has allowed this methodology, as shown in Scheme 5, to be used for the synthesis of biologically important cis-α-α'-dialkylated pyrrolidines and piperidines [37] (25) by the nucleophilic ring opening of 24 followed by the N-dealkylation reaction.

The dependence of the α-deprotonation site on the kinetic acidity of the planar amine radical cation is best demonstrated by the sensitized ET photoreaction of N-benzyl tertiary amines and N-alkyl anilines in the presence of ^{1}DCN*, which leads to the efficient N-debenzylation [38] and N-dealkylation [39], respectively. These results are explained by considering the hydrolysis of the respective regioselective iminium cations produced during the reaction. The utility of these studies is demonstrated in the N-debenzylation [38] reaction of benzyl-protected tertiary amines and in the understanding of the mechanism of oxidative N-dealkylation [39] of tertiary amines by cytochrome P-450-dependent monooxygenases.

Further utility of the regiospecific iminium cation generation is shown [40] by the synthesis of 28 as the substrate for the enantioselective synthesis of α-

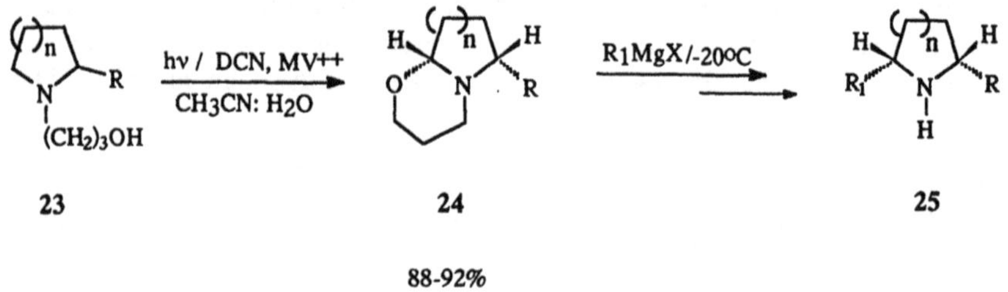

n = 1&2, R = Me, n-Bu

Scheme 5

amino acids and their *N*-methyl derivatives (29). The precursor 28 is obtained in a 11.5 : 1 diastereomeric ratio by the intramolecular cyclization of iminium cation intermediate 27, generated by the usual PET reaction of 26 using ^{1}DCN* as an electon acceptor. The nucleophilic ring opening of 28 followed by relevent chemical manipulations produced α-amino acids and/or their derivatives (29) as shown in Scheme 6. In this methodology (L)-prolinol (30) acts as a recyclable chiral auxiliary. Another interesting application of such cyclizations may be cited [41] where compound 33 is designed as the precursor (Scheme 7) for the enantioselective alkylation of cyclic amines at the α-position—an important and challenging problem in organic synthesis. Optically, active mandelic acid acts as a recyclable chiral auxiliary in this strategy. Both (*R*) and (*S*) alkylated products (35) have been obtained from pyrrolidine, piperidine, and tetrahydro-isoquinoline in 92–98% *ee* depending on the configuration of mandelic acid.

Intermolecular trapping of the iminium cation, generated by the sensitized PET reaction of tertiary amine moiety using *N*,*N*′-dimethyl-2,7-diazapyrelium-bis tetrafluoroborate salt as an electron acceptor, by Me$_3$SiCN is reported by

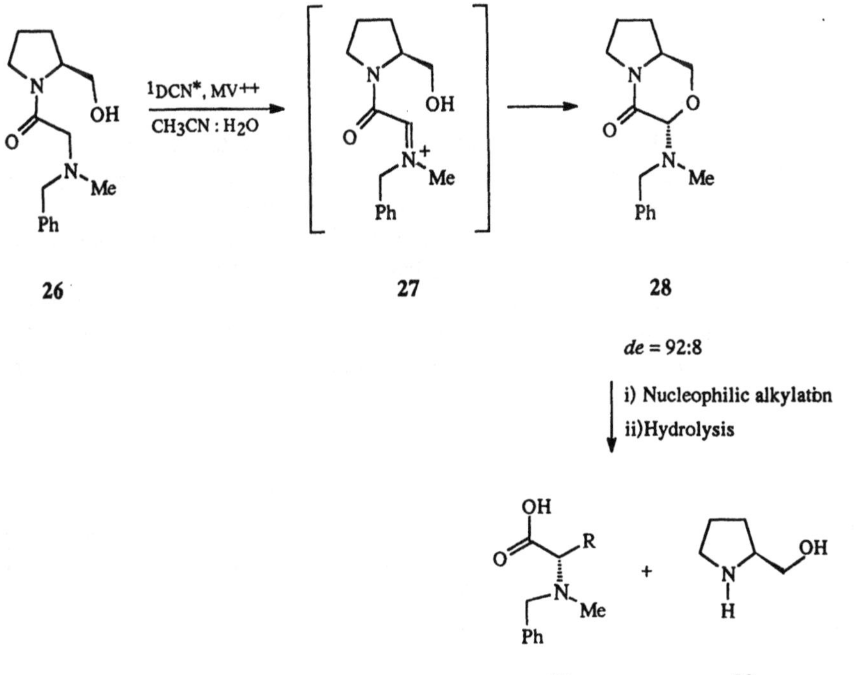

Scheme 6

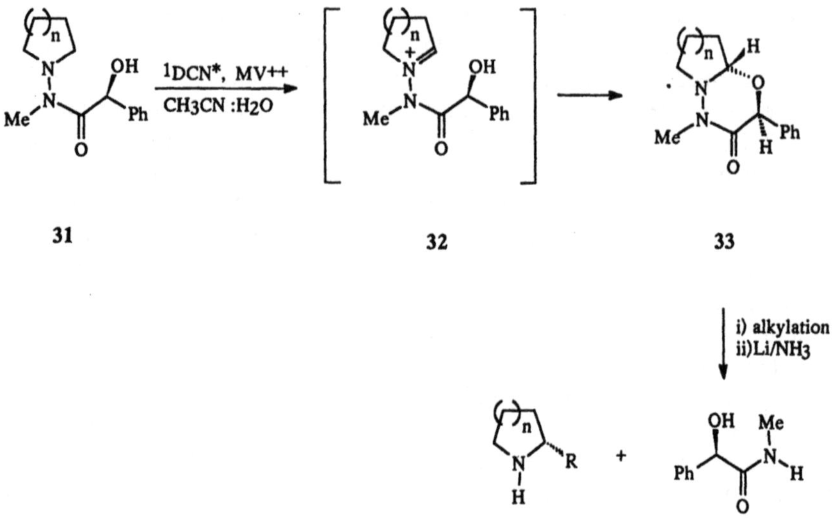

31   **32**   **33**

i) alkylation
ii)Li/NH3

35   +

ee = 92-98%   **34**

Scheme 7

Santamaria et al [42] for synthesizing α-amino nitriles in the alkaloid field and
also for preparing 6-cyano-1,2,3,6-tetrahydropyridine from corresponding pyridine nucleus. A similar approach has also been used by Sundberg et al.
[43a,43b] for the cyanation of Catharanthine alkaloids. In situ trapping of the
iminium cation (37) by allyltrimethylsilanes or silyl enol ethers is also shown
[44] recently as a direct —C–C— bond formation methodology at the α-position of tertiary amines (Scheme 8). The success of this reaction is based on the
comparative correlation of ion-pair yield with the ΔG_{et} values from amines and
enol ethers.

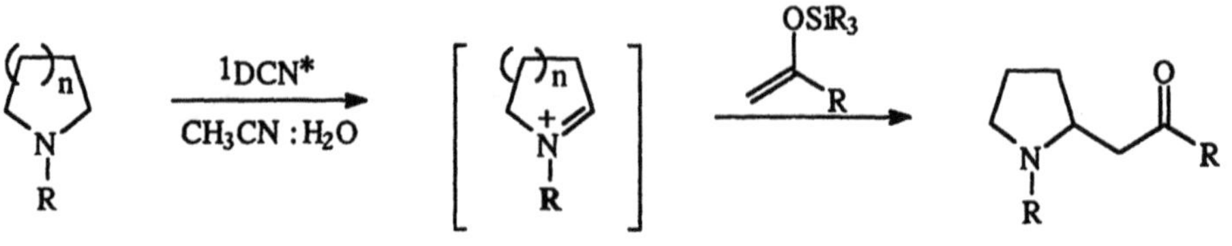

36   **37**   **38**
Scheme 8

2. Carbon–Metal Bond Dissociation

Photoelectron spectroscopic studies of group 4A organometallics (R_4M or $R'MR_3$; M= Si, Ge, Sn, and Pb) have revealed that the ionization of these compounds is associated with the electrons relatively close to the central atom (i.e., from —C–M bonding orbitals [45,46]. Ionization potentials of these compounds are found to be in the range of typical σ-donors effective in charge-transfer-(CT) complex systems and many such organometallics are shown [47] to form ground stae CT complexes with electron-deficient acceptors such as tetracyanoethylene (TCNE). The chemical bond insertion reaction resulting from the thermally induced ET bond cleavage of resultant $R_4M^{+\cdot}$ has been established by Kochi et al. [48,49] in these systems. ESR annealing studies [50,51] have suggested that the cleavage of $R_4M^{+\cdot}$ species produces a carbon-centered radical and M^+. Subsequently, however, Eaton [52] has demonstrated that there could be a dual pathway for the cleavage of the benzyltrimethylstannane radical cation, providing both a carbon-centered radical and carbocationic products. Eaton [53] in the meantime reported on the efficient PET from benzyltrialkyl stannane (39) to the singlet excited state of DCA and the formation of photoadducts 41–43 (Scheme 9). In mechanistic terms, it is explained that, initially, product 40 is formed by the recombination of DCA$^{-\cdot}$ with the benzyl radical and the tributyltin cation, formed by the —C–Sn— bond dissociation step of $39^{+\cdot}$ in the solvent cage. ET involving a nonemitting CT complex is suggested on the basis of the detailed fluorescence quenching study of ^{1}DCA* with benzyltrialkylstannane and other kinetic parameters. Fragmentation of the —C–M— bonds from their corresponding radical cations of group 4A organometallics usually generates carbon-centered radicals which apparently adds to the reduced form of the acceptor to give photoaddition products. Photoalkylation of cyanoaromatics often results when these are irradiated in the presence of organometallic compounds [54]. Indirect alkylation of dicyanobenzene (DCNB) is also achieved [55] by activating RMR'_3 (Scheme 10) by phenanthrene sensitization. Comparative analysis of the rate of $RMR'_3]^{-\cdot}$ cleavage suggests the $Si > Ge > Sn$ order. Recently, Mella's group [56] have reported a method of radical addition to the activated olefins which is based on the carbon-centered radical generation by the fragmentation of the tetraalkylstannane radical cation generated by the sensitized PET reaction using cyanoaromatics as light-absorbing electron acceptors. In several cases, the alkylation is shown to occur more efficiently in the presence of a secondary electron donor (phenanthrene or bibenzyl). Several groups have also utilized this cleavage pattern of $R'MR_3]^{+\cdot}$ to produce carbon-centered radicals for the photoallylation/benzylation of quinones [57], N-methyl arene carboxamides [58], and α-diketones [59]. In all of these reactions the principal mechanism appears to be the same, where an excited triplet ketone acs as an electron acceptor from $R'MR_3$ and the radical generated

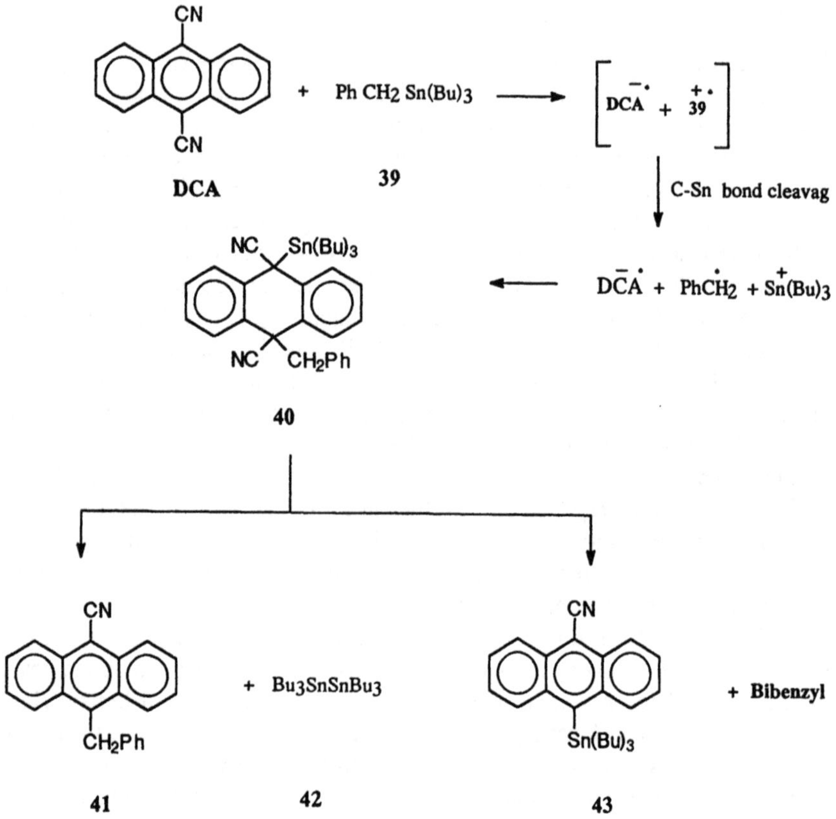

Scheme 9

after the mesolysis of the corresponding radical cation adds to the reduced ketone in a 1,2-fashion (Fig. 4).

Carbon–silicon bond heterolysis from the radical cation of trimethylsilyl-substituted ethers, thioethers, and amines, generated by ET to ^{1}DCA*, is reported by Mariano et al. [60] to produce a corresponding methylene radical which ultimately combines with reduced DCA$^-$ to yield the net photoaddition products 45 and 46, respectively. Photoaddition of $Me_3SiN(Et)_2$ initially produces corresponding dihydroanthracene adduct (47), which spontaneously dehydrocyanates under the reaction condition to give (48) (Scheme 11).

Mariano's group [61] have extensively investigated the synthetic potentials of the cleavage of the —C–Si— bond from allyl and benzyl silane radical cations produced by the photoreaction of electron-deficient iminium salts. Mechanistically, this result is interpreted by considering one electron oxidation

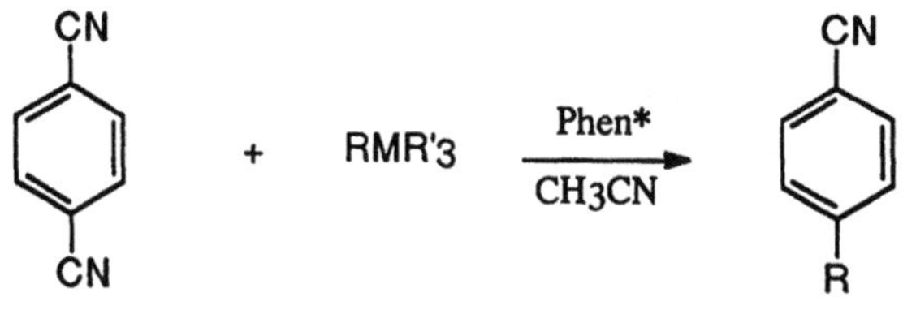

DCNB   44

R = allyl or benzyl
R' = Me
M = Si, Ge, Sn
Phen = Phenanthrene

Mechanism:

$$Phen \xrightarrow{h\nu} Phen^*$$

$$Phen^* + DCNB \longrightarrow [\, Phen \overset{+\bullet}{......} DCNB \overset{-\bullet}{}\,] \longrightarrow Phen \overset{+\bullet}{} + DCNB \overset{-\bullet}{}$$

$$Phen \overset{\bullet}{} + RMR'_3 \longrightarrow Phen + [RMR'_3] \overset{+\bullet}{}$$

$$[RMR'_3] \overset{+\bullet}{} \longrightarrow R \overset{\bullet}{} + R'_3M \overset{+}{}$$

$$DCNB \overset{-\bullet}{} + R \overset{\bullet}{} \longrightarrow 44$$

Scheme 10

of the silane moiety (50) to the excited iminium salt (49) to produce a radical/radical cation pair (51-52). Subsequent —C–Si— bond fragmentation from the silane radical cation (52) by the electrofugal loss of TMS^+ forms an allyl/benzyl radical, which, upon coupling with 53 in the solvent cage, provides 54 (Scheme 12). Nucleophile-assisted elimination of TMS^+ is suggested as the key factor in these reactions. The detailed study of the fluorescence quenching of iminium salt with silanes with a rate near diffusion control and the evaluation of other relevant kinetic parameters has formed the basis of a ET mechanism in these reactions. A comparative study [62] using benzyltrimethylsilanes and corresponding stannanes has shown similar chemistry. The —C–Si— bond cleavage chemistry in this fashion has been exploited extensively for synthetic purposes.

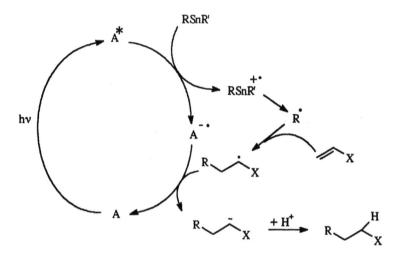

A = Cyanoaromatics

X = COOEt, CN

Figure 4 Radical addition to alkenes via ET photosensitization.

Among the most prominent is the synthesis of (±)-xylopinine [63] (56), a protoberberine alkaloid, and the alkaloids possessing an erythrane ring [64] (59), by the intramolecular PET cyclization of 55 and 57, respectively (Scheme 13).

   Dissociation of —C–Si— bond from the α-silylmethylamine radical cation (63), formed by ET to a triplet excited enone (60) is described by Mariano et al. [65] to produce a "free α-amino radical" (65) which adds to the β-ketyl radical (64) to produce a conjugate addition product (66). This cleavage mode is shown to be highly solvent dependent. For example, fragmentation of —C–Si— bond occurs only in alcoholic solvent, whereas α-deprotonation is the predominant pathway in acetonitrile solvent (Scheme 14). An interesting explanation to this observation is made by implicating the intermediacy of FRIP in the former case, whereas CIP is important in acetonitrile. Subsequently, 63 generated under a photosensitized ET reaction condition using ^{1}DCA* as an elecron acceptor has been shown to undergo exclusive —C–Si— bond cleavage [66] independent of the solvent polarity possibly through FRIP. The photosensitized ET reaction from α-trimethylsilylmethyl amine compounds tethered with activated olefin are shown to undergo efficient intramolecular cyclizations, and a variety of useful compounds such as fused as well as spiro N-heterocycles [67], the highly functionalized E-ring of Yohimbane alkaloids [68], and stereosubstituted piperidines [69] are synthesized using this methodology.

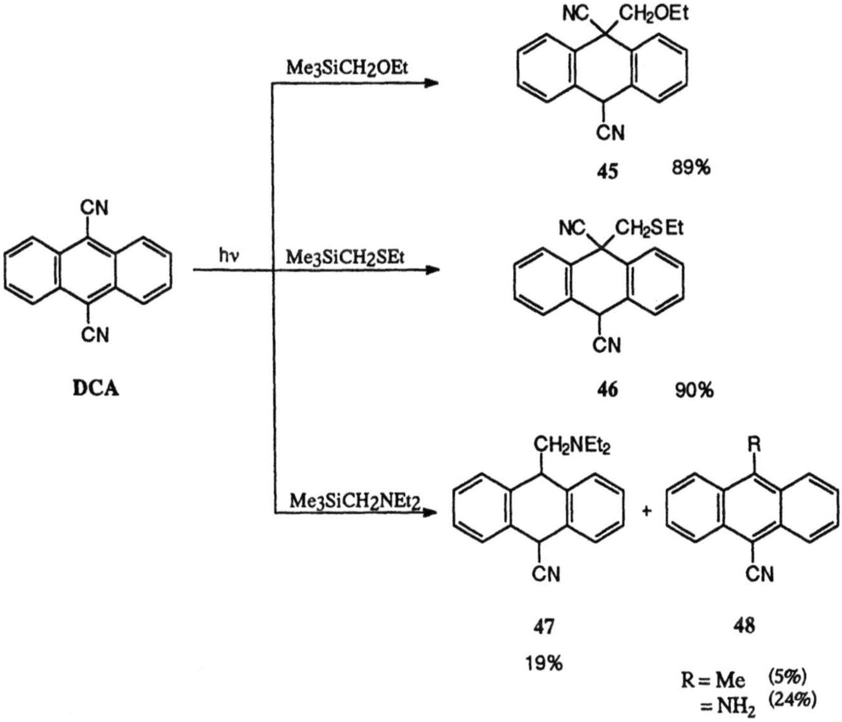

Scheme 11

Simultaneous with Mariano et al.'s [65] first publication on the desilylation reaction from α-trimethylsilylamine radical cation, we reported [70] photocyclization of silylated methylamine of type 69 by sensitized photoreaction using ^{1}DCN* as an electron acceptor in i-PrOH to produce N-heterocycles

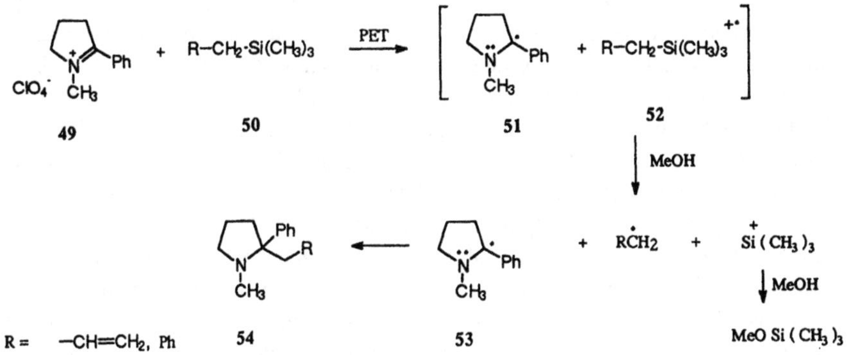

Scheme 12

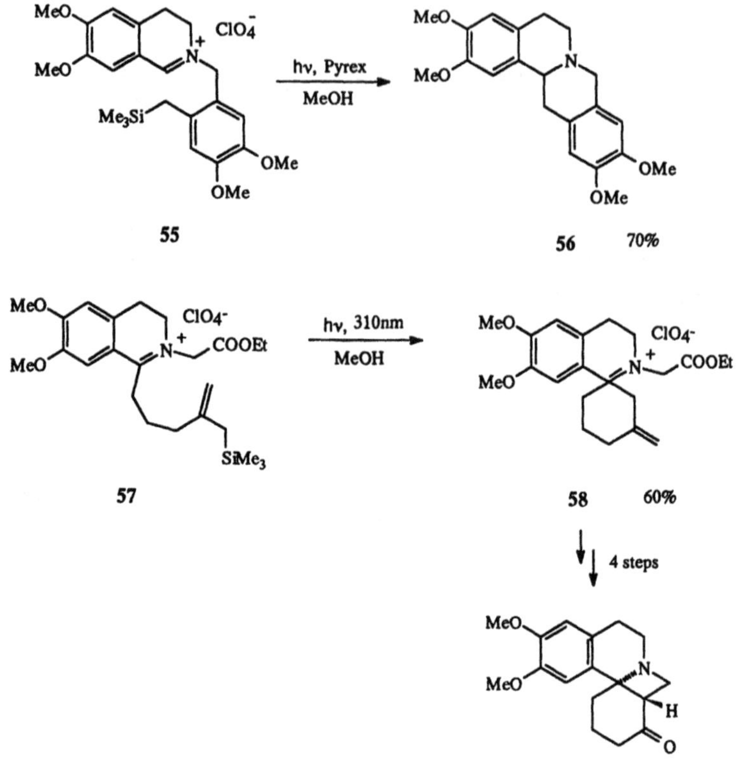

Scheme 13

(73). In sharp contrast to Mariano et al.'s [66] suggestion on the involvement of the "free α-amino radical" in these cyclizations, we have pointed out that the cleavage of the —C—Si— σ-bond from the delocalized trimethylsilyl methyl amine radical cation (71), produced by the vertical overlap of the —C—Si— bond and empty p-orbital of nitrogen, is assisted by the π orbitals of the olefin [71] (Scheme 15). Convincing support of this arguement is provided on the basis of extensive experimentation and drawing parallels from literature reports. The stereochemical aspect of these cyclizations is also investigated [71] to support our hypothesis by studying the cyclization of (74) which is found to be nonstereospecific (1 : 1 mixture of 75 and 76) unlike the 3-substituted carbon-centered radical cyclisations [72] (Scheme 16). A plausible explanation to this observed difference is advanced [71] by considering the lower energy barrier between the two possible "chair" conformations of the transition state due to

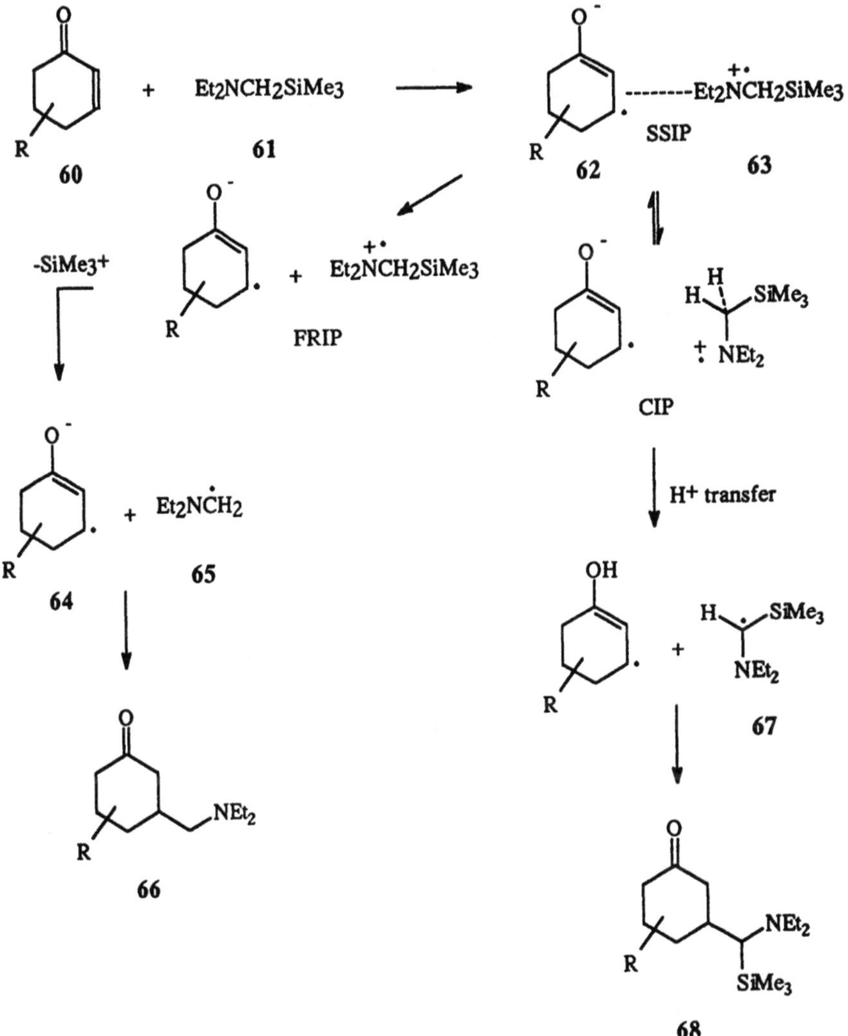

Et2NCH2SiMe3

61

60

62 SSIP

-------Et2NCH2SiMe3

63

-SiMe3+

+• Et2NCH2SiMe3

FRIP

H H SiMe3

+ NEt2

CIP

+ Et2NCH2

65

64

H+ transfer

OH

H SiMe3

NEt2

+

67

NEt2

R

66

O

NEt2

R SiMe3

68

Scheme 14

the lone pair flipping on the nitrogen. This argument is further supported [73] by studying the cyclization stereochemistry of the cyclic analogs (77), where lone pair flipping is expected to be restricted and gave diastereoselective 1-azabicyclo [m.n.o] alkanes (78 and 79) (Scheme 17). The stereochemistry of these cyclizations are dependent on the size of the new ring formed (e.g., 1,5-*cis* and 1,6-*trans*). An interesting application of observed stereoselectivity is

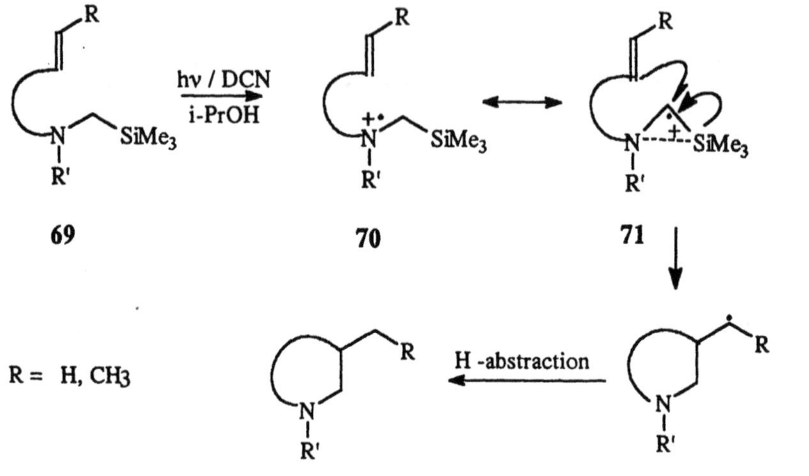

Scheme 15

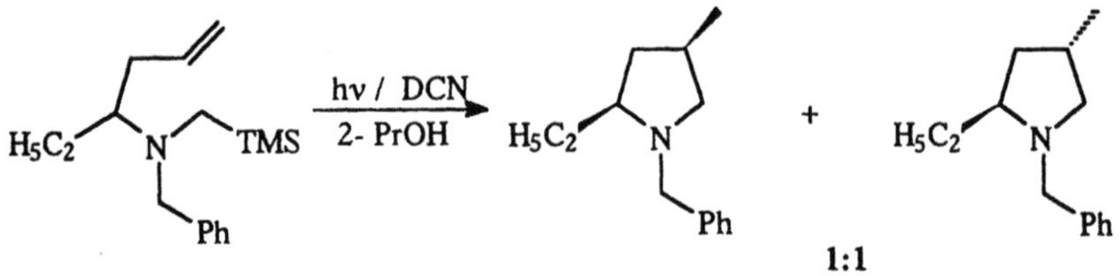

Scheme 16

extended for the synthesis of (±)-iso-retronecanol and (±)-epilupinine and other related alkaloids [74].

Immediately after our detailed mechanistic disclosure [71] on these cyclizations, Mariano's group [75] countered, surprisingly, that cyclization of (77i) to (78i), as reported by Pandey and Reddy [73], is not reproducible and insisted that such cyclizations occur only when tethered olefin is activated. The reason for their failure in reproducing the cyclization of (77i) by following the reaction condition in Ref. 73 is not very clear; however, to reconfirm the earlier experimental observation [73], optically active (−)-retronecanol (82), an important necine base in the *senecio* series of alkaloids, is synthesised [76] by the PET cyclization of (81) (Scheme 18). The precursor compound 81 is obtained by the silylation of N-BOC-derivatives of (−)-3-hydroxypyrrolidine (80) employ

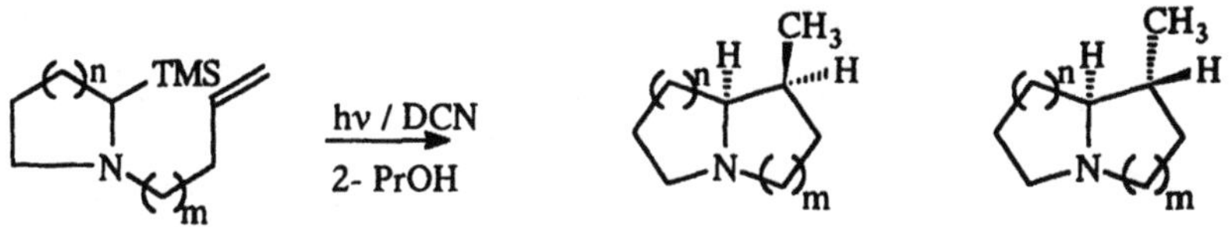

77 78 79

	78/ 79	% Yield
i) n = m =1	97:3	90
ii) n = m = 2	0:100	88
iii) n = 1, m = 2	2:98	85
iv) n = 2, m = 1	95:5	87

Scheme 17

ing Beak's [77] metallation procedure. In a related study, the sensitized PET reaction of α-trimethylsilylmethyl amine tethered with the aldehyde group is observed to produce corresponding cycloalkanols in the identical stereochemical pattern as found in the case of (77). (+)-Castenospermine [78] (88), an important glycosidase inhibitor and potential anti-HIV agent, it synthesized [79]

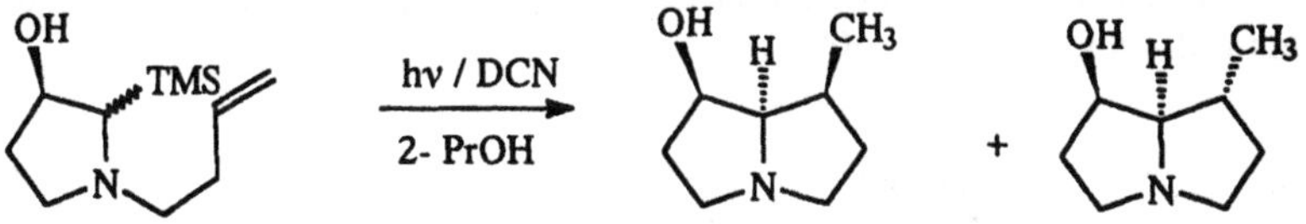

81 82 83

major minor

(-) - Retronecanol

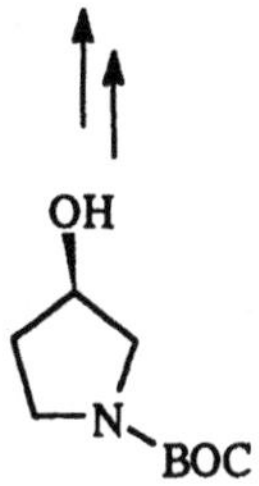

80

Scheme 18

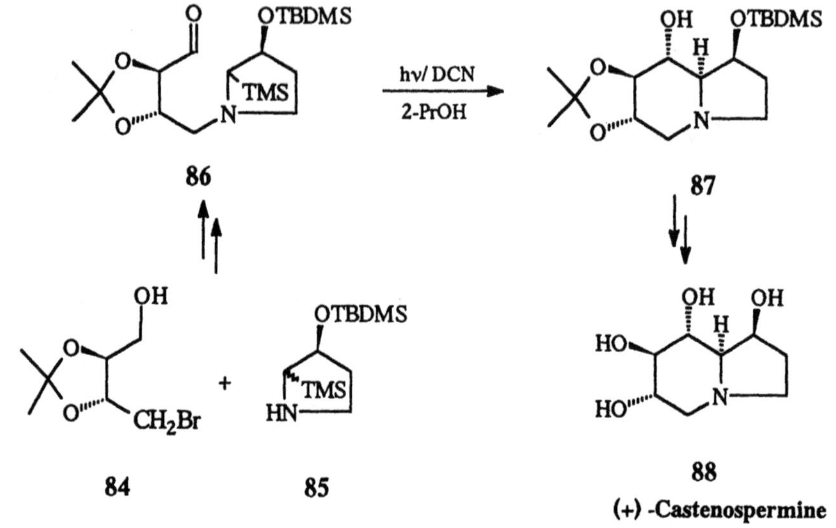

Scheme 19

by PET cyclization of 86 as shown in Scheme 19. The PET generation of a radical cation from N,N'-bis(trimethylsilylmethyl)alkylamines (89) in acetonitrile solvent is observed [80] to undergo a sequential double desilylation (silylation–electron–desilylation) reaction to generate nonstabilized azomethine ylide (92) (1,3-dipole) efficiently, in an identical manner as reported [35] during the transformation of N-hydroxylamines to nitrones, which, on cycloaddition with various activated olefins, produces corresponding pyrrolidines (93) in good yields (60–95%) (Scheme 20).

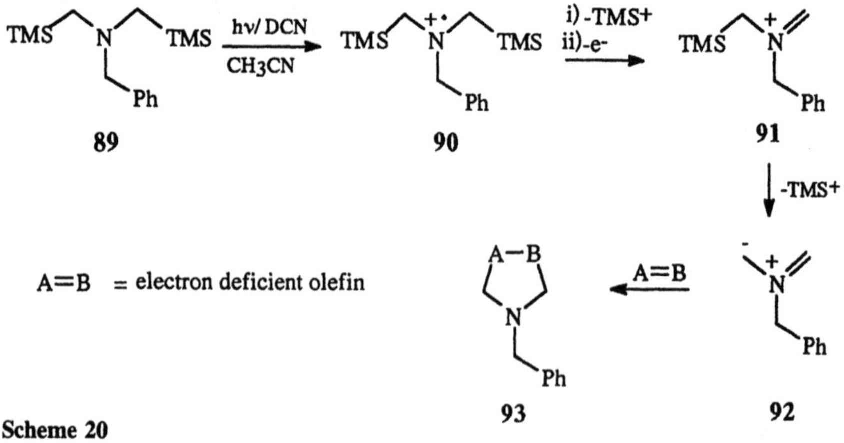

A=B = electron deficient olefin

Scheme 20

It is apparent from the above examples that a vast majority of radical cations generated photochemically from group 4A organometallics cleave to produce carbon-centered radicals. However, a contrasting observation is made [81] during the cleavage of —C–Se—]$^{+\cdot}$, generated by the sensitized ET photoreaction of organoselenium substrates in the presence of ^{1}DCN* in polar solvents which produced carbocationic species. These carbocationic species are normally terminated by the nucleophilic addition of solvents or other nuclophiles. The FRIP produced in polar solvent appears to undergo efficient —C–Se— bond disproportionation, producing carbocationic species, and PhSe$^{\cdot}$, which has provided unique opportunity for utilizing this reaction for the deselenylation reaction [81,82] from organoselenium compounds (e.g., 94→95) without compromising the functionality aspect (Scheme 21). Another interesting application of this activation pattern of the —C–Se— bond has led to the development of a unique —C–C— bond formation methodology [83] by the coupling of nonactivated alkylselenides (96) with silyl enolethers (97) (Fig. 5).

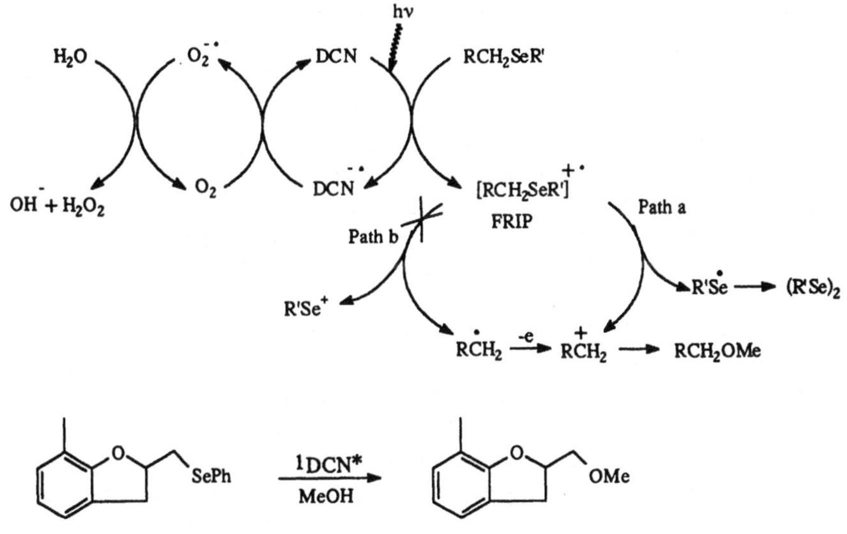

Scheme 21

3. Metal–Metal Bond Dissociation

Photoelectron spectroscopic studies of group 4A organodimetallics (R_3M–$M'R_3$, M–M'; Si, Sn, and Ge) indicate the low energy of ionization associated with the M–M bond rather than adjacent —C–M— bond [84]. Two independent groups [85,86] have observed CT-complex formation between organodimetallics

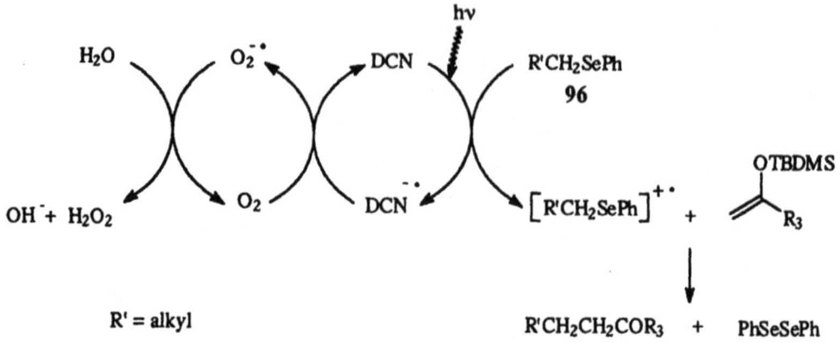

Figure 5 Unique —C–C— bond formation methodology.

(permethylpolysilanes) and tetracyanoethylene (TCNE) even though silanes lack either a metallic bond character or lone pair electrons. The —Si–Si— bond dissociation reaction is observed [86] when these CT complexes are irradiated in the ESR cavity at room temperature. The electron transfer processes in these systems have been confirmed by ESR detection of TCNE⁻·; however, the corresponding radical cation from the —Si–Si— bond could not be detected. Radical cations could only be noticed [87] when these complexes were irradiated using ^{90}Co γ-rays at frozen temperatures in Freon. These studies, in fact, have established that in organodimetallics, the hole lies between two metal atoms, which weakens the bond by stretching and allowing two metal atoms to attain planarity, culminating in fragmentation to provide stable radical and cationic species.

Sakurai et al. [88] have demonstrated, possibly for the first time, the ET-induced photodissociation of —Si–Si— bond from 98, by irradiating the CT complex with TCNE, to produce silyl extrusion product 102 (Scheme 22). Subsequently, the same group has also reported [89] chlorinative cleavage of the —Si–Si— bond of permethyl polysilanes initiated by ET using DCA as the sensitizer. The cleavage is suggested to proceed via a transient polysilyl radical cation intermediate. Usually, during the sensitized cleavage of the —Si–Si— bond, the nucleophilic trapping of the cationic site has been rather difficult under normal reaction conditions, partly due to the short lifetime of radical cation and/or rapid back electron transfer from DCA⁻·. However, the trapping of a cationic site could finally be accomplished [90] by the PET reaction of a molecule having a tethered nucleophile such as polysilyl alkanols (103) to produce cyclic product 104 (Scheme 23). The —Si–Si— bond cleavage from $Me_3SiSiMe_3$ is also reported by Fukuzumi et al. [91] when it is photolyzed in the presence of the 10-methyl acridinium ion (AcrH⁺). The AcrH⁺, an analog of NAD⁺, acts as an electron acceptor in this reaction. The cleavage of

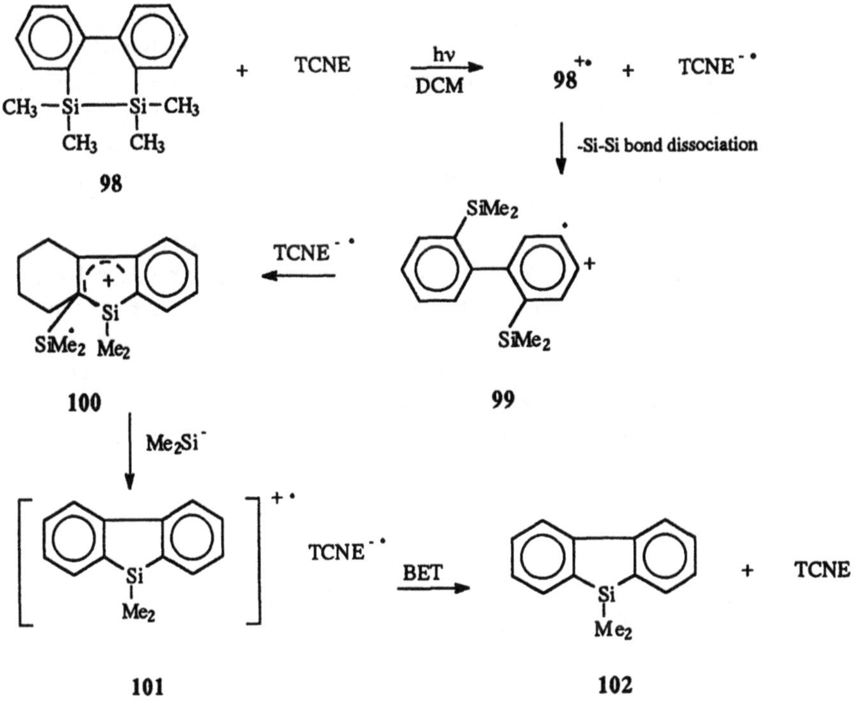

Scheme 22

other group 4A organodimetallics ($R_3MM'R_3$; $M = M' = Si$, Sn, Ge) is also described [92] via their corresponding radical cations, generated by ET from dimetallic compounds to the excited $AcrH^+$. However, if $M = M' = Si$, two electron reductions of $AcrH^+$ is normally observed [93] because of the lower reducing ability of $Me_3Si^{\cdot}$ compared with $Me_3Sn^{\cdot}$ and $Me_3Ge^{\cdot}$.

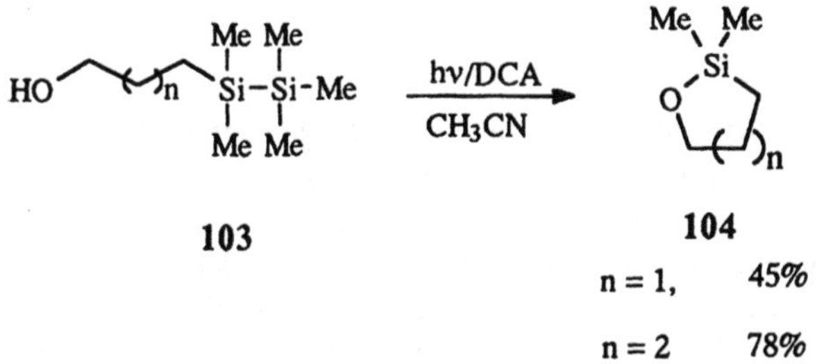

n = 1,   45%

n = 2   78%

Scheme 23

Selenium-selenium (—Se–Se—) bond dissociation from PET-generated PhSeSePh]$^{+ \cdot}$ using DCN as an electron acceptor is reported [94,95] by the author's group to produce an electrophilic selenium (PhSe$^+$) species which has been suggested to be useful in carrying out a variety of selenoetherification reactions (Scheme 24). Diffusion-controlled fluorescence quenching of DCN with PhSeSePh and the evaluation of other relevent kinetic and thermodynamic parameters has formed the basis of implicating ET processes in this reaction. The utility of electrophilic selenium species generated in this fashion is further demonstrated [96] to bring about enyne (106) cyclizations to produce 108 and 109 (Scheme 25). TBAB has been used as a substrate to enhance the nucleophilicity of acetylenic moiety toward the episelenonium cation intermediate (107) required for the cyclization step. The formation of unexpected product (109) in this cyclization has indicated the possible participation of initially formed PhSeSePh]$^{+ \cdot}$ itself as an electrophilic selenium species either in full or in part [97].

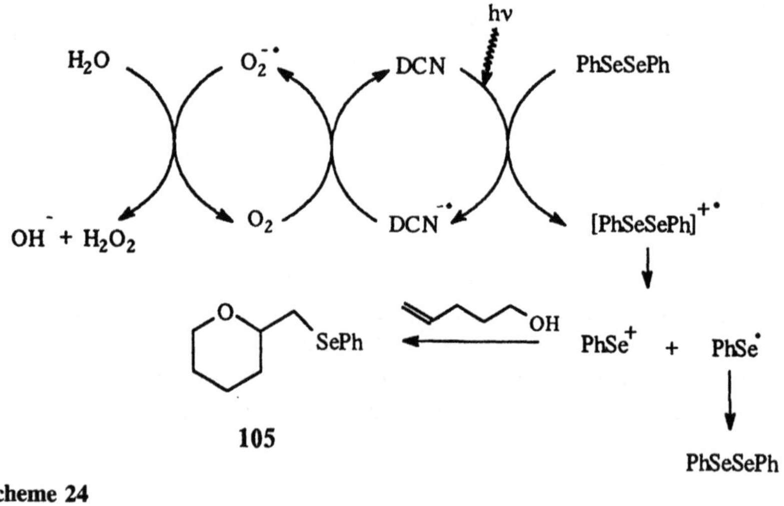

105

Scheme 24

A novel intra-ion-pair electron(*3*) transfer cleavage of the —C–B— bond is reported by Chatterjee et al. [98] from the photoreaction of triphenyl alkyl borate salts of cyanine dye (110) to generate a carbon-centered radical (113) whose utility for initiating a free radical polymerization reaction has been demonstrated (Scheme 26). An intramolecular concerted bond cleavage/coupling process is described [99] from phenyl anthracene sulphonium salt derivatives as a part of designing a novel "Photoacid" system.

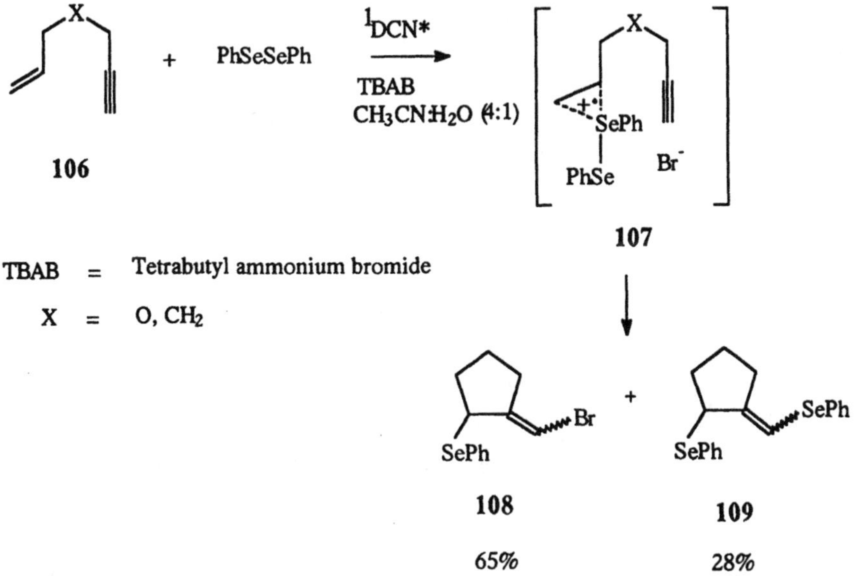

TBAB = Tetrabutyl ammonium bromide

X   =   O, CH$_2$

108
65%

109
28%

Scheme 25

4. Carbon–Carbon Bond Dissociation

Arnold's [100–102] pioneering work on the oxidative —C–C— bond cleavage reaction from the corresponding radical cation of 2,2-diphenylether and 1,2-diphenylethane, initiated by the sensitized photoreaction, using DCN as the electron acceptor is very well recognized. The fragmentation of the —C–C— bond is governed by the stabilities of radicals and cations in competition with other possible deactivation pathways such as proton loss, nucleophilic additions, or further electron transfer. Griffin et al. [103,104] have provided strong evidence of the ET mechanism from laser flash photolysis studies for the cleavage of diphenylethane and benzpinacols. Dissociation of the activated pinacols to generate corresponding ketones is reported by Sankaraman et al. [105] from the photolysis of the CT complex of pinacols with chloranil. Analogous cleavage of the methoxy bicumins is also demonstrated [106] from the photolysis of the CT complexes of bicumins and chloranil. Photolysis of the CT complex of 4-methoxy-4'-X-bicumenes (115, c: Me, d: CF$_3$, e: CN) with tetranitromethane (TNM) leads [107] to the formation of 119 and 120 via —C–C— bond scission of the initially formed 115$^{+\cdot}$ and —C–NO$_2$— bond dissociation of TNM$^-$· followed by radical coupling reactions, as shown in the Scheme 27. The nature of the X and the solvent polarity greatly influence the fragmentation

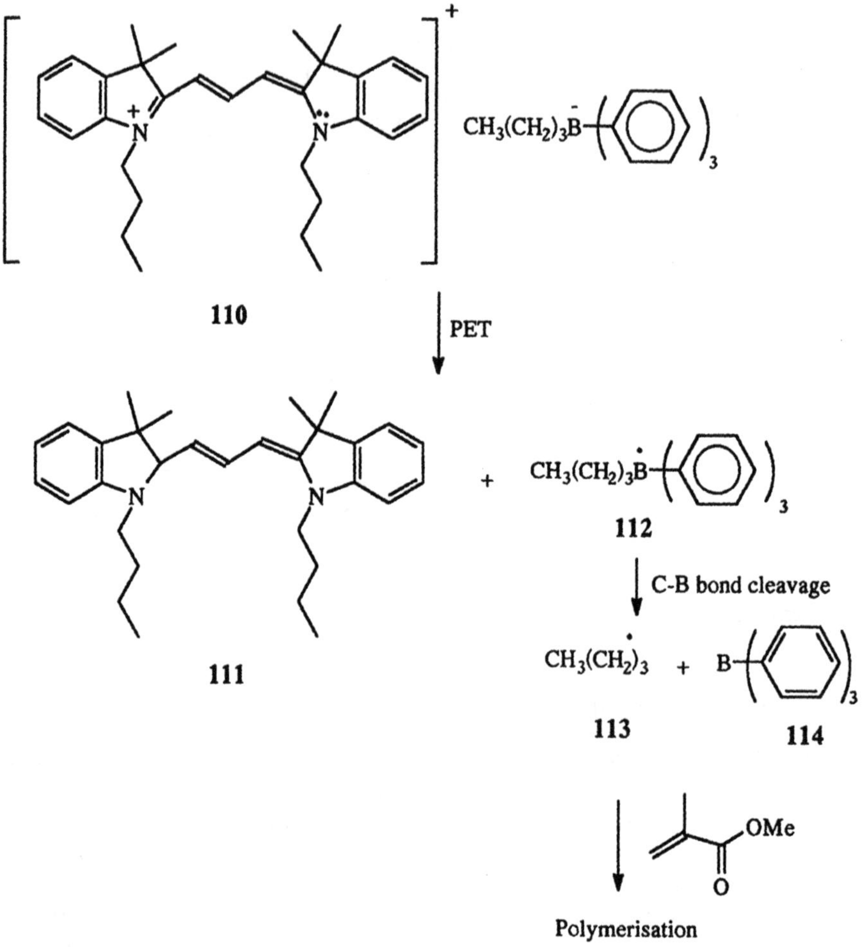

Scheme 26

processes. The cumyl cation (116) produced by the cleavage of $115^{+\cdot}$ undergoes trinitromethylation at CIP, whereas 118 produced from the thermal oxidation of 117 is trapped by SSIP.

Whitten and co-workers [108] have observed C_α-C_β bond fragmentation from various tertiary amine radical cations possessing at least one radical stabilizing substituent attached α to nitrogen center. The primary step in all these cleavage reactions involved ET from amine to excited acceptor generating amine radical cation followed by —C–C— bond fragmentation α to nitrogen. For example, when 122 possessing a radical stabilizing substituent attached to nitrogen center, is subjected to a PET reaction [109] using visible-light-absorb-

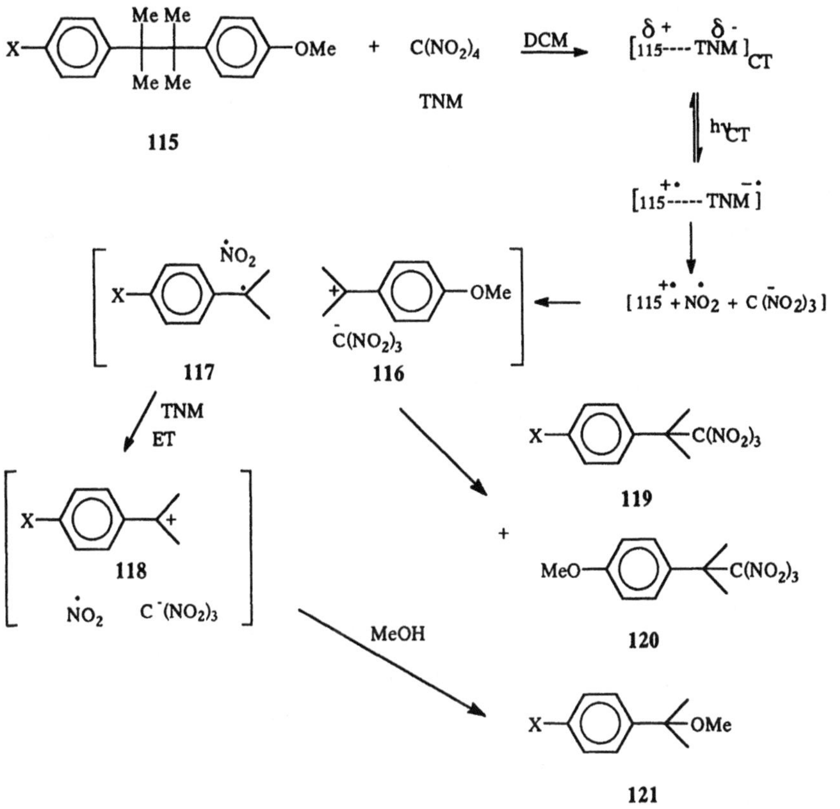

Scheme 27

ing electron acceptors such as thioindigo(TI) or metal complexes of ruthenium (RuL_3^{+2}), it resulted into the fragmentation of $123^{+\cdot}$ with the release of a radical (124) and cation 125 (Scheme 28). Facile cleavage of $123^{+\cdot}$ is explained by implicating the stability of the iminium ion (125) and radical (124) released during the process. C_α–C_β bond dissociation reactions from β-amino alcohols are also reported [110–112] by the same group. Extensive mechanistic studies have suggested that the fragmentation of β-amino alcohol radical cations depend on the basicity of anion radicals and also on the solvent polarity [111]. For a set of acceptors, the rate of fragmentation varied in the order of basicity of anion radicals [Ti > DCA > DCN] in a solvent of constant polarity. DCA sensitized the photofragmentation of amino ketone (126) where the cooperative reactivity of a donor radical cation (128) and a radical anion (127) have lead [113] to the scission of the C_α–C_β bond of 128 in CIP (Scheme 29). Product

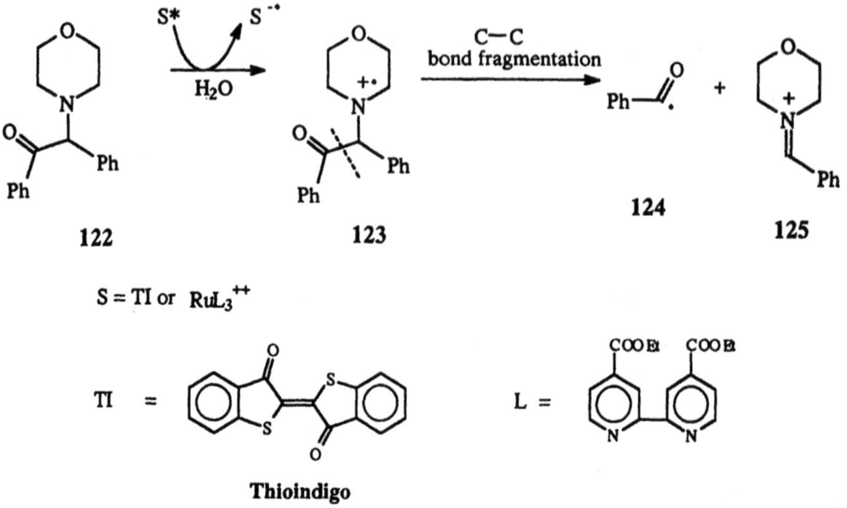

S = TI or RuL₃++

TI =

Thioindigo

L =

Scheme 28

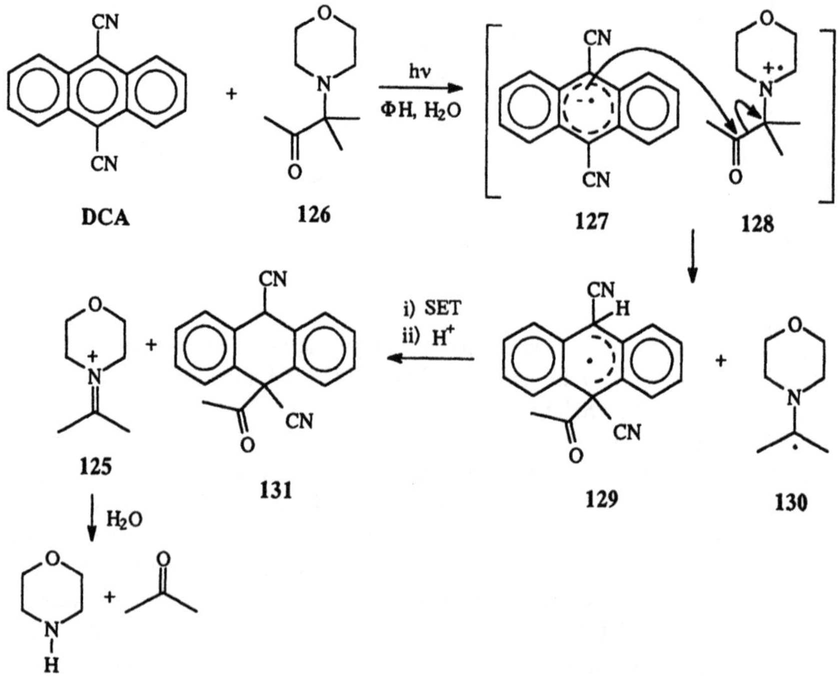

DCA   126   127   128

125

131

129   130

Scheme 29

studies have revealed that the radical anion functions as a nucleophile in assisting the fragmentation. For an increase in the efficiency of net fragmentation of such types of reactions, both a fragmentable donor and fragmentable acceptor has been suggested [114].

Electron transfer photoionization of strained ring compounds generally leads to ring opening reactions in order to release the strain, and the resultant radical cationic species undergo a variety of reactions such as isomerization, rearrangement, and nucleophilic and/or cycloaddition reactions. From the many reviews [115] of the chemistry originating from the cleavage of strained ring radical cations, only few selected examples will be highlighted in this chapter.

Substituted cyclopropanes [116] and oxiranes [117] are shown to undergo efficient cis, trans-isomerization by ET-sensitized photoreactions. Normally, cyclopropane radical cations undergo ring opening followed by rearrangement or nucleophilic additions, depending on the nature of the solvent [118,119]. Rao and Hixon [120] has rationalized the PET addition of methanol to cyclopropanes by postulating the nucleophilic capture of a ring closed cyclopropane radical cation; however, a nucleophilic addition to completely —C–C— bond dissociated species is also suggested [121]. Recently, ET-mediated ring opening of cyclopropane radical cations strategy has been used for —C–C— bond formation reaction by the photoreaction of cyclopropane acetals in the presence of triplet excited states of ketone [122].

An interesting application of the PET reaction of cyclopropanes has been described [123] for preparing 3,5-diaryl-2-isoxazolines in good yields (91%) by direct NO insertion reaction into the 1,2-diaryl cyclopropane radical cation. [3+2]-cycloaddition-type products are obtained [124] from the PET reaction of 1,1,2-triaryl cyclopropane (132) and vinyl ether in the presence of DCNB as an electron acceptor. The detailed reaction pathways are shown in Scheme 30. PET-initiated ring opening strategy of cyclopropanes has also been utilized for the ring expansion reaction by Gassman and Burns [125] (Scheme 31). Clawson et al. [126] have utilized this methodology earlier to other derivatives as well.

Another important application of DCN-sensitized photodissociation of strained ring compounds has been demonstrated by Muller and Mattay [127] for synthesizing N-substituted imidazoles (147) by the [3+2]-cycloaddition of the 2-azaallenyl radical cation (144), produced by the cleavage of corresponding radical cation from azirine (143), with imines. This strategy is further extended [128] for the synthesis of pyrrolophane 3,4-dimethyl ester (152) by the ring opening cycloaddition reaction of (148) with dimethyl acetylene dicarboxylate (Scheme 32).

Cyclobutanes are also known [129–131] to undergo similar types of ring opening reaction, and several examples on this topic along with their applications may be found in the comprehensive review article of Mattay [6].

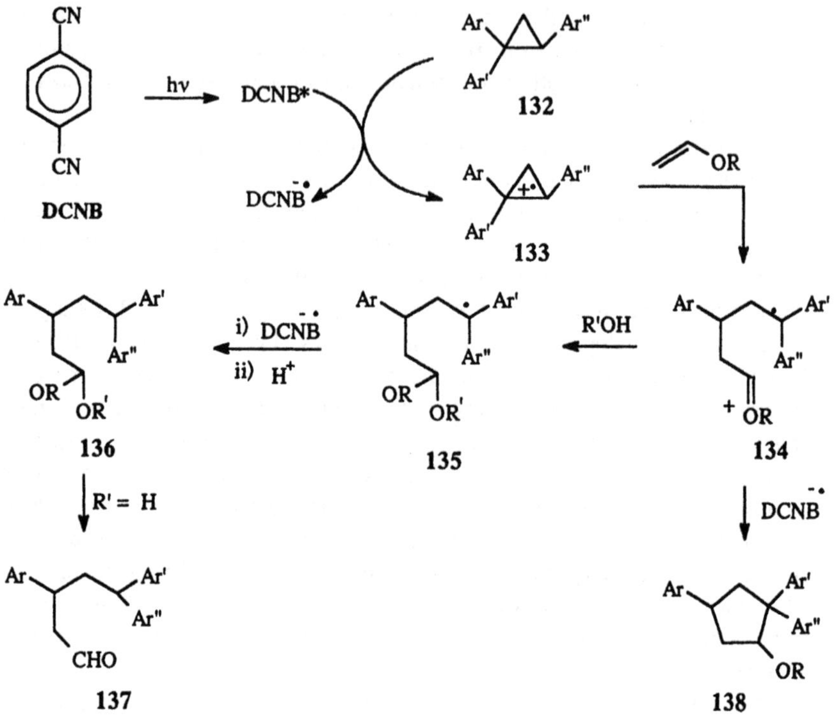

Scheme 30

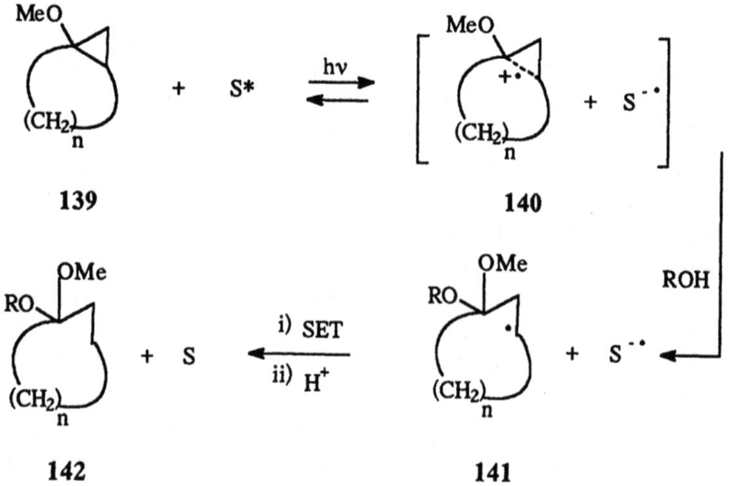

Scheme 31

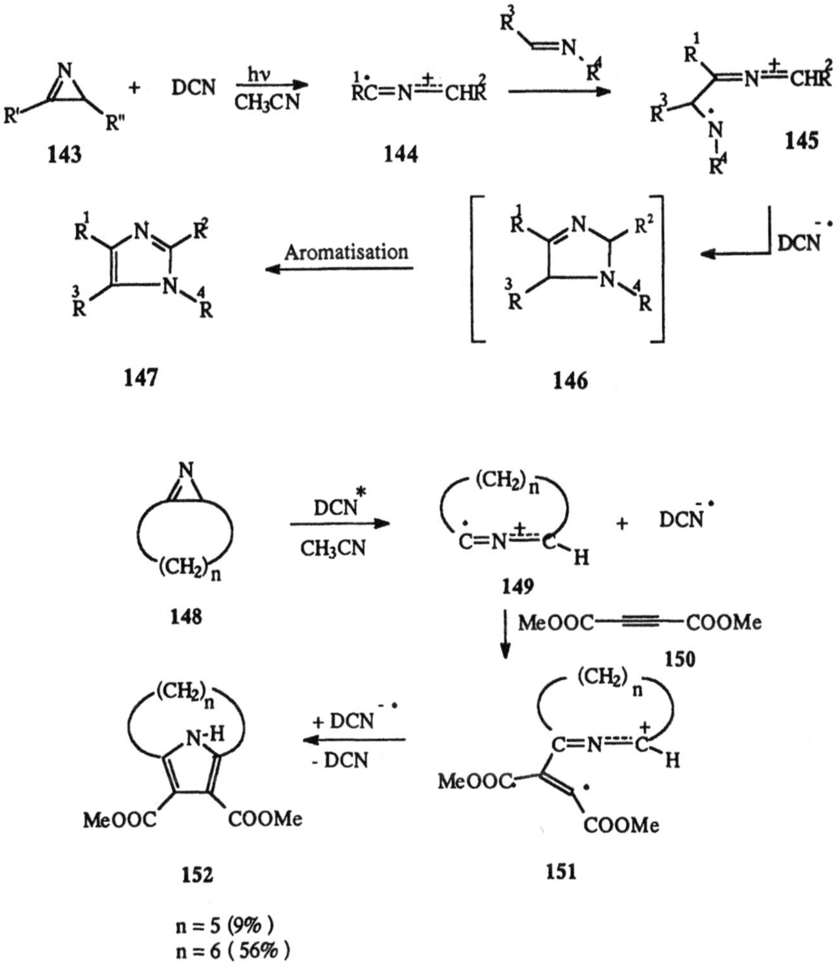

R'—[143]—R" + DCN $\xrightarrow[CH_3CN]{h\nu}$ $R\overset{\cdot}{C}=N\overset{\pm}{=}CH\overset{2}{R}$ [144] $\longrightarrow$

$R^1\underset{R^3}{\overset{}{C}}=N\overset{\pm}{=}CH\overset{2}{R}$ [145]

[147] $\xleftarrow{\text{Aromatisation}}$ [146] $\xleftarrow{}$ $DCN^{-\cdot}$

[148] $\xrightarrow[CH_3CN]{DCN^*}$ $\underset{H}{\overset{(CH_2)_n}{C}}=N\overset{\pm}{=}\overset{}{C}$ + $DCN^{-\cdot}$ [149]

[149] $\downarrow$ MeOOC—≡—COOMe [150]

[151] $\xrightarrow[-DCN]{+DCN^{-\cdot}}$ [152]

MeOOC COOMe [152]   MeOOC COOMe [151]

n = 5 (9%)
n = 6 (56%)

Scheme 32

B. Nucleophilic Addition Reactions

Nucleophilic addition to organic radical cations is one of the most common available reaction pathways. Usually, alkene radical cations give anti-Markonikov addition products with a variety of nucleophiles. In this context, the pioneering work of Arnold et al. [132,133] for the addition of alcohol or cyanide ion to conjugated alkenes may be cited.

Gassman and Bottorff [134] have utilized this strategy for synthesising lactones (158) by the intramolecular addition of carboxylic acid to alkene radical cation (154) generated from the PET reaction of 153 using 1-cyanonaphthalene as an electron acceptor. Improvement [135] in the cyclization yield of 158 is further suggested by employing sterically hindered cyanoaromatics as a photosensitizer (Scheme 33).

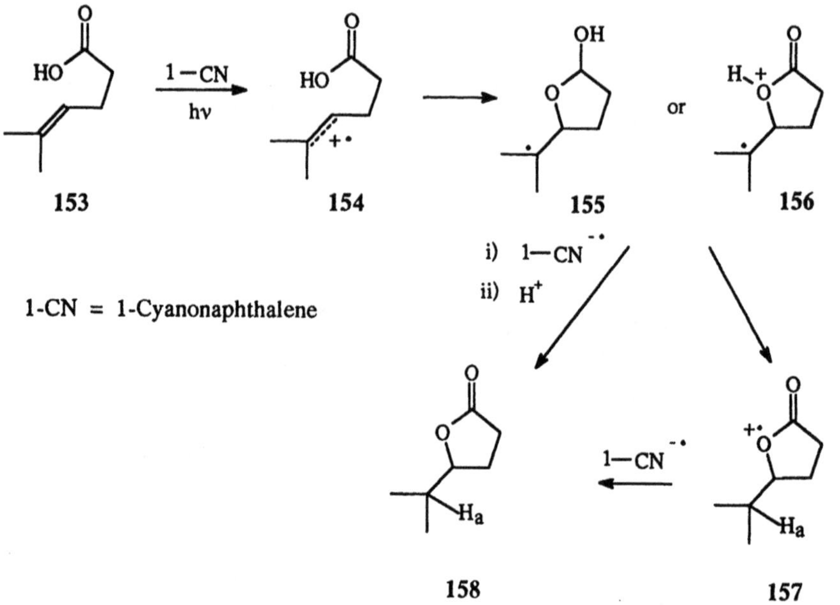

1-CN = 1-Cyanonaphthalene

Scheme 33

An indirect PET methodology known as "redox photosensitization" has been developed by Majima et al. [136] and Tazuke and Kitamura [137] for achieving higher yields of nucleophilic addition products to alkene radical cations. One interesting example of this approach may be cited by illustrating anti-Markonikov alcohol addition to nonconjugated olefin (159) using biphenyl as cosensitizer [138]. The complete reaction sequence is shown in Scheme 34. More examples on this topic may be found in Refs. 4 and 5.

The PET-generated arene radical cations also undergo nucleophilic substitution via the σ-complex. Photocyanation of arenes may be cited in this context as a very early example [139], where hydrogen served as the group undergoing displacement. This concept is further extended [140] for the direct amination of polynuclear aromatic hydrocarbons with ammonia or primary amines via the arene radical cation produced by irradiating arenes in the presence of DCNB. Another potentially useful application of this methodology is

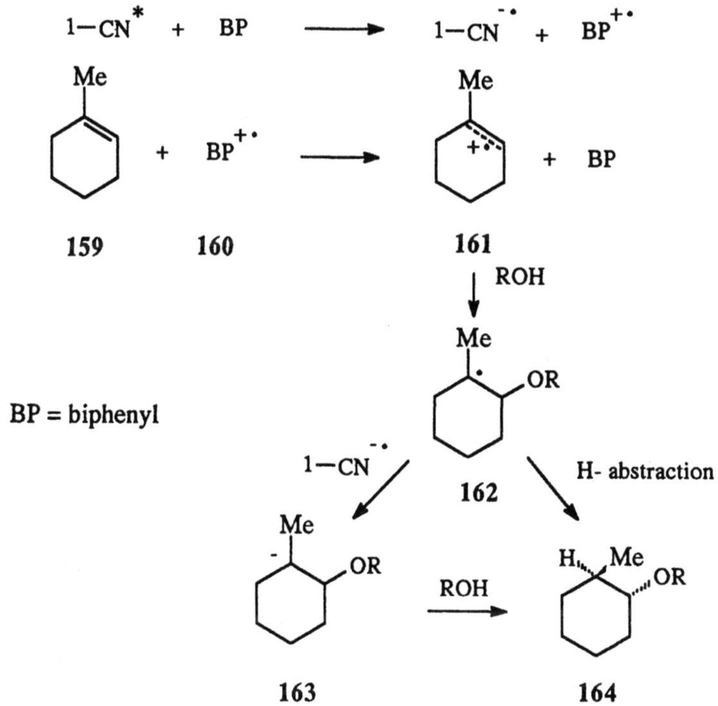

Scheme 34

known [141] for quite sometime for the reduction of electron-rich arenes to corresponding dihydroderivatives (e.g., Birch-type reduction) using NaBH$_4$ as an hydride donor to an arene radical cation. An important aspect of this methodology lies in the selective reduction of electron-donating substituted rings of unsymmetrically substituted arenes, in contrast to Birch-type procedures that favor the reduction of less-electron-rich aromatic rings.

Intramolecular trapping of arene radical cations from methoxy-substituted aromatic compounds have led to the development [142] of an interesting methodology for preparing variety of oxygen, nitrogen, and carbocyclic aromatic compounds. The arene radical cations in this strategy have been generated by ET from excited states of methoxy-substituted aromatic molecules to the ground state of DCN. The sensitized PET processes are depicted in Fig. 6. One interesting application of this methodology may be exemplified by citing the synthesis of coumarins [143] (170, 70–90% yield) directly by the PET cyclization of corresponding cinnamic acids (165, Scheme 35). Several observations

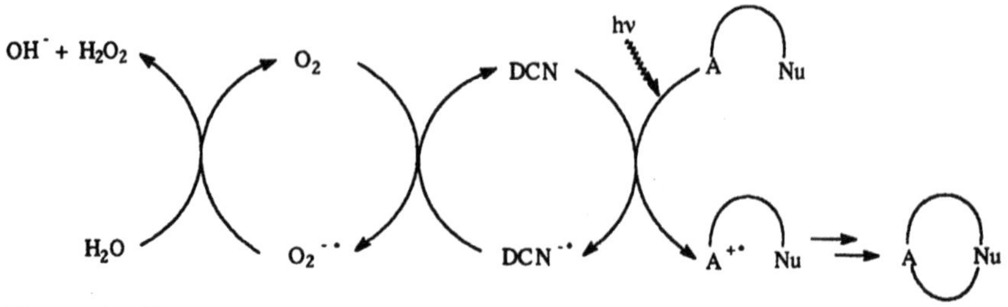

Figure 6 The PET processes for generating arene radical cations

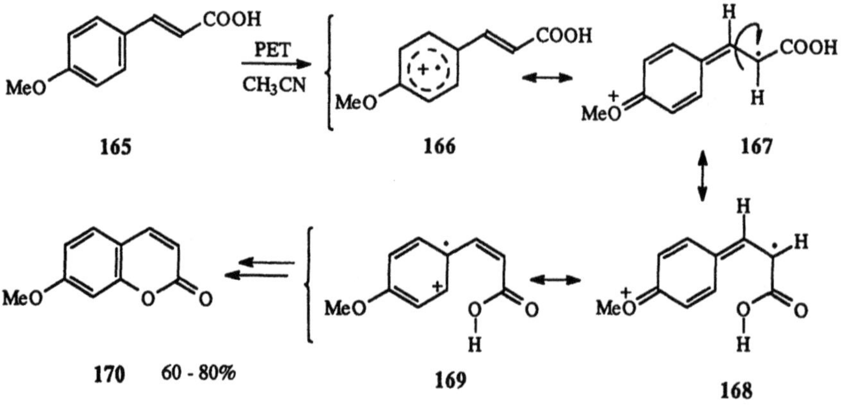

Scheme 35

such as diffusion-controlled fluorescence quenching of 165 with DCN and exergonic values for the free energy change (ΔG_{et}) suggest the ET pathways [142] for this reaction. The regiospecificity of the cyclization mode is in accord with the calculated electron densities (Huckel or MNDO) at different carbons of the HOMO of the arene radical cation. Precocenes-I [144], a potent antijuvenile hormone compound, and several of its analogs, 2-alkylated dihydrobenzofurans [145], believed to possess antifungal and phytoalexin properties are also prepared using this methodology.

The success of the strategy is further applied for the synthesis of carbo- and spiro-annulated aromatic compounds [146,147] by the intramolecular cyclization of silyl enolethers to PET-generated arene radical cations. Two types of carbocyclic compounds (170 and 173), varying in ring sizes, may be synthesized [146] starting from the same ketone (i.e., 169), as two types of silyl enol ethers can be produced using either thermodynamic or kinetic enolisation procedures. The core spiro structure (177) of the anticancer antibiotic fredericamycin is also prepared [147] by the PET cyclization of 176 (Scheme 36).

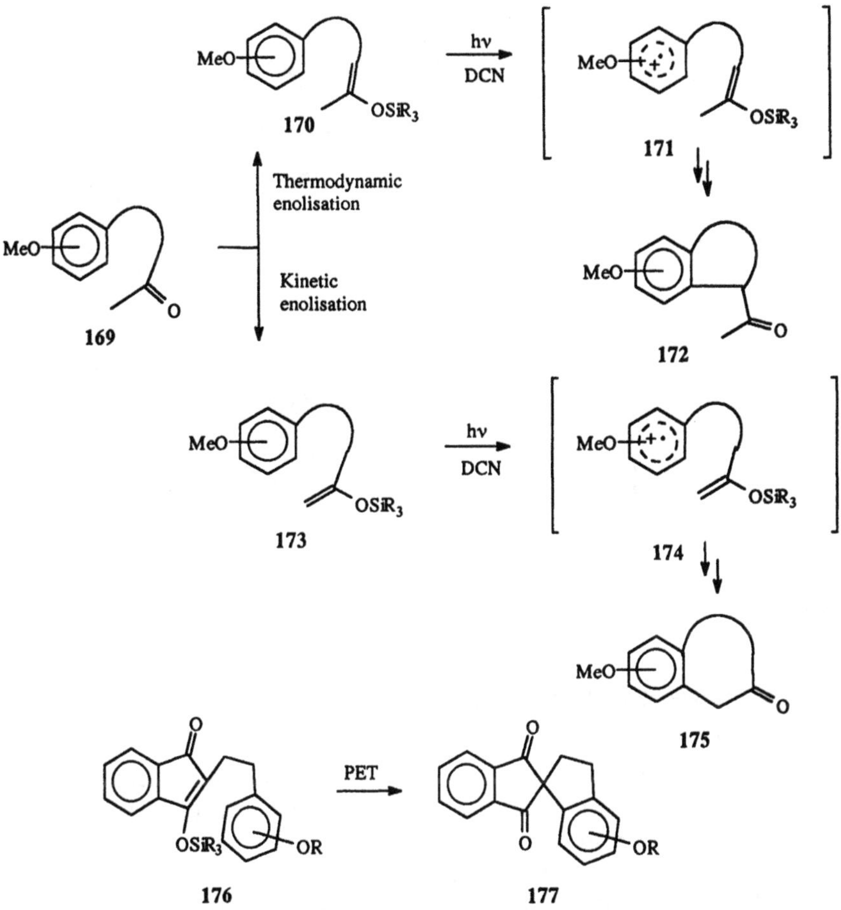

Scheme 36

The synthetic utility of these cyclizations is further demonstrated [148] for the efficient and regiospecific cyclisations of β-arylethylamines to produce substituted dihydroindoles. A unique combination of two independent PET operating reactions discovered by Pandey and co-workers [35,148] has been used for realizing one pot "wavelength switch" approach for synthesising benzopyrrolizidines (182) related to mitomycin skeleton starting from 178 (Scheme 37). The strategy involved the construction of a dihydroindole ring (180) via intramolecular cyclization of an arene radical cation intermediate (179) by the photolysis of 178 in the presence of DCA at >300 nm (light absorbed by 178 only), followed by building the pyrrolizidine moiety of 182 via iminium cation intermediate (181) cyclizations, as shown in Scheme 37.

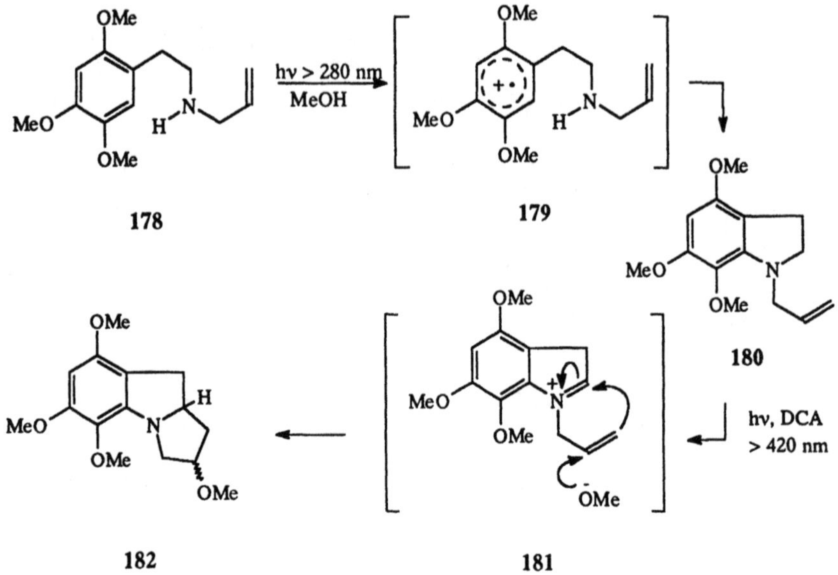

Scheme 37

C. Isomerizations, Dimerizations, Rearrangements, and Cycloadditions

These types of reaction are generally associated with the alkene radical cations which have incidentally formed one of the important areas of PET-related research activities over the years, and many good reviews have been written on this subject. Radical cation catalyzed Diels–Alder cycloaddtions, especially where both dienes and dipolarophiles are electron rich, have emerged as one of the most interesting and useful reactions from such research activities. Readers are advised to consult the recent review article [7] on these topics for detailed examples.

III. CHEMISTRY FROM RADICAL ANIONS

Unlike radical cations, the quantum of chemistry originating from PET-generated radical anions is still limited possibly due to the impending development of a suitable photosystem to initiate photosensitized one-elecron redox reactions in wide array of functionalities. Nevertheless, the radical anion chemistry follows, more or less, the analogous pattern of bond dissociation and addition (electrophilic/radical) reactions as observed for the radical cations. As there are not many examples to describe the separate categories, this section is subdivided into only two parts, where selected examples from bond cleavage and addition reactions are summarized.

A. Bond Cleavage Reactions

Efficient carbon–halogen bond cleavage, both from the arene ring as well as from the benzylic position of the corresponding haloarene radical anions, has been observed [6] to produce aryl or benzyl radicals. Aryl radicals may react with nucleophiles to give $S_{RN}1$-type substitution products, whereas benzyl radicals can participate in intramolecular —C–C—bond formation reactions with tethered π systems. Analogous photodissociation of benzyl esters and benzyl-sulfones is also reported [3]. Photoreduction of carboxylic acid ester of secondary alcohols to corresponding alkanes in high yields (84%) by irradiating the esters in the presence of HMPT at 254 nm is described [149]. The reaction is initiated by the transfer of an electron from HMPT to carboxylic esters followed by —C–C— bond fragmentation and the termination of the resultant radical by H-abstraction. Based on a similar mechanistic paradigm, a practical method of photodecarboxylation is reported [150] by the PET cleavage of the —N–O— bond of carboxylic acid derivatives N-acyloxyphthalimides using 1,6-bis(dimethyl amino) pyrene (BDMP) as an electron donor. However, the exact fate of BDMP after electron donation is not very clear. The same decarboxylation strategy is extended for generating a carbon-centered radical [151] for utilization in radical-type Michael addition reactions, although the PET strategy in this case employed $Ru(biPy)_3Cl_2$ as a light-absorbing electron acceptor from BNAH initially. The in-situ-generated Ru(I) species ultimately became the electron donor to initiate cleavage chemistry. The exact photosensitization mechanism, however, is also not very clear here.

Hamada et al. [152] have described a novel approach for the photodissociation of sulphonamides for detosylation purposes from corresponding radical anions of amine to sylates, generated by the PET reaction employing dimethoxy benzene as an electron donor and in the presence of a reductant such as $NaBH_4$, ascorbic acid, hydrazine, and so forth. A similar methodology [153] is also utilized for the deprotection of aryl sulphonate esters from carbohydrate substrates utilizing tertiary amines (DABCO or Et_3N) as the electron donor. Improvements in these methodologies are subsequently reported by Nishida et al. [154] and Art et al. [155]. Saito et al. [156] have demonstrated selective photodissociation of secondary alcohol aroyl esters via their corresponding radical anion, produced by ET from N-methyl carbazole. The application of this cleavage methodology is demonstrated for the selective deoxygenation of secondary alcohols (91%) in general and 2,3-deoxygenation of ribonucleoside in particular.

Dissociation of the —C–NO_2 bond from the tetranitromethane (TNM) radical anion, produced by the visible-light irradiation of the CT complex of anthracene-TNM pair is reported [157]. The formation of products occur by the cage recombination of geminate ions and radical pairs. A novel application of this reaction for the carboxylation of *p*-dimethoxy benzene is demonstrated [158] by the same group. Recently [159], the same strategy is extended for the α-nitration of ketones via silyl enolethers. Organoselenium radical anions [(—C—Se—$^{-}$)], generated through a photosystem comprised of DMN as a light-harvesting electron donor and ascorbic acid as a sacrificial electron donor, has been found [160] (Fig. 7) to undergo a heterolytic bond cleavage reaction to product carbon-centered radicals and PhSeSePh. Use of this cleavage pattern is suggested for the unimolecular group transfer radical chain reactions [161] (e.g., 183→184) (Scheme 38). Support for the ET mechanism from excited DMN to —C–Se— moiety to generate —C-Se]$^{-}$ is provided by studying the diffusion-controlled fluorescence quenching of DMN by organoselenium substrates, estimation of negative ΔG_{et} values, and the evaluation of other kinetic parameters [161]. The same photosystem is further extended to initiate cleavage of PhSeSiR$_3$ via its radical anion to provide silicon radical fragment ($^{\cdot}$SiR$_3$) and PhSeSePh [162]. Owing to the better halophilicity of $^{\cdot}$SiR$_3$ than conventionally utilized tin radicals, and the radical trapping ability of PhSeSePh, this strategy of PhSeSiR$_3$ dissociation is utilized for the group transfer radical chain reactions[162] (Fig. 8, 185→186). A slight modification, utilizing pyrene instead

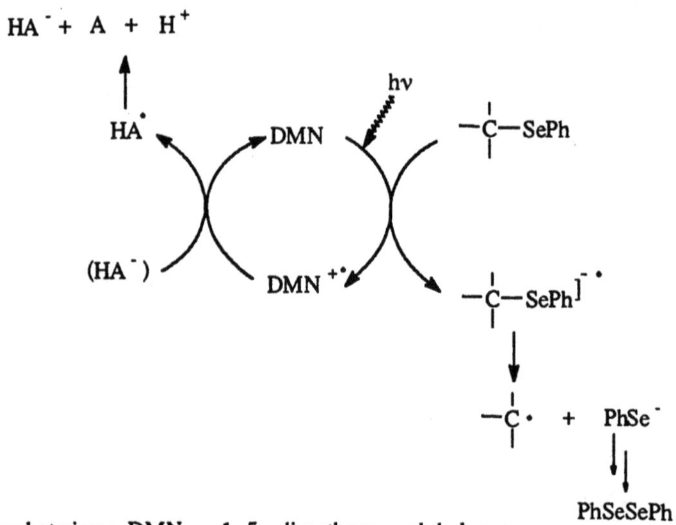

HA^{-} = Ascorbate ion ; DMN = 1, 5 - dimethoxy naphthalene

A = Dehydro ascorbic acid

Figure 7 PET Generation and cleavage of —C—Se]$^{-}$.

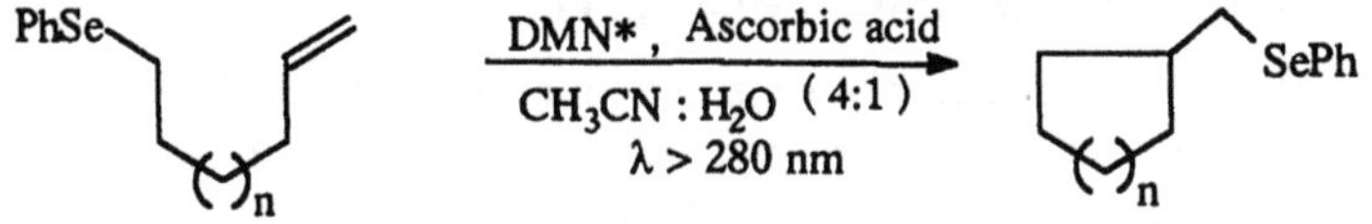

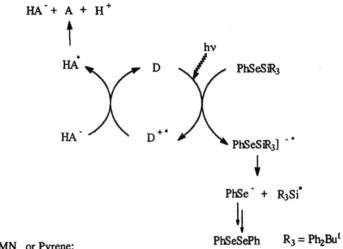

183

n = 1 or 2

184

yield ~ 75.1%

Scheme 38

of DMN as a visible-light-absorbing (366 nm) electron donor is also suggested recently in the existing photosystem [163]. Simultaneous addition of silyl radicals and PhSeSePh to olefinic bond is utilized for the preparation of vinylsilanes [164] and diene cyclizations as well.

D = DMN or Pyrene;
HA = Ascorbate ion; A = Dehydroascorbic acid

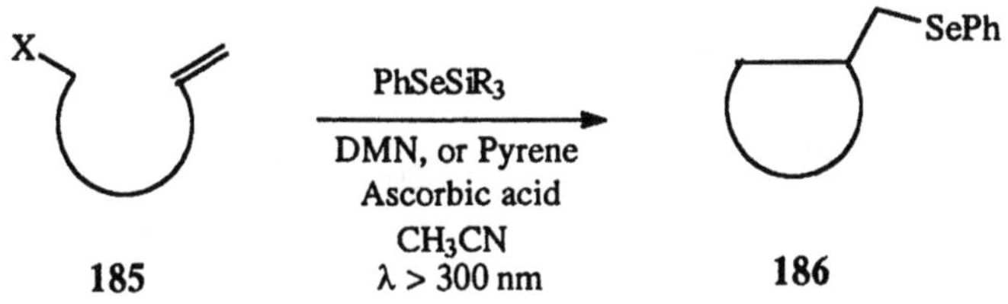

X = Cl, Br.

Yield = ~ 81.1%

185

186

Figure 8 Strategy of $PhSeSiR_3$ dissociation for group transfer radical chain reactions.

B. Electrophilic/Radical Additon Reactions

Arene radical anions, particularly from polynuclear aromatic hydrocarbons
(e.g., phenanthrene, anthracene, and pyrene), generated by ET using amines
as the electron donor has been shown [165] to undergo carboxylation reaction
(e.g., Phen→189) by the electrophilic addition of CO_2, followed by the termination of the resultant radical species by H-abstraction from the solvent (Scheme
39). A laser flash photolysis study [166] has recently confirmed the involvement of arene radical anions in this reaction.

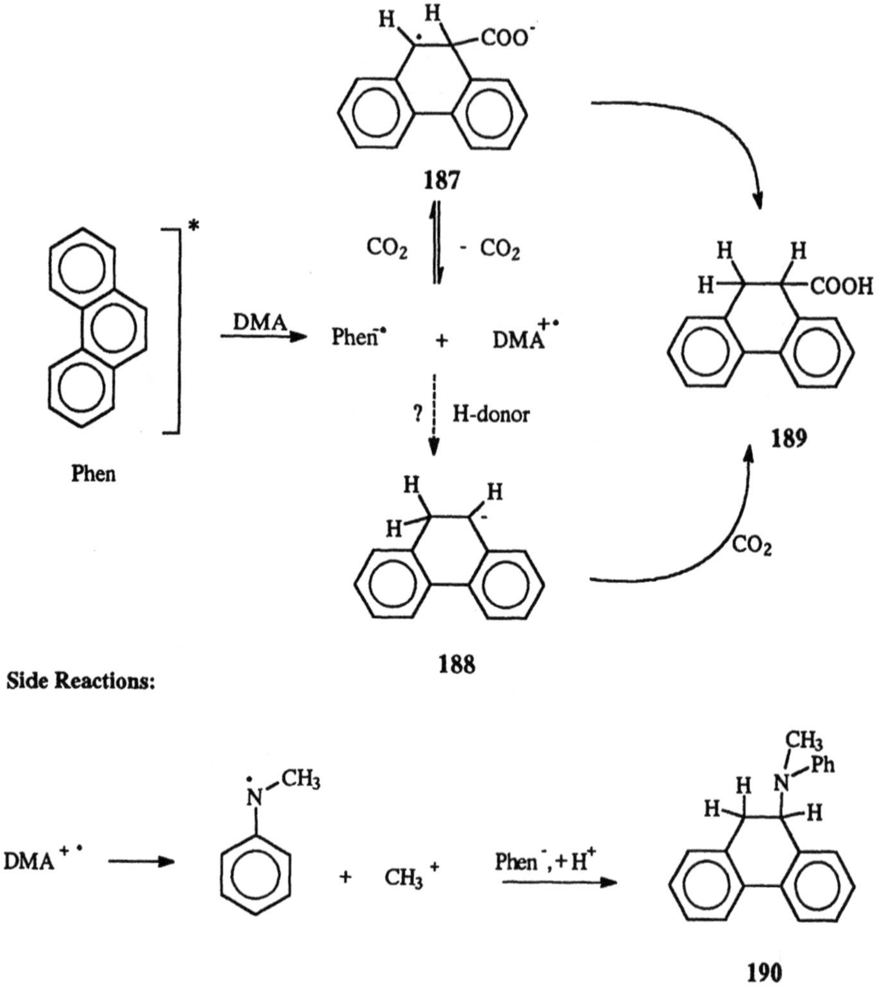

Side Reactions:

DMA = N, N- dimethylaniline

Scheme 39

Intra-molecular as well as intermolecular addition of an activated olefin at the β position of the α,β-unsaturated ketone radical anion (192) to a olefinic π bond is reported by Pandey et al. [167] as a new strategy for —C-C— bond formation reaction. The radical anions (192) are generated by the secondary [168] ET from the in-situ-generated DCA$^-$ through a photosystem comprised of DCA as a light-harvesting electron acceptor and triphenylphosphine as a sacrificial electron donor [169] (Fig. 9). The thermodynamic feasibility of ET at each step is supported by the evaluation of various kinetic parameters. This unique mode of activation of enones is applied for the construction of many cyclic compounds [170]. Further improvements in the photosystem are also suggested [171] by utilizing DMN as the primary electron donor to ^{1}DCA* and ascorbic acid as the sacrificial electron donor (Fig. 10). The efficient intermolecular coupling of activated alkenes and alkynes at the β position of cyclic α,β-unsaturated ketones has led to the development [172] of novel strategy toward the synthesis of prostaglandin skeleton (196, Scheme 40). This photosensitized ET methodology of initiating one electron reduction processes is extended further to carbonyl functionalities to generate ketyl radicals [173] for intramolecular cyclization reactions with olefins to synthesize substituted cycloalkanols (e.g., 198→199, Scheme 41). An interesting application of ketyl radical cyclization methodology is demonstrated by the synthesis of C-furanoside [174] (203, Scheme 42). ET selectivity studies with enones versus ketones has suggested that the former is much more prone for reduction than ketones due to a greater

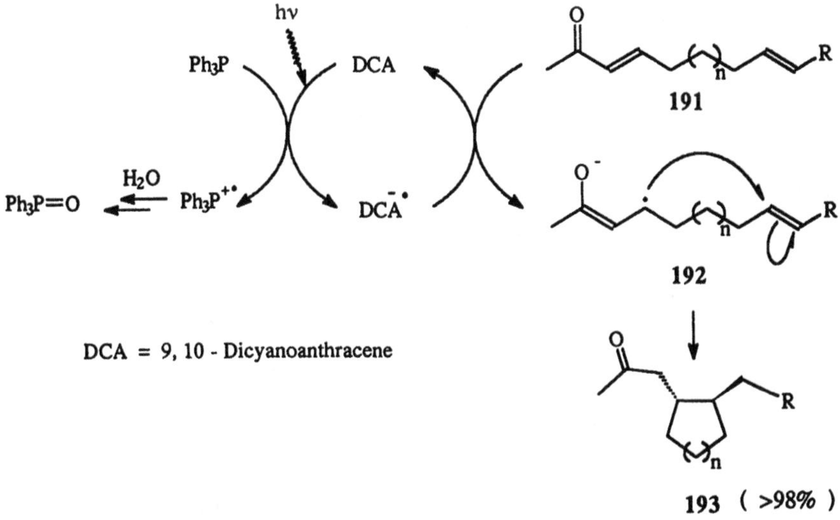

Figure 9 Photosystem comprised of a light-harvesting electron acceptor and a sacrificial electron donor.

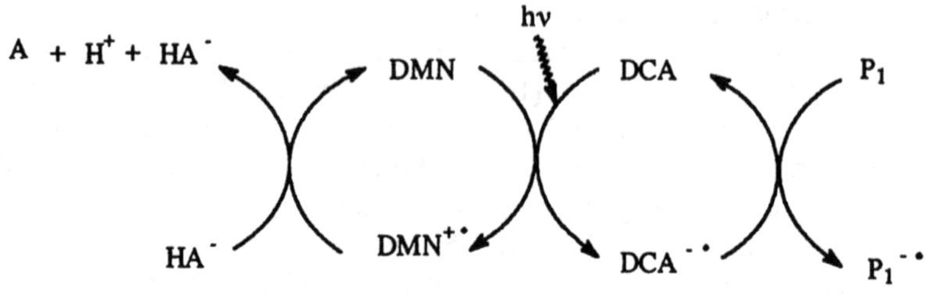

HA$^-$ = Ascorbate ion ; DMN = 1, 5- Dimethoxynaphthalene

DCA = 9, 10- dicyanoanthracene; P$_1$ = Saturated / unsaturated carbonyl compounds

Figure 10   Photosystem comprised of a primary electron donor and a sacrificial electron donor.

magnitude of free energy change associated with the electron transfer processes [175].

IV. CONCLUSIONS

This chapter could be concluded with a note that the consequence of PET transformations from organic substrates may be characterized by evaluating the

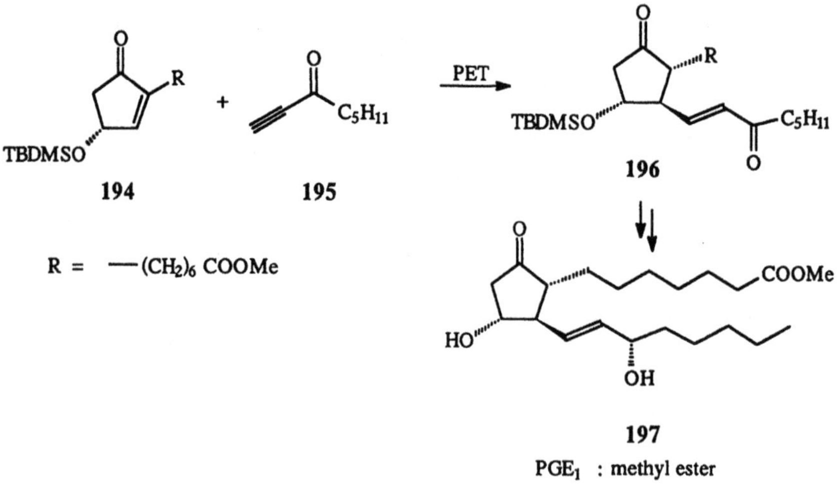

R = —(CH$_2$)$_6$ COOMe

PGE$_1$: methyl ester

Scheme 40

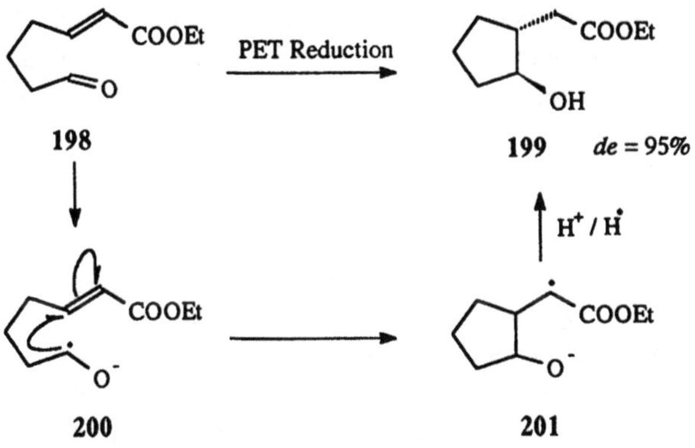

Scheme 41

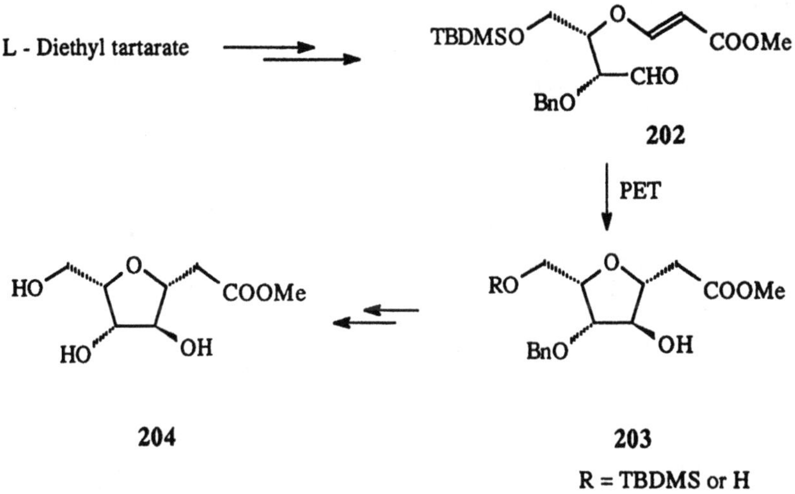

Scheme 42

reactivity profiles of initially produced radical ions. The selected examples presented in this chapter possibly indicate a general reactivity pattern of these reactive intermediates which is expected to encourage synthetic organic chemists to design newer synthetic strategies. Although we are a long way from making predictions in such reactions, the author is of the opinion that frequent use of these reactions in selective transformations during complex molecule synthesis will prevail. It may be essential to collect more kinetic data for the chemical reactions of photogenerated radical ions that may help to develop an understanding of structure–activity relationships, competing reaction pathways, and the relationship between photochemical and nonphotochemical methods of generating radical ions. The study of electron transfer processes on semiconductor surfaces and in restricted environments will strongly influence further development in this exciting area.

ACKNOWLEDGMENTS

The author is thankful to all his colleagues whose names have figured in this article for their dedication and intellectual contributions. I am indebted to Dr. (Mrs.) Smita Gadre, Mr. Anjan Ghatak, and Mr. Debasis Chakrabarti for their untiring help in preparing this chapter. I am also grateful to CSIR, DST, New Delhi and BRNS, BARC, Bombay for financing our research in this area over the past several years.

REFERENCES

1. Fox, M. A.; Chanon, M., Eds. *Photoinduced Electron Transfer Reactions*; Elsevier: Amsterdam, 1988; Parts A–D.
2. Davidson, R.S. *Adv. Phys. Org. Chem.* 1983, *19*, 1. (b) Lewis, F. D. *Adv. Photochem.* 1986, *13*, 165.
3. Kavarnos, G.J.; Turro, N. J. *Chem. Rev.* 1986, *86*, 401.
4. Matts, S. L.; Farid, S. *Org. Photochem.* 1983, *6*, 223.
5. Mariano, P. S.; Stavinoha, J. L. *Synthetic Organic Photochemistry*; M. W. Horspool, Ed.; Plenum Press: London, 1983; pp. 145.
6. Mattay, *J. Synthesis.* 1989, 233. (b) Mattay, J. *Angew Chem. Int. Ed. Engl.* 1987, *26*, 825.
7. Pandey, G. *Top. Curr. Chem.* 1993, *168*, 175.
8. Maslak, P.; Chapman, W. H.; Vallombroso Jr., T. M.; Watson, B. A. *J. Am. Chem. Soc.* 1995, *117*, 12380.
9. Beens, H.; Weller, A. *Organic Molecular Photophysics*; J. D. Birks, Ed.; Wiley: London, 1975; Vol. 2; Chap. 4. (b) Mataga, N.; Ottolenghi, M. *Molecular Association*; R. Foster, Ed.; Academic Press: London, 1975; Vol. 2; Chap. 1.
10. Rehm, D.; Weller, A. *Isr. J. Chem.* 1970, *8*, 259.
11. Gould, I. R.; Ege, D.; Moser, J. E.; Farid, S. *J. Am. Chem. Soc.* 1990, *112*, 4290.
12. Masuhara, H.; Mataga, N. *Acc. Chem. Res.* 1981, *14*, 312. (b) Mataga, N.; Okada, T.; Kanda, Y.; Shioyama, H. *Tetrahedron* 1986, *42*, 6143.
13. Kellett, M. A.; Whitten, D. G.; Gould, I. R.; Bergmark, W. R. *J. Am. Chem. Soc.* 1991, *113*, 358.
14. Zhang, X. M.; Bordwell, F. G. *J. Am. Chem. Soc.* 1994, *116*, 904, and 4251. (b) Dinnocenzo, J. P.; Banach, T. E. *J. Am. Chem. Soc.* 1989, *111*, 8646.
15. Nicholas, A. M. D. P.; Arnold, D. R. *Can. J. Chem.* 1982, *60*, 2165.
16. Albini, A.; Spreti, S. *Tetrahedron* 1984, *40*, 2975.
17. Albini, A.; Fasani, E.; Sulpizio, A. *J. Am. Chem. Soc.* 1984, *106*, 3562.

18. Lewis, F. D.; Petisce, J. R. *Tetrahedron* 1986, *42*, 6207.
19. Albini, A.; Fasani, E.; Mella, M. *Top. Curr. Chem.* 1993, *168*, 143.
20. Santamaria, J.; Jroundi, R.; Rigaudy, J. *Tetrahedron Lett.* 1989, *30*, 4677.
21. Pandey, G.; Krishna, A. *Synth. Commun.* 1988, *18*, 2309.
22. Nishida, A.; Oishi, S.; Yonemitsu, O. *Chem. Pharm. Bull.* 1989, *37*, 2266.
23. Mella, M.; Freccero, M.; Albini, A. *J. Chem. Soc., Chem. Commun.* 1995, 41.
    (b) Mella, M.; Freccero, M.; Albini, A. *Tetrahedron* 1996, *52*, 5533 and 5549.
24. Barltrop, J. A. *Pure Appl. Chem.* 1973, *33*, 179.
25. Bellas, M.; Bryce-Smith, D.; Clarke, M. T.; Gilbert, A.; Klunkin, G.;
    Krestonosich, S.; Manning, C.; Wilson, S. *J. Chem. Soc., Perkin Trans. 1* 1977,
    2571.
26. Yang, N. C.; Shold, D. M.; Kim, B. *J. Am. Chem. Soc.* 1976, *98*, 6587.
27. Cohen, S. G.; Parola, A.; Parsons Jr., G. H. *Chem. Rev.* 1973, *73*, 141.
28. Lewis, F. D.; Ho, T-I; Simpson, J. T. *J. Am. Chem. Soc.* 1982, *104*, 1924. (b)
    Hub, W.; Schneider, S.; Dorr, F.; Simpson, J. T.; Oxman, J. D.; Lewis F. D.
    J. Am. Chem. soc. 1982, *104*, 2044.
29. Lewis, F. D. *Acc. Chem. Res.* 1986, *19*, 401. (b) Lewis, F. D.; Ho, T-I;
    Simpson, J. T. *J. Org. Chem.* 1981, *46*, 1077.
30. Pandey, G. *Synlett* 1992, 546.
31. Griller, D.; Lossing, F. P. *J. Am. Chem. Soc.* 1981, *103*, 1586.
32. Chow, Y. L.; Danen, W. C.; Nelsen, S. F.; Rosenblatt, D. H. *Chem. Rev.* 1978,
    78, 243.
33. Reddy, P. Y. Ph. D. Thesis, Osmania University, Hyderabad, India, 1994.
34. Pandey, G.; Kumaraswamy, G.; Krishna, A. *Tetrahedron Lett.* 1987, *28*, 4615.
35. Pandey, G.; Kumaraswamy, G. *Tetrahedron Lett.* 1988, *29*, 4153.
36. Pandey, G.; Kumaraswamy, G.; Reddy, P. Y. *Tetrahedron* 1992, *48*, 8295.
37. Pandey, G.; Reddy, P. Y.; Bhalerao, U. T. *Tetrahedron Lett.* 1991, *32*, 5147.
38. Pandey, G.; Sudha Rani, K. *Tetrhedron Lett.* 1988, *29*, 4157.
39. Pandey, G.; Sudha Rani, K.; Bhalerao, U. T. *Tetrahedron Lett.* 1990, *31*, 1199.
40. Pandey, G.; Reddy, P. Y.; Das, P. *Tetrahedron Lett.* 1996, *37*, 3175.
41. Pandey, G.; Das, P. Unpublished work.
42. Santamaria, J.; Kaddachi, M. T.; Rigaudy, J. *Tetrahedron Lett.* 1990, *31*, 4735.
43. Sundberg, R. J.; Hunt, P. J.; Desos, P.; Gadamasetti, K. G. *J. Org. Chem.*
    1991, *56*, 1689. (b) Sundberg, R. J.; Desos, P.; Gadamasetti, K. G.; Sabat, M.
    Tetrahedron Lett. 1991, *32*, 3035.
44. Pandey, G.; Sudha Rani, K.; Lakshmaiah, G. *Tetrahedron Lett.* 1992, *33*, 5107.
45. Evans, S.; Green, J. C.; Joachim, P. J.; Orchard, A. F.; Turner, D. W.; Maier,
    J. P. *J. Chem. Soc., Faraday Trans. 2* 1972, *68*, 905.
46. Boschi, R.; Lappert, M. F.; Pedley, J. B.; Schmidt, W.; Wilkins, B. T. *J.
    Organomet. Chem.* 1973, *50*, 69.
47. Reutov, O. A.; Rozenberg, V. I.; Gavrilova, G. V.; Nikanorov, V. A. *J.
    Organomet. Chem.* 1979, *177*, 101 and references cited therein.
48. Gardner, H. C.; Kochi, J. K. *J. Am. Chem. Soc.* 1976, *98*, 2460.
49. Kochi, J. K., Ed. *Organometallic Mechanisms and Catalysis*; Academic Press:
    New York, 1978; pp. 445.

50. Symons, M. C. R. *Chem. Soc. Rev.* 1984, *13*, 393.
51. Walther, B. W.; Williams, F.; Lau, W.; Kochi, J. K. *Organometallics* 1983, *2*, 688.
52. Eaton, D. F. *J. Am. Chem. Soc.* 1980, *102*, 3278.
53. Eaton, D. F. *J. Am. Chem. Soc.* 1980, *102*, 3280; 1981, *103*, 7235.
54. Kyushin, S.; Otani, S.; Takahashi, T.; Nakadaira, Y.; Ohashi, M. *Tetrahedron* 1990, *44*, 6395. (b) Mizuno, K.; Terasaka, K.; Yasueda, M.; Otsugi, T. *Chem. Lett.* 1988, 145.
55. Mizuno, K.; Nakanishi, K.; Otsuji, Y. *Chem. Lett.* 1988, 1833.
56. Fagnoni, M.; Mella, M.; Albini, A. *J. Am. Chem. Soc.* 1995, *117*, 7877.
57. Maruyama, K.; Imahori, H.; Osuka, A.; Takuwa, A.; Tagawa, H. *Chem. Lett.* 1986, 1719.
58. Kubo, Y.; Imaoka, T.; Shiragami, T.; Araki, T. *Chem. Lett.* 1986, 1749.
59. Takuwa, A.; Nishigaichi, Y.; Yamashita, K.; Iwamoto, H. *Chem. Lett.* 1990, 639. (b) Takuwa, A.; Tagawa, H.; Iwamoto, H.; Soga, O.; Maruyama, K. *Chem. Lett.* 1987, 1091.
60. Hasegawa, E.; Brumfield, M. A.; Mariano, P. S.; Yoon, U-C. *J. Org. Chem.* 1988, *53*, 5435.
61. Mariano, P. S. *Acc. Chem. Res.* 1983, *16*, 130.
62. Borg, R. M.; Mariano, P. S. *Tetrahedron Lett.* 1986, *27*, 2821.
63. Cho, I-S.; Mariano, P. S. *J. Org. Chem.* 1988, *53*, 1590.
64. Ahmed-Schofield, R.; Mariano, P. S. *J. Org. Chem.* 1987, *52*, 1478.
65. Hasegawa, E.; Xu, W.; Mariano, P. S.; Yoon, U-C.; Kimg, J-U. *J. Am. Chem. Soc.* 1988, *110*, 8099. (b) Zhang, X-M.; Mariano, P. S. *J. Org. Chem.* 1991, *56*, 1655.
66. Jeon, Y. T.; Lee, C-P.; Mariano, P. S. *J. Am. Chem. Soc.* 1991, *113*, 8847.
67. Xu, W.; Zhang, X-M.; Mariano, P. S. *J. Am. Chem. Soc.* 1991, *113*, 8863.
68. Khim, S. K.; Mariano, P. S. *Tetrahedron Lett.* 1994, *35*, 999.
69. Jung, Y. S.; Mariano, P. S. *Tetrahedron Lett.* 1993, *34*, 4611.
70. Pandey, G.; Kumaraswamy, G.; Bhalerao, U. T. *Tetrahedron Lett.* 1989, *30*, 6059.
71. Pandey, G.; Reddy, G. D.; Kumaraswamy, G. *Tetrahedron* 1994, *50*, 8185.
72. Beckwith, A. L. J.; Schiesser, C. H. *Tetrahedron* 1985, *41*, 3925.
73. Pandey, G.; Reddy, G. D. *Tetrahedron Lett.* 1992, *33*, 6533.
74. Pandey, G.; Reddy, G. D.; Chakrabarti, D. *J. Chem. Soc., Perkin Trans. 1* 1996, 219.
75. Hoegy, S. E.; Mariano, P. S. *Tetrahedron Lett.* 1994, *35*, 8319.
76. Pandey, G.; Chakrabarti, D. *Tetrahedron Lett.* 1996, *37*, 2285.
77. Beak, P.; Lee, W. K.; *J. Org. Chem.* 1993, *58*, 1109.
78. Karpas, A.; Fleet, G. W. J.; Dwek, R. A., Petursson, S.; Namgoong, S.K; Ramsden, N. G.; Jacob, G. S.; Rademacher, T. W. *Proc. Natl. Acad. Sci. U.S.A.* 1988, *85*, 9229.
79. Pandey, G.; Sahoo, A. K.; Rao, K. V. N.; Bhagwat, B. V. Unpublished work.
80. Pandey, G.; Lakshmaiah, G.; Kumaraswamy, G. *J. Chem. Soc., Chem. Commun.* 1992, 1313.

81. Pandey, G.; Soma Sekhar, B. B. V.; Bhalerao, U. T. *J. Am. Chem. Soc.* 1990, *112*, 5650.
82. Pandey, G.; Soma Sekhar, B. B. V. *J. Org. Chem.* 1994, *59*, 7367.
83. Pandey, G.; Sochanchingwung, R. *J. Chem. Soc., Chem. Commun.* 1994, 1945.
84. Szepes, L.; Koranyi, T.; Naray-Szabo, G.; Modelli, A.; Distefano, G. *J. Orgmet. Chem.* 1981, *217*, 35.
85. Traven, V. F.; West, R. *J. Am. Chem. Soc.* 1973, *95*, 6824.
86. Sakurai, H.; Kira, M.; Uchida, T. *J. Am. Chem. Soc.* 1973, *95*, 6826.
87. Wang, J. T.; Williams, F. *J. Chem. Soc., Chem. Commun.* 1981, 666. (b) Symons, M. C. R. *J. Chem. Soc., Chem. Commun.* 1982, 869. (c) Walther, B. W.; Williams, F. *J. Chem. Soc., Chem. Commun.* 1982, 270.
88. Sakurai, H.; Sakamoto, K.; Kira, M. *Chem. Lett.* 1984, 1213.
89. Nakadaira, Y.; Komatsu, N.; Sakurai, H. *Chem. Lett.* 1985, 1781.
90. Nakadaira, Y.; Sekiguchi, A.; Funada, Y.; Sakurai, H. *Chem. Lett.* 1991, 327.
91. Fukuzumi, S.; Kitano, T.; Mochida, K. *Chem. Lett.* 1989, 2177.
92. Fukuzumi, S.; Kitano, T.; Mochida, K. *J. Chem. Soc., Chem. Commun.* 1990, 1236.
93. Fukuzumi, S.; Kitano, T. *J. Am. Chem. Soc.* 1990, *112*, 3246.
94. Pandey, G.; Rao, V. J.; Bhalerao, U. T. *J. Chem. Soc., Chem. Commun.* 1989, 416.
95. Pandey, G.; Soma Sekhar, B. B. V. *J. Org. Chem.* 1994, *59*, 7367.
96. Pandey, G.; Soma Sekhar, B. B. V. *J. Chem. Soc., Chem. Commun.* 1993, 780.
97. Pandey, G.; Soma Sekhar, B. B. V. *Tetrahedron* 1995, *51*, 1483.
98. Chatterjee, S.; Gottschalk, P.; Davis, P. D.; Schuster, G. B. *J. Am. Chem. Soc.* 1988, *110*, 2326.
99. Saeva, F. D.; Breslin, D. T.; Luss, H. R. *J. Am. Chem. Soc.* 1991, *113*, 5333.
100. Arnold, D. R.; Maroulis, A. J. *J. Am. Chem. Soc.* 1976, *98*, 5931.
101. Okamoto, A.; Snow, M. S.; Arnold, D. R. *Tetrahedron* 1986, *42*, 6175.
102. Popielarz, R.; Arnold, D. R. *J. Am. Chem. Soc.* 1990, *112*, 3068.
103. Reichel, L. W.; Griffin, G. W.; Muller, A. J.; Das, P. K.; Ege, S. N. *Can. J. Chem.* 1984, *62*, 424.
104. Davis, H. F.; Das, P. K.; Reichel, L. W.; Griffin, G. W. *J. Am. Chem. Soc.* 1984, *106*, 6968.
105. Sankararaman, S.; Perrier, S.; Kochi, J. K. *J. Am. Chem. Soc.* 1989, *111*, 6448.
106. Maslak, P.; Chapman Jr., W. H. *J. Chem. Soc., Chem. Commun.* 1989, 1809.
107. Maslak, P.; Chapman, Jr., W. H. *J. Org. Chem.* 1990, *55*, 6334.
108. Whitten, D. G. *Photoinduced Electron Transfer Reactions*; M. A. Fox and M. Chanon, Eds.; Elsevier: Amsterdam, 1988; Part C; pp. 553.
109. Lee, L. Y. C.; Ci, X.; Giannotti, C.; Whitten, D. G. *J. Am. Chem. Soc.* 1986, *108*, 175.
110. Ci, X.; Lee, L. Y. C.; Whitten, D. G. *J. Am. Chem. Soc.* 1987, *109*, 2536.
111. Ci, X.; Whitten, D. G. *J. Am. Chem. Soc.* 1987, *109*, 7215.
112. Ci, X.; Whitten, D. G. *J. Am. Chem. Soc.* 1989, *111*, 3459.
113. Bergmark, W. R.; Whitten, D. G. *J. Am. Chem. Soc.* 1990, *112*, 4042.

114. Chen, L.; Farahat, M. S.; Gan, H.; Farid, S.; Whitten, D. G. *J. Am. Chem. Soc.* 1995, *117*, 6380.
115. Gassman, P. G. *Photoinduced Electron Transfer Reactions*; M. A. Fox and M. Chanon, Eds.; Elsevier: Amsterdam, 1988; Part C.
116. Wong, P. C.; Arnold, D. R. *Tetrahedron Lett.* 1979, 2101.
117. Albini, A.; Arnold, D. R. *Can. J. Chem.* 1978, *56*, 2985.
118. Roth, H. D. *Acc. Chem. Res.* 1987, *20*, 343.
119. Arnold, D. R.; Humphreys, R. W. R. *J. Am. Chem. Soc.* 1979, *101*, 2743. (b) Gassman, P. G.; Olson, K. D.; Walter, L.; Yamaguchi, R. *J. Am. Chem. Soc.* 1981, *103*, 4977. (c) Ipaktschi, J. *Tetrahedron Lett.* 1970, 3183.
120. Rao, V. R.; Hixson, S. S. *J. Am. Chem. Soc.* 1979, *101*, 6458.
121. Dinnocenzo, J. P.; Todd, W. P.; Simpson, T. R.; Gould, I. R. *J. Am. Chem. Soc.* 1990, *112*, 2462.
122. Abe, M.; Oku, A. *J. Org. Chem.* 1995, *60*, 3065.
123. Ichinose, N.; Mizuno, K.; Yoshida, K.; Otsuji, Y. *Chem. Lett.* 1988, 723.
124. Tomioka, M.; Kobayashi, D.; Hashimoto, A.; Murata, S. *Tetrahedron Lett.* 1989, *30*, 4685.
125. Gassman, P. G.; Burns, S. J. *J. Org. Chem.* 1988, *53*, 5576.
126. Clawson, P.; Lunn, P. M.; Whiting, D. A. *J. Chem. Soc., Chem. Commun.*, 1984, 134.
127. Muller, F.; Mattay, J. *Angew. Chem. Int. Ed. Engl.* 1991, *30*, 1336.
128. Muller, F.; Mattay, J. *Angew. Chem. Int. Ed. Engl.* 1992, *31*, 209.
129. Kaupp, G.; Laarhoven, W. H. *Tetrahedron Lett.* 1976, *941*. (b) Kadota, S.; Tsubono, K.; Makino, K.; Takeshita, M.; Kikuchi, T. *Tetrahedron Lett.* 1987, *28*, 2857. (c) Pac, C. *Pure Appl. Chem.* 1986, *58*, 1249. (d) Pac, C.; Fukunaga, T.; Go-An, Y.; Sakae, T.; Yanagida, S. *Photochem. Photobiol. (A)* 1987, *41*, 37.
130. Harm, W. *Biological Effects of Ultraviolet Radiations*; Cambridge University Press: Cambridge, 1980. (b) Beddard, G. *Light, Chemical Change and Life;* J. D. Coyle, R. R. Hill, and D. R. Roberts, Eds.; The Open University Press: Walton Hall, Milton Keynes, U.K., 1982; p. 178.
131. Hartman, R. F.; Camp, J. R. V.; Rose, S. D. *J. Org. Chem.* 1987, *52*, 2684.
132. Shigemitsu, Y.; Arnold, D. R. *J. Chem. Soc., Chem. Commun.* 1975, 407. (b) Maroulis, A. J.; Arnold, D. R. *Synthesis* 1979, 819.
133. Maroulis, A. J.; Shigemitsu, Y.; Arnold, D. R. *J. Am. Chem. Soc.* 1978, *100*, 535.
134. Gassman, P. G.; Bottorff, K. J. *J. Am. Chem. Soc.* 1987, *109*, 7547.
135. Gassman, P. G.; De Silva, S. A. *J. Am. Chem. Soc.* 1991, *113*, 9870.
136. Majima, T.; Pac, C.; Nakasone, A.; Sakurai, H. *J. Am. Chem. Soc.* 1981, *103*, 4499.
137. Tazuke, S.; Kitamura, N. *J. Chem. Soc., Chem. Commun.* 1977, 515.
138. Gassman, P. G.; Bottorff, K. J. *Tetrahedron Lett.* 1987, *28*, 5449. (b) Gassman, P. G.; Olson, K. D. *Tetrahedron Lett.* 1983, *24*, 19.
139. Mizuno, K.; Pac, C.; Sakurai, H. *J. Chem. Soc., Chem. Commun.* 1975, 553.

(b) Yasuda, M.; Pac, C.; Sakurai, H. *J. Chem. Soc., Perkin Trans. 1* 1981, 746.

(c) Bunce, N. J.; Bergsma, J. P.; Schmidt, J. *J. Chem. Soc., Perkin Trans. 2* 1981, 713.

140. Yasuda, M.; Yamashita, T.; Shima, K. *J. Org. Chem.* 1987, *52*, 753.
141. Yasuda, M.; Pac, C.; Sakurai, H. *J. Org. Chem.* 1981, *46*, 788.
142. Krishna, A. Ph.D. Thesis, Osmania University, Hyderabad, India, 1988.
143. Pandey, G.; Krishna, A.; Rao, J. M. *Tetrahedron Lett.* 1986, *27*, 4075.
144. Pandey, G.; Krishna, A. *J. Org. Chem.* 1988, *53*, 2364.
145. Pandey, G.; Krishna, A.; Bhalerao, U. T. *Tetrahedron Lett.* 1989, *30*, 1867.
146. Pandey, G.; Krishna, A.; Girija, K.; Karthikeyan, M. *Tetrahedron Lett.* 1993, *34*, 6631.
147. Pandey, G.; Karthikeyan, M. Unpublished work.
148. Pandey, G.; Sridhar, M.; Bhalerao, U. T. *Tetrahedron Lett.* 1990, *31*, 5373.
149. Portella, C.; Deshayes, H.; Pete, J. P. ; Scholler, D. *Tetrahedron* 1984, *40*, 3635.
150. Okada, K.; Okamoto, K.; Oda, M. *J. Am. Chem. Soc.* 1988, *110*, 8736.
151. Okada, K.; Okamoto, K.; Morita, N.; Okubo, K.; Oda, M. *J. Am. Chem. Soc.* 1991, *113*, 9401.
152. Hamada, T.; Nishida, A.; Yonemitsu, O. *J. Am. Chem. Soc.* 1986, *108*, 140.
153. Masnovi, J.; Koholic, D. J.; Berki, R. J.; Binkley, R. W. *J. Am. Chem. Soc.* 1987, *109*, 2851.
154. Nishida, A.; Hamada, T.; Yonemitsu, O. *J. Org. Chem.* 1988, *53*, 3386.
155. Art, J. F.; Kestemont, J. P.; Soumillion, J. P. *Tetrahedron Lett.* 1991, *32*, 1425.
156. Saito, I.; Ikehira, H.; Kasatani, R.; Watanabe, M.; Matsuura, T. *J. Am. Chem. Soc.* 1986, *108*, 3115.
157. Masnovi, J. M.; Kochi, J. K.; Hilinski, E. F.; Rentzepis, P. M. *J. Am. Chem. Soc.* 1986, *108*, 1126.
158. Sankararaman, S.; Kochi, J. K. *Rec. Trav. Chim.* 1986, *105*, 278.
159. Rathore, R.; Kochi, J. K. *J. Org. Chem.* 1996, *61*, 627.
160. Pandey, G.; Rao, K. S. S. P.; Soma Sekhar, B. B. V. *J. Chem. Soc., Chem. Commun.* 1993, 1636.
161. Pandey, G.; Rao, K. S. S. P. Unpublished work.
162. Pandey, G.; Rao, K. S. S. P. *Angew. Chem. Int. Ed. Engl.* 1995, *34*, 2669.
163. Pandey, G.; Rao, K. S. S. P. Unpublished work.
164. Pandey, G.; Rao, K. S. S. P. Bhagwat, B. V. Unpublished work.
165. Tazuke, S.; Kazama, S.; Kitamura, N. *J. Org. Chem.* 1986, *51*, 4548.
166. Nikolaitchik, A. V.; Rodgers, M. A. J.; Neckers, D. C. *J. Org. Chem.* 1996, *61*, 1065.
167. Pandey, G.; Hajra, S.; Ghorai, M. K. *Pure Appl. Chem.* 1996, *68*, 653.
168. Pandey, G.; Hajra, S. *Angew. Chem. Int. Ed. Engl.* 1994, *33*, 1169.
169. Pandey, G.; Pooranchand, D.; Bhalerao, U. T. *Tetrahedron* 1991, *47*, 1745.
170. Pandey, G.; Hajra, S.; Ghorai, M. K. *Tetrahedron Lett.* 1994, *35*, 7837.
171. Pandey, G.; Hajra, S.; Ghorai, M. K. Unpublished work.
172. Pandey, G.; Hajra, S.; Ghorai, M. K. Unpublished work.

173. Pandey, G.; Hajra, S.; Ghorai, M. K. Unpublished work.
174. Pandey, G.; Hazra, S.; Ghorai, M. K. Unpublished work.
175. Pandey, G.; Hazra, S.; Ghorai, M. K. Unpublished work.

8

Photochemical Reactions on Semiconductor Particles for Organic Synthesis

Yuzhuo Li
Clarkson University, Potsdam, New York

I. INTRODUCTION

Semiconductor photocatalysis has recently become one of the most active interdisciplinary research areas, attracting efforts from photochemists, photophysicists, environmental engineers, and scientists in related fields. The initial driving force for these efforts was the search for alternative energy sources due to the oil crisis in early 1970s. In the past three decades, there has been extensive research on semiconductor particle-based photoelectrochemical cells [1–3], photoelectrochemical production of hydrogen [4–6], photocatalytic degradation of air/water pollutants [7–10], organic functional group transformations, and metal recovery [11–16]. Through these investigations, the fundamental processes of photocatalysis are now much better understood [17–20]. High-efficiency photochemical solar cells have been constructed [3]. A wide range of setups based on heterogeneous photocatalysis for air and water treatment are on the brink of commercialization [9,21].

The potential applications of semiconductor-catalyzed organic functional group transformations for organic synthesis have been studied actively with two objectives [15,22,23]. The first is to explore any new photochemical reactions that are unique due to the redox microenvironment on semiconductor surfaces. The second is to search for reactions that can serve as superior alternatives for conventional chemical processes. As photocatalysts, semiconductor particles are usually inexpensive and nontoxic [8]. The use of semiconductor photocatalytic reactions may eliminate the need for some unsafe reducing or oxidizing reagents and wastes such as metal/metal oxides in conventional chemical processes. This would reduce the production of pollutants and hazardous materials at their sources—a strategy that many chemical industries and federal agencies have adopted recently [24,25].

This chapter will give an overview on the factors that significantly influence the efficiency of a semiconductor-mediated photocatalytic process. Representative organic functional group transformations that have synthetic importance will be reviewed in the order of oxidative, reductive, and combined redox transformations. A case study for the synthesis of benzimidazoles will be followed by a discussion on future challenges and opportunities in this area.

II. SEMICONDUCTOR PHOTOCATALYSIS

Photochemical reactions on semiconductor particles can be divided into two categories: semiconductor-mediated photocatalysis and direct photolytic transformation of substrates adsorbed on semiconductor particles. Semiconductor photocatalysis begins with photoexcitation of semiconductor particles, whereas direct photolytic reactions start with excitation of the organic substrate. This chapter deals with the former. The efficiency of a semiconductor-mediated photocatalytic reaction is determined by the properties of the semiconductor, type of substrate, amount of competition from the solvent, and the experimental setup. This section will discuss these four factors in terms of band-gap and band-edge positions of semiconductors, competitive adsorption on semiconductor surfaces, electron and electron-hole scavenging, light sources, and reactors. For a more complete examination of factors that have significant impact on heterogeneous photocatalysis, readers are referred to several excellent articles [12–15].

A. Semiconductor Particles

1. Band-Gap and Band-Edge Positions of Semiconductor Particles

A semiconductor is commonly characterized by the energy gap between its electronically populated valence band and its largely vacant conduction band [26]. The band gap determines the wavelengths that can cause excitation or charge separation in the semiconductor (SC). After excitation, the conduction-band electron, $SC(e^-)$, and the valence-band electron hole, $SC(h^+)$, may undergo either electron transfer reactions with the adsorbed substrates or recombine [17,18]. For large particles, when the solid surface is in contact with an electrolyte solution, the bands bend to form a Schottky surface charge layer, which causes the electron and electron hole to flow in different directions. For n-type semiconductor particles, the conduction-band electrons flow into the bulk of the particle and the electron holes flow toward the interface. For nanoparticles, the degree of such bending is negligible [15–18]. The electron transfer processes among a semiconductor particle and two neutral substrates are illustrated in Fig. 1. A represents an electron acceptor and D denotes an electron donor. This simplified illustration is also applicable for charged substrates.

The efficiency of the electron transfer reactions illustrated in Fig. 1 governs a semiconductor's ability to serve as a photocatalyst for a redox reaction. This efficiency is, in turn, a function of the positions of the semiconductor's conduction and valence bands (band-edge positions) relative to the redox potentials of the adsorbed substrates. For a desired electron transfer reaction to occur, the potential of the electron acceptor should be below (more positive than) the conduction band of the semiconductor, whereas the potential of the electron donor is preferred to be above (more negative than) the valence band of the semiconductor. For an efficient organic synthesis via oxidative photocatalysis, the substrates must have potentials more negative than the valence band of the semiconductor. For an efficient organic synthesis via reductive photocatalysis, it is the reverse.

A number of semiconductors have been investigated for use in heterogeneous photocatalysis. Their band-gap and band-edge positions have been summarized in several well-known reviews [8,12,20,27,28]. Semiconductor particles can be divided into categories according to their ability to split water: for oxidative or O-type, which includes WO_3 and Fe_2O_3, the oxidation power is strong enough to oxidize water, but the reduction power is not strong enough to reduce water. For reductive or R-type, which includes Si and CdSe, the re-

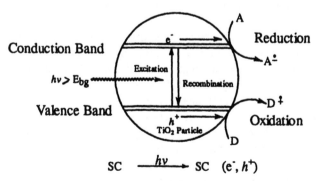

Figure 1 Photoexcitation and electron transfer reactions on an SC particle.

duction power is strong enough to reduce water but the oxidation power is not strong enough to oxidize water. The OR-type (redox) includes TiO_2 and CdS, which have strong enough oxidizing and reducing powers to split water [27]. For example, TiO_2 has a valence-band potential of $+3.1$ V (versus saturated calomel electrode (SCE)) and a conduction band potential of -0.1V (versus SCE). Many organic compounds have potentials above that of the TiO_2 valence band and therefore can be oxidized by TiO_2. By contrast, fewer organic compounds can be reduced by TiO_2 because a smaller number has a potential below that of the TiO_2 conduction band [29].

In addition to band-gap and band-edge positions, some criteria for the selection of a good semiconductor include its chemical and photochemical stability and its environmental impact. TiO_2 is the most popular semiconductor because of its resistivity to strong acids and bases and its stability under illumination [15,17,18]. ZnO, although its band-edge positions are very similar to those of TiO_2 [30], is less desirable due to photocorrosion induced by self-oxidization. CdS has limited potential for practical use despite its attractive spectral response to solar radiation because CdS decomposes to release environmentally harmful Cd^{2+} [15].

2. Preparation and Modification of Semiconductor Particles

Commercially available Degussa P-25 is the most commonly used form of TiO_2 in semiconductor photocatalysis [9]. The advantages of using Degussa P-25 include its high photoactivity, high surface area, consistency in physical and chemical properties, and ready availability [31]. It has even been suggested that the photocatalytic degradation of phenol on unmodified Degussa P-25 should be used as a standard reaction for heterogenous photocatalytic efficiency measurements [32]. Degussa P-25 is a mixture of anatase and rutile titania. Its surface area ranges from 50 to 100 m^2/g^{-1}. Another catalyst, the recently introduced Hombikat UV100, is entirely anatase and has a higher surface area (~ 250 m^2/g^{-1}) [33]. There have been limited reports on the use of this catalyst so far [34].

Titanium dioxide particles can also be prepared easily through a hydrolysis of various titanium alkoxides. The most commonly used alkoxide is isopropoxide. Titanium dioxide with a wide range of particle sizes, morphologies, and surface areas have been prepared [35–37]. It is relevant that semiconductor photoactivity is a function of particle morphology. Characterization of the particles should include the determination of surface ion content and isoelectric points. These properties have a direct impact on particles' electron transfer efficiencies and affinities to organic substrates. It is equally important to realize that the quantum size effect may become significant for small semiconductor particles. Although some observations for TiO_2 are still inconclusive

[38], the quantum size effect has been well documented for CdS and other semiconductors [39–40].

A considerable amount of effort has been made in modifying semiconductor particles. The key objective is to increase quantum efficiency through schemes that prolong charge separation and retard electron–electron-hole recombination after photoexcitation. Efforts also have been made to use the visible region of the sunlight spectra with modified particles. Modifications of semiconductor particles include composition, metal coating, transition metal doping, dye sensitization, and surface hydrophobic treatment.

A composite semiconductor may be a pair of coupled photocatalysts [41–45]. For example, when TiO_2 particles are coupled with smaller-band-gap CdS particles, photoexcited CdS transfer an electron from the conduction band of CdS to the conduction band of TiO_2 [41–43]. As a result, conduction-band electrons on TiO_2 can be made available by using photons with less energy than the TiO_2's band gap. In addition, the charge separation in the composite lasts longer than that in TiO_2 particles alone. The oxidizing power, in this case, is determined by the electron hole of CdS, which is slightly less powerful than that of TiO_2.

Homogeneous doping of semiconductor particles with a small amount of metal ions such as Fe^{3+} and V^{4+} prolongs the electron–electron-hole separation and hence increases the photocatalytic efficiency [46,47]. However, doping of TiO_2 with metal ions such as Cr^{3+} and Sb^{5+} creates electron acceptor and donor centers that accelerate the charge recombination—an undesirable result for photocatalysis [48–50].

The addition of noble metals as small "dots" or "islands" on the surface of semiconductor particles can suppress electron–electron-hole recombination by pulling conduction-band electrons onto the metal islands [51–54]. In addition to prolonging the charge separation, these noble metals have their own catalytic abilities such as assembling and disassembling H_2 for the reduction of organic substrates. These metal islands may lower the conduction-band edge and thus lower the reducing power of the semiconductor [55].

Surface sensitization of a semiconductor via chemisorbed or physisorbed dyes is another efficient method to expand the wavelength range of response (e.g., from UV to visible light) and prolong the electron–electron-hole separation [56–64]. A potential problem associated with a dye sensitization system is the instability of the organic dyes in the presence of strong oxidants such as hydroxyl radicals and electron holes [65]. However, it has been demonstrated that some organic dyes complexed with metals are stable under redox conditions [66]. After dye sensitization, the oxidation power is determined by the highest occupied molecular orbital (HOMO) of the organic dye, which is usually less powerful as an oxidant than the electron hole.

B. Substrates and Solvents

1. Competitive Surface Adsorption

The influence of substrate concentration on the kinetics of photocatalysis has been successfully explained by the Langmuir–Hinshelwood equation and its variations [13,67–71]. In addition to the reactant, many other molecules (solvent, reaction intermediate, reaction product, and other nonreactive solution components) may simultaneously be adsorbed onto the semiconductor surface. As in more classical areas of heterogeneous catalysis, this simultaneous adsorption influences the observed reaction kinetics. In most experiments, the concentration of a reactant ranges in the millimolar. At this concentration, competition from reaction intermediates and/or products can be ignored. Increasing the concentration of a substrate may increase competition between the reactant and intermediate/product for adsorption sites and/or electrons or electron holes, as well as the competition between organic substrates and semiconductor particles for photons [72,73]. This will lower the quantum efficiency and product selectivity. For waste degradation, the objective is to completely convert the starting substrates to products such as water and carbon dioxide. An enhanced surface adsorption for starting material and the reactive intermediates translates to a higher reactivity for the substrate and is always more desirable. For organic synthesis, the goal is controlled functional group transformation. Strong interactions between the semiconductor particles and the starting material, reactive intermediates, and products may not be equally important.

2. Competitive Electron and Electron-Hole Scavenging

To suppress the electron and electron-hole recombination and continuously supply valence-band electron holes for an oxidative process, a sacrificial electron acceptor must be used to scavenge the electron in the conduction band [15,20]. Oxygen is usually used in photocatalytic oxidation because of its ability to capture conduction-band electrons from most semiconductors [8]. The oxygen radical anion formed after the quenching of an electron is an additional oxidant, which is useful for oxidative waste degradation, but may be a source of side reactions in organic synthesis. To reduce such side reactions, the use of other electron acceptors such as methyl viologen and N_2O has been reported [74].

To reduce an organic substrate which cannot compete efficiently with oxygen for electrons, oxygen must be removed from the system. This is usually accomplished by purging the reaction mixture with nitrogen or argon. In addition, an electron donor, such as an alcohol or an amine, is often used to scavenge the electron holes so the conduction-band electrons are available for the reduction [22,23].

C. Experimental Setups

1. Light Sources and Reactors

Two types of light sources have been used often in semiconductor-mediated photocatalysis—medium-pressure mercury arc lamps and phosphor-coated low-pressure mercury lamps [75,76]. A medium-pressure arc lamp has high photon flux and a wide spectrum of photons (200 nm—visible). This type of light source can lead to both direct photolysis of organic substrates and semiconductor photocatalysis. The reaction mechanism in this case can be extremely complicated, as an intermediate of one process can become the substrate for another. For waste degradation, this may be an advantage [7]. However, a direct photolytic reaction would be detrimental to organic synthesis. A phosphor-coated lamp with a much narrower emission band matching the absorption spectrum of the semiconductor particles is thus recommended for organic synthesis. The most commonly used phosphor-coated lamp is a Rayonet lamp that emits light centered at 350 nm (RPR-3500A). Most organic substrates studied in a semiconductor photocatalysis absorb weakly at wavelengths above 350 nm. Particles of a semiconductor, such as TiO_2, shield organic substrates from receiving photons. However, some aromatic compounds do have strong absorptions above 350 nm. For these compounds, a lamp with an even narrower emission band or one equipped with optical filters is recommended. The concentration of the organic substrate should also be low enough to avoid competition with SC particles for photons. However, a low concentration may translate into an unrealistically large amount of solvent for a small amount of organic substrates—a key challenge that prevents using photocatalysis in synthetic applications.

The use of sunlight in waste degradation has been investigated extensively [77–80]. Although typical complaints about sunlight are that it has low intensity in the UV range and that the visible (bulk) portion of the sunlight is useless in exciting semiconductor particles, experimental results on the degradation of organic wastes using sunlight have been quite successful. There has been no report on using solar energy driven semiconductor photocatalysis for synthetic scale organic functional group transformations.

With a low-intensity light source, the number of electrons and electron holes available on the semiconductor particles increases linearly with the increase of photon flux. With a high-intensity light source, however, the number of electrons and electron holes available on the semiconductor particles increases with the square root of photon flux because of electron–electron-hole recombination and the dynamic shortage of organic substrates to scavenge electrons and/or electron holes [81–83]. Therefore, the quantum yield of a reaction catalyzed by semiconductor particles generally decreases with the increase of photon flux. In some cases, the efficiency of organic waste degradation can be increased significantly by turning the light source on and off repeatedly [84]. However, this method has not been used for organic synthesis.

At an analytical scale, an experiment can be carried out in a small Shell vial. A suspension of semiconductor particles in a solution of an organic substrate can be placed in the vial and exposed to a light source. Postreaction analysis can be carried out after filtration with a syringe filter [22,23]. For synthetic purposes, a larger photochemical reactor is required. There has been a considerable amount of effort devoted to the reactor design for semiconductor-mediated photocatalysis. These reactors are designed mainly for water or air treatment [8,85–88]. A flow photochemical reactor used in the author's research lab consists of a series of reaction chambers, a reservoir, and a variable tubing pump. A three-neck round flask serves as the reaction reservoir. The construction of each reaction chamber is similar to that of a reflux condenser [89]. A 15-W lamp (GE F15T8 BL) is located inside of each reaction chamber. Each reaction chamber has an outside tube with a 33.0-mm inner diameter and an inside tube with a 28.0-mm inner diameter, both of which are 400 mm long. The suspension of the semiconductor particles with an organic substrate in solvent is pumped through the jackets of the reaction chambers during the irradiation. A typical flow rate is ~ 20 ml/min^{-1}. After the reaction, the suspension is filtered and the semiconductor particles can be reused. This type of batch-flow reactor can handle up to several liters of the reaction mixture at once. The reaction chamber is made of Pyrex because the light source does not emit much light below 300 nm.

2. Suspended Versus Immobilized Semiconductor Particles

Semiconductor suspensions provide a large total particulate surface area and, therefore, a higher efficiency for photocatalysis. The major disadvantage of a slurry system is that the catalyst must be removed after the reaction. Because most semiconductors used in photocatalysis are submicron-sized particles, an inexpensive and efficient separation is almost impossible. To surmount this disadvantage, immobilized catalyst configurations have been investigated extensively and applied to the construction of photochemical flow reactors for water/air treatment [85–88,90]. In general, these efforts involve adhering the semiconductor particles to a rigid supporting material, such as glass, fiberglass, ceramic, or metal. Some of these processes have already been patented and commercialized [90]. The main problems with the existing immobilization techniques are the inconsistency and lack of durability of the coating. A new particle-coating technique has been developed in the author's lab for the construction of a photochemical reactor [89]. Prior to the particle coating, the glass surface is sandblasted. Sandblasting makes the coating more even and more durable by increasing the roughness of the glass surface. Thus, more particles can be loaded onto the glass, and the reactor activity is significantly enhanced [89]. In addition, the coating of semiconductor particles on a sandblasted surface is much more durable than that in other methods and gives much more consistent results for up to several hundred hours of use.

3. A Semiconductor-Based Actinometer

An actinometer is often necessary to determine the quantum efficiency of a photochemical process, especially when it involves sunlight, which fluctuates in strength at different times and locations. A conventional actinometer involves either a photolabile compound, such as potassium ferrioxalate, or a photoelectrometer [91,92]. For TiO_2-mediated photocatalysis, a disadvantage of these conventional methods is that the absorption spectrum of an actinometer is generally wider than the TiO_2 adsorption profile. The results, therefore, must be carefully corrected for the actual overlap between the two spectra. This difficulty may be overcome by using an actinometer based on the photochemical properties of TiO_2 particles [93]. The dosimeter works because in the presence of a good electron donor such as an alcohol, $TiO_2(e^-)$ can be readily trapped [6,94–97]. To a certain extent, the number of electrons trapped in the matrix is proportional to the number of photons absorbed by TiO_2. The trapped electrons produce a visible color change under UV radiation, which can revert to original transparency upon exposure to oxygen. The absorption spectrum of the trapped electrons, therefore, can be used for actinometry. A clear advantage of the new method is that it measures the light with wavelengths covered by and potentially utilized by TiO_2 particles in photocatalysis. In principle, dosimeters based on other semiconductor particles also can be constructed [94–97].

III. REPRESENTATIVE ORGANIC FUNCTIONAL TRANSFORMATIONS

Essentially, all semiconductor-mediated photocatalytic transformations begin with a redox reaction. In some cases, a substrate either donates an electron to or receives an electron from an SC directly. In others, a reactant is subject to an oxidant or reducing agent generated through a redox reaction. This chapter will discuss these initial steps, beginning with the least complex. Furthermore, a fundamental difference between photocatalytic degradation and organic synthesis is that only those reactions that can be halted at a desired stage, with a high yield, have synthetic value. For this reason, the following discussion is restricted to photoreactions that are "controllable" in product formation.

A. Isomerization

The simplest reaction of an organic substrate (S) on illuminated SC particles involves single electron transfer and back electron transfer only. If the intermediate (radical cation or anion) is thermodynamically and kinetically stable, no net transformation occurs [Eq. (1)]. When the intermediate is unstable, a product (E) may form [Eq. (2)]:

$$S \xrightarrow[\text{or } e^-]{h^+} S^{\ddagger} \text{ or } S^{\div} \xrightarrow[\text{or } h^+]{e^-} S \tag{1}$$

$$S \xrightarrow{h^+} S^{\ddagger} \xrightarrow{\text{isomerization}} E^{\ddagger} \longrightarrow E \tag{2}$$

One of the simplest transformations of a radical cation or anion is structural isomerization. A simple alkene such as *cis*-2-butene is capable of donating an electron to SC(h^+) on an illuminated SC, such as TiO$_2$, ZnS, or CdS, to yield a chemisorbed radical cation which then undergoes cis-trans isomerization. This radical cation takes back an electron from SC to give a mixture of cis- and trans-isomers [98–103]. The ratio of the two isomers is determined by the thermodynamic stability of their radical cations. 1,2-Diarylcyclopropanes also undergo cis-trans isomerization, using ZnO as a photocatalyst [Eq. (3)] [104]:

$$\text{ZnO, } h\nu \longrightarrow \text{ cis- and trans-diarylcyclopropanes} \tag{3}$$

a: R^1 = H, R^2 = OCH$_3$; b: R^1 = R^2 = OCH$_3$

However, unlike simple alkenes, this reaction produces predominately trans-isomers since the trans-isomer is much more stable that its cis-isomer. A third example is the valence isomerization of quadricyclene through a similar mechanism, which yields norbornadiene on irradiated SC particles, including Cds, TiO$_2$, and ZnO [Eq. (4)] [105]:

$$\xrightarrow[\text{CH}_2\text{Cl}_2]{\text{CdS, } h\nu} \tag{4}$$

Quadricyclene   Norbornadiene

Because the initial step in isomerization is electron donation from a substrate to SC(h^+), the isomerization reactions can be quenched readily by a competitive electron donor such as 1,4-diazobicycloctane, diisopropylamine, or 1,2,4-trimethoxybenzene [104]. The presence of methyl viologen dictation (a conduction-band electron quencher) or diphenylamine (an electron-hole relay) enhances the reaction efficiency [105]. Although oxygen is an effective scavenger for conduction-band electrons and capable of increasing the electron-hole availability, the presence of oxygen reduces product yields because oxygen quenches radical cations so effectively.

When a radical cation intermediate undergoes electron transfer with the reactant, a chain reaction may take place [Eq. (5)].

$$S \xrightarrow{h^+} S^{\ddagger} \longrightarrow E^{\ddagger} \xrightarrow{\overset{S}{\underset{}{\bigvee}} S^{\ddagger}} E \tag{5}$$

For example, the quantum yield for a CdS-, TiO_2-, or ZnO-mediated valence isomerization of hexamethyl-dewar benzene to hexamethylbenzene is greater than unity [106]. A cation radical chain reaction mechanism accounts for this observation (Fig. 2).

Although many radical cation-mediated structural isomerizations on an SC have been documented, isomerizations via a radical anion pathway have not been reported. Potential candidates for such an isomerization would be electron-deficient alkenes or cycloalkanes, as they are more likely to accept conduction-band electrons.

B. Dimerization and Polymerization

A radical cation generated form an alkene on an SC surface may also bond with another molecule of the reactant, leading to dimer products, D:

$$S \xrightarrow{h^+} S^{\ddagger} \xrightarrow{S} D^{\ddagger} \xrightarrow{e^-} D \tag{6}$$

For example, *trans*- and *cis*-1,2-diphenoxycyclobutane can be formed from phenylvinyl ether on an irradiated semiconductor such as ZnO, TiO_2, or

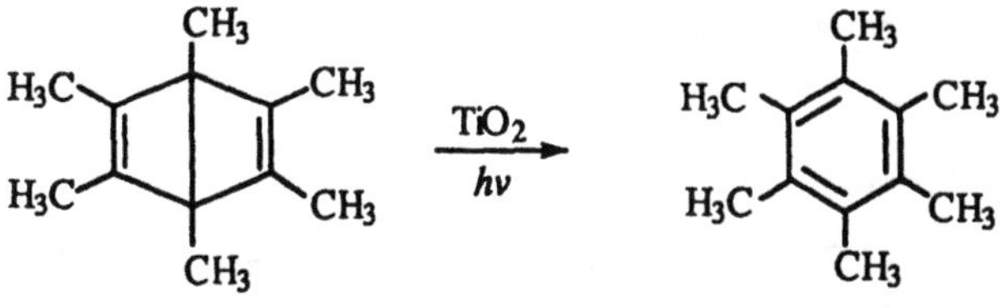

$$\text{HMDB} + h^+ \longrightarrow \text{HMDB}^{\ddagger}$$

$$\text{HMDB}^{\ddagger} \longrightarrow \text{HMB}^{\ddagger}$$

$$\text{HMDB} + \text{HMB}^{\ddagger} \longrightarrow \text{HMDB}^{\ddagger} + \text{HMB}$$

$$\text{HMB}^+ + e^- \longrightarrow \text{HMB}$$

Figure 2 Isomerization of hexamethyl-dewarbenzene (HMDB).

CdS [Eq. (7)] [107–109]. The ratio of trans- to cis- products is determined by the thermodynamic stability of the two dimers. In another example, CdS-mediated photodimerization of N-vinyl carbazole leads to *trans*-cyclobutane dimer exclusively [Eq. (8)] [110].

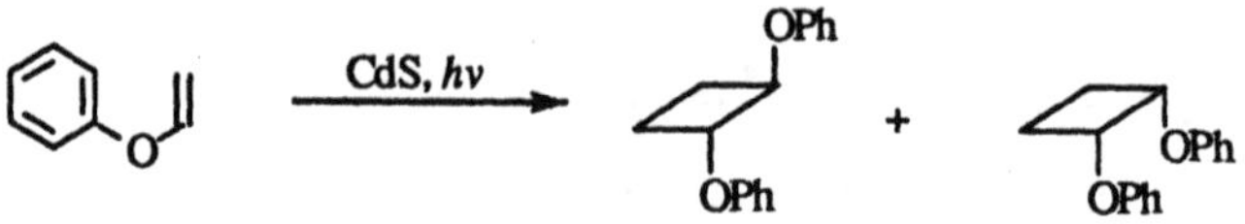

$$ \text{(7)} $$

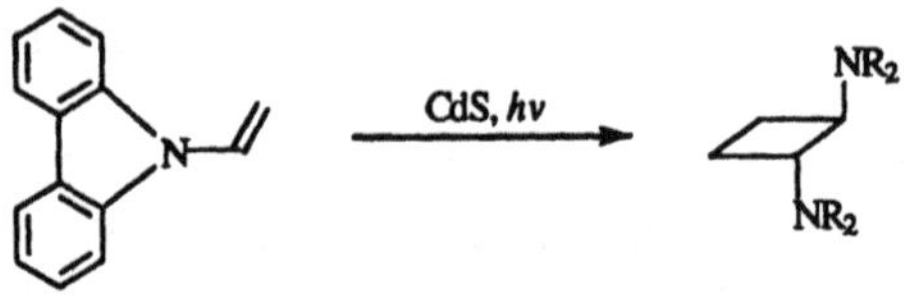

$$ \text{(8)} $$

Polymerization occurs when a radical cation generated from an organic substrate reacts with more than one reactant molecule (monomer) before picking up an electron:

$$ S \xrightarrow{h^+} S^{+} \xrightarrow{nS} P^{+} \xrightarrow{e^-} P \tag{9} $$

For example, irradiation of a suspension of TiO_2 in 1,3,5,7-tetramethylcyclotetrasiloxane (TMCTS) initiates a ring opening polymerization which gives poly(methylsiloxane) (PMS) [111]. A radical cation mechanism with ring opening steps has been proposed for this process (Fig. 3), in which the radical cation of TMCTS undergoes electron transfer with monomers. Similarly, the synthesis of polypyrrole films on illuminated TiO_2 particles has been documented [112]. There has been no report on a dimerization or polymerization via a reductive pathway.

C. Photo-Kolbe Reactions

After donating an electron to $SC(h^+)$, an organic substrate may become unstable and decompose. One well-known example of such a reaction is the Photo-Kolbe

$$ TMCTS + h^+ \longrightarrow TMCTS^+ \xrightarrow{\text{ring opening}} TMTS^+ $$

$$ TMTS^+ + nTMCTS \longrightarrow PMS^+ $$

$$ PMS^+ + e^- \longrightarrow PMS $$

Figure 3 Polymerization of TMCTS on TiO_2 surface.

reaction, in which an alkyl carboxylate anion donates an electron to SC (h^+), followed by decarboxylation to yield an alkyl radical:

$$S \xrightarrow{h^+} S^{\ddagger} \xrightarrow{-X} R\cdot \tag{10}$$

The alkyl radical may dimerize, hydrogen abstract, disproportionate, or initiate polymerization. One such example is the polymerization of methacrylic acid on CdS, CdS/HgS, or CdS/TiO$_2$ particles [Eq. (11)] [113–116]. Through a similar mechanism, benzyl radicals can be obtained from the decomposition of benzyltrimethylsilanes on illuminated TiO$_2$. Dimerization of benzyl radicals gives diarylethanes in a high yield [Eq. (12)] [117].

$$CdS(e^-, h^+) + RCOOH \xrightarrow[-CO_2]{} CdS\,(e^-)\,H^+ + R\cdot \tag{11}$$

$$ArCH_2SiMe_3 \xrightarrow[MeCN]{TiO_2,\, h\nu/Ag_2SO_4} ArCH_2CH_2Ar \tag{12}$$

D. Substitution Reactions

The radical cation of a reactant formed via electron-hole oxidation may also be subject to nucleophilic attack. For example, the radical cation of p-dimethoxybenzene is attacked by cyanide and leading the formation of cyanoanisole [Eq. (13)] [118]. Similarly, selective fluorination of triphenylmethane on irradiated TiO$_2$ in the presence of AgF has been reported [Eq (14)] [119]. A stable carbocation, which is formed after a sequential electron transfer and proton elimination from the reactant, is key for successful fluorination (Fig. 4). Phogocatalytic fluorination employs safe and easy-to-handle reagents and eliminates the need for toxic fluorine gas or other problematic fluorination reagents.

$$(13)$$

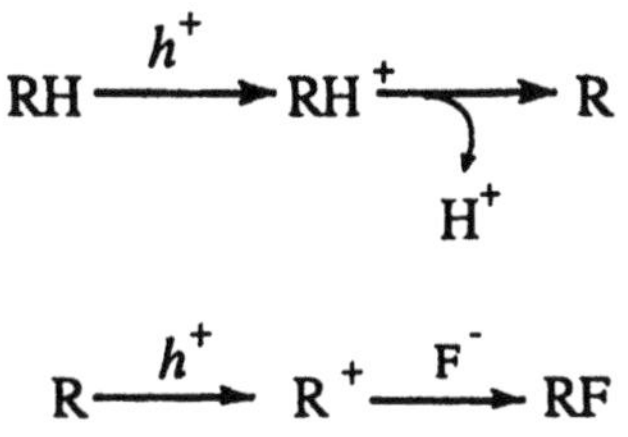

$$RH \xrightarrow{h^+} RH^{+\cdot} \xrightarrow{} R$$
$$\downarrow$$
$$H^+$$

$$R \xrightarrow{h^+} R^+ \xrightarrow{F^-} RF$$

Figure 4 Substitution reaction with AgF on TiO$_2$ surface.

$$(C_6H_5)_3CH \xrightarrow[\substack{AgF \\ CH_3CN}]{TiO_2, h\nu} (C_6H_5)_3CF \qquad (14)$$

E. Oxidations

A wide range of organic compounds can be oxidized on illuminated semiconductor particles [8,9]. Only a few of the reactions, however, can be stopped easily at the desired stage. In addition to the electron hole (the sole oxidant in reactions described above), other oxidants such as oxygen, oxygen radical anion [formed through $O_2 + SC(e^-)$], and hydroxyl radical (formed after electron-hole oxidation of hydroxide or water) may also aid or initiate oxidation of organic substrates on illuminated SC particles (Fig. 5) [120,121]. When the number of oxidants and the complexity of reaction increases, chemical selectivity and product yield often decrease significantly. For some compounds containing heteroatomic sites with lone pair electrons (e.g., $-XH$ or $-CH_2-X-$, where $X=O$, S, and N), such as alcohols, amines, and sulfides, the product yields and chemical selectivities are much higher than those simple hydrocarbons, because of the heteroatoms' ability to stabilize a radical or radical cation intermediate.

Oxidation of some sulfides by semiconductor photocatalysis produces high yields of sulfoxides [Eq. (15)] [122]. It is proposed that the radical cation of the sulfide reacts with oxygen, and then undergoes elimination to yield the sulfoxide. The reaction efficiency of substituted aryl sulfides can be enhanced by an electron-donating substituent in the order of $OCH_3 > CH_3 > H > Cl$, Br.

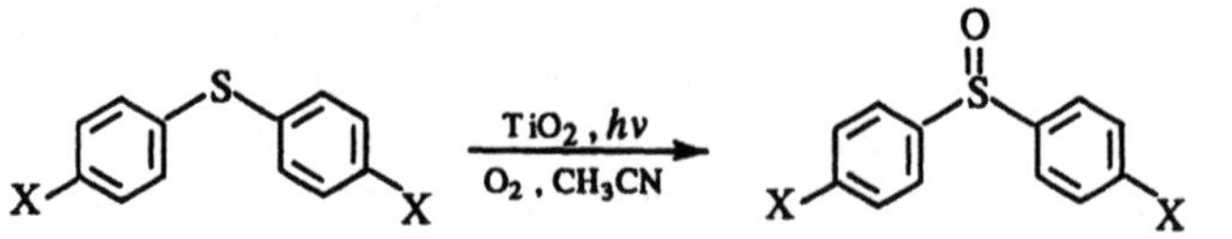

$$(15)$$

High yield and selective formation of carbonyl compounds from aromatic olefins have been obtained in air-saturated acetonitrile solution [Eq.(16)] [123–125]. The combination of a radical cation of the alkene and an oxygen radical anion gives a dioxane intermediate which then undergoes cleavage. For substituted naphthalenes, the oxidative cleavage can be stopped at the side chain of one benzene ring [Eq. (17)] [126].

$$TiO_2 \xrightarrow{h\nu} TiO_2\,(e^-, h^+)$$

$$S \xrightarrow{h^+} S^{\cdot +}$$

$$O_2 \xrightarrow{e^-} O_2^{\cdot -}$$

$$H_2O \xrightarrow{h^+} HO\cdot + H^+$$

Figure 5 TiO_2-mediated oxidation of organic substrates.

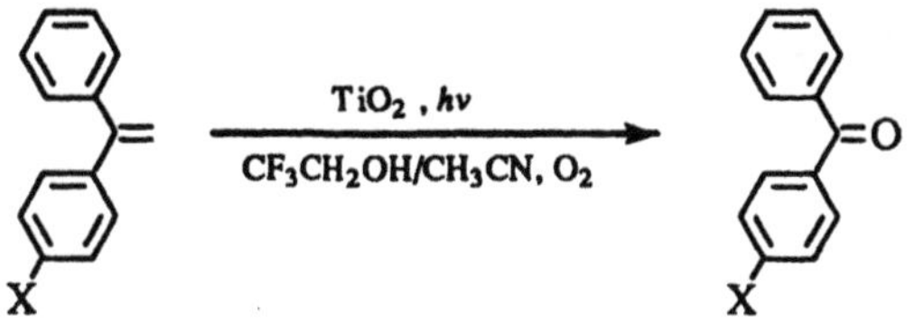

$$\text{(16)}$$

$$X = H, OMe, Me, Cl, NO_2$$

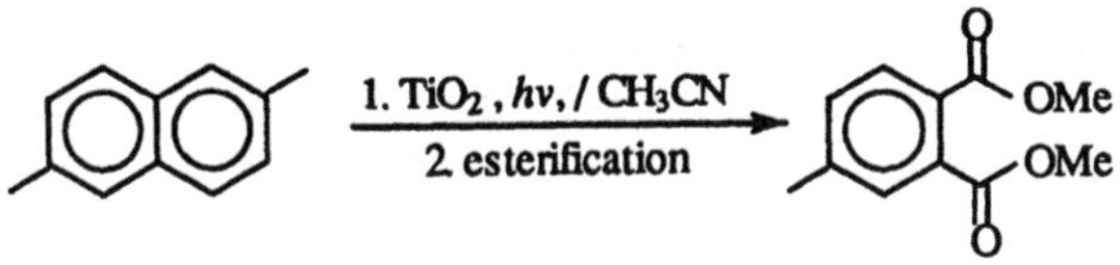

$$\text{(17)}$$

One of the most studied oxidation reactions on semiconductor particles is the photoinduced dehydrogenation of alcohols to aldehydes or ketones [127–136]. Generally, the oxidation is initiated by an electron transfer from the alcohol to the electron hole (Fig. 6). In the presence of adsorbed water, the oxidation can also be initiated by a hydroxyl radical through hydrogen abstraction. Hydrogen is formed when conduction-band electrons or electrons collected on metal islands reduce H^+. An alkoxy radical may also eliminate hydrogen atoms to yield hydrogen molecules. An aldehyde or ketone product can neither compete with the alcohol for h^+ nor compete with O_2 for e^- efficiently, so further oxidation to corresponding carboxylic acids can be prevented.

Highly regioselective oxidation of the carbon next to a heteroatom such as oxygen and nitrogen has been accomplished on illuminated semiconductor particles. One example is the formation of benzoate esters from the photooxidation of benzyl ethers on TiO_2 [Eq. (18)] [137]. The selectivity is the result of a stable benzyl radical cation.

$$PhCH_2OR \xrightarrow[CH_3CN, O_2]{TiO_2, h\nu} PhCO_2R \tag{18}$$

$$R = CH_3, (CH_2)_5CH_3, C(CH_3)_3, (CH_2)_3Ph,$$

Similarly, imides can be synthesized through the photooxidation of lactams and N-acylamines in aqueous solutions in the presence of oxygen [138]:

$$1/2\ H_2 + R_2CHO^- \xleftarrow{\ e^-\ } R_2CHOH \xrightarrow{\ h^+\ } R_2CHOH^{\ddagger} \longrightarrow R_2C{=}O + H^+ + 1/2\ H_2$$

Figure 6 Oxidation of alcohols on semiconductor particles.

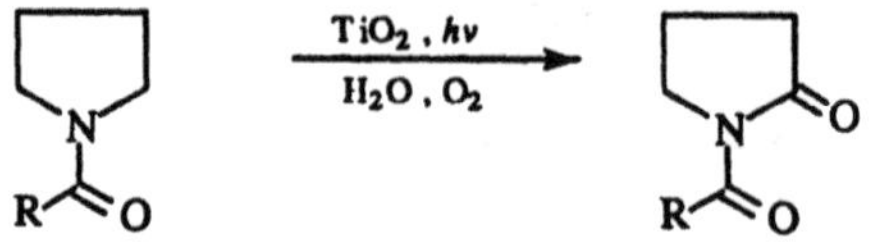

$$\text{(structure)} \xrightarrow[\text{H}_2\text{O},\ \text{O}_2]{\text{TiO}_2\,,\ h\nu} \text{(structure)} \tag{19}$$

$$R = CH_2CH_3\ ,\ CH_3$$

Saturated hydrocarbons have the lowest reactivities on illuminated semi-conductor particles because of their inability to donate electrons to the electron hole. With a few exceptions, such as the oxidation of cyclohexane to cyclohexanone and toluene to benzaldehyde [139–142], the oxidations of hydrocarbons are usually nonselective and synthetically useless [143–145].

F. Reductions

Although valence-band electron holes on most semiconductors are strong oxidants, their conduction-band electrons are not very powerful. In spite of persistant efforts, success in some reductive semiconductor-mediated photocatalytic processes, such as fixation of carbon dioxide and nitrogen, has been limited. However, reductive degradation of some electron-deficient compounds such as polyhalogenated hydrocarbons and metal recovery have shown promising results [84]. In addition, a reductive process is easier to control and thus more suitable for organic synthesis.

Generally, a semiconductor-catalyzed reduction involves of electron transfer, protonation, and sometimes dehydration (Fig. 7) [22,23]. Usually, an alcohol is the source of protons as well as the electron donor to suppress electron–electron-hole recombination. Because most organic substrates cannot compete with oxygen efficiently, oxygen must be removed from the reaction. The hydrogen generated at the semiconductor reduction sites may also be a reducing agent (Fig. 7) [146].

Aromatic aldehydes can be reduced to their corresponding alcohols over irradiated TiO_2 [22]. The reaction involves the formation of an α-hydroxyl radical via a single electron transfer from excited TiO_2 to the aldehyde followed by protonation of the radical anion. The hydroxyl radical then is reduced by a

$$TiO_2 \xrightarrow{h\nu} TiO_2(e^-, h^+)$$

$$S \xrightarrow{e^-} S^{\cdot-} \xrightarrow{H^+} \dot{S}H \xrightarrow{e^-} \xrightarrow{H^+} SH_2$$

$$S + H_2\text{(ad)} \longrightarrow SH_2$$

Figure 7   Reduction of an organic substrate on semiconductor particles.

second electron transfer from TiO_2, followed by a second protonation. This is consistent with the fact that irradiation of TiO_2-benzaldehyde in the presence of O-deuterated ethanol (C_2H_5OD) gives a product labeled with two deuterium atoms [22]:

$$\begin{array}{c} CHO \\ \text{\bigcirc} \end{array} \xrightarrow[CH_3CH_2OD]{TiO_2,\,h\nu} \begin{array}{c} CHDOD \\ \text{\bigcirc} \end{array} \qquad (20)$$

Reduction of a multisubstituted aromatic compound is functional group selective, where selectivity is based on the reduction potentials. For example, the aldehyde group is reduced in p-cyanobenzaldehyde and p-acetobenzaldehyde, whereas cyano and aceto groups are not. The nitro group, however, is reduced more readily than the aldehyde group when p-nitrobenzaldehyde is employed as a substrate [23]. The reduction efficiency for an aliphatic aldehyde is much less than that for an aromatic one because of the difference in reduction potentials.

When illuminated TiO_2 is used as a catalyst, nitro compounds are reduced to their amino derivatives:

$$X = CH=CH_2,\ COCH_3 \qquad (21)$$

$$X = H,\ CHO,\ COCH_3,\ CN,\ CH_3,\ OCH_3 \qquad (22)$$

A six-electron reduction mechanism involving repeated steps of electron transfer, protonation, and dehydration has been proposed (Fig. 8) [24]. A conventional reduction of a halogenated aromatic nitro compound with metals in acidic media is often accompanied by dehalogenation [147]. The semiconductor-mediated reduction of the compounds does not cause dehalogenation.

Although both ZnO and TiO_2 have very similar band-gap and band-edge positions, their isoelectric points (IEP) are very different, ~ 6.0 for TiO_2 and 8.0 for ZnO [148]. In a electrokinetic measurement, IEP is the pH at which the particles are neutral. A lower IEP value means the particles need a lower pH solution to neutralize, hence a more acidic surface surface [149]. It is found

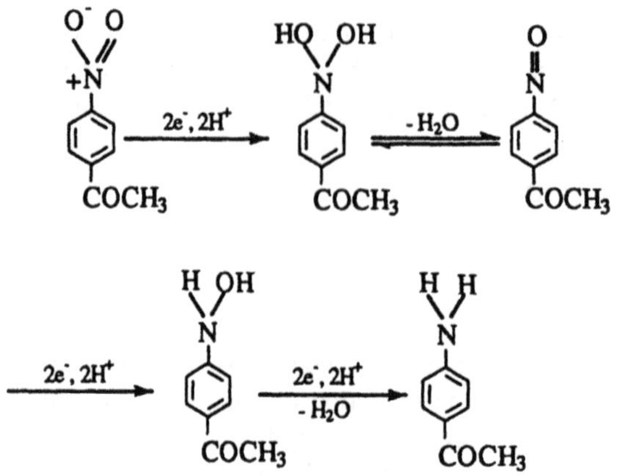

Figure 8 Reduction of aromatic nitro compounds on semiconductor particles.

that surface acidity of a semiconductor particle could have a significant impact on the reaction outcome. For example, hydroxylamino compounds are generally formed as intermediates during the reduction of nitro compounds to amino compounds [150]. When TiO_2 is used as the catalyst, the reduction cannot be stopped at the hydroxylamino stage. In comparison, hydroxylamino derivatives can be synthesized selectively through the photoinduced reduction of nitro compounds over ZnO [151]:

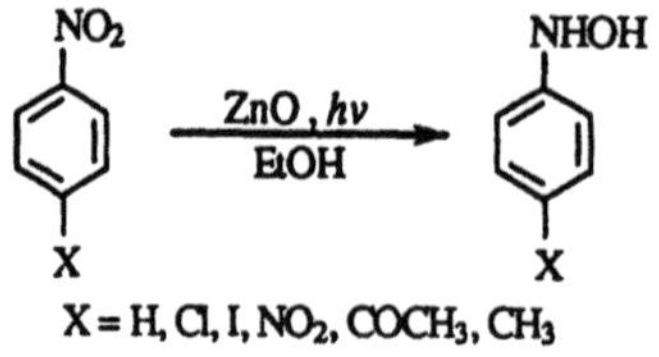

$$X = H, Cl, I, NO_2, COCH_3, CH_3$$

(23)

The lack of further reduction of hydroxylamino intermediate is due to the limited proton source on a basic ZnO surface.

Carbon–carbon multiple bonds can be hydrogenated using illumination of semiconductor particles as catalysts [152,153]. One example is the reduction of pyruvate to lactate under illumination on aqueous suspension of TiO_2 particles [154]. Triethanolamine increases the reaction efficiency by donating electrons to h^+. Alkyne reduction, however, usually results in a mixture of alkenes and alkanes [152].

Semiconductor-mediated photocatalysis also can be used in deoxygenating sulphoxides to sulfides, pyridine-N-oxide to pyridine, and azoxybenzene to azobenzene [Eqs. (24)–(26)] [151].

$$\text{(pyridine N-oxide)} \xrightarrow[\text{EtOH}]{hv,\ TiO_2} \text{(pyridine)} \qquad (24)$$

$$\text{(PhS(O)CH}_3) \xrightarrow[\text{EtOH}]{hv,\ TiO_2} \text{(PhSCH}_3) \qquad (25)$$

$$\text{(azoxybenzene)} \xrightarrow[\text{EtOH}]{hv,\ TiO_2} \text{(azobenzene)} \qquad (26)$$

In a similar process, 1-azidoadmatane is reduced to 1-adamantamine via a single electron transfer followed by elimination of nitrogen:

$$\text{1-azido- damantane} \xrightarrow{TiO_2,\ hv} \text{1-adamantamine} \qquad (27)$$

1-azido: damantane   1-adamantamine

The yield of a 1-adamantamine decreases in the presence of a more polar compound, such as cyclohexanol, which adsorbs more readily onto TiO_2 particles than 1-azidoadamantane. Because azides can be produced readily from alkyl halides, photoreduction of azides on SC provides an easier alternative to amine synthesis [151].

Unlike oxidations, reduction efficiency for organic compounds decreases when metal-coated TiO_2 such as Au/TiO_2 or Pt/TiO_2 is the catalyst [151]. This is because these noble metals lower the conduction-band position and hence lower the reducing power of the TiO_2 particles. Also, the reduction efficiency of CdS is much lower than that of TiO_2 for most polar compounds, although the reducing power of CdS should be higher than that of TiO_2 according to their relative band-edge positions. The inefficiency may be due to poor surface adsorptivity of the organic substrates on CdS. Furthermore, hydrophobic surface treatment (methylation of surface OH groups) for TiO_2 increases its reduction efficiency for less polar or nonpolar compounds, such as azidoadmantane, and decreases its reduction efficiency for polar compounds, such as aldehydes and nitro compounds [89].

G. Combined Redox Reactions

A primary amine can be photocatalytically converted to a secondary amine on TiO_2 particles [Eq. (28)]. The transformation involves both oxidative and reductive steps, as shown in Fig. 9, in which an oxidative process generates a Schiff-base intermediate, which then reduces to the secondary amine [155]. α,ω-Diamines give cyclic amines through a similar but intramolecular process [Eq. (29)] [156].

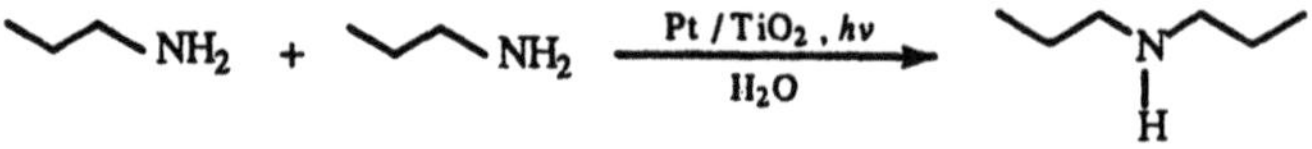

$$\text{\raisebox{0pt}{$\diagdown$}}NH_2 \; + \; \text{\raisebox{0pt}{$\diagdown$}}NH_2 \xrightarrow[\text{H}_2\text{O}]{\text{Pt / TiO}_2 \, , \, hv} \text{\raisebox{0pt}{$\diagdown$}}\underset{\underset{H}{|}}{N}\text{\raisebox{0pt}{$\diagup$}} \tag{28}$$

$$H_2N\text{\raisebox{0pt}{$\diagdown$}}NH_2 \xrightarrow[\text{H}_2\text{O}]{\text{Pt / TiO}_2 \, , \, hv} \underset{\underset{H}{|}}{N} \tag{29}$$

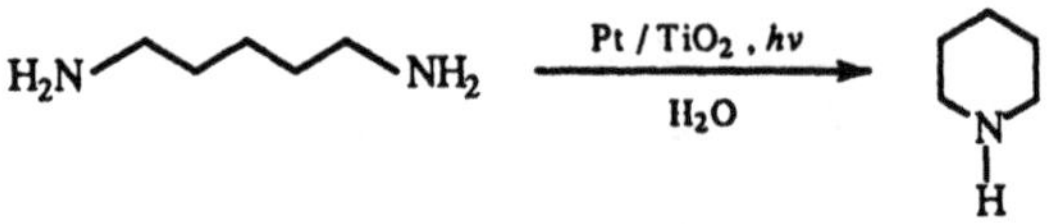

In an alcoholic solution of a primary amine, the alcohol is oxidized to an aldehyde or ketone, which couples with the amine to form a Schiff-base by N-alkylation. The intermediate is then reduced to a secondary amine (Fig. 10) [146]. Similar reactions can yield tertiary amines.

Photocatalytic reduction of o-dinitrobenzene and its derivatives on TiO_2 particles in the presence of a primary alcohol gives benzimidazoles in high yield [157]. This transformation is also an example of combined redox processes (Fig. 11).

First, the solvent is oxidized to aldehyde **I** while the dinitro compound is reduced to 2-nitroaniline, **II**. The condensation between aldehyde **I** and the

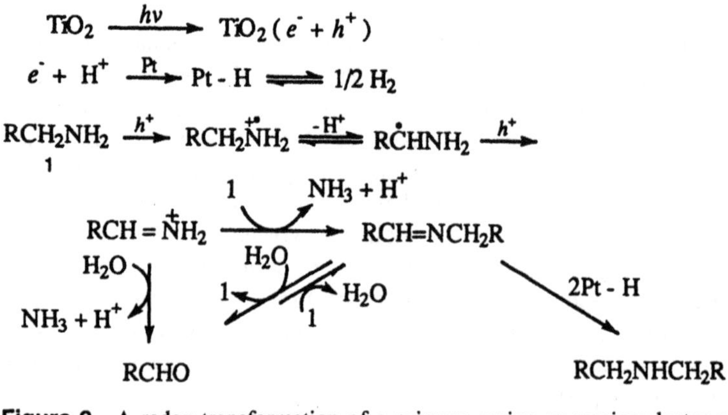

Figure 9   A redox transformation of a primary amine on semiconductor particles.

$$RR'CHOH \longrightarrow RR'C=O + H_2$$

$$R''NH_2 + RR'C=O \longrightarrow (R''NHC(OH)RR' \xrightarrow[-H_2O]{} R''N=CRR') \xrightarrow{H_2/Pt} R''NHCHRR'$$

Figure 10 A redox transformation of a primary amine/alcohol solution on SC.

amino group of **II** forms intermediate **III**. At room temperature and in the presence of a considerable amount of water, **III** is in equilibrium with the aniline and aldehyde. However, the further reduction of **III** into hydroxylamine, **IV**, push the equilibrium away from **III**. The nitrogen atom in the hydroxylamine group attacks the imino carbon to close the ring. Further dehydration leads to a benzimidazole. The reaction is applicable to a number of dinitro compounds and primary alcohols with yields ranging from 70% to 95%.

A number of experiments have been conducted to verify this reaction mechanism. A TiO$_2$-catalyzed reaction of 3,4-dinitrotoluene with intentionally added 1-propanal gives 2-ethyl-6-methylbenzimidazole, as expected from the reaction mechanism above. This finding supports the involvement of aldehyde in the formation of benzimidazole. Irradiation of a 2-nitroaniline solution containing TiO$_2$ produces benzimidazole as well. This is consistent with the fact that 2-nitroaniline has been found in the formation of benzimidazole as a transient product. Irradiation of o-phenylenediamine in ethanol with TiO$_2$, however, has failed to produce 2-methylbenzimidazole. Although monoanil and dianil might have been formed in this reaction mixture, they could not have proceeded further to form benzimidazoles in the Philips synthesis [158] due to the lack

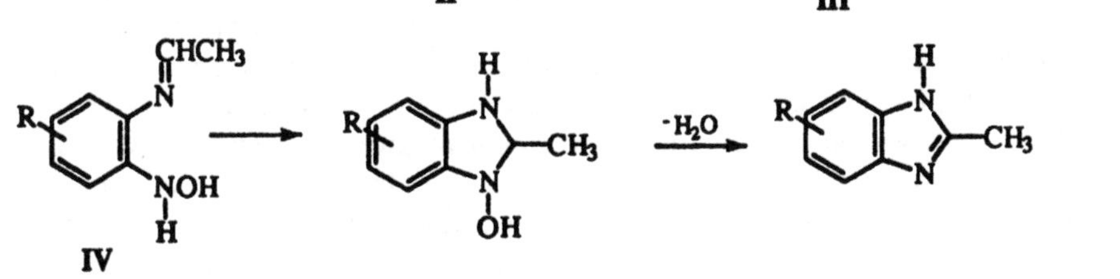

R= H,CH$_3$,OCH$_2$CH$_3$,COOCH$_2$CH$_3$,Cl

Figure 11 Formation of benzimic azoles on TiO$_2$ particles.

of oxidizing reagents. Because irradiation of nitro compounds with ZnO under UV light generates hydroxylamine as a final product [151] and irradiation of *o*-nitroaniline in ethanol with ZnO yields 2-methylbenzimidazole, the second nitro group must not be reduced to an amino group after **II** forms. When 1,2-dinitrobenzene is used, 2-methylbenzimidazole is also found on the irradiated ZnO particles. This appears inconsistent with the reaction mechanism illustrated in Fig. 11, as an amino group is required for the first step, and hydroxylamine is the only expected product with ZnO. The 1,2-dihydroxylaminobenzene, however, may undergo an intramolecular redox reaction in which one hydroxylamino group is oxidized to a nitro group, whereas the other is reduced to an amino group. The nitroaniline is then converted to 2-methylbenzimidazole.

The reaction product and reaction rate are strongly solvent dependent. When ethanol is used as a solvent, 2-methylbenzimidazoles are formed, although when 1-propanol is used, 2-ethylbenzimidazoles are produced. When a secondary alcohol such as 2-propanol is used as the solvent, no benzimidazole products are obtained. The oxidation of 2-propanol generates acetone instead of acetaldehyde. Due to the steric effect, acetone is much less likely to condense with nitroaniline to form an intermediate like **III**.

IV. CHALLENGES AND OPPORTUNITIES

With the examples illustrated in this chapter, semiconductor-mediated photocatalysis has shown its potential applications in organic synthesis. These organic functional group transformations, operated through redox reactions, cover a wide range of organic compounds. Although most of these transformations are possible using conventional reagents and processes, one obvious advantage of semiconductor-mediated photocatalysis is its potential to provide an environmentally benign alternative.

In the past several decades, significant progress has been made in semiconductor particle preparation and modifications to enhance surface electron-electron-hole separation. The progress includes the preparation of quantum size particles, dye sensitization, doping/coating semiconductor particles, and the assembly of semiconductor thin films. A key difference between semiconductor photocatalysis and electrochemical processes is the physical distance between the electrodes. For semiconductor-mediated reactions, both reduction and oxidation sites are on the same particle. This may have disadvantages, such as charge recombination between conduction-band electrons and valence-band electron holes, or back electron transfer reactions between SC particles and organic substrates. The close proximity between redox sites can also be an advantage. As shown by the reactions described in the coupled redox reaction section, intermediates generated from one site can be substrates of another. With a careful design, chemical transformations that are impossible or difficult to carry out in other environments can be achieved easily via semiconductor photocatalysis.

There has been a wide range of semiconductor-mediated organic functional group transformations based on an oxidative step (electron-hole oxidation). However, reactions based on a reductive process (conduction-band electron reduction) are still limited. There are some excellent opportunities for new chemical transformations initiated with conduction-band electron transfer. Despite progress made in semiconductor particle immobilization and reactor design, fundamental challenges still await in conducting semiconductor-mediated photocatalysis at a synthetic scale. One key issue is competitive surface adsorption. Competitive adsorption among the reactants, intermediate, and products influences the reactivity and reaction outcome and prevents the use of a high reactant concentration. Another obstacle to using a high reactant concentration is competition between the semiconductor particles and organic substrates for photons. A semiconductor system such as a carefully designed dye sensitization or composite semiconductor system that retains the redox desired potentials may tolerate a higher reactant concentration. Semiconductor modifications also may have surface properties for selective adsorption of the target reactant and/or intermediates. Modifications on the semiconductor particles must be coupled with a reactor design in which a large surface area of semiconductor particles is available to the organic substrate. Overall, there are tremendous challenges and opportunities for research and development in the area of nanocrystalline semiconductor particles for organic synthesis.

ACKNOWLEDGMENTS

Financial support from the National Science Foundation and the Petroleum Research Fund is gratefully acknowledged. Thanks are also due to Ms. Lijuan Wang of Clarkson University for her assistance in the preparation of the manuscript.

REFERENCES

1. Grätzel, M. (ed.), *Energy Resources through Photochemistry and Catalysis,* Academic Press: New York, 1983.
2. Vlachopoulos, N.; Liska, P.; Augustynski, J.; Grätzel, M. *J. Am. Chem. Soc.,* 1988, *110*, 1216-1220.
3. Grätzel, M. Research Opportunities in Photochemical Sciences, Workshop Proceedings, Estes Park, Colorado, February 5-8, 1996; pp. 1–22.
4. Kamat, P. V.; Dimitrijevic, N. M. *Solar Energy,* 1990, *44*, 83-98.
5. Serpone, N. in *Photochemical Energy Coversion,* J. R. Norris, and D. Meisel, Eds; Elsevier Science Publishers B.V.: Amsterdam, The Netherlands, 1989; pp. 297–315.

6. Grätzel, M. in *Homogeneous and Heterogeneous Photocatalysis*, E. Pelizzetti, and N. Serpone, Eds; D. Reidel Publishing Company: Dordrecht, The Netherlands, 1986; pp. 91-100.

7. Legrini, O.; Oliveros, E.; Braun, A. M. *Chem. Rev.*, 1993, *93*, 671-698.

8. Mills, A.; Davies, R. H.; Worsley, D. *Chem. Soc. Rev.*, 1993, 22, 417-425.

9. Blake, D. M. NREL/TP-430-6084, National Renewable Energy Laboratory, 1994.

10. Hoffmann, M. R.; Martin, S. T.; Choi, W.; Bahnemann, D. W. *Chem. Rev.*, 1995, *95*, 69-96.

11. Fox, M. A.; *ACS Symposium Series 278*, American Chemical Society: Washington, D.C. 1985; pp. 43-55.

12. Fox, M. A. in *Photocatalysis - Fundamentals and Applications*, N. Serpone, and E. Pelizzetti, Eds. John Wiley & Sons: New York, 1989; pp. 420-455.

13. Al-Ekabi, H. in *Photochemistry in Organized and Constrained Media*, V. Ramamurthy, Ed.; VCH Publishers, Inc.: New York, 1991; pp. 495-534.

14. Fox, M. A. *Top. Curr. Chem.*, 1991, *142*, 71-100.

15. Fox, M. A.; Dulay, M. T. *Chem. Rev.*, 1993, *93*, 341-357.

16. Borgarello, E.; Serpone, N.; Barbeni, M.; Pelizzetti, E.; Pichat, P.; Herrmann, J. M.; Fox, M. A. *J. Photochem.*, 1987, *10*, 373.

17. Serpone, N.; Pelizzetti, E. in *Homogeneous and Heterogeneous Photocatalysis*, E. Pelizzetti, and N. Serpone, Eds., D. Reidel Publishing Company: Dordrecht, The Netherlands, 1986; pp. 51-89.

18. Weisbuch, C.; Vinter, B. *Quantum Semiconductor Structures*, Academic Press, Inc.: New York, 1991.

19. Miller, R. J. D.; McLendon, G. L.; Nozik, A. J.; Schmickler, W.; Willig, F. *Surface Electron Transfer Processes*, VCH Publishers, Inc.: New York, 1995.

20. Linsebgler, A. L.; Lu, G.; Yates Jr., J. T. *Chem. Rev.*, 1995, *95*, 735-758.

21. *The Second International Conference on TiO₂ Photocatalytic Purification and Treatment of Water and Air*, Cincinnati, Ohio, October 22-29, 1996.

22. Joyce-Pruden, C.; Pross, J. K.; Li, Y. *J. Org. Chem.*, 1992, *57*, 5087-5091.

23. Mahdavi, F.; Bruton, T. C.; Li, Y. *J. Org. Chem.*, 1993, *58*, 744-746.

24. Illman, D. L. *C&EN*, September 5, 1994; pp. 22-27.

25. Anastas, P. T.; Farris, C. A. (ed.), *Benign by Design, ACS Symposium Series 577*, American Chemical Society: Washington, D.C., 1994.

26. Lewis, N. A.; Rosenbluth, M. L. in *Photocatalysis - Fundamentals and Applications*, N. Serpone, and E. Pelizzetti, Eds., John Wiley & Sons: New York, 1989; pp. 45-98.

27. Sakata, T. in *Photocatalysis - Fundamentals and Applications*, N. Serpone, and E. Pelizzetti, Eds., John Wiley & Sons: New York, 1989, pp. 311-338.

28. Pelizzetti, E., Barbeni, M.; Pramauro, E.; Erbs, W.; Borgarello, E.; Jamieson, M. A.; Serpone, N. *Quimica Nova*, 1985, 288-302.

29. Murov, S. L.; Carmichael, I.; Hug, G. L. *Handbook of Photochemistry, 2nd ed.*, Marcel Dekker, Inc.: New York, 1993, pp. 269-278.

30. Kakuta, N.; White, J. M.; Bard, A. J.; Campion, A.; Fox, M. A.; Webber, S. E.; Finlayson, M. *J. Phys. Chem.*, 1985, *89*, 48.

31. Koth, D.; Ferch, H. *Chem. Ing. Techn.*, 1980, *52*, 628.
32. Matthews, R. W.; McEvoy, S. R. *J. Photochem. Photobiol. A: Chem.*, 1992, *66*, 355.
33. Tahiri, H.; Serpone, N.; Levanmao, R. *J. Photochem. Photobiol. A: Chem.*, 1996, 93, 199-203.
34. *The First International Conference on Advanced Oxidation Technologies for Water and Air Remediation*, London, Ontario, Canada, June 25-30, 1994.
35. Hisanaga, T.; Tanaka, K. *Mizu Shori Gijutsu*, 1993, *34*, 13-18.
36. Kato, K.; Tsuzuki, A.; Torri, Y.; Taoda, H.; Kato, T.; Butsugan, Y. *Nagoya Kogyo Gijutsu Kenkyusho Hokoku*, 1994, *61*, 876-877.
37. Pelizzetti, E.; Minero, C.; Borgarello, E.; Tinucci, L.; Serpone, N. *Langmuir*, 1993, *9*, 2995-3001.
38. Serpone, N.; Lawless, D.; Khairutdinov, R. *J. Phys. Chem.*, 1995, *99*, 16646-16654.
39. Grätzel, M. in *Photocatalysis - Fundamentals and Applications*, N. Serpone, and E. Pelizzetti, Eds., John Wiley & Sons: New York, 1989; pp. 123-157.
40. Henglein, A. *70th Colloid and Surface Science Symposium*, Clarkson University, New York, June 16-19, 1996; p. 8.
41. Sclafani, A.; Mozzanega, M.-N.; Pichat, P. *J. Photochem. Photobiol. A: Chem.*, 1991, *59*, 181.
42. Gopida, K. R.; Bohorquez, M.; Kamat, P. V. *J. Phys. Chem.*, 1990, *94*, 6435-6440.
43. Spanhel, L.; Weller, H.; Henglein, A. *J. Am. Chem. Soc.*, 1987, *109*, 6632.
44. Bahnemann, D. W.; Bockelmann, D.; Goslich, R.; Hilgendorff, M. in *Aquatic and Surface Photochemistry*, G. R. Helz, R. G. Zepp, D. G. Crosby, Eds., Lewis Publishers: Ann Arbor, 1994; pp. 349-367.
45. Li, S.; Lu, G. *New J. Chem.*, 1992, *16*, 517-519.
46. (a) Cunningham, J.; Srijaranai, S. *J. Photochem. Photobiol. A:* Chem., 1988, *43*, 329. (b) Grätzel, M.; Howe, R. F. *J. Phys. Chem.*, 1990, *94*, 2566.
47. Escribano, V. S.; Busca, G.; Lorenzelli, V. *J. Phys. Chem.*, 1991, *95*, 5541.
48. Palmisano, L.; Augugliaro, V.; Sclafani, A.; Schiavello, M. *J. Phys. Chem.*, 1988, *92*, 6710.
49. Mu, W.; Herrmann, J. M.; Pichat, P. *Catal. Lett.*, 1989, *3*, 73.
50. Day, V. W.; Klemperer, W. G.; Main, D. *J. Inorg. Chem.*, 1990, *29*, 2345.
51. Kraeutler, B.; Bard, A. J. *J. Am. Chem. Soc.*, 1978, *100*, 2239.
52. Kraeutler, B.; Bard, A. J. *J. Am. Chem. Soc.*, 1978, *100*, 4317-4318.
53. Sato, A. *J. Catal.*, 1985, *92*, 11-16.
54. Fox, M. A. *Nouveau Journal de Chimie*, 1987, *11*, 129-133.
55. Wagner, J.; Alamo, J. *J. Appl. Phys.*, 1987, *63*, 425.
56. Nakahira, T.; Grätzel, M. *Makromol. Chem., Rapid Commun.*, 1985, *6*, 341-347.
57. Kamat, P. V. *J. Phys. Chem.*, 1989, *93*, 859-864.
58. Moser, J.; Grätzel, M. *J. Am. Chem. Soc.*, 1984, *106*, 6557-6564.
59. Borgarello, E.; Kiwi, J.; Pelizzetti, E.; Visca, M.; Grätzel, M. *Nature*, 1981, *289*, 158-160.

60. Kamat, P. V. *Langmuir*, 1990, *6*, 512-513.
61. Kamat, P. V.; Fox, M. A. *Chem. Phys. Lett*, 1983, *102*, 397.
62. Patrick, B.; Kamat, P. V. *J. Phys. Chem.*, 1992, *96*, 1423.
63. Blachopoulos, N.; Liska, P.; Augustynski, J.; Grätzel, M. *J. Am. Chem. Soc.*, 1988, *110*, 1216.
64. Desilvestro, J.; Grätzel, M.; Kavan, K.; Moser, J.; Augastynski, J. *J. Am. Chem. Soc.*, 1985, *107*, 2988.
65. Vinodgopal, K.; Kamat, P. V. *Environ. Sci. Technol.*, 1995, *29*, 841-845.
66. Regan, O.; Grätzel, M. *Nature*, 1991, 353.
67. Pichat, P.; Herrmann, J.-M. in *Photocatalysis - Fundamentals and Applications*, N. Serpone, and E. Pelizzetti, Eds., John Wiley & Sons: New York, 1989; pp. 217-250.
68. Turchi, C. S.; Ollis, D. F. *J. Catal.*, 1990, *122*, 178.
69. Jenny, B.; Pichat, P. *Langmuir*, 1991, *7*, 947.
70. Langmuir, I. *Trans. Faraday Soc.*, 1921, *17*, 621.
71. Matthews, R. W. *J. Chem. Soc., Faraday Trans. I*, 1989, *85*, 1291.
72. Hidaka, H.; Nohara, K.; Zhao, J.; Serpone, N.; Pelizzetti, E. *J. Photochem. Photobiol. A: Chem.*, 1992, *64*, 247-254.
73. Sato, S. *Langmuir*, 1988, *4*, 1156.
74. Serpone, N.; Maruthanuthu, P.; Pichat, P.; Pelizzetti, E.; Hidaka, H. *J. Photochem. Photobiol. A. Chem.*, 1995, *85 (3)*, 247-255.
75. Murov, S. L.; Carmichael, I.; Hug, G. L. *Handbook of Photochemistry*, 2nd ed., Marcel Dekker, Inc.: New York, 1993; pp. 325-337.
76. Gould, I. R. in *Handbook of Organic Photochemistry*, J. Scaiano, Ed.; CRC Press: 1989; pp. .155-196.
77. Wilkins, F. W.; Blake, D. M. *Chemical Engineering Progress*, June 1994.
78. Matthews, R. W. *Solar Energy*, 1987, *38*, 405-413.
79. Anderson, J. V.; Clyne, R. J. SERI/TP-250-4474, UC Category: 234 DE91015007, Solar Energy Research Institute, 1991.
80. Blake, D. M. NREL/TP-430-5594, UC Category: 237 DE93010033, National Renewable Energy Laboratory, 1993.
81. Cunningham, J.; Hodnett, B. K. *J. Chem. Soc., Faraday Trans. I*, 1981, *77*, 2777.
82. Harvey, P. R.; Rudham, R.; Ward, S. *J. Chem. Soc., Faraday Trans. I*, 1979, *75*, 2507.
83. Davidson, R. S.; Slater, R. M.; Meck, R. R. *J. Chem. Soc., Faraday Trans. I*, 1983, *79*, 1391.
84. Ollis, D. F. Research Opportunities in Photochemical Sciences, Workshop Proceedings, Estes Park, Colorado, February 5-8, 1996; pp. 87-94.
85. Matthews, R. W. AU Patent B 76028, 1987.
86. Turchi, C. S.; Mehos, M. S. in *Chemical Oxidation: Technologies for the Nineties*, Technomic Publishing Company: Lancaster, 1992, pp. 155-161.
87. Al-Ekabi, H.; Edwards, G.; Holden, W.; Safarazadey-Amiri, A.; J. Story, J. in *Chemical Oxidation: Technologies for the Nineties*, Technomic Publishing Company: Lancaster, 1992, pp. 254-261.

88. Hofstadler, K.; Bauer, R.; Novalic, S.; Heisier, G. *Environ. Sci. Technol.*, 1994, *28*, 670–674.
89. Wang, M.; Li, Y. unpublished results.
90. Robertson, M. K.; Headerson, R. B. US Patent, 1990, 4,892,712.
91. Murov, S. L.; Carmichael, I.; Hug, G. L. *Handbook of Photochemistry*, 2nd ed., Marcel Dekker, Inc: New York, 1993; pp. 299-313.
92. Diffey, B. L. in *Sunscreens, Development, Evaluation, and Regulatory Aspects*, N. J. Lowe and N. A. Shaath, Eds., Marcel Dekker: New York, 1990, pp. 93–123.
93. Huang, J.; Ding, H.; Dodson, W. S.; Li, Y. *Analytica Chimica Acta*, 1995, *311*, 115-122.
94. Warman, J. M.; de Haas, M. P.; Pichat, P.; Serpone, N. *J. Phys. Chem.*, 1991, *95*, 8858.
95. Vinodgopal, K.; Bedja, I.; Hotchandani, S.; Kamat, P. V., *Langmuir*, 1994, *10*, 1767.
96. Kamat, P. V., *Chem. Rev.*, 1993, *93*, 267.
97. Serpone, N.; Lawless, D.; Disdier, J.; Herrman, J. M., *Langmuir*, 1994, *10*, 643.
98. Yanagida, S.; Mizumoto, K.; Pac, C. *J. Am. Chem. Soc.*, 1986, 108, 647-654.
99. Anpo, M.; Yabuta, M.; Kodama, S.; Kubokawa, Y. *Bull. Chem. Soc. Jpn.*, 1986, *59*, 259-264.
100. Kodama, S.; Nakaya, H.; Anpo, M.; Kubokawa, Y. *Bull. Chem. Soc. Jpn.*, 1985, *58*, 3645-3646.
101. Al-Ekabi, H.; Mayo, P. *J. Phys. Chem.*, 1985, *89*, 5815-5821.
102. Kodama S.; Yagi, S. *J. Phys. Chem.*, 1989, *93*, 4556-4561.
103. Anpo, M.; Sunamoto, M. *J. Phys. Chem.*, 1989, *93*, 1187-1189.
104. Carson, P. A.; Mayo, P. *Can. J. Chem.*, 1987, *65*, 976-979.
105. Ikezawa, H.; Kutal, C. *J. Org. Chem.*, 1987, *52*, 3299-3303.
106. Al-Ekabi, H.; Mayo, P. *J. Phys. Chem.*, 1986, *90*, 4075-4080.
107. Barber, R. A.; Mayo, P.; Okada, K. *J. Chem. Soc. Commun.*, 1982, 1073-1074.
108. Draper, A. M.; Ilyas, M.; Mayo, P.; Ramamurthy, V. *J. Am. Chem. Soc.*, 1984, *106*, 6222.
109. Ilyas, M.; Mayo, P. *J. Am. Chem. Soc.*, 1985, *107*, 5093.
110. Al-Ekabi H.; Mayo, P., *Tetrahedron*, 1986, *42*, 6277.
111. Hiroaki, T.; Hyodo, M.; Kawahara, H. *J. Phys. Chem.*, 1991, *95*, 10185-10188.
112. Kawai, K.; Mihara, N.; Kuwabata, S.; Yoneyama, H. *J. Electrochem. Soc.*, 1990, *137*, 1793-1796.
113. Popovic, I. G.; Katsikas, L.; Weller, H. *Polymer Bull.*, 1994, *32*, 597-603.
114. Krauetler, B.; Bard, A. J. *J. Am. Chem. Soc.*, 1978, *100*, 5985-5992.
115. Izumi, I.; Fan, F.; Bard, A. J. *J. Phys. Chem.*, 1981, *85*, 218-223.
116. Yoneyama, H.; Takao, Y.; Tamura, H.; Bard, A. J. *J. Phys. Chem.*, 1983, 87, 1417-1422.
117. Bachiocchi, E.; Rol, C.; Rosato, G.; Sebastiani, G. V. *J. Chem. Soc., Chem. Commun.*, 1992, 59-60.

118. Maldotti, A.; Amadelli, R.; Bartocci, C.; Carassiti, V. *J. Photochem. Photobiol. A: Chem.*, 1990, *53*, 263.

119. Wang, C. M.; Mallouk, T. E. *J. Am. Chem. Soc.*, 1990, *112*, 2016-2018.

120. Fox, M. A.; Pettit, T. *J. Org. Chem.*, 1985, *50*, 5013-5015.

121. Blake, D. M.; Webb, J.; Turchi, C.; Magrini, K. *Solar Energy Materials*, 1991, *24*, 584-593.

122. Fox, M. A.; Abdel-Wahab, A. A. *Tetrahedron Lett.*, 1990, *231*, 4533-4536.

123. Fox, M. A.; Chen, C.-C. *J. Am. Chem. Soc.*, 1981, *103*, 6757-6759.

124. Fox, M. A.; Chen, C.-C. *Tetrahedron Lett.*, 1983, *224*, 547-550.

125. Kanno, T.; Oguchi, T.; Sakuragi, H.; Tokumaru, K. *Tetrahedron Lett.*, 1980, *21*, 467-470.

126. Fox, M. A.; Chen, C.-C.; Younathan, J. N. N. *J. Org. Chem.*, 1984, *49*, 1969-1974.

127. Miyake, H.; Suganuma, F.; Matsumoto, T.; Kamiyama, H. *Bulletin of Japan Petroleum Institute*, 1974, *16*, 50-54.

128. Hussein, F. H.; Pattenden, G.; Rudham, R.; Russell, J. J. *Tetrahedron Lett.*, 1984, *225*, 3363-3364.

129. Pichat, P.; Mozzanega, M.-N.; Courbon, H. *J. Chem. Soc., Faraday Trans. I*, 1987, *83*, 697-704.

130. Fox, M. A.; Abdel-Wahab, A. A. *J. Catal.*, 1990, *126*, 693-696.

131. Enea, O.; Ali, A.; Duprez, D. *Int. J. Hydrogen Energy*, 1988, *13*, 569-572.

132. Harvey, P. R.; Rudham, R.; Ward, S. *J. Chem. Soc., Faraday Trans. I*, 1983, *79*, 1381-1390.

133. Borgarello E.; Pelizzetti, E. *La Chimica E L'Industria*, 1983, 65, 474-478.

134. Fraser, I. M.; MacCallum, J. R. *J. Chem. Soc. Faraday Trans. I*, 1986, *82*, 2747-2754.

135. Fraser, I. M.; MacCallum, J. R. *J. Chem. Soc. Faraday Trans. I*, 1986, *82*, 607-615.

136. Kawai, M.; Kawai, T.; Naito, S.; Tamaru, K. *Chem. Phys. Lett.*, 1984, *110*, 58-62.

137. Pincock, J. A.; Pincock, A. L.; Fox, M. A., *Tetrahedron*, 1985, *41*, 4107-4117.

138. Pavlik, J. W.; Tantayanon, S. *J. Am. Chem. Soc.*, 1981, 103, 6755-6757.

139. Weng, Y.; Wang, F.; Lin, L. *Acta Energiae Solaris Sinica*, 1989, *10*, 259-264.

140. Fujihira, M.; Satoh, Y.; Osa, T. *J. Electroanal. Chem*, 1980, *126*, 277.

141. Fujihira, M.; Satoh, Y.; Osa, T. *Nature*, 1981, *293*, 206-207.

142. Fujihira, M.; Satoh, Y.; Osa, T. *Bull. Chem. Soc. Jpn.*, 1982, *55*, 666-671.

143. Izumi, I.; Dunn, W. W.; Wilbourn, K. O.; Fan, F.-R. R.; Bard, A. J. *J. Phy. Chem.*, 1980, *84*, 3207-3210.

144. Morris, S. Photocatalysis for Water Purification, Ph.D. Thesis, University of Wales, 1992, Table 1.1 and references.

145. Ollis, D. F. in *Photocatalysis and Environment: Trends and Applications*, M. Schiavello, Ed., Kluwer Academic Publishers: Dordrecht, 1988; pp. 663.

146. Ohtam, B.; Osaki, H.; Nishimoto, S.; Kagiya, T. *J. Am. Chem. Soc.*, 1986, *108*, 308-310.

147. Isobe, K.; Nakano, Y.; Fujise, M. JP Paten 50059333 750522, Chem. Abstr. 83:113919.

148. Augustinski, J. *Struct. Bonding*, 1988, *69*, 1.
149. Mattson, S.; Pugh, A. J. *Soil Sci.*, 1934, *38*, 229.
150. Zuman, P.; Fijalek, Z. *J. Electroanal. Chem.*, 1990, *196*, 583–588.
151. Dodson, W. S.; Bao, R.; Bruton, T. C.; Li, Y. *J. Org. Chem.*, submitted.
152. Frank, A. J.; Goren, Z.; Willner, I. *J. Chem. Soc., Chem. Commun.*, 1985 1029–1030.
153. Yamataka, H.; Seto, N.; Ichihara, J.; Hanafusa, T.; Teratani, S. *J. Chem. Soc., Chem. Commun.*, 1985, 788–789.
154. Lin, L.; Kuntz, R. R. *Langmuir*, 1992, *8*, 870–875.
155. Nishimoto, S.; Ohtani, B.; Yashikawa, T.; Kagiya, T. *J. Am. Chem. Soc.*, 1983, *105*, 7180–7182.
156. Fox, M. A.; Chen, M.-J. *J. Am. Chem. Soc.*, 1983, *105*, 4497–4499.
157. Wang, H.; Li, Y.; *J. Org. Chem.*, submitted.
158. Philips, M. A. *J. Chem. Soc.*, 1928, 2393.

9

Photophysics and Photochemistry of Fullerene Materials

Ya-Ping Sun
Clemson University, Clemson, South Carolina

Since large quantities of fullerenes became readily available, there has been great scientific interest in the understanding of their properties. Fullerenes are highly colored materials, naturally attracting the attention of molecular spectroscopists and photochemists. The photophysics and photochemistry of fullerenes have been studied quite extensively, with objectives ranging from a quantitative understanding of their electronic structures and photoexcited states to photochemical preparations of novel fullerene-containing materials and to potential applications of fullerenes in nonlinear optical devices. For such a rapidly developing research field, a critical review of all issues under investigation is rather difficult. Thus we will focus our discussion on the topics that are either widely studied and therefore relatively well understood or related to our own research interest in fullerenes. Our purpose is to provide a representative picture of the current understanding of fullerene photoexcited state related properties and also a sense of future directions in this exciting research field.

I. PHOTOPHYSICS OF MONOMERIC FULLERENES

Electronic absorption and emission properties of fullerenes have been investigated extensively. A historic account for the study of fullerene photophysics has been given by Foote [1]. Here, the focus of our discussion will be of the current understanding of the photophysical and related properties of [60]fullerene (C_{60}), [70]fullerene (C_{70}), and high fullerenes in solution and in low-temperature glass.

A. Absorption and Fluorescence of C_{60} and C_{70}

The absorption spectra of C_{60} and C_{70} in room-temperature toluene have been reported by several groups [2–4]. The molar absorptivities at the first absorption band maxima are 940 $M^{-1}cm^{-1}$ and 21,000 $M^{-1}cm^{-1}$ for C_{60} and C_{70}, respectively (Fig. 1). In general, the absorption spectra of C_{60} and C_{70} are not sensitive to solvent environment. However, dramatic solution color changes were reported for C_{60} in the solvent series of benzene, toluene, o-xylene, 1,2,4-trimethylbenzene, and 1,2,3,5-tetramethylbenzene [5,6]. These solvents are in

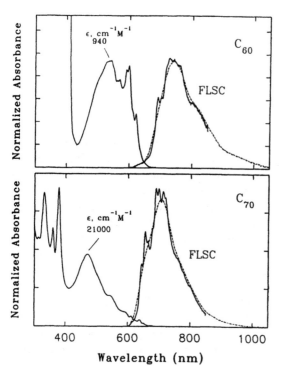

Figure 1 Absorption and fluorescence spectra of C_{60} and C_{70} in room-temperature hexane. The fluorescence spectra were obtained by using a regular detector and small slit (——) and a near-infrared-sensitive detector and large slit (–···–···–).

fact benzenes with different numbers of methyl substitution. The solution color changes are due to significant shifts of the strong second absorption band, while the weak longer wavelength first absorption band of C_{60} remains the same (Fig. 2) [6,7]. The absorption spectra of C_{60} in solvents of naphthalene derivatives are also different [6,7]. There are in fact similar absorption spectral changes for C_{70} in the solvent series of methyl substituted benzenes (Fig. 3). However, color changes in C_{70} solutions are not obvious because molar absorptivities in the first absorption band of C_{70} are much larger than those of C_{60}, which makes the solvent induced C_{70} absorption spectral changes relatively less significant [7]. A satisfactory explanation for the dramatic solvent dependence of fullerene electronic absorption in a specific series of aromatic solvents remains to be found.

In 77K solvent glass, the absorption spectra of C_{60} and C_{70} become much better resolved (Figs. 4 and 5) [8–13]. The 0-0 bands are weak for both C_{60} and C_{70}, indicative of weakly allowed electronic transitions.

Fullerenes C_{60} and C_{70} are only weakly fluorescent in room-temperature solution. Fluorescence spectra of C_{60} and C_{70} in different solvents have been reported by several groups [3,4,7–10,14–16]. Because of low fluorescence

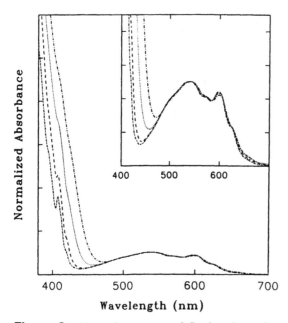

Figure 2 Absorption spectra of C_{60} in toluene (–···–···–), o-xylene (- - -), 1,2,4-trimethylbenzene (· · · · ·), and 1,2,3,5-tetramethylbenzene (–··–··–) at room temperature.

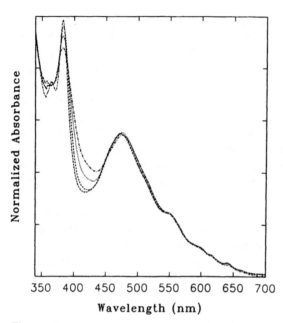

Figure 3 Absorption spectra of C_{70} in toluene (—···—···—), *o*-xylene (- - -), 1,2,4-trimethylbenzene (· · · · ·), and 1,2,3,5-tetramethylbenzene (—·—·—) at room temperature.

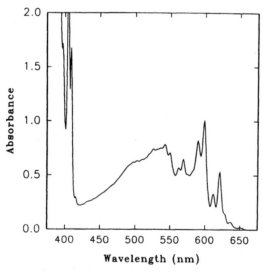

Figure 4 Absorption spectrum of C_{60} in 3-methylpentane-methylcyclohexane glass at 77K.

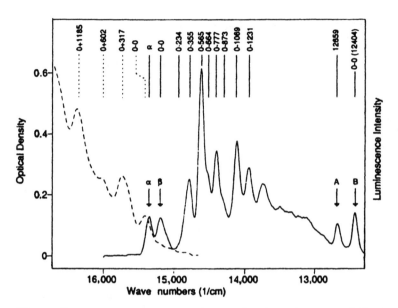

Figure 5 Fluorescence (————) and absorption (- - - -) spectra of C_{70} in methyl-cyclohexane at 77K. (From Ref. 11.)

quantum yields of the fullerenes, special efforts are required to obtain structurally resolved fluorescence spectra [14–16]. Recently, emission measurements with a near-infrared-sensitive detector have shown that the fluorescence spectrum of C_{60} is broader than that of C_{70}, extending further into the near-infrared region (Fig. 1) [7]. Without taking the near-infrared region ($>$ 830 nm) into account, the fluorescence quantum yields of C_{60} and C_{70} in toluene are $\sim 2 \times 10^{-4}$ and $\sim 5 \times 10^{-4}$, respectively [17], though other values have also been reported. The true fluorescence quantum yields over the entire emission wavelength region (600–1200 nm) are 3.2×10^{-4} for C_{60} and 5.7×10^{-4} for C_{70} in room-temperature toluene [7]. According to these results, C_{60} is not so much less fluorescent than C_{70} after all. For both C_{60} and C_{70}, the fluorescence quantum yields are excitation wavelength independent and the fluorescence excitation spectra are in excellent agreement with the absorption spectra [14,15]. Observed fluorescence intensities are not affected by dissolved oxygen in air-saturated solution. The fluorescence quantum yields are only weakly solvent dependent, even in the solvent series of methyl substituted benzenes. Variations in observed fluorescence quantum yields of C_{60} and C_{70} with solvent changes follow the same trend, so that the yield ratios $\Phi_{F,C70}/\Phi_{F,C60}$ in different solvents are similar (Table 1). For most applications, an average fluorescence quantum

Table 1   Fluorescence Quantum Yields of C_{60} and C_{70} and Yield Ratios ($\Phi_{F,C70}/\Phi_{F,C60}$) under Different Solvent Conditions

Solvent	n	ε	$\Phi_{F,C60}(10^4)$	$\Phi_{F,C70}(10^4)$	$\Phi_{F,C70}/\Phi_{F,C60}$
Hexane	1.375	1.890	3.3	5.9	1.8
Dichloromethane	1.424	9.080	3.3	5.4	1.6
Chlorobenzene	1.524	5.710	2.8	5.3	1.9
o-Dichlorobenzene	1.552	9.930	3.0	4.8	1.6
Carbon Disulfide	1.632	2.641	2.6	4.2	1.6
Toluene	1.496	2.379	3.2	5.7	1.8
o-Xylene	1.506	2.568	2.9	5.7	2.0
1,2,4-Trimethylbenzene	1.505		3.1	5.8	1.9
1,2,3,5-Tetramethylbenzene	1.513		3.2	5.7	1.8
1-Methylnaphthalene	1.618	2.920	3.7	6.2	1.7

Source: Ref. 7.

yield ratio $\Phi_{F,C70}/\Phi_{F,C60}$ of 1.8 can be used as a solvent independent constant [7].

The fluorescence spectral profile of C_{70} is strongly solvent dependent, with vibronic structures changing systematically with solvent polarity and polarizability [18]. For example, observed fluorescence spectra of C_{70} in different hexane–THF mixtures are apparently different (Fig. 6). The spectrum in hexane is well resolved with eight peaks, corresponding to at least the same number of vibronic bands. As THF volume fraction in the solvent mixtures increases, the fluorescence vibronic profile changes significantly. The most visible is a continuous enhancement in the intensity of the third vibronic peak, which is accompanied by a slight increase in the fourth peak and decreases in the first and second peaks. However, intensities of the four longer wavelength bands remain essentially unchanged. The solvent dependence of observed fluorescence spectra has also been observed for C_{70} in other solvent mixtures including hexane–acetone, hexane–dichloromethane, hexane–chloroform, hexane–ethanol, and hexane–toluene [18]. The results have been correlated with the solvent polarity scale of pyrene, namely the intensity ratio of the first and the third fluorescence vibronic bands of pyrene (Fig. 7) [18,19]. The solvent polarity dependence has been rationalized in terms of the Ham effect in molecular spectroscopy, which was first established in the study of the electronic transitions of benzene and its derivatives [20–22]. Such a rationalization is supported by the results obtained at low temperatures [11].

A similar solvent dependence of C_{70} phosphorescence spectra has been examined systematically over the temperature range 4–200K (Fig. 8) [11]. The

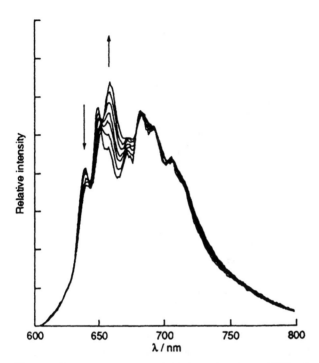

Relative intensity

λ / nm

Figure 6 Fluorescence spectra of C_{70} in hexane–THF mixtures with THF volume fractions (in the direction of arrows) of 0, 4, 7, 10, 14, 19, and 25%. (From Ref. 18.)

observed phosphorescence spectra were analyzed by focusing on the three low-energy vibronic bands (A, B, and C in Fig. 8). The relative intensities of these bands are strongly dependent on solvent and temperature. By using the line width of the vibronic bands as a measurement of the strength of solvent–solute interactions, the polarity effect in terms of Onsager solvent reaction field was found to be insignificant [11]. Thus, it has been concluded that the extreme solvent sensitivity exhibited by the luminescence spectra of C_{70} has the same origin as that observed in benzene and pyrene (Ham effect) [11,20–22].

Fluorescence spectra of C_{60} and C_{70} in solvent glass at low temperature are much better resolved than the spectra in room-temperature solution (Figs. 5 and 9) [8,10–13]. The spectrum of C_{70} is particularly well-resolved, making it possible to assign vibronic structures [11]. For C_{60}, the broadness in the observed emission spectrum is partially due to a severe overlap between fluorescence and phosphorescence (Fig. 9) [10]. According to an early estimate, [8]; the fluorescence quantum yields of C_{60} and C_{70} in 77K hydrocarbon sol-

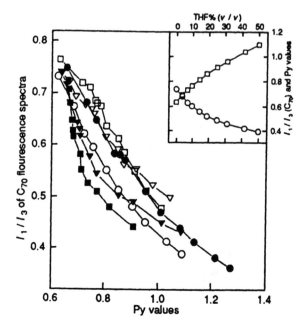

Figure 7   Plots of I_1/I_3 ratios in C_{70} fluorescence spectra vs. Py values in hexane–TFH (o), hexane–acetone (●), hexane–chloroform (∇), hexane–dichloromethane (▼), hexane–ethanol (□), and hexane–toluene (■) mixtures with the second component volume fraction from 0 to 50%. Inset: Plots of I_1/I_3 ratios in C_{70} fluorescence spectra (O) and Py values (□) as a function of THF volume fraction in hexane–THF mixtures. (From Ref. 18.)

vent glass are smaller than those in room-temperature solution. The results are rather unusual, but a search for an explanation may be premature until more accurate values of fluorescence quantum yields are determined. For fluorescence quantum yields at different excitation wavelengths, recent results show no excitation wavelength dependence of the relative fluorescence yields of C_{70} in 3-methylpentane-methylcyclohexane glass at 77K [12]. The fluorescence excitation spectrum of C_{60} in decaline-cyclohexane (3:1 v/v) at 1.2 K is in excellent agreement with the absorption spectrum obtained in 77K solvent glass (Fig. 10), which also indicates the excitation wavelength independence of fluorescence quantum yields.

Fluorescence lifetimes of C_{60} and C_{70} in solution at room temperature have been measured by several groups using both fluorescence decay (time-correlated single photon counting) and transient absorption methods [23-28]. Selected results are summarized in Table 2. The reported lifetime values are rather diverse, which is probably caused by more than experimental uncertain-

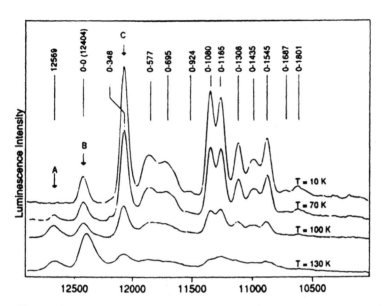

Figure 8 Temperature variation of the C_{70} phosphorescence spectrum in methylcyclohexane. (From Ref. 11.)

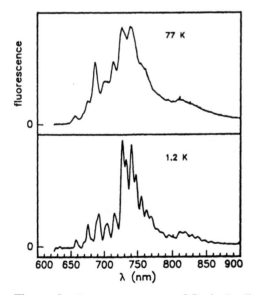

Figure 9 Fluorescence spectra of C_{60} in decaline-cyclohexane (3:1 v/v) at 77K and 1.2K. (From Ref. 10.)

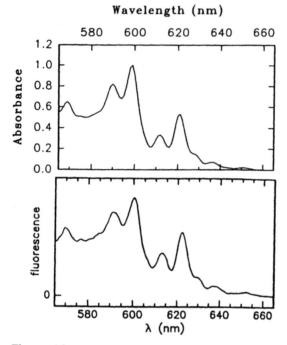

Figure 10 Fluorescence excitation spectrum of C_{60} in decaline-cyclohexane (3:1 v/v) at 1.2K (bottom, from Ref. 10) is compared with the red portion of the absorption spectrum shown in Fig. 4.

Table 2 Fluorescence Lifetimes of C_{60} and C_{70} in Room-Temperature Solution

Solvent	Reference	τ_F (ps)	k_F (s^{-1})
C_{60}			
Benzene	25	1200[a,b]	2.7 × 10^5 [d]
Toluene	23	650 ± 100[b]	4.9 × 10^5 [d]
Toluene	28	1170 ± 20[c]	2.7 × 10^5 [d]
	17	1245 ± 120[e]	2.6 × 10^5 [d]
C_{70}			
Toluene	26	607[a,b]	8.5 × 10^5 [f]
Toluene	28	660 ± 20[c]	8.6 × 10^5 [f]
Benzene	27	627[c]	9.1 × 10^5 [f]
	17	630 ± 75[e]	9.0 × 10^5 [f]

[a]From the decay of singlet transient absorption.
[b]From the rising of triplet transient absorption.
[c]From time-correlated single photon counting.
[d]Calculated using the Φ_F value of 3.2 × 10^{-4}.
[e]Average values.
[f]Calculated using the Φ_F value of 5.7 × 10^{-4}.

ties. One factor that has not been examined carefully is a possible fullerene concentration dependence of observed fluorescence lifetimes. For the excited triplet states of the fullerenes, self-quenching and excimer formation have been observed in dilute C_{60} and C_{70} solutions [29–31]. Although the excited singlet states are much shorter-lived, the consideration of possible concentration related self-quenching should also take into account the fact that fullerene molecules have a strong tendency to aggregate in room-temperature solution [32]. In this regard, the fluorescence decay measurements using very dilute fullerene solutions should be more reliable. It seems likely that the correct fluorescence lifetimes are ~ 1.2 ns for C_{60} and ~ 0.65 ns for C_{70} in room-temperature toluene.

As first pointed out by Hochstrasser and coworkers [23], there are significant discrepancies between the electronic transition probabilities obtained experimentally and calculated from observed absorption and fluorescence spectra in terms of the Strickler–Berg equation [7,14,15,23,28,33]. The experimental fluorescence radiative rate constants $k_{F,e}$ of C_{60} and C_{70} can be obtained from observed fluorescence quantum yields Φ_F and lifetimes τ_F,

$$k_{F,e} = \frac{\Phi_F}{\tau_F} \tag{1}$$

With the known Φ_F and τ_F results of C_{60} (3.2×10^{-4}, 1.2 ns) and C_{70} (5.7×10^{-4}, 0.65 ns), $k_{F,e}$ values are 2.7×10^5 s^{-1} for C_{60} and 8.8×10^5 s^{-1} for C_{70} in room-temperature toluene. Theoretical fluorescence radiative rate constants $k_{F,c}$ can be calculated from the observed absorption and fluorescence spectra of C_{60} and C_{70} in terms of the Strickler–Berg equation [33].

$$k_{F,c} = 2.880 \times 10^{-9} n^2 \langle v_F^{-3} \rangle_{AV}^{-1} \left(\frac{g_0}{g_e} \right) \int \left(\frac{\varepsilon}{v} \right) dv \tag{2}$$

$$\langle v_F^{-3} \rangle_{AV}^{-1} = \frac{\int I_F(v) dv}{\int v^{-3} I_F(v) dv} \tag{3}$$

where g denotes state degeneracies and n represents the refractive index of the solvent. By including the entire observed first absorption band in the calculation of $\int (\varepsilon/v) dv$ in Eq. (2). the $k_{F,c}$ values for C_{60} and C_{70} in room-temperature toluene are $3.3 \times 10^6 (g_0/g_e)$ s^{-1} and $8.1 \times 10^7 (g_0/g_e)$ s^{-1}, respectively. g_0 is unity for both C_{60} and C_{70}. If $g_e = 3$ for C_{60} and $g_e = 1$ for C_{70} are assumed on the basis of computational results [34–37], the $k_{F,c}$ values are significantly larger than the $k_{F,e}$ values, with $k_{F,c}/k_{F,e}$ ratios of 4 for C_{60} and 92 for C_{70}.

The difference between $k_{F,c}$ and $k_{F,e}$ has been attributed to an overestimation of $k_{F,c}$, namely that the observed first absorption bands of C_{60} and C_{70} are in fact contributed by more than the lowest excited states of the fullerene molecules [7,14,15,23,28]. With an assumption that there is a mirror image relationship between the fluorescence and the absorption associated with the lowest excited state, the term $\int(\varepsilon/v)dv$ Eq. (2) corresponding to the transition to the lowest excited singlet state may be estimated (Fig. 11) [7,14,15]. For C_{60}, the estimated absorption band due to the lowest electronic transition yields a $k_{F,c}$ value of $1.45 \times 10^6/g_e$ s^{-1} in room-temperature toluene. The $k_{F,c}/k_{F,e}$ ratio becomes $5.4/g_e$, or 1.8 for $g_e = 3$. In order to have a unity $k_{F,c}/k_{F,e}$ ratio, the $\int(\varepsilon/v)dv$ term in Eq (2) must be even smaller.

The assumed g_e values for the lowest excited singlet states of C_{60} and C_{70}, which are based largely on theoretical calculations [34–37]; may also be questioned. The degeneracies in the excited states might be more than what have been predicted theoretically. There could even be pseudo-degeneracies such that the emissions are from a set of low-lying excited states of very close energies. It is quite possible that both the overestimation of the $\int(\varepsilon/v)dv$ term and the selection of g_e values contribute to the observed discrepancies between $k_{F,e}$ and $k_{F,c}$. For C_{60}, other g_e values have been suggested also on the basis of theoretical calculations. If a g_e value of 5 for C_{60} is used, an agreement between $k_{F,e}$ and $k_{F,c}$ can be achieved by considering only the portion of the observed first absorption band that is a mirror image of the fluorescence spectrum (Fig. 11) in the Strickler–Berg equation calculation [7]. A similar treatment may also

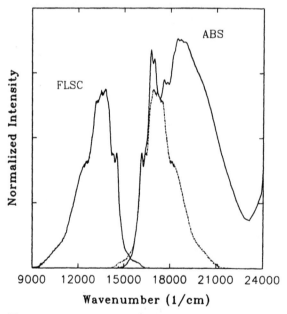

Figure 11 An estimation of the absorption band due to the lowest excited states of C_{60} by assuming a mirror image relationship with the observed fluorescence spectrum.

be performed for C_{70}, though estimating the portion of the observed first absorption band corresponding to the lowest-lying electronic transition is more difficult because of a less clear mirror image relationship between the absorption and emission spectra of C_{70}.

Absorption and fluorescence properties of C_{60} and C_{70} have also been studied in the solid state [38] and in other environments [39–41], such as low-temperature matrix [40] and cyclodextrin [41].

B. Excited Triplet States of C_{60} and C_{70}

The lowest-lying triplet states of C_{60} and C_{70} play dominating roles in their photochemical processes. Upon direct photoexcitation, C_{60} undergoes intersystem crossing quantitatively, with a yield of unity [3]. The intersystem crossing yield of C_{70} is also very high (0.86) [4,42].

The triplet state of C_{60} has a strong and broad transient absorption spectrum over the entire visible region (Fig. 12) [23,43–47]. The transient absorption peaks near 720–750 nm, depending on the solvent environment. The reported transient rise time values vary from 650 ps to 1.2 ns [23,25]. Since the intersystem crossing yield of C_{60} is unity and the excited singlet state lifetime from fluorescence decay measurements is ~1.2 ns, the triplet–triplet absorption rise time of 1.2 ns should be correct. However, somewhat puzzling is the fact that significantly shorter rise time values were obtained for the triplet–triplet absorption of C_{60} in some rather careful transient absorption measurements. Again, the substantial difference in the observed lifetime values may not be attributed simply to experimental uncertainties.

The triplet–triplet transient absorption of C_{70} has also been studied extensively [45,48,49]. Shown in Fig. 13 is the transient absorption spectrum of C_{70} in room-temperature benzene [48].

The triplet state decay lifetimes of C_{60} and C_{70} have been reported by several groups [24,29–31,45,47]. The results are very diverse, which are likely due to effects of fullerene solution concentrations. Weisman and coworkers demonstrated unambiguously that the observed triplet lifetimes are strongly dependent on fullerene solution concentrations as a result of self-quenching [29,30]. As illustrated in Fig. 14, the concentration dependence or the C_{70} excited triplet state decay is obvious. On the basis of careful triplet–triplet absorption measurements, Weisman and coworkers reported that in the absence of bimolecular self-quenching, the correct excited triplet state lifetimes of C_{60} and C_{70} in room-temperature toluene are 133 µs and ≥12 ms, respectively, and the self-quenching rate constants for the excited triplet states of C_{60} and C_{70} are 1.5×10^7 M^{-1}s^{-1}, and 1×10^8 M^{-1}s^{-1}, respectively [29,30]. In addition, by use of a method called high-definition transient absorption kinetics, in which small induced absorptions are measured with high precision and accuracy and

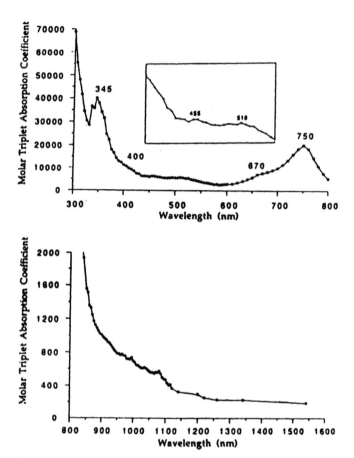

Figure 12 Corrected triplet–triplet absorption spectrum of C_{60} in benzene solution. (From Ref. 43.)

the resulting kinetic traces are then fitted precisely to simulated traces computed from detailed kinetic models, they found that the concentration dependent triplet kinetics of C_{60} and C_{70} are surprisingly complex, which strongly suggests the facile formation of triplet excimers through encounters of triplet and ground state fullerene molecules [30,31]. For C_{70}, the rate constants of triplet excimer formation, dissociation, and deactivation are approximately 10^9 $M^{-1}s^{-1}$, 10^5 s^{-1} and 10^4 s^{-1}, respectively, on the basis of a kinetic model in which triplet monomers rapidly preequilibrate with shorter-lived excimers [30].

Phosphorescence of C_{60} and C_{70} can be observed in solvent glass and solid matrix at low temperature [8–13,24]. The phosphorescence spectrum of C_{60} overlaps with the fluorescence spectrum. By using zinc tetraphenylporphyrin as

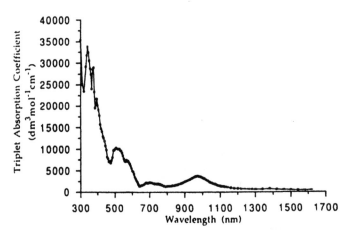

Figure 13 Corrected triplet–triplet absorption spectrum of C_{70} in benzene solution. (From Ref. 48.)

a reference, the phosphorescence quantum yield of 1.3×10^{-3} for C_{70} in toluene containing 10% poly(α-methylstyrene) as a glassing agent at 77K was obtained [24]. In the same glassy medium, the C_{70} phosphorescence lifetime of 53.0 ± 0.1 ms was determined, which is in good agreement with the result of 51 ± 2 ms obtained from the decay of the ESR signal (Fig. 15) [24].

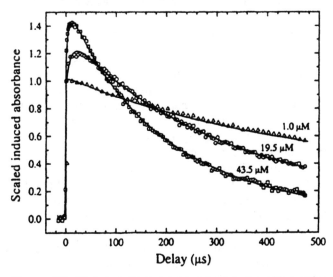

Figure 14 Variation of induced absorption kinetics with sample concentration. Room-temperature absorptions of C_{70} having concentrations of 1.0, 19.5, and 43.5 μM probed at 758 nm.

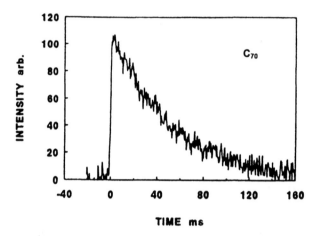

Figure 15 Decay of the EPR signal at 9K for triplet C_{70} at 3240 G (ν = 9.248 GHz). (From Ref. 24.)

The excited triplet states of C_{60} and C_{70} are very efficient sensitizers for the generation of singlet molecular oxygen, with yields of unity and 0.81, respectively [3,4,51]. The photodynamic activities of C_{60} and C_{70} and their potential biological applications have received much recent attention [51–53].

C. Absorption and Excited States of High Fullerenes

The absorption spectra of the soluble high fullerenes [76]fullerene (C_{76}), [78]fullerene (C_{78}), [82]fullerene (C_{82}), [84]fullerene (C_{84}), [90]fullerene (C_{90}), and [96]fullerene (C_{96}) have been reported (Fig. 16) [54–64], though the quality of the reported spectra may require further improvements. The high fullerene samples used in photophysical measurements are mixtures of isomeric species. For C_{84}, the presence of two isomers C_{84}-D_2 and C_{84}-D_{2d} has been proposed, and their relative populations of 2:1 have been estimated on the basis of ^{13}C NMR results [54,56–58,64,65]. There are also experimental results that suggest other compositions for the isomeric mixture of C_{84} [59–63]. While the absorption spectrum of the C_{84} isomeric mixture extends well into the near-infrared wavelength region, the molar absorptivities are relatively small (Fig. 17) [66].

, For the high fullerenes in 77K solvent glass, the resolution in observed absorption spectra is hardly improved. For example, the spectra of C_{76} in 3-methylpentane-methylcyclohexane at 77K and room temperature are essentially the same (Fig. 18) [67].

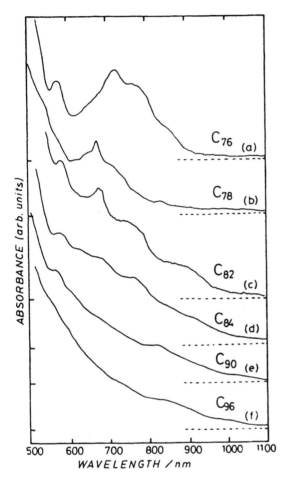

Figure 16 Absorption spectra of high fullerenes in benzene solution at room temperature. (From Ref. 55.)

The detection of emissions from the higher fullerene molecules has been pursued vigorously with the purpose of understanding their excited state properties [66,67]. Fluorescence measurements were carried out, without success, for C_{76}, C_{84}, and a C_{76}–C_{78} mixture (30% C_{78} according to the result of matrix-assisted laser desorption mass spectroscopy) in room-temperature toluene by use of both regular (up to 850 nm) and near-infrared-sensitive (up to 1150 nm) emission detectors. Under the same experimental conditions, the fluorescence of C_{60} can easily be detected with an excellent signal/noise ratio. Oxygen

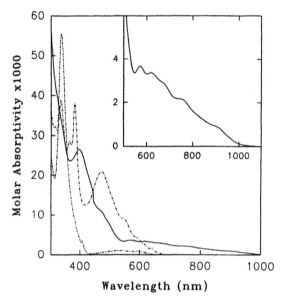

Figure 17 The absorption spectrum of C_{84} in toluene (———) compared with those of C_{60} (–··–··–) and C_{70} (–·–·–).

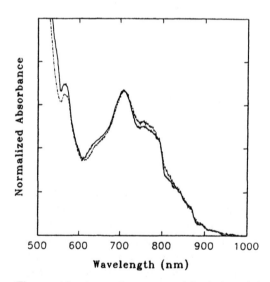

Figure 18 Absorption spectra of C_{76} in 3-methylpentane-methylcyclohexane at room temperaure (–··–··–) and 77K (———).

quenching plays no role in the absence of emissions from the high fullerenes. The same measurements were carried out, also without success, for the high fullerenes in 3-methylpentane-methylcyclohexane glass at 77K [67]. Thus, it may be concluded that the fluorescence quantum yields of the high fullerene molecules are at least an order of magnitude smaller than that of C_{60} (3.2×10^{-4}).

An interesting question is what are the efficient competing decay pathways in the excited singlet states of the high fullerene molecules. For C_{84}, it has been shown [68] that the intersystem crossing following direct photoexcitation is insignificant and that the generation of the excited triplet state requires a sensitizer. The very low intersystem crossing yield is likely responsible for the absence of singlet oxygen generation with the photoexcitation of C_{84} even in an oxygen-saturated toluene solution [69]. Thus, the internal conversion processes must be very efficient for the excited singlet states of the high fullerene molecules.

The excited triplet state of C_{84} can be generated through sensitization. In a transient absorption study by Kamat and coworkers [68], $^3C_{84}^*$ was generated by the triplet–triplet energy transfer method using pulse radiolytically produced biphenyl triplet $^3BP^*$ as a sensitizer.

$$^3BP^* + C_{84} \rightarrow BP + {}^3C_{84}^* \qquad (4)$$

The excited triplet state of C_{84} has a weak absorption in the UV (difference absorption maxima at 310 and 345 nm) but no significant absorption in the visible (Fig. 19). C_{84} can also be sensitized by the excited triplet state of C_{60} [68].

$$^3C_{60}^* + C_{84} \rightarrow C_{60} + {}^3C_{84}^* \qquad (5)$$

The rate constants for energy transfer from $^3BP^*$ and $^3C_{60}^*$ are 4×10^9 and 4.7×10^9 M^{-1}s^{-1}, respectively, in room-temperature benzene.

D. Absorption and Emission Properties of Fullerene Derivatives

There have been significant advances in the derivatization of fullerenes [70]. Electronic absorption spectra of many functionalized C_{60} molecules have been reported. In general, C_{60} derivatives are stronger absorbers in the visible region, largely due to enhanced transition probabilities as a result of symmetry reduction in the functionalization of the C_{60} cage. The absorption spectra are also broader, extending to longer wavelengths. Methano (1) and pyrrolidino derivatives of C_{60} in which the fullerene cage is monofunctionalized exhibit a somewhat structured band at ~700 nm [17,70–72]. Shown in Fig. 20 is a typical example for the absorption spectra of C_{60} derivatives.

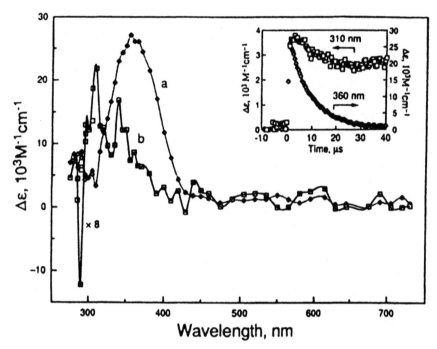

Figure 19 Transient absorption spectra obtained (a) 1 μs (^{3}BP*) and (b) 35 μs (^{3}C$_{84}$*) after pulse radiolysis of deaerated benzene solution containing 0.05 M biphenyl and 10 μM C$_{84}$. Inset shows normalized decay traces recorded at 360 and 310 nm corresponding to the transients ^{3}BP* and ^{3}C$_{84}$*, respectively. (From Ref. 68.)

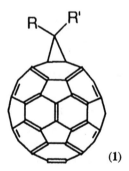

(1)

The functionalization of fullerenes is predominantly through addition reactions [70]. With many double bonds in a fullerene molecule (30 for C$_{60}$), multiple additions to the same fullerene cage are quite common in fullerene derivatization. For C$_{60}$ derivatives, there is a suggestion that the monoadducts

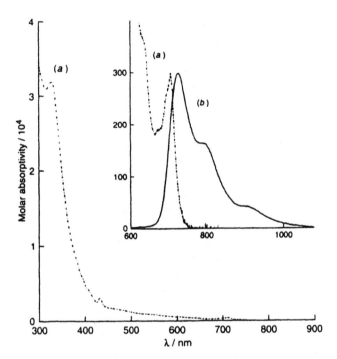

Figure 20 The absorption (*a*) and the uncorrected fluorescence (*b*) spectra of *N*-ethyl-*trans*-2′,5′-dimethyl-pyrrolidino[3′,4′:1,2][60]fullerene. (From Ref. 71.)

(equivalent to the consumption of one double bond on the C_{60} cage) have a sharp absorption peak in the 400–450 nm region as a fingerprint for mono-functionalization [73–75]. The suggestion is obviously empirical, based on a limited number of experimental observations. Indeed, many monofunctionalized C_{60} molecules exhibit a sharp absorption peak in the 400–450 nm region (Fig. 20, for example). However, the suggested characteristic feature can hardly be used as a general criterion for the monofunctionalization of a C_{60} cage. For example, the absorption spectrum of the monoadduct of C_{60} and morpholine does not show such a peak [76], nor does the dimer 2 in which both C_{60} cages are monofunctionalized [77]. Thus, while the presence of a sharp peak at 400–450 nm in the observed absorption spectrum may be an indication that the underlying compound is probably a monofunctionalized C_{60} derivative, the absence of such an absorption feature is not enough for the conclusion that the C_{60} cage in the derivative is multiple-functionalized.

There have been only a few reports of emission properties of functionalized fullerenes. A systematic fluorescence study of four C_{60} derivatives

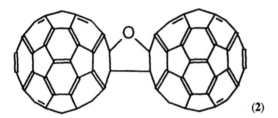

(2)

(3–6) has been reported by Lin et al. [72]. A noticeable common feature in the observed fluorescence spectra of these monofunctionalized C_{60} derivatives is a peak at ~700 nm, which is a mirror image of the 0-0 absorption band (Fig. 21). There is also a shoulder at the blue onset of the observed fluorescence spectra [72]. The same spectral features are present in the fluorescence spectra of pyrrolindine derivatives of C_{60} [17,71]. The shoulder is absent in the low-temperature spectra of the fullerene derivatives in solvent glass. Because of the temperature dependence, the shoulder may be assigned to hot emissions [17].

Lin et al. have also measured the fluorescence lifetimes of the mono-functionalized C_{60} derivatives. The observed fluorescence lifetimes are 1.1–1.6 ns for 3–6, which are similar to that of C_{60} [72]. A more comprehensive fluorescence lifetime study of a series of C_{60} derivatives has recently been carried out by Schuster and coworkers [78]. The results show that the fluorescence lifetimes of the C_{60} derivatives are only marginally different from that of C_{60}.

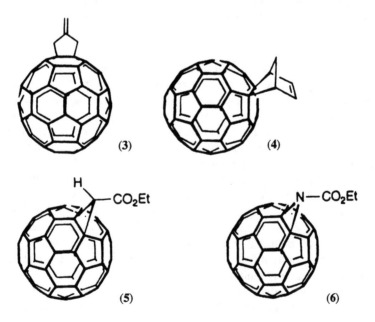

(3)   (4)

(5)   (6)

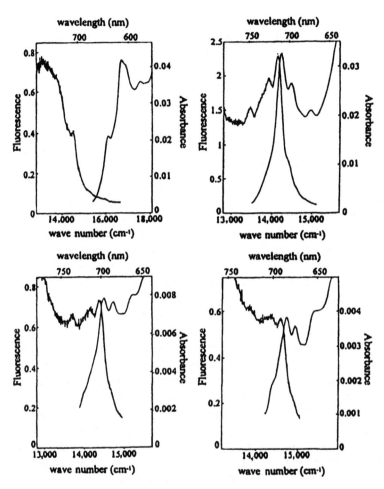

Figure 21 Absorption and fluorescence spectra of C_{60} (upper left), 3 (upper right), 5 (bottom left), and 6 (bottom right) in methylcyclohexane at room temperature. (From Ref. 72.)

II. PHOTOPHYSICS OF FULLERENE AGGREGATES

There seems to be a general tendency for fullerene molecules to aggregate in solution. Slow aggregation of C_{60} in a room-temperature benzene solution has been reported [32]. However, the C_{60} aggregates thus formed are not stable, and can be dispersed simply by shaking the solution by hand [32]. The effects of such aggregation on the photophysical properties of C_{60} are not clear.

Photophysical properties of fullerenes are significantly changed upon the formation of stable aggregates in room-temperature binary solvent mixtures [79–81]. In fact, the formation of stable fullerene aggregates was first observed when the addition of acetonitrile to a C_{70} solution in toluene resulted in dramatic solution color changes [79]. The unusual solvatochromism due to the formation of microscopic aggregates or clusters of C_{70} molecules has been studied systematically under different concentration and solvent conditions [79,80]. Experimental evidence for the cluster formation includes results from dynamic light-scattering analysis based on photon correlation spectroscopy of quasi-elastic light scattering (PCS-QELS) [81]. The binary mixture that facilitates the cluster formation always consists of solvents in which the fullerene has very different solubilities [81,82]. The mixture of toluene (fullerene-soluble) and acetonitrile (fullerene-insoluble) serves as a classical example in this regard.

The cluster formation is strongly dependent on the methods by which solvent mixtures are prepared [80]. One method is to add acetonitrile to a toluene solution of fullerene drop by drop, which is referred to as the slow method. The other is the fast method, in which acetonitrile is poured into the toluene solution as rapidly as possible. For example, at a C_{70} concentration of 1.4×10^{-5} M in a toluene–acetonitrile mixture containing 76% (v/v) acetonitrile, the slow and fast methods yield solutions with very different colors. The slow solution has a faint bluish color and the fast solution is a strong pinkish purple, both differing from the reddish orange color of C_{70} in neat toluene. The clusters formed with the slow method are much closer to fullerene solid particles, which can be precipitated either in storage or more efficiently in a centrifuge field [80]. However, the cluster solution generated by the fast method is stable (Fig. 22), showing no signs of precipitation over a long period of time. It is also stable with respect to deoxygenation by continuously bubbling dry nitrogen gas through the cluster solution. The stability is very important in the use of fullerene clusters as a vehicle for efficient photochemical preparations of polyfullerenes [81,82].

The cluster formation is strongly dependent on the concentration of fullerene and the solvent mixture composition in a correlated fashion. For fullerene in toluene–acetonitrile mixtures, the higher the fullerene concentration, the lower the acetonitrile composition required for the observation of cluster formation [80,81].

The cluster formation is also reversible [80]. After the solvatochromic changes take place at a given acetonitrile composition, decreasing the acetonitrile composition by adding toluene to the mixture can change the solution color and absorption spectrum back to those of the fullerene monomer. The absorption spectrum of the cluster solution also changes upon dilution. A gradual

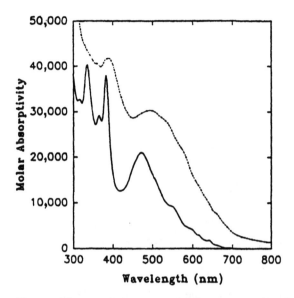

Figure 22 Absorption spectra of C_{70} monomer in toluene (———) and C_{70} clusters in a toluene–acetonitrile mixture with C_{70} concentration of 1.4×10^{-6} M and acetonitrile volume fraction of 76% (-··–··-). (From Ref. 81.)

dilution will eventually result in an absorption spectrum almost the same as that of the fullerene monomer (Fig. 23) [79].

The stable fullerene clusters obtained from the fast method have higher molar absorptivities than the fullerene monomers, corresponding to larger transition probabilities [80]. As the acetonitrile composition in acetonitrile-toluene mixtures increases, the absorption of the fullerene clusters quickly predominates, resulting in a significant increase in integrated molar absorptivities (Fig. 24).

Fluorescence and fluorescence excitation spectra, and fluorescence quantum yields of the fullerene clusters are different from those of the fullerene monomers. The excitation and absorption spectra of the fullerene clusters are in good agreement [80]. The observed fluorescence yields of fullerenes decrease significantly upon the formation of clusters. This is opposite to the trend of increasing transition probabilities as measured by the integrated molar absorptivities. The lower fluorescence yields for the fullerene clusters are likely due to more efficient excited state nonradiative decays. One possible mechanism is the quenching of photoexcited fullerenes by neighboring fullerene molecules in the clusters with excimer-like interactions [82]. However, no fullerene–fullerene excimer emissions have been observed.

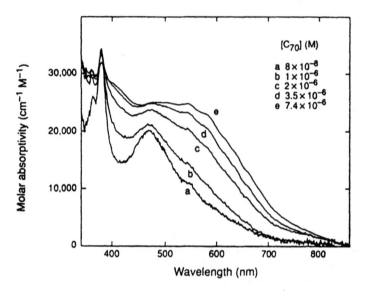

Figure 23   Absorption spectra of C_{70} in acetonitrile (70%)–toluene (30%) mixture as a function of C_{70} concentration. (From Ref. 79.)

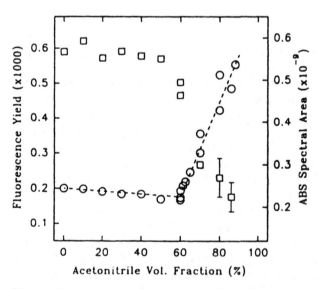

Figure 24   Fluorescence quantum yield (□) and integrated absorption spectral area (○) of C_{70} in toluene–acetonitrile mixtures as a function of acetonitrile composition. (From Ref. 80.)

III. ELECTRON TRANSFER

Fullerenes are excellent electron acceptors. The early examples for the high electron affinity of fullerenes include efficient nucleophilic addition reactions of fullerenes with electron donors such as primary and secondary amines. Since then, there have been many studies of electron transfer interactions and reactions involving fullerene molecules. It is now well established that both ground and excited state fullerene molecules can form charge transfer complexes with electron donors. The photochemically generated fullerene radical anions as a result of excited state electron transfers serve as precursors for a wide range of functionalizations and other reactions.

A. Ground State Charge Transfer Complexes

Fullerene ground state charge transfer complexes with N,N-dimethylaniline (DMA) and N,N-diethylaniline (DEA) were first reported by Hochstrasser and coworkers [83] and Wang [8], respectively. Due to contributions of the charge transfer complexes, the observed absorption spectra of the fullerenes become broader in the presence of DMA [83,84]. Mittal, Rao, and coworkers extended the study of fullerene–amine charge transfer complexes to other aromatic amines including diphenylamine (DPA) and triphenylamine (TPA) [85,86]. Absorption measurements were also carried out for C_{60} and C_{70} solutions in the presence of an aliphatic amine triethylamine (TEA) [85]. New absorption bands in the blue region (<450 nm) of the observed spectra were initially attributed to contributions of fullerene–TEA ground state charge transfer complexes. The formation constants of the assumed fullerene–TEA complexes are much larger that those of the fullerene–aromatic amine complexes [85]. However, results from subsequent experimental investigations have shown that the new absorption bands in the spectra of fullerene–TEA solutions are in fact due to products from efficient chemical reactions between the fullerene molecules and TEA [87]. There is no valid spectroscopic evidence for the formation of fullerene–TEA ground state charge transfer complexes.

As shown in Fig. 25, the C_{60}–DMA charge transfer complex is a stronger absorber than free C_{60} in the visible wavelength region. The pronounced changes in the observed C_{60} absorption spectra with increasing DMA concentration are due largely to the fact that molar absorptivities of free C_{60} are weak in the visible region. For C_{70}, absorption spectral changes upon complexation with DMA (Fig. 26) are less significant because absorptivities of free C_{70} are much higher [88].

The formation of fullerene–amine 1:1 ground state complexes was examined by use of a chemometric method principal component analysis (PCA) [89]. With the digitization of n observed absorption spectra to form a data matrix $\mathbf{Y}$

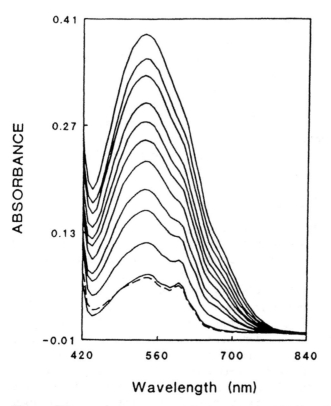

Figure 25 Absorption spectra of C_{60} in mixtures of DMA and toluene. The DMA concentrations are, from top to bottom, 7.887 (neat DMA), 7.10, 6.39, 5.52, 4.81, 4.02, 3.24, 2.37, 1.66, 0.83, and 0.072 M. The dashed line is the spectrum of C_{60} in neat toluene. (From Ref. 83.)

with the spectra being n row vectors, the PCA method can be used to determine the number of independent components k in the n observed absorption spectra in the data matrix. Mathematically, the data matrix $\mathbf{Y}$ has only k significant eigenvalues and their corresponding eigenvectors. An observed spectrum as a row vector Y_i in the matrix $\mathbf{Y}$ can be represented as a linear combination of the significant eigenvectors V_j [89].

$$Y_i = \sum_{j=1}^{k} \xi_{ij} V_j \tag{6}$$

For the C_{60}–DMA complex formation in room-temperature toluene as an example, the data matrix $\mathbf{Y}$ consists of observed absorption spectra of C_{60} in

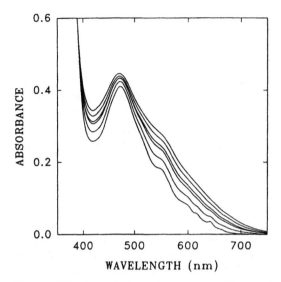

Figure 26 Selected absorption spectra of C_{70} as a function of DMA concentration in room-temperature toluene. The DMA concentrations are (for spectra in the order of increasing absorbance) 0, 1.6, 3.4, 4.7, 6.3, and 7.9 (neat DMA) M. (From Ref. 88.)

toluene–DMA mixtures at different DMA volume fractions [88]. PCA treatment of the data matrix yields only two significant eigenvalues, indicative of the presence of only two independent absorbers (C_{60} monomer and C_{60}–DMA 1:1 complex) contributing to the observed absorption spectra. By assuming a thermodynamic equilibrium between the monomer and complex with equilibrium constant K, there is the following relationship [88,90].

$$\frac{A_{OBS,\lambda} - A_{M,\lambda}}{b} = \frac{cK(\varepsilon_{C,\lambda} - \varepsilon_{M,\lambda})[Q]}{1 + K[Q]} \tag{7}$$

where b is the optical path length, ε represents molar absorptivities, c is the total fullerene concentration, the subscripts M and C denote monomer and complex, respectively, and Q represents the complexing agent. Eq. 7 can be rearranged to the linear Benesi–Hildebrand plot [91].

$$b\Delta A_\lambda^{-1} = (cK\Delta\varepsilon_\lambda)^{-1}[Q]^{-1} + (c\Delta\varepsilon_\lambda)^{-1} \tag{8}$$

where $\Delta A_\lambda = A_{OBS,\lambda} - A_{M,\lambda}$, $\Delta\varepsilon_\lambda = \varepsilon_{C,\lambda} - \varepsilon_{M,\lambda}$. The equilibrium constant K can be obtained from the ratio intercept/slope. In many cases, however, a determination of K at a single wavelength λ is quite sensitive to the effects of uncertainties in the spectra obtained at low complexing agent concentrations. Thus, a global treatment based on multiple wavelengths may be more accurate [92].

$$b\left(\frac{\Delta A_\lambda}{\Delta A_{Q,\lambda}}\right)^{-1} = \left(\frac{cK\Delta\varepsilon_\lambda}{\Delta A_{Q,\lambda}}\right)^{-1}[Q]^{-1} + \left(\frac{c\Delta\varepsilon_\lambda}{\Delta A_{Q,\lambda}}\right)^{-1} \tag{9}$$

where $\Delta A_{Q,\lambda} = \Delta A_{Q,\lambda} - A_{M,\lambda}$, with $A_{Q,\lambda}$ being the absorbance of the fullerene in neat complexing agent. Interestingly, K values for different charge transfer complexes of C_{60}, C_{70}, and C_{84} with aromatic amines are rather similar, varying in a narrow range of 0.04–0.2 M^{-1} [83–86,88,90,92].

The molar absorptivity of the fullerene–amine complex at a given wavelength $\varepsilon_{C,\lambda}$ can also be obtained from Eq. (7).

$$\varepsilon_{C,\lambda} = \left(\frac{\Delta A_\lambda}{b}\frac{1 + K[Q]}{cK[Q]}\right) + \varepsilon_{C,\lambda} \tag{10}$$

For the C_{60}–DEA complex the molar absorptivity at the maximum is $\sim 20,000$ $M^{-1}cm^{-1}$ [90], which is much larger than that of the C_{60} monomer. In principle, Eq. (10) can be applied at all wavelengths to obtain the pure absorption spectrum of the complex (Fig. 27), though the results may be affected by spectra noise at wavelengths where absorptions are relatively weak [91,92]. Alternatively, the pure absorption spectrum of the complex can be obtained by using the eigenvectors from the PCA method. The complex absorption spectrum S_C can be represented by the two largest eigenvectors V_1 and V_2 [93].

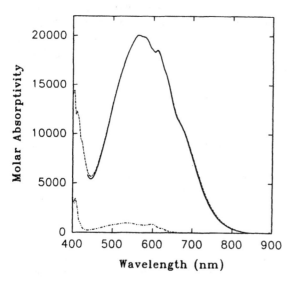

Figure 27 Absorption spectra of the C_{60} monomer (–··–··–) and C_{60}–DEA complex obtained from Benesi–Hildebrand plots (–···–···–) and from the PCA treatment (——).

$$S_C = \xi_{C1}V_1 + \xi_{C2}V_2 \tag{11}$$

where ξ_{C1} and ξ_{C2} are combination coefficients, which can be determined by use of the known condition that the molar absorptivity of the C_{60}–DEA complex at 564 nm is 20,000 $M^{-1}cm^{-1}$. For C_{60}–DEA, the two methods apparently yield similar results (Fig. 27) [90].

Fullerene–amine charge transfer complex formation has also been studied at different temperatures to determine thermodynamic parameters of the equilibria. For C_{84}–DEA as an example, observed absorbances decrease systematically with increasing temperature, which is due to a shift in the complex formation equilibrium with temperature, because molar absorptivities of C_{84} in toluene are essentially temperature independent [92].

$$\ln K = \frac{-\Delta H}{RT} + \frac{\Delta S}{R} \tag{12}$$

where ΔH and ΔS are enthalpy and entropy changes for the complex formation. From Eqs. (7) and (12),

$$\ln\left[\left(\frac{bc\Delta\varepsilon_\lambda}{\Delta A_\lambda}\right) - 1\right] = \frac{\Delta H}{RT} - \left(\frac{\Delta S}{R} + \ln[Q]\right) \tag{13}$$

The right side of the equation is wavelength independent, which allows a global treatment of spectra data at multiple wavelengths. For the C_{84}–DEA complex formation, a global plot based on data at 15 different wavelengths yields a ΔH value of –1.14 kcal/mol and a ΔS value of –9.17 cal $mol^{-1}K^{-1}$ (Fig. 28) [92]. Results for C_{60} and C_{70} charge transfer complex formations with DMA and DEA are similar [90].

B. Electron Transfer Quenching and Exciplexes

The photoexcited states of fullerenes are quenched efficiently by electron donors. Wang first reported [8] that fluorescence intensities of C_{70} in methylcyclohexane at room temperature and 77K glass are quenched by DEA. Similar electron transfer quenchings of C_{70} fluorescence intensities by other donors including DMA, 1,2,4-trimethoxybenzene (TMB), and N,N-dimethyl-p-toluidine (DMPT) in room-temperature benzene were reported by Williams and Verhoeven [27]. Stern–Volmer plots for the fluorescence quenchings by DMA and DMPT are nonlinear (Fig. 29). The upward curvatures were attributed to static quenching, which was modeled by assuming contributions due to the excitation of ground state 1:1 C_{70}–donor complexes [27]. The modified Stern–Volmer equation used to fit the experimental fluorescence quenching results is as follows [27].

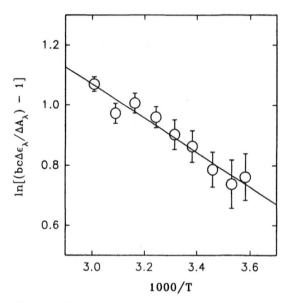

Figure 28 A plot based on Eq. (13) for the temperature dependence of the C_{84} –DEA complex formation in neat DEA. (From Ref. 92.)

$$\frac{I_o}{I} = (1 + K[Q])(1 + k_q \tau_F[Q]) \qquad (14)$$

where I_o and I are unquenched and quenched fluorescence intensities, respectively, [Q] is the quencher concentration, k_q is quenching rate constant, τ_F is the unquenched fluorescence lifetime, and K is the equilibrium constant of the ground state complex. The experimental results can be fitted well using Eq. (14) (Fig. 29). However, equilibrium constants K of 4.2 M^{-1} for DMA and 11.2 M^{-1} for DMPT thus obtained are orders of magnitude larger than the actual equilibrium constants from direct measurements of the ground state C_{70}–donor complexes [27]. The much larger K values from fluorescence results were interpreted in terms of a Perrin-type long-range electron transfer mechanism, in which all excited species that have a quencher within a distance Δr undergo static quenching [27]. Δr values of 5.5 and 9.7 Å for DMA and DMPT, respectively, were obtained from the K values. These distances are sufficiently large to infer that even quencher molecules separated from the fullerene surface by one or two solvent (benzene) molecules can contribute to the quenching [27]. The authors then went on to discuss the special role of benzene in mediating such efficient static quenching due to electron transfer [27].

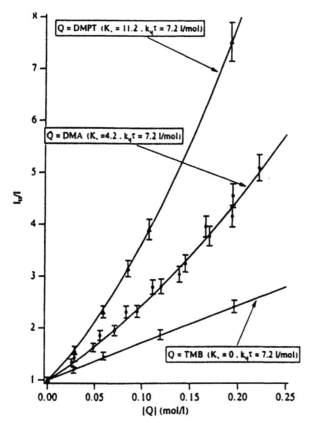

Figure 29 Dependence of C_{70} fluorescence intensity on quencher concentration in room-temperature benzene solution. The curves represent best fits obtained with a combined static and dynamic quenching model Eq. (14) using the parameters indicated. (From Ref. 27.)

A similar treatment in terms of a Perrin-type formulation for static quenching was applied to the fullerene–donor systems without invoking contributions due to the excitation of ground state fullerene–donor complexes [88]. In room-temperature toluene, upward curvatures were observed in quenchings of C_{60} and C_{70} fluorescence intensities by DMA and DEA (Fig. 30). The upward curvatures were also treated by including static fluorescence quenching in terms of the equation as follows [88,94].

$$\frac{\Phi_F^0}{\Phi_F} = (1 + K_{SV}[Q]) \exp(Nv[Q]) \tag{15}$$

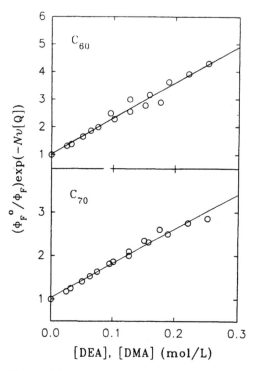

Figure 30 A treatment for the results of fluorescence quenching of C_{60} and C_{70} by DEA and DMA in room-temperature toluene based on a model in which both static and dynamic quenchings are considered Eq. (15). (From Ref. 88.)

where N is Avogadro's number and v is the static quenching volume. By treating the results of DEA and DMA in a single plot (eq 15) because quenchings by DEA and DMA are virtually the same, K_{SV} and v of 13 M^{-1} and 5300 Å^3 for C_{60}, and 8.1 M^{-1} and 5800 Å^3 for C_{70}, respectively, were obtained from least-squares fits [88]. The ratio of the Stern–Volmer constants for C_{60} and C_{70} is 1.6, close to the ratio of C_{60} and C_{70} fluorescence lifetimes. The fluorescence quenchings are diffusion-controlled with the rate constant of 1.2×10^{10} M^{-1} s^{-1} [88]. With an assumption that the static quenching volume is spherical, the v values correspond to static fluorescence quenching radii of 10.8 Å and 11.1 Å for C_{60} and C_{70}, respectively, comparable to the sum of molecular radii of the fullerene–quencher pairs [88].

In saturated hydrocarbon solvents such as methylcyclohexane and hexane, quenchings of fullerene fluorescence by aromatic amines result in the formation of emissive fullerene–amine exciplexes [8,84,88,95]. The exciplex fluorescence spectra are red-shifted from those of free fullerenes (Fig. 31). The

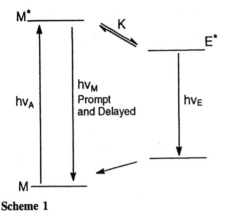

Scheme 1

exciplex formation has been modeled in terms of a mechanism illustrated in Scheme 1 [88,96,97]. The fluorescence quantum yield ratio of the equilibrating excited monomer M* ($\Phi_{F,MD}$) and exciplex E* ($\Phi_{F,E}$) can be expressed as follows.

$$\frac{\Phi_{F,E}}{\Phi_{F,MD}} = \frac{\Phi_{F,E}/\Phi_{F,M}}{(1 - \Phi_{F,MP}/\Phi_{F,M})} = K_1\left(\frac{k_{F,E}}{k_{F,M}}\right)[Q] \qquad (16)$$

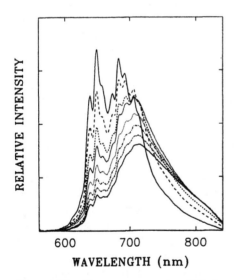

Figure 31 Fluorescence spectra of C_{70} in room-temperature hexane at different DEA concentrations. The DEA concentrations are (in the order of decreasing intensity at the 650 nm peak of the spectra) 0, 0.025, 0.050, 0.075, 0.101, 0.126, 0.151, and 0.176 M. (From Ref. 88.)

where $\Phi_{F,MP}$ is the yield of monomer prompt fluorescence, K_1 is the equilibrium constant for $M^* + Q \rightleftharpoons E^*$, and $k_{F,M}$ and $k_{F,E}$ are fluorescence radiative rate constants of the monomer and exciplex, respectively. The term $\Phi_{F,MP}/\Phi_{F,M}$ can be expressed as

$$\frac{\Phi_{F,MP}}{\Phi_{F,M}} = \left(\frac{\Phi^o_{F,M}}{\Phi_{F,M}}\right)\bigg/\left(\frac{\Phi^o_{F,M}}{\Phi_{F,MP}}\right) \tag{17}$$

where $\Phi^o_{F,M}$ is the monomer fluorescence yield in the absence of the quencher. Both $(\Phi^o_{F,M}/\Phi_{F,MP})$ and $(\Phi^o_{F,M}/\Phi_{F,M})$ are quencher concentration dependent. The former can be written as Eq. (15), and the latter can be obtained from the fractional contributions of monomer fluorescence in observed fluorescence spectra [88].

Thus

$$\frac{(\Phi_{F,E}/(\Phi_{F,M}))}{[Q]} = \left[1 - \frac{\Phi^o_{F,M}/\Phi_{F,M}}{(1 + K_{SV}[Q])\exp(Nv[Q])}\right]^{-1} = K_1\left(\frac{k_{F,E}}{k_{F,M}}\right) \tag{18}$$

Obviously, the right side of the equation is a constant independent of quencher concentration. The left side of the equation is also independent of quencher concentration with correct quenching parameters k_q and v. The k_q value in room-temperature hexane is 2.2×10^{10} M^{-1}s^{-1} [88]. The static fluorescence quenching volume v for C_{70} in hexane has been determined as a parameter in such a way that the variation on the left side of Eq. (18) is forced to a minimum. The v value optimized for both C_{70}–DEA and C_{70}–DMA is 8100 Å^3 [88]. Interestingly, the static quenching radius of C_{70} in hexane (12.4 Å) is somewhat larger than that in toluene. Also interesting is that even with contributions of the delayed monomer emission (which accounts for ~1/3 of the observed monomer fluorescence yield at [DEA] = 0.17 M) the decrease in monomer fluorescence yields with increasing quencher concentration is faster in hexane than in toluene (Fig. 32) [88].

This was clearly a simplified treatment. Other possibilities for the observed upward curvatures in $\Phi_{F,E}/\Phi_{F,M}$ vs. [Q] plots, including, for example, the quenching of exciplex fluorescence and/or the formation of fullerene-quencher 1:2 exciplexes, may also be considered. In fact, substantial exciplex-quencher interactions were observed in anthracene–DMA and related systems [88–100]. For C_{60}–DEA/DMA, to which Eq. (18) is not applicable, the situation is more complicated. As shown in Fig. 33, the observed monomer fluorescence quantum yield changes with quencher concentration in an unusual pattern [88]. At low quencher concentrations, the observed yields are actually

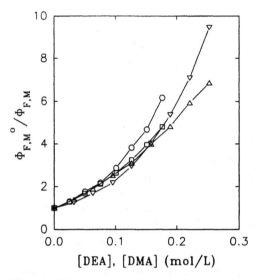

Figure 32 The quenching ratio for C_{70} monomer fluorescence in hexane (DEA, O and DMA, ∇) and toluene (DEA, □ and DMA, $\triangle$). (From Ref. 88.)

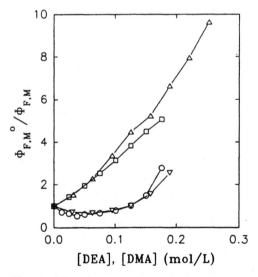

Figure 33 The quenching ratio for C_{60} monomer fluorescence in hexane (DEA, O and DMA, ∇) and toluene (DEA, □ and DMA, $\triangle$). (From Ref. 88.)

larger than those in the absence of quenchers. Such an increase of monomer fluorescence yield with increasing quencher concentration cannot be explained in the context of the simple mechanism (Scheme 1) [96,97]. Further experimental investigations are required to allow a rigorous treatment in which prompt fluorescence, static quenching, ground state complexation, and exciplex-quencher interactions are considered simultaneously.

The extreme solvent sensitivity of the exciplex fluorescence is very interesting. Fullerene–amine exciplex emissions observed in saturated hydrocarbon solvents are absent in solvents such as benzene and toluene (27,84,88,101), which has been explained in terms of solvent polarizability effects [101]. However, there has also been an explanation [84] that the formation of exciplexes in a solvent such as benzene is hindered by specific solute–solvent interactions that result in complexation between the fullerene and solvent molecules. The two explanations are fundamentally different. In the former, the exciplex state is effectively quenched through a radiationless decay pathway facilitated by a stronger dielectric field of the solvent. However, the latter assumes that the ground state fullerene–solvent complexation prevents the formation of fullerene-donor exciplexes. In order to understand whether the extreme solvent sensitivity is solvent specific (limited to benzene, toluene, and other aromatic solvents) or solvent property specific (solvent polarity and polarizability), fluorescence spectra of C_{70}–DEA were measured systematically in mixtures of hexane and a polar solvent (acetone, THF, or ethanol) with volume fraction up to 10% [101]. The results are consistent with the explanation of solvent polarity and polarizability effects.

Solvent dependence of exciplex decay is known for many nonfullerene systems [97,98,102]. Exciplex emissions are often quenched in polar solvents. Generally, the decrease of exciplex emission with increasing solvent polarity is due to a more efficient radiationless transition or exciplex dissociation into an ion pair. A polar solvent environment stabilizes the exciplex state, thus reducing the excited state–ground state energy gap and increasing the rate of radiationless transition [103,104]. This is hardly the case for the quenching of the fullerene exciplex emission because the substantial reduction of exciplex fluorescence yield is not accompanied by bathochromic shifts of the exciplex emission band [101]. In polar solvents, more efficient formation of solvated ions as a competing process to the exciplex fluorescence has been used to account for decreases in the fluorescence quantum yield and lifetime of anthracene–DEA exciplex with increasing solvent polarity [98,105,106]. It was suggested [105,106] that the initial charge transfer quenching of the monomer excited state results in the formation of an intermediate denoted as encounter complex (or excited quencher complex). The solvated ions can also be formed directly from the encounter complex, in competition with the formation of exciplexes. The dependence of C_{70}–DEA exciplex fluorescence on solvent polarity and polarizability is a typical example of such a mechanism. However, the solvent sensitivity for the fullerene–donor systems is to the extreme. The exciplex emissions are completely quenched as the solvent is changed from a nonpolar alkane (hexane or methylcyclohexane) to a still nonpolar but more polarizable hydrocarbon (benzene or toluene) [84,101]. The extreme sensitivity is probably due to an intrinsic tendency for the fullerenes to form solvated anions. The large surface area of the fullerenes (on the order of 700 $Å^2$) can accommodate a relatively large number of solvent molecules in the inner solvation shell, resulting in the formation of solvated ions in a polarizable or slightly polar solvent environment.

In both room-temperature solution and low-temperature solvent glass, fullerene radical ions and fullerene–donor ion pairs as a result of photoinduced electron transfer have been observed in transient absorption experiments [85,107].

(7)

C. Photoinduced Intramolecular Electron Transfer

There have been several investigations of intramolecular electron transfers in bridged fullerene–donor systems. Williams, et al. [17] prepared a C_{60}–bridge–DMA compound (7) by functionalizing a pyrrolindine derivative of C_{60} using a procedure slightly modified from the one first reported by Maggini et al. [108]. It was shown that the excited singlet state lifetime of 7 is strongly dependent on the solvent polarity. The observed fluorescence lifetime is much shorter for 7 in a polar solvent (Fig. 34) due to intramolecular electron transfer quenching. The electron transfer process also competes effectively with intersystem crossing in a polar solvent, reducing significantly the triplet population of 7 [17].

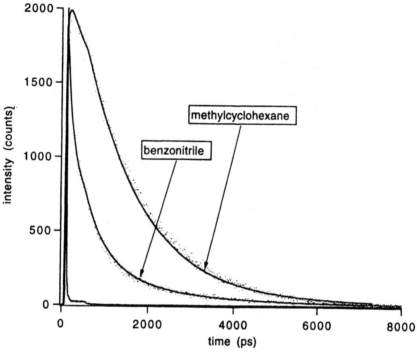

Figure 34 Fluorescence decays of compound **7** in methylcyclohexane and in benzonitrile. (From Ref. 17.)

Dyads of free base and Zn-complexed porphyrins and C_{60} (**8**) were prepared by Gust and coworkers [109]. When the porphyrin moiety in the dyads is photoexcited, there is intramolecular energy transfer from the excited porphyrin to the fullerene moiety, generating the excited singlet state of C_{60}. For the dyad consisting of Zn-complexed porphyrin, the C_{60} excited singlet accepts an electron from the porphyrin, resulting in the formation of C_{60} radical anion and porphyrin radical cation [109].

Photophysical properties of porphyrin–C_{60} dyads (**9**) have also been studied systematically by Imahori, et al. [110]. In addition, a picosecond transient absorption study of a series of ferrocene–C_{60} dyads for distance dependent photoinduced electron transfer has been carried out by Guldi et al. [111].

Sariiftci et al. prepared a supramolecular dyad consisting of ruthenium(II) tris(bipyridine) functionalized C_{60} (**10**) [112]. While the supramolecule shows no interaction between donor and acceptor moieties in the ground state with no charge transfer band in the observed absorption spectrum, there is clearly photoinduced electron transfer in **10** according to results from transient absorption and time-resolved luminescence measurements [112].

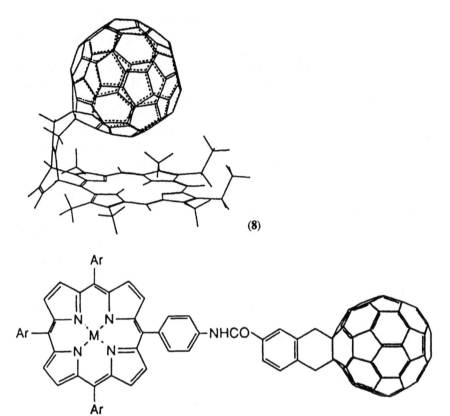

(8)

M=H$_2$
M=Zn
Ar=3,5-(t-Bu)$_2$C$_6$H$_3$

(9)

(10)

Photoinduced intramolecular electron transfer has also been observed in polymer-bound fullerene materials [12,113]. Pendant C_{60}–poly(ethyleneimine) polymer (C_{60}–PEI) was prepared in a photochemical "buckyball-fishing" reaction [113,114]. The primary structure of the pendant polymer is such that each C_{60} cage is functionalized by two amino groups on the PEI polymer chain through either 1,2- or 1,4- connections. Observed fluorescence quantum yields of the pendant C_{60}–PEI polymer are strongly solvent dependent (Table 3). The yield is lower in a more polar solvent environment, which has been attributed to electron transfer quenching of excited C_{60} chromophores by amine units in the polymer structure [12,113]. The intrapolymer electron transfer is suppressed when the amine groups in the pendant polymer are protonated in the presence of 1% trifluoroacetic acid (Table 3). There are two possible intrapolymer electron transfer fluorescence quenching processes. One is through the C_{60}–amine linkage, and the other is the quenching of excited C_{60} chromophores by neighboring amine units, which likely involves diffusional reorientation of the polymer backbone. The latter is similar to the intermolecular quenching of monomeric C_{60} fluorescence by diethylamine, which is somewhat dependent on solvent viscosity. For the pendant C_{60}–PEI polymer, the result of effective fluorescence quenching in a more viscous solvent mixture of 50% (v/v) DMSO–chloroform is in favor of the mechanism of intrapolymer electron transfer through the C_{60}–amine linkage [12,113].

IV. FULLERENE PHOTOCHEMISTRY

Fullerenes have very rich photochemistry. Here, several commonly observed photochemical reactions of fullerenes including photoadditions through an electron transfer–proton transfer mechanism and excited state [2+2] cycloadditions

Table 3 The Dependence of Fluorescence Quantum Yields of the Pendant C_{60}–Poly(ethyleneimine) Polymer on Solvent Environment

Solvent	Dielectric constant[a]	Relative yield[b]
Chloroform	4.81	1.0
Chloroform with 1% TFAA[c]		0.99
50% Toluene–Chloroform	2.38	1.0
50% Acetonitrile–chloroform	35.9	0.61
50% Acetonitrile–chloroform with 1% TFAA		1.09
50% DMSO–chloroform	45	0.48

[a]Dielectric constant of the component added to chloroform in solvent mixtures.
[b]excited at 550 nm. The yield in chloroform is 3×10^{-3}.
[c]TFAA = trifluoroacetic acid.
Source: Ref. 12.

will be discussed in more detail, and some other photochemical reactions will be highlighted.

A. Photoadditions of Amines to Fullerenes

Aliphatic primary and secondary amines react with C_{60} and C_{70} rather efficiently, typically forming multiple addition products [76,115,116]. One of a few examples in which the reaction products are well characterized is the nucleophilic additions of morpholine and piperidine to C_{60} in the presence of oxygen, resulting in the formation of products 11–13 [76].

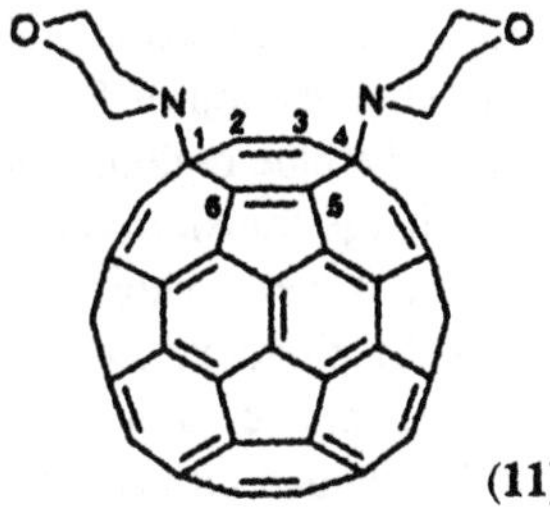

(11)

X = O
X = CH₂   (12)

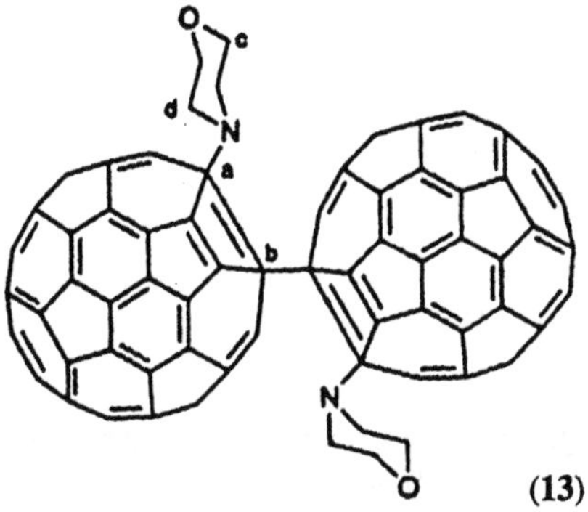

(13)

In the photoexcited states, fullerenes undergo addition reactions with aliphatic amines at much lower amine concentrations. For example, the fluorescence of C_{60} is strongly quenched by diethylamine at a nearly diffusion-controlled rate (Fig. 35) [117]. The quenching results in the bleaching of ground state C_{60}, which may be attributed to the functionalization of C_{60} through diethylamine additions. However, similar to thermal nucleophilic addition reactions, photochemical additions of diethylamine to C_{60} also yield complicated mixtures of adducts [117]. This is most likely due to the fact that the absorption spectrum of the C_{60}–diethylamine monoadduct overlaps with that of C_{60}, therefore competing for excitation photons to making further photochemical additions to the monoadduct possible.

Ground state C_{60} and C_{70} also react readily with a tertiary aliphatic amine triethylamine (TEA) at high TEA concentrations [87]. The reaction of C_{60} and TEA results in the formation of a new absorption band in the blue region, which was initially mistaken as the absorption of a C_{60}–TEA charge transfer complex [85,87]. The reaction products appear to be complicated as well, whose separations and identifications remain to be completed. In the photoexcited states of fullerenes, however, reactions with TEA are more efficient even at low TEA concentrations [66,71,118]. The reaction mixture can be divided into two fractions in terms of the solubility in toluene. The relative quantities of the two fractions are somewhat dependent on irradiation time.

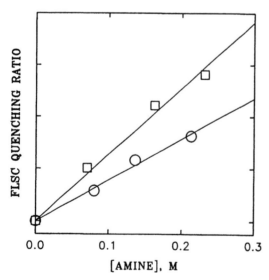

Figure 35   Stern–Volmer plots for the quenching of fluorescence quantum yields of C_{60} by diethylamine in toluene (O, slope = 1.57 M^{-1}) and in a mixture of 30% (v/v) acetonitrile-toluene (□, slope = 2.54 M^{-1}).

The toluene-soluble fraction consists of a major product that has been identified as a C_{60}–TEA monocycloadduct N-ethyl-trans-2',5'-dimethyl-pyrrolidino[3',4':1,2][60]fullerene (**14**) by use of matrix-assisted laser desorption ionization mass spectroscopy and NMR methods [71]. It is interesting that the photochemical reaction actually results in the formation of a cycloadduct. This is unique to the fullerene system because there have been no reports of cycloadducts in reactions involving nonfullerene acceptors [119–122], such as trans-stilbene. In the context of the classical photoinduced electron transfer-proton transfer mechanism [124], a two-step process for the formation of the cycloadduct has been proposed [71].

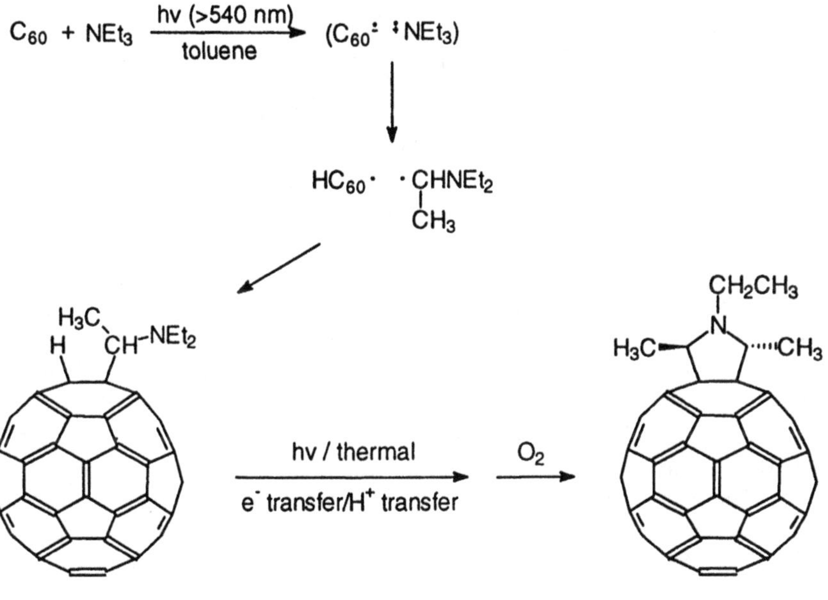

14'

Scheme 2 (14)

It is argued that a special property of fullerenes is that the monofunctionalized C_{60} can continue to serve as an electron acceptor, making it possible to undergo a further electron transfer reaction intramolecularly. The transfer might even be a thermal process because of the known dark reactions between C_{60} and TEA [71]. Recently, in a similar photochemical reaction of C_{60} and trimethylamine (TMA), Liou and Cheng isolated the C_{60}–TMA monoadduct (**15**) [123]. They also demonstrated that the monoadduct **15** can undergo further photochemical reaction to form the C_{60}–TMA monocycloadduct **16**. The results show that the C_{60}-tertiary amine cycloadducts can indeed be generated in a two-step photochemical process within the context of the classical photoinduced electron transfer–proton transfer mechanism. However, mechanistic details for the absence of methine hydrogens on the C_{60} cage are still unknown. It has been suggested [71] that the methine hydrogens might have been eliminated in an oxidation process similar to the one observed in the formation of phenanthrene from dihydrophenanthrene in the photocyclization reaction of *cis*-stilbene [124–126].

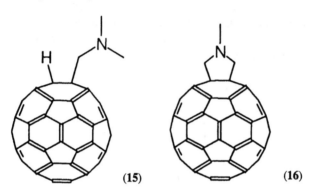

(15)   (16)

Polymeric materials were found in the toluene-insoluble fraction [127]. The materials have no solubility in common organic solvents or so-called fullerene solvents but are partially soluble in DMSO. Evidence for the polymeric nature of the materials includes results of dynamic scattering based on photon correlation spectroscopy of quasi-elastic light scattering (PCS-QELS) [128,129] and gel permeation chromatography using DMSO as mobile phase [117]. The structural characterization of the polymers remains a challenge.

Photochemical reactions of C_{60} and amines have also been used in the preparation of fullerene amino acid derivatives. In the reactions, C_{60} was irradiated in the presence of amino acid esters to form fulleropyrrolidines [130,131].

A complicated mechanism has been proposed to explain the reactions [131]. The mechanism is based on the formation of two carbon-centered radicals due to oxidation by singlet molecular oxygen, which is produced presumably by C_{60} excited triplet state sensitization. As discussed by the authors, several steps of the proposed mechanism are unprecedented, which they credit to the unique properties of C_{60}. However, the proposed mechanism and related speculations may not be evaluated by use of the reported results because of their poor quality for a mechanistic elucidation. For example, the sample solution was allowed to be heated by an uncooled light source during photoirradiation [131], so that in fact the photochemical reactions were carried out in the presence of thermal processes. In addition, the proposed role of singlet molecular oxygen is highly questionable because the excited singlet and triplet states of

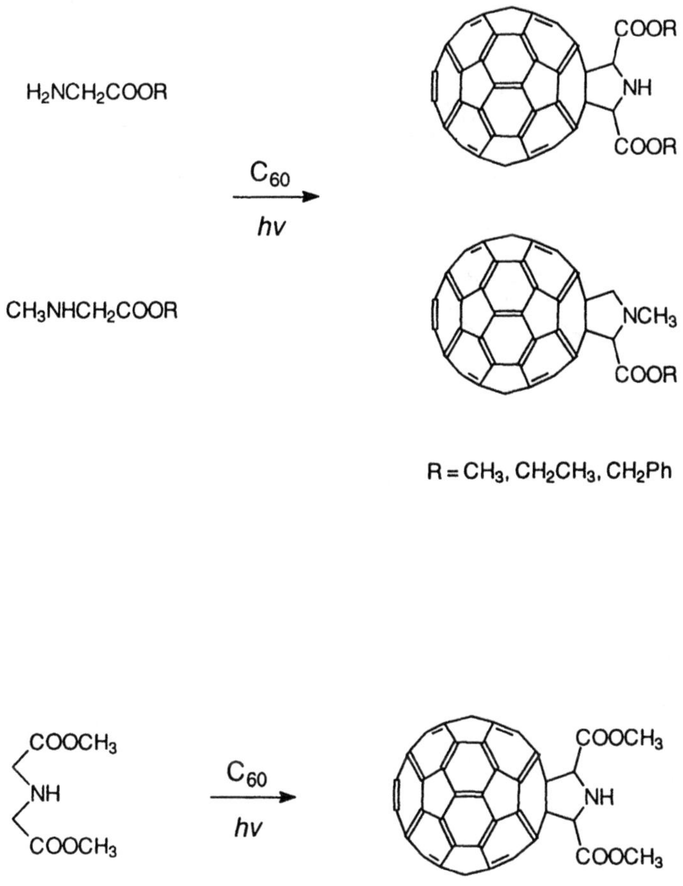

H2NCH2COOR
CH3NHCH2COOR
C60
hv
COOR
NH
COOR
NCH3
COOR
R = CH3, CH2CH3, CH2Ph
COOCH3
NH
COOCH3
C60
hv
COOCH3
NH
COOCH3

Scheme 3

C_{60} should be effectively quenched by the amino groups in the solution. Photochemistry experiments under better controlled conditions are required for a mechanistic understanding of these interesting and useful reactions.

Photoinduced electron transfer–proton transfer reactions between the fullerene cage and polymeric secondary amine units serve as an efficient and selective "buckyball fishing" method for the preparation of pendant fullerene aminopolymers. As a novel approach to carry highly hydrophobic fullerene molecules into aqueous solution, pendant C_{60}-poly(propionylethyleneimine-co-ethyleneimine) polymer (**17**) has been prepared by the photochemical method [113,114]. The pendant polymer is highly water-soluble, with the equivalent aqueous solubility of the polymer-bound C_{60} units (10–12 mg/mL) much higher than the solubility of C_{60} in the so-called fullerene solvent toluene (2.9 mg/mL) [113].

$$-\left[CH_2CH_2NCH_2CH_2N\right]_y-\left[CH_2CH_2N\right]_z-$$

$$\underset{\underset{C_2H_5}{|}}{\overset{\overset{O}{||}}{C}}$$

(17)

B. Photoinduced [2+2] Cycloadditions

A number of photoinduced cycloaddition reactions have been reported [132–134]. For example, Schuster, Wilson, and coworkers reported [2+2] photocycloaddition reactions of C_{60} and enones [133,134].

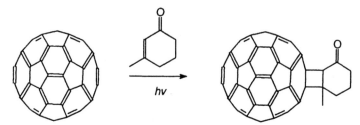

Scheme 4

They have also separated the enantiomers of the mixture of the cis and trans photocycloadducts by use of HPLC with a chiral column [135].

Photoinduced [2+2] cycloaddition reactions through the excited triplet states of fullerenes are believed to be responsible for the photopolymerization of fullerene molecules in the solid state [136,137] and in solution [81,82]. Evidence for the involvement of the excited triplet states is primarily from the oxygen quenching results.

The dimers may then be linked through further [2+2] cycloadditions into polyfullerenes. Although all-carbon polyfullerenes have been obtained (see below), fullerene dimers as the species that are most important to the proposed [2+2] cycloaddition mechanism for fullerene photopolymerization are yet to be isolated and characterized.

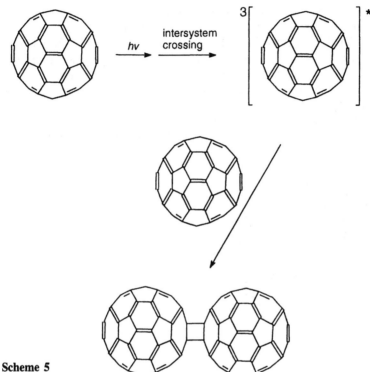

Scheme 5

C. Other Photochemical Reactions

Photochemical fulleroid to methanofullerene conversion via di-π-methane rearrangement has been reported by Wudl and coworkers [138].

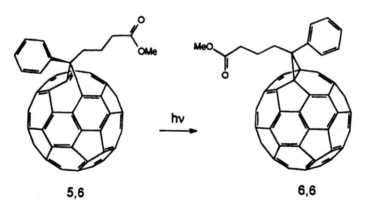

5,6   6,6

Scheme 6

The photochemical conversion is efficient and quantitative, serving as an excellent method for the preparation of methanofullerenes.

Cox, Smith, and coworkers have prepared photochemically the fullerene epoxide $C_{60}O$, [139], which has since proven to be an important starting material for further functionalizations [77,140]. The photooxidation was effected by irradiating an oxygenated benzene solution of C_{60} at room temperature for 18 hr, resulting in $C_{60}O$ in 7% yield. The photooxidation rate and yield were significantly enhanced in the presence of benzil in the solution for photo-irradiation [139]. It has been shown that the C_{60} epoxide thus prepared has a 6,6-closed structure (**18**).

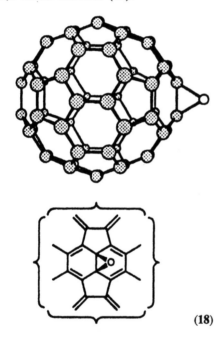

(**18**)

Photochemistry has also been used in many other functionalizations of fullerenes [141,142] such as the preparation of fullereneacetates through the addition of ketene silyl acetals to excited triplet C_{60} [142].

V. PHOTOCHEMICAL PREPARATIONS OF POLYFULLERENES

Fullerene-containing polymeric materials have received much attention for their potential technological applications [143–158]. The polymeric fullerene materials under active investigations can roughly be classified into three categories. One category includes copolymers of fullerenes and comonomers [148–154], such as fullerene–styrene and fullerene–methyl methacrylate copolymers. With the potential use of a variety of comonomers, copolymerization represents an effective method to incorporate a large amount of fullerene molecules into polymeric structures. However, fullerene copolymers are typically complicated systems, so that their structural characterizations are quite challenging. The second category consists of pendant fullerene polymers, in which fullerene molecules are attached to groups or branches on a polymer chain often through so-called "buckyball fishing" reactions [75,113,155–157]. In the absence of multiple functionalization of a single fullerene cage, pendant fullerene polymers are referred to as "fullerene charm-bracelet polymers". However, extraordinary effort is often required in order to prevent multiple functionalization of a single fullerene cage in "fishing" reactions. The third category is represented by polyfullerenes, which are all-carbon polymers from covalently linking fullerene molecules [81,136,137,158]. As a new class of polymeric carbon materials in addition to graphite and diamond, polyfullerenes are being actively studied for their potential applications. For example, highly hydrogenated polyfullerenes have been prepared [159]. As relative light materials, fullerene polymers may be used as sponges for chemically storing a large weight percentage of hydrogen. Polyfullerenes are also being considered for lithographic applications [160].

Preparation of polymeric fullerene materials through the linking of neighboring fullerene molecules in the solid state has been reported [136]. By irradiating oxygen-free, face-centered-cubic C_{60} films with visible or ultraviolet light, C_{60} molecules in the films are linked into oligomers/polymers [136]. According to results from laser desorption mass spectroscopy, these oligomers/polymers in the phototransformed films consist of up to 21 C_{60} units [161]. The phototransformed films were characterized using electron microscopy, thermal, and spectroscopic methods [136,161–164]. In particular, changes in Raman frequencies upon phototransformation of C_{60} solid films are characteristic. The pentagonal pinch mode (A_g) of C_{60} as a narrow line at ~ 1470 cm^{-1}, which is characteristic of the van der Waals bonded face-center-cubic C_{60} solid, shifts to a lower frequency and becomes broader as a result of photopolymerization [161–166]. The same research group also reported [166,167] that a similar phototransformation of solid C_{70} films is considerably more difficult. It was suggested that the much lower phototransformation yield for solid C_{70} films is a result of a smaller number of C=C double bonds that can participate in the fullerene-fullerene linking since a specific alignment of two neighboring C_{70} molecules is required [167]. In the context of the proposed explanation, the required molecular orientation must be specific for C_{70} solid films because solution-based photopolymerization of C_{70} is as efficient as that of C_{60} [81,82].

Recently, a solution-based method for photochemical preparation of polyfullerences has been discovered [81,82]. In the photopolymerization, carefully deoxygenated fullerene cluster solutions are used. As discussed in the previous section, C_{60} and C_{70} form microscopic aggregates or clusters in room-temperature solvent mixtures [79–81]. The fullerene clusters are solidlike species, amenable to photopolymerization. Yields of the photopolymerization reactions vary from 25% to 75%, depending on experimental conditions. For poly[60]fullerene, effects of reaction conditions on polymerization yields have been examined systematically [82]. With the same C_{60} concentration and the same solvent mixture composition, a longer photoirradiation time results in a higher polyfullerene yield. However, when the photoirradiation time is kept constant, the observed photopolymerization yields decrease with increasing C_{60} concentration in the cluster solution [82].

The linking of fullerene molecules inside the clusters likely also occurs through photoinduced cycloaddition reactions in the excited triplet states of the fullerenes [81]. For comparison, two solutions of C_{60} in toluene–acetonitrile mixtures with 15% and 85% acetonitrile (v/v) were irradiated under the same reaction conditions. In the solution with high acetonitrile composition, C_{60} clusters are formed, and so are poly[60]fullerene polymers after photoirradiation. However, the other solution is homogeneous without C_{60} clusters, resulting in no photopolymerization [82]. The photopolymerization of C_{60} is also absent in other deoxygenated homogeneous C_{60} solutions, such as C_{60} in toluene and 1,2-dichlorobenzene with higher C_{60} concentrations [82]. These results are very interesting with respect to the mechanism for photopolymerization. According to the known rate constants of fullerene self-quenching and the triplet excimer formation, there should be significant interactions between the excited triplet and ground state fullerene molecules in the photoirradiation of concentrated homogeneous fullerene solutions. The interactions apparently result in no formation of fullerene dimers or polymers. However, polyfullerenes are obtained when the excited and ground state fullerene molecules are confined in the environment of fullerene clusters. If the photopolymerization of fullerenes is indeed through the excited triplet states of fullerenes, a possible explanation is that the restrictive environment in fullerene clusters may be able to shift the formation–dissociation equilibria significantly toward the formation of fullerene dimers and polymers.

Preparations of polyfullerene materials using other methods have also been reported [168–176]. A quantitative comparison of the polyfullerene polymers obtained photochemically with the materials from other methods will prove to be very interesting.

VI. PHOTOPHYSICS OF POLYMERIC FULLERENE MATERIALS

Studies of polymer-based fullerene materials have been centered on systems in which fullerene molecules are blended into polymer hosts [177–180]. Here, however, the discussion will be focused on the photophysics of covalently linked polymer–fullerene materials, fullerene–styrene copolymers in particular.

Fullerene-styrene copolymers have been prepared in radical initiated and thermal polymerization reactions [148–151]. In radical copolymerizations of C_{60} and styrene, copolymers with C_{60} contents up to 50% (wt/wt) can be obtained [150]. Electronic absorption spectra of the copolymers are very different from that of monomeric C_{60} (Fig. 36). The absorptivities per unit weight concentration of the copolymers $\varepsilon_{w,i}$ increase with increasing C_{60} contents in the copolymers in a nearly linear relationship (Fig. 37). Fluorescence spectra of the C_{60}–styrene copolymers, blue-shifted from the spectrum of monomeric C_{60}, are dependent on excitation wavelengths in a systematic fashion [149]. Interestingly, the observed absorption and fluorescence spectral profiles of C_{60}-styrene and C_{70}-styrene copolymers are very similar, even though the spectra of monomeric C_{60} and C_{70} are very different. The absorption and fluorescence spectra of the fullerene–styrene copolymers are also similar to those of the pendant C_{60}–polystyrene polymer (**19**) prepared in a Friedel–Crafts type reaction [150,156].

The observed fluorescence quantum yields of C_{60}–styrene copolymers are strongly dependent on the C_{60} contents (Fig. 38) [150]. For the copolymer with a very low C_{60} content, the observed fluorescence quantum yield is significantly larger than that of monomeric C_{60}. The enhanced fluorescence yield is likely

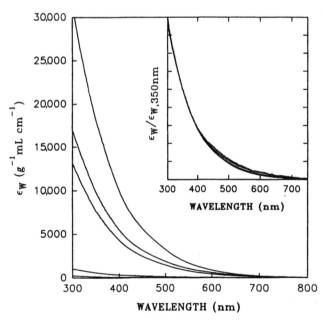

Figure 36 Absorption spectra of the C_{60}-styrene copolymers containing (in the direction of increasing ε_w values) ~0.4%, ~1.7%, ~14%, ~30%, and ~50% (wt/wt) in dichloromethane. Shown in the insert are the spectra normalized at 350 nm.

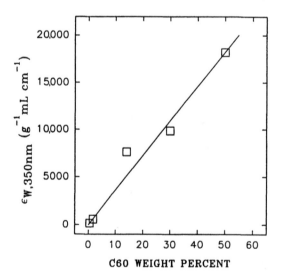

(19)

due to effects associated with symmetry reduction in substituted C_{60} cages in the copolymer. The rapid decreases in observed fluorescence quantum yields with increasing C_{60} contents in the copolymers may be attributed to quenching by neighboring C_{60} cages, probably through excited state intrapolymer complexation. The proposed excited state fullerene–fullerene interactions in the C_{60}–styrene copolymers should be phenomenonlogically similar to photoinduced charge transfers in bichromophoric systems. Such interactions might be facilitated by the copolymer structures, which are determined by the copolymerization mechanism.

Figure 37 A plot of absorptivities per unit weight of the C_{60}–styrene copolymers at 350 nm as a function of C_{60} contents in the copolymers.

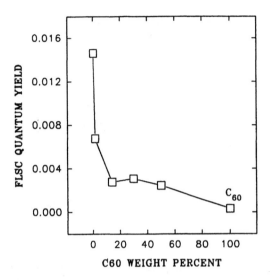

Figure 38 Fluorescence quantum yields of the C_{60}-styrene copolymers as a function of C_{60} contents at 400 nm excitation.

In a logical extrapolation of the classical mechanism of radical polymerization, the copolymerization of C_{60} and styrene has been described as follows [150,178].

$$BP\cdot + S \rightarrow BP\text{-}S\cdot \tag{19a}$$

$$BP\cdot + C_{60} \rightarrow BP\text{-} C_{60}\cdot \tag{19b}$$

$$BP\text{-}S\cdot + S \rightarrow PS\cdot \tag{20a}$$

$$BP\text{-} C_{60}\cdot + S \rightarrow BP\text{-}C_{60}\text{-}S\cdot \tag{20b}$$

$$PS\cdot \text{ (or } PS'\text{-} C_{60}\text{-}PS\cdot\text{) } + C_{60} \rightarrow PS\text{-} C_{60}\cdot \text{ (or } PS'\text{-} C_{60}\text{-}PS\text{-} C_{60}\cdot\text{) } \tag{20c}$$

$$PS\text{-}C_{60}\cdot + PS''\cdot \rightarrow PS\text{-}C_{60}\text{-}PS'' \tag{21a}$$

$$PS\text{-}C_{60}\cdot + PS''\text{-}C_{60}\cdot \rightarrow PS\text{-}C_{60}\text{-}C_{60}\text{-}PS'' \tag{21b}$$

$$PS'\text{-}C_{60}\text{-}PS\text{-}C_{60}\cdot + PS''\cdot \rightarrow PS'\text{-}C_{60}\text{-}PS\text{-}C_{60}\text{-}PS'' \tag{21c}$$

However, because the propagation of $C_{60}\cdot$ radicals is slow in general as a result of their relatively high stability [178,179] the formation of true C_{60}-styrene copolymers (Eq. (21c)) with high C_{60} contents is likely insignificant. Instead, it has been suggested that copolymers containing more than two C_{60} cages may be generated in what are essentially termination reactions [180].

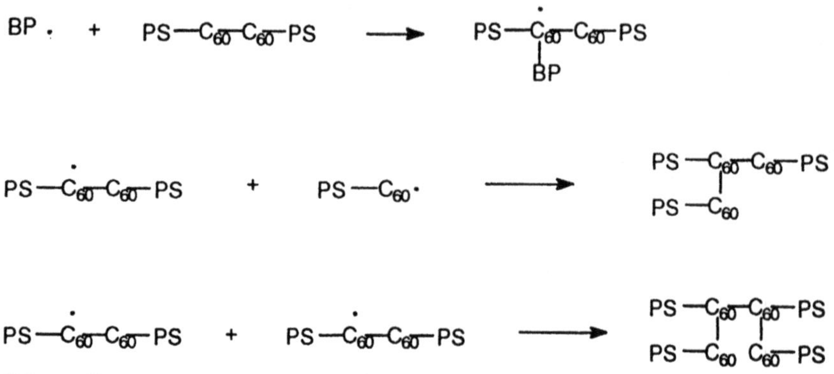

Scheme 7

Because each C_{60} can have more than one polystyrene arm, the copolymers may have a starlike structure conceptually similar to that of the so-called flagellenes [154], namely multiple polystyrene arms are attached to a core of fullerene cages (**20**). The high concentration of fullerene units in the core probably makes the excimer-like interactions possible.

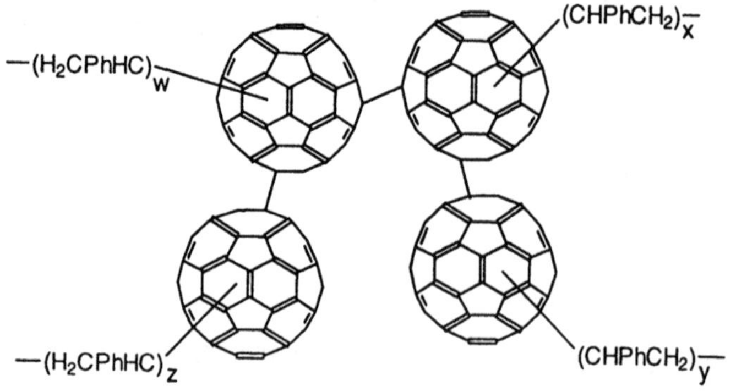

(20)

The proposed excited state fullerene–fullerene interactions are consistent with the results of picosecond transient absorption measurements, in which excimer-like transients were observed [181]. The transient absorption is probably responsible for the significant optical limiting effects of the C_{60}–styrene copolymers [182].

ACKNOWLEDGMENT

Financial support from the National Science Foundation (CHE-9320558) is gratefully acknowledged.

REFERENCES

1. Foote, C. S. in *Topics in Current Chemistry: Electron Transfer I*; Mattay, J., ed.; Springer-Verlag: Berlin, 1994, p. 347.

2. (a) Kroto, H. W.; Allaf, A. W.; Balm, S. P. *Chem. Rev.* 1991, *91*, 1213. (b) Ajie, H.; Alvarez, M. M.; Anz, S. J.; Beck, R. D.; Diederich, F.; Fostiropoulos, K.; Huffman, D. R.; Kratschmer, W.; Rubin, Y.; Schriver, K. E.; Sensharma, D.; Whetten, R. L. *J. Phys. Chem.* 1990, *94*, 8630. (c) Hare, J. P.; Kroto, H. W.; Taylor, R. *Chem. Phys. Lett.* 1991, *177*, 394. (d) Leach, S.; Vervloet, M.; Despres, A.; Breheret, E.; Hare, J. P.; Dennis, T. J.; Kroto, H. W.; Taylor, R.; Walton, D. R. M. *Chem. Phys.* 1992, *160*, 451. (e) Catalán, J. *Chem. Phys. Lett.* 1994, *223*, 159.

3. Arbogast, J. W.; Darmanyan, A. P; Foote, C. S.; Rubin, Y.; Diederich, F. N.; Alvarez, M. M.; Whetten, R. B. *J. Phys. Chem.* 1991, *95*, 11.

4. Arbogast, J. W.; Foote, C. S. *J. Am. Chem. Soc.* 1991, *113*, 8886.

5. Scrivens, W. A.; Tour, J. M. *J. Chem. Soc. Chem. Commun.* 1993, 1207.

6. Catalán, J.; Saiz, J. L.; Laynez, J. L.; Jagerovic, N.; Elguero, J. *Angew. Chem. Int. Ed. Engl.* 1995, *34*, 105.

7. Ma, B.; Sun, Y.-P, *J. Chem. Soc. Perkin Trans. 2* 1996, 2157.

8. Wang, Y. *J. Phys. Chem.* 1992, *96*, 764.

9. Sibley, S.P.; Argentine, S. M.; Francis, A. H. *Chem. Phys. Lett.* 1992, *188*, 187.

10. van den Heuvel, D. J.; van den Berg, G. J. B.; Groenen, E. J. J.; Schmidt, J. *J. Phys. Chem.* 1995, *99*, 11644.

11. Argentine, S. M.; Kotz, K. T.; Francis, A. H. *J. Am. Chem. Soc.* 1995, *117*, 11762.

12. Bunker, C. E.; Rollins, H. W.; Ma, B.; Sun, Y.-P. in *Fullerenes, Recent Advances in the Chemistry and Physics of Fullerenes and Related Materials*; Kadish, K. M.; Ruoff, R. S., Eds., Electrochemical Society, 1996, *3*, 308

13. Ma, B.; Sun, Y.-P., unpublished results.

14. Sun, Y.-P.; Wang, P.; Hamilton, N. B. *J. Am. Chem. Soc.* 1993, *115*, 6378.

15. Sun, Y.-P.; Bunker, C. E. *J. Phys. Chem.* 1993, *97*, 6770.

16. Catalán, J.; Elguero, J. *J. Am. Chem. Soc.* 1993, *115*, 9249.

17. Williams, R. M.; Zwier, J. M.; Verhoeven, J. W. *J. Am. Chem. Soc.* 1995, *117*, 4093.

18. Sun, Y.-P.; Ma, B.; Bunker, C. E. *J. Chem. Soc. Chem. Commun.* 1994, 2099.

19. Reichardt, C. *Solvent and Solvent Effects in Organic Chemistry*; VCH: Weinheim, 1988.

20. Ham, J. S. *J. Chem. Phys.* 1953, *21*, 756.

21. Leach, S.; Lopez-Delgado, R. in *Advances in Molecular Spectroscopy, Vol. 1*; Mangini A., Ed., Pergamon Press: Oxford, 1962.

22. Nakajima, A. *Bull. Chem. Soc. Jpn.* 1971, *44*, 3272.

23. Sension, R. J.; Phillips, C. M.; Szarka A. Z.; Romanow, W. J.; McGhie, A. R.; McCauley, J. P. Jr., Smith, A. B. III, Hochstrasser, R. M. *J. Phys. Chem.* 1991, *69*, 6075.

24. Wasielewski, M. R.; O'Neil, M. P.; Lykke, K. R.; Pellin, M. J.; Gruen, D. M. *J. Am. Chem. Soc.* 1991, *113*, 2774.

25. Ebbesen, T. W.; Tanigaki, K.; Kuroshima, S. *Chem. Phys. Lett.* 1991, *181*, 501.
26. Tanigaki, K.; Ebbesen, T. W.; Kuroshima, S. *Chem. Phys. Lett.* 1991, *185*, 189.
27. Williams, R. M.; Verhoeven, J. W. *Chem. Phys. Lett.* 1992, *194*, 446.
28. Kim, D.; Lee, M.; Suh, Y. D.; Kim, S. K. *J. Am. Chem. Soc.* 1992, *114*, 4429.
29. Fraelich, M. R.; Weisman, R. B. *J. Phys. Chem.* 1993, *97*, 11145.
30. Etheridge H. T. III; Weisman, R. B. *J. Phys. Chem.* 1995, *99*, 2782.
31. Wiseman, R. B., Ausman, K. D.; et al, in *Fullerenes Recent Advances in the Chemistry and Physics of Fullerenes and Related Materials*; Kadish, K. M.; Ruoff, R. S., Eds., Electrochemical Society, 1996, *3*, 276.
32. Ying, Q.; Marecek, J.; Chu, B. *Chem. Phys. Lett.* 1994, *219*, 214.
33. Strickler, S. J.; Berg, R. A. *J. Chem. Phys.* 1962, *37*, 814.
34. Haddon, R. C.; Brus, L. E.; Raghavachari, K. *Chem. Phys. Lett.* 1986, *125*, 459.
35. Kataoka, M.; Nakajima, T. *Tetrahedron* 1986, *42*, 6437.
36. (a) Fowler, P. W.; Woolrich, J. *Chem. Phys. Lett.* 1986, *127*, 78. (b) Fowler, P. W.; Lazzeretti, P.; Malagoli, M.; Zanasi, R. *Chem. Phys. Lett.* 1991, *179*, 174.
37. (a) Negri, F.; Orlandi, G.; Zerbetto, F. *Chem. Phys. Lett.* 1988, *144*, 31. (b) Negri, F.; Orlandi, G.; Zerbetto, F. *J. Chem. Phys.* 1992, *97*, 6496.
38. (a) Shin, E. J.; Park, J.; Lee, M.; Kim, D.; Suh, Y. D.; Yang, S. I.; Jin, S. M.; Kim, S. K. *Chem. Phys. Lett.* 1993, *209*, 427. (b) Andreoni A.; Bondani, M.; Consolati, G. *Phys. Rev. A.* 1994, *50*, 317. (c) Capozzi, V.; Casamassima, G.; Minafra, A.; Piccolo, R.; Trovato, T.; Valentini, A. *Synth. Metals* 1996, *77*, 3. (d) Byrne, H. J.; Maser, W.; Ruhle, W. W.; Mittelbach, A.; Honle, W.; Von Schnering, H. G.; Movaghar, B.; Roth, S. *Chem. Phys. Lett.* 1993, *204*, 461. (e) Pichler, K.; Graham, S.; Gelsen, O. M.; Friend, R. H.; Romanow, W. J.; McMauley, J. P., Jr,; Coustel, N.; Fischer, J. E.; Smith, A. B. III, *J. Phys. Condens. Matter* 1995, *3*, 9259.
39. Eastoe, J.; Crooks, E. R.; Beeby, A.; Heenan, R. K. *Chem. Phys. Lett.* 1995, *245*, 571.
40. Fulara, J.; Jakobi, M.; Maier, J. P. *Chem. Phys. Lett.* 1993, *206*, 203.
41. Andersson, T.; Nilsson, K.; Sundahl, M.; Westman, G.; Wennerstrom, O. *J. Chem. Soc., Chem. Commun.* 1992, 604.
42. Hung, R. R.; Grabowski, J. J. *Chem. Phys. Lett.* 1992, *192*, 249.
43. Bensasson, R. V.; Hill, T.; Lambert, C.; Land, E. J.; Leach, S.; Truscott, T. G. *Chem. Phys. Lett.* 1993, *201*, 326.
44. Kajii, Y.; Nakagawa, T.; Suzuki, S.; Achiba, Y.; Obi, K.; Shibuya, K. *Chem. Phys. Lett.* 1991, *181*, 100.
45. Leach, S. in *Physics and Chemistry of the Fullerenes;* Prassides, K., Ed.; Kluwer Academic Publishers: Netherlands, 1994; p. 117.
46. Watanabe, A.; Ito, O.; Watanabe, M.; Saito, H.; Koishi, M. *J. Chem. Soc. Chem. Commun.* 1996, 117.
47. Sauvé, G.; Dimitrijevic, N. M.; Kamat, P. *J. Phys. Chem.* 1995, *99*, 1199.
48. Bensasson, R. V.; Hill, T.; Lambert, C.; Land, E. J.; Leach, S.; Truscott, T. G. *Chem. Phys. Lett.* 1993, *206*, 197.

49. Lee, M.; Song, O. K.; Seo, J. C.; Kim, D.; Suh, Y. D.; Jin, S. M.; Kim, S. K. *Chem. Phys. Lett.* 1992, *196*, 325.

50. Haufler, R. E.; Wang, L. S.; Chibante, L. P. F.; Jin, C.; Conceicao, J.; Chai, Y.; Smalley, R. E. *Chem. Phys. Lett.* 1991, *179*, 449.

51. Foote, C. S. *ACS Symp. Ser. 616 (Light Activated Pest Control)*, 1995, 17.

52. Tokuyama, H.; Yamago, S.; Nakamura, E.; Shiraki, T.; Sugiura Y. *J. Am. Chem. Soc.* 1993, *115*, 7918.

53. Nakajima, N.; Nishi, C.; Li, F. M.; Ikada, Y. *Fullerene Sci. Technol.* 1996, *4*, 1.

54. Diederich, F.; Ettl, R.; Rubin, Y.; Whetten, R. L.; Beck, R.; Alvarez, M.; Anz, S.; Sensharma D.; Wudl, F.; Khemani, K. C.; Koch, A. *Science* 1991, *252*, 548.

55. (a) Kikuchi, K.; Nakahara, N.; Honda, M.; Suzuki, S.; Saito, K.; Shiromaru, H.; Yamauchi, K.; Ikemoto, I.; Kuramochi, T.; Hino, S.; Achiba, Y. *Chem. Lett.* 1991, 1607. (b) Kikuchi, K.; Nakahara, N.; Wakabayashi, T.; Honda, M.; Matsumiya, H.; Moriwaki, T.; Suzuki, S.; Shiromaru, H.; Saito, K.; Yamauchi, K.; Ikemoto, I.; Achiba, Y.; *Chem. Phys. Lett.* 1992, *188*, 177.

56. (a) Thilgen, C.; Diederich, F.; Whetten, R. L. in *Buckminsterfullerenes*; Billups, W. E.; Ciufolini, M. A., Eds., VCH: New York, 1993, p. 59. (b) Diedrich, F.; Whetten, R. L. *Acc. Chem. Res.* 1992, *25*, 119.

57. Kikuchi, K.; Nakahara, N.; Wakabayashi, T.; Suzuki, S.; Shiromaru, H.; Miyake, Y.; Saito, K.; Ikemoto, I.; Kainosho, M.; Achiba, Y. *Nature* 1992, *357*, 142.

58. Taylor, R.; Langley, G. J.; Avent, A. G.; Dennis, T. J. S.; Kroto, H. W.; Walton, D. R. M. *J. Chem. Soc. Perkin Trans 2* 1993, 1029.

59. Saunders, M.; Jiménez-Vásquez, H. A.; Cross, R. J.; Billups, W. E.; Gesenberg, C.; Gonzalez, A.; Luo, W.; Haddon, R. C.; Diederich, F.; Herrmann, A. *J. Am. Chem. Soc.* 1995, *117*, 9305.

60. Hawkins, J. M.; Nambu, M.; Meyer, A. *J. Am. Chem. Soc.* 1994, *116*, 7642.

61. Boulas, P.; Jones, M. T.; Kadish, K. M.; Ruoff, R. S.; Lorents, D. C.; Tse, D. S. *J. Am. Chem. Soc.* 1994, *116*, 9393.

62. Tanaka, K.; Zakhidov, A. A.; Yoshizawa, K.; Okahara, K.; Yamabe, T.; Kikuchi, K.; Suzuki, S.; Ikemoto, I.; Achiba, Y. *Solid State Commun.* 1993, *85*, 69.

63. Balch, A. L.; Ginwalla, A. S.; Lee, J. W.; Noll, B. C.; Olmstead, M. M. *J. Am. Chem. Soc.* 1994, *116*, 2227.

64. Manolopoulos, D. E.; Fowler, P. W.; Taylor, R.; Kroto, H. W.; Walton, D. R. M. *J. Chem. Soc. Faraday Trans.* 1992, *88*, 3117.

65. (a) Manolopoulos, D. E.; Fowler, P. W. *J. Chem. Phys.* 1992, *96*, 7603. (b) Bakowies, D.; Kolb, M.; Thiel, W.; Richard, S.; Ahlrichs, R.; Kappes, M. M. *Chem. Phys. Lett.* 1992, *200*, 411.

66. Sun, Y.-P. in *Fullerenes, Recent Advances in the Chemistry and Physics of Fullerenes and Related Materials*; Kadish, K. M.; Ruoff, R. S., Eds., Electrochemical Society, 1995, p. 510.

67. Ma, B.; Bunker, C. E.; Sun, Y.-P., unpublished results.

68. Sauvé, G.; Kamat, P. V.; Ruoff, R. S. *J. Phys. Chem.* 1995, *99*, 2162.

69. Dabestani, R.; Sun, Y.-P., unpublished results.
70. (a) Diederich, F.; Thilgen, C. *Science* 1996, *271*, 317. (b) Hirsch, A. *The Chemistry of Fullerenes*; Thiemes: Stuttgart, 1994.
71. Lawson, G. E.; Kitaygorodskiy, A.; Ma, B.; Bunker, C. E.; Sun, Y.-P. *J. Chem. Soc. Chem. Commun.* 1995, 2225.
72. Lin, S.-K.; Shiu, L.-L.; Chien, K. M.; Luh, T.-Y.; Lin, T.-I. *J. Phys. Chem.* 1995, *99*, 105.
73. Suzuki, T.; Li, Q.; Khemani, K. C.; Wudl, F.; Almarson, Ö. *Science* 1991, *254*, 1186.
74. Hirsch, A.; Soi, A.; Karfunkel, H. R. *Angew. Chem. Int. Ed. Engl.* 1992, *31*, 766.
75. Geckeler, K. E.; Hirsch, A. *J. Am. Chem. Soc.* 1993, *115*, 3850.
76. Schick, G.; Kampe, K.-D.; Hirsch, A. *J. Chem. Soc. Chem. Commun.* 1995, 2023.
77. Smith, A. B., III; Tokuyama, H.; Strongin, R. M.; Furst, G. T.; Romanow, W. J. *J. Am. Chem. Soc.* 1995, *117*, 9359.
78. Schuster, D. I. et al., unpublished results.
79. Sun, Y.-P.; Bunker, C. E. *Nature* 1993, *365*, 398.
80. Sun, Y.-P.; Bunker, C. E. *Chem. Mater.* 1994, *6*, 578.
81. Sun, Y.-P.; Ma, B.; Bunker, C. E.; Liu, B. *J. Am. Chem. Soc.* 1995, *117*, 12705.
82. Ma, B.; Guduru, R.; Sun, Y.-P., unpublished results.
83. Sension, R. J.; Szarka, A. Z.; Smith, G. R.; Hochstrasser, R. M. *Chem. Phys. Lett.* 1991, *185*, 179.
84. (a) Seshardi, R.; D'Souza, F.; Krishnan, V.; Rao, C. N. R. *Chem Lett.* 1993, 217. (b) Sibley, S. P.; Campbell, R. L.; Silber, H. B. *J. Phys Chem.* 1995, *99*, 5274.
85. Palit, D. K.; Ghosh, H. N.; Pal, H.; Sapre, A. V.; Mittal, J. P.; Seshadri, R.; Rao, C. N. R. *Chem. Phys. Lett.* 1992, *198*, 113.
86. Seshadri, R.; Rao, C. N. R.; Pal, H.; Mukherjee, T.; Mittal, J. P. *Chem. Phys. Lett.* 1993, *205*, 395.
87. Sun, Y.-P.; Ma, B.; Lawson, G. E. *Chem. Phys. Lett.* 1995, *233*, 57.
88. Sun, Y.-P.; Bunker, C. E.; Ma, B. *J. Am. Chem. Soc.* 1994, *116*, 9692.
89. Malinowski, E. R. *Factor Analysis in Chemistry*, 2d ed., Wiley: New York, 1991.
90. Rollins, H. W.; Bunker, C. E.; Sun, Y.-P., unpublished results.
91. Benesi, H.; Hildebrand, J. H. *J. Am. Chem. Soc.* 1949, *71*, 2703.
92. Bunker, C. E.; Rollins, H. W.; Sun, Y.-P. *J. Chem. Soc. Perkin Trans. 2* 1996, 1307.
93. Lawton, W. H.; Sylvestre, E. A. *Technometrics* 1971, *13*, 617.
94. Sun, Y.-P.; Wallraff, G. M.; Miller, R. D.; Michl, J. *J. Photochem. Photobil. A: Chem.* 1991, *62*, 333.
95. Caspar, J. V.; Wang, Y. *Chem Phys. Lett.* 1994, *218*, 221.
96. Hui, M.-H.; Ware, W. R. *J. Am. Chem. Soc.* 1976, *98*, 4718.
97. Birks, J. B. *Photophysics of Aromatic Molecules;* Wiley-Interscience: London, 1970.

98. Bhattacharyya, K.; Chowdhury, M. *Chem. Rev.* 1993, *93*, 507.
99. (a) Yang, N. C.; Shold, D. M.; Kim, B. *J. Am. Chem. Soc.* 1976, *98*, 6587. (b) Saltiel, J.; Townsend, D. E.; Watson, B. D.; Shannon, P.; Finson, S. L. *J. Am. Chem. Soc.* 1977, *99*, 884.
100. (a) Pearson, J. M.; Turner, S. R. In *Molecular Association;* Forster, R., Ed.; Academic Press: London, 1979; Vol. 2, p. 79. (b) Davison, R. S. in *Advances in Physical Organic Chemistry*; Gold, V.; Bethell, D., Eds.; Academic Press: New York, 1983; Vol. 19, p. 1.
101. Sun, Y.-P.; Ma, B. *Chem. Phys. Lett.* 1995, *236*, 285.
102. (a) Knibbe, H.; Röllig, K.; Schäfer, F. P.; Weller, A. *J. Chem. Phys.* 1967, *47*, 1184. (b) Mataga, N.; Okada, T.; Yamamoto, N. *Chem. Phys. Lett.* 1967, *1*, 119.
103. (a) Siebrand, W. *J. Chem. Phys.* 1967, *46*, 440. (b) Freed, K. F.; Jortner, J. *J. Chem. Phys.* 1970, *52*, 6272. (c) Englman, R.; Jortner, J. *Mol. Phys.* 1970, *18*, 145.
104. (a) Caspar, J. V.; Kober, E. M.; Sullivan, B. P.; Meyer, T. J. *J. Am. Chem. Soc.* 1982, *104*, 630. (b) Caspar, J. V.; Meyer, T. J. *J. Am. Chem. Soc.* 1983, *105*, 5583. (c) Caspar, J. V.; Meyer, T. J. *J. Phys. Chem.* 1983, *87*, 952.
105. (a) McDonald, R. J.; Selinger, B. K. *Molec. Photochem.* 1971, *3*, 99. (b) Selinger, B. K.; McDonald, R. J. *Aust. J. Chem.* 1972, *25*, 897.
106. Hirata, Y.; Kanda, Y.; Mataga, N. *J. Phys. Chem.* 1983, *87*, 1659.
107. (a) Ghosh, H. N.; Pal, H.; Sapre, A. V.; Mittal, J. P. *J. Am. Chem. Soc.* 1993, *115*, 11722. (b) Park, J.; Kim, D.; Suh, Y. D.; Kim, S. K. *J. Phys. Chem.* 1994, *98*, 12715. (c) Ito, O.; Sasaki, Y.; Yoshikawa, Y.; Watanabe, A. *J. Phys. Chem.* 1995, *99*, 9838. (d) Sasaki, Y.; Yoshikawa Y.; Watanabe, A.; Ito, O. *J. Chem. Soc. Faraday Trans.* 1995, *91*, 2287.
108. Maggini, M.; Scorrano, G.; Prato, M. *J. Am. Chem. Soc.* 1993, *115*, 9798.
109. Liddell, P. A.; Sumida, J. P.; Macpherson, A. N.; Noss, L.; Seeley, G. R.; Clark, K. N.; Moore, A. L.; Moore, T. A.; Gust, D. *Photochem. Photobiol.* 1994, *60*, 537.
110. Imahori, H.; Hagiwara, K.; Akiyama, T.; Taniguchi, S.; Okada, T.; Sakaya, Y. *Chem. Lett.* 1995, 265.
111. Kamat, P. V.; Guldi, D. in *Fullerenes, Recent Advances in the Chemistry and Physics of Fullerenes and Related Materials*; Kadish, K. M.; Ruoff, R. S., Eds., Electrochemical Society, 1996, *3*, 254.
112. Sariciftci, N. S.; Wudl, F.; Heeger, A. J.; Maggini, M.; Scorrano, G.; Prato, M.; Bourassa, J.; Ford, P. C. *Chem Phys. Lett.* 1995, *247*, 510.
113. Sun, Y.-P.; Liu, B.; Bunker, C. E. *Macromolecules* 1996, submitted.
114. Lawson, G. E.; Liu, B.; Bunker, C. E.; Sun, Y.-P. in *Fullerenes, Recent Advances in the Chemistry and Physics of Fullerenes and Related Materials;* Kadish, K. M.; Ruoff, R. S., Eds., Electrochemical Society, 1996, *3*, 1218.
115. Wudl, F.; Hirsch, A.; Khemani, K. C.; Suzuki,T.; Allemand, P.-M.; Koch, A.; Eckert, H.; Srdanov, G.; Webb, H. M. in *Buckminsterfullerenes*; Billups, W. E.; Cinfolini, M. A., Eds., VCH: New York, 1993, Chapter 11.
116. Kampe, K. D.; Egger, N.; Vogel, M. *Angew. Chem., Int. Ed. Engl.* 1993, *32*, 1174.

117. Bunker, C. E.; Sun, Y.-P., unpublished results.
118. (a) Kajii, Y.; Takeda, K.; Shibuya, K. *Chem. Phys. Lett.* 1993, *204*, 283. (b) Pola, J.; Darwish, A. D.; Jackson, R. A.; Kroto, H. W.; Meidine, M. F.; Abdul-Sada, A. K.; Taylor, R; Walton, D. R. M. *Full. Sci. Tech.* 1995, *3*, 299. (c) Takeda, K.; Kejeii, Y.; Shibuya, K. *J. Photochem. Photobiol. A: Chem.* 1995, *92*, 69.
119. Barltrop, J. A. *Pure Appl. Chem.* 1973, *33*, 179.
120. Bryce-Smith, D.; Gilbert, A. *Tetrahedron.* 1977, *33*, 2459.
121. (a) Lewis, F. D.; Ho, T.-I. *J. Am. Chem. Soc.* 1977, *99*, 7991. (b) Lewis, F. D.; Ho, T.-I.; Simpson, J.-T. *J. Organ. Chem.* 1981, *46*, 1077. (c) Lewis, F. D.; Ho, T.-I.; Simpson, J.-T. *J. Am. Chem. Soc.* 1982, *104*, 1924. (d) Hubb, W.; Schneider, S.; Dörr, F.; Oxman, J. D.; Lewis, F. D. *J. Am. Chem. Soc.* 1984, *106*, 708.
122. Lewis, F. D. *Acc. Chem. Res.* 1986, *19*, 401.
123. Liou, K. F.; Cheng, C. H. *J. Chem. Soc., Chem. Commun.* 1996, 1423.
124. (a) Saltiel, J.; Charlton, J. L. in *Rearrangements in Ground and Excited States*; de Mayo, P., Ed., Academic Press: New York, 1980, p. 25. (b) Mazzucato, U.; *Pure Appl. Chem.* 1982, *54*, 1705.
125. Mallory, F. B.; Mallory, C. W. *Organic Reactions* 1984, *30*, 1.
126. Laarhoven, W. H. in *Organic Photochemistry*; Padwa, A., Ed., Marcel Dekker: New York, 1989, Vol. 10, p. 163.
127. Ma, B.; Lawson, G. E.; Bunker, C. E.; Kitaygorodskiy, A.; Sun, Y.-P. *Chem. Phys. Lett.* 1995, *247*, 51.
128. Phillies, G. D. J. *Anal. Chem.* 1990, *62*, 1049A.
129. Weiner, B. B. in *Modern Methods of Particle Size Analysis*; Barth, H. G., Ed., John Wiley: New York, 1984, Chap. 3.
130. Zhou, D.; Tan, H.; Luo, C.; Gan, L.; Huang, C.; Pan, J.; Lu, M.; Wu, Y. *Tetrahedron Lett.* 1995, *36*, 9169.
131. Gan, L.; Zhou, D.; Luo, C.; Tan, H.; Huang, C.; Lu, M.; Pan, J.; Wu, Y. *J. Organ. Chem.* 1996, *61*, 1954.
132. (a) Beer, E.; Feurerer, M.; Knorr, A.; Daub, J. *Angew. Chem. Int. Ed. Engl.* 1994, *33*, 1087 (b) Banks, M. R.; Cadogan, J. I. G.; Gosney, I.; Hodgson, P. K. G.; Langridge-Smith, P. R. R.; Millar, J. R. A.; Parkinson, J. A.; Sadler, I. H.; Taylor, A. T. *J. Chem. Soc., Chem Commun.* 1995, 1171. (c) Wilson, S. R.; Cao, J.; Lu, Q.; Wu, Y.; Kaprinidas, N.; Lem, G.; Saunders, M.; Jimenéz-Vásquez, H. A.; Schuster, D. I., *Mat. Res. Soc. Proc.* 1995, *359*, 357. (d) Arce, M. J.; Viado, A. L.; An, Y. Z.; Khan, S. I.; Rubin, Y. *J. Am. Chem. Soc.* 1996, *118*, 3775. (e) Averdung, J.; Mattay, J. *Tetrahedron* 1996, *52*, 5407.
133. Wilson, S. R.; Kaprinidis, N.; Wu, Y. H.; Schuster, D. I. *J. Am. Chem. Soc.* 1993, *115*, 8495.
134. Wilson, S. R.; Cao, J.; Lu, Q.; Wu, Y.; Kaprinidis, N.; Lem, G.; Saunders, M.; Jiménez-Vásquez, H. A.; Schuster, D. I. *Mat. Res. Soc. Symp. Proc.* 1995, *359*, 357.
135. Wilson, S. R.; Wu, Y. H.; Kaprinidis, N. A.; Schuster, D. I. *J. Org. Chem.* 1993, *58*, 6548.

136. Rao, A. M.; Zhou, P.; Wang, K-A.; Hager, G. T.; Holden, J. M.; Wang, Y.; Lee, W.-T.; Bi, X.-X; Eklund, P. C.; Cornett, D. S.; Duncan, M. A.; Amster, I. J. *Science* 1993, *259*, 955.

137. Zhou, P.; Dong, Z.-H.; Rao, A. M.; Eklund, P. C. *Chem. Phys. Lett.* 1993, *211*, 337.

138. (a) Janssen, R. A. J.; Hummelen, J. C.; Wudl, F. *J. Am. Chem. Soc.* 1995, *117*, 544. (b) Gonzalez, R.; Hummelen, J. C.; Wudl, F. *J. Org. Chem.* 1995, *60*, 2618.

139. Creegan, K. M.; Robbins, J. L.; Win, K.; Millar, J. M.; Sherwood, R. D.; Tindall, P. J.; Cox, D. M.; Smith, A. B., III; McCauley, J. P., Jr.; Jones, D. R.; Gallagher, R. T. *J. Am. Chem. Soc.* 1992, *114*, 1103.

140. Smith, A. B., III; Strongin, R. M.; Bard, L.; Furst, G. T.; Romanow, W. J.; Owens, K. G.; King, R. C. *J. Am. Chem. Soc.* 1993, *115*, 5829. (b) Smith, A. B., III; Strongin, R. M.; Bard, L.; Furst, G. T.; Romanow, W. J.; Owens, K. G.; Goldschumidt, R. J.; King, R. C. *J. Am. Chem. Soc.* 1995, *117*, 5492.

141. (a) Smith A. B. III; Strongin, R. M.; Bard, L.; Furst, G. T.; Atkins, J. H.; Romanow, W. J. *J. Org. Chem.* 1996, *61*, 1904. (b) An, Y. Z.; Viado, A. L.; Arce, M. J.; Rubin, Y. *J. Org. Chem.* 1995, *60*, 8330. (c) Westmeyer, M. D.; Galloway, C. P.; Rauchfuss, T. B. *Inorg. Chem.* 1994, *33*, 4615. (d) Yan, M.; Cai, S. X.; Keana, J. F. W. *J. Org. Chem.* 1994, *59*, 5951. (e) Zhang, Y. K.; Janzen, E. G.; Kotake, Y. *J. Chem. Soc., Perkin Trans.* 1996, *2*, 1191. (f) Taliani, C.; Ruani, G.; Zamboni, R.; Danieli, R.; Rossini, S.; Denisov, V. N.; Burlakov, V. M.; Negri, F. *J. Chem Soc., Chem. Commun.* 1993, 220. (g) Hummelen, J. C.; Prato, M.; Wudl, F. *J. Am. Chem. Soc.* 1995, *117*, 7003. (h) Arce, M. J.; Viado, A. L.; An, Y. Z.; Khan, S. I.; Rubin, Y. *J. Am. Chem. Soc.* 1996, *118*, 3775. (i) Ohno, T.; Martin, N.; Knight, B.; Wudl, F.; Suzuki, T.; Yu, H. *J. Org. Chem.* 1996, *61*, 1306. (j) Kusukawa, T.; Shike, A.; Ando, W. *Tetrahedron* 1996, *52*, 4995.

142. Mikami, K.; Matsumoto, S.; Ishida, A.; Takamuku, S.; Suenobu, T.; Fukuzumi, S. *J. Am. Chem. Soc.* 1995, *117*, 11134.

143. Fischer, J. E. *Science* 1994, *264*, 1548.

144. Hirsch, A. *Adv. Mater.* 1993, *5*, 859.

145. Geckeler, K. E. *Trends Polym. Sci.* 1994, *2*, 355.

146. (a) Hawker, C. J.; Wooley, K. L.; Fréchet, J. M. J. *J. Chem. Soc. Chem. Commun.* 1994, 925. (b) Chiang, L. Y.; Wang, L. Y.; Tseng, S.-M.; Wu, J.-S.; Hsieh, K.-H. *J. Chem. Soc. Chem. Commun.* 1994, 2675. (c) Wooley, K. L.; Hawker, C. J.; Fréchet, J. M. J.; Wudl, F.; Srdanov, G.; Shi, S.; Li, C.; Kao, M. *J. Am. Chem. Soc.* 1993, *115*, 9836. (d) Wang, J.; Javahery, G.; Petrie, S.; Bohme, D. K. *J. Am. Chem. Soc.* 1992, *114*, 9665. (e) Guhr, K. I.; Greaves, M. D.; Rotello, V. M. *J. Am. Chem. Soc.* 1994, *116*, 5997. (f) Fey, H.; Weis, C.; Friedrich, C.; Mulhaupt, R. *Macromolecules* 1995, *28*, 403. (g) Shi, S.; Khemani, K. C.; Li, Q. C.; Wudl, F. *J. Am. Chem. Soc.* 1992, *114*, 10656. (h) Hawker, C. J. *Macromolecules* 1994, *27*, 4836. (i) Berrada, M.; Hashimoto, Y.; Miyata S. *Chem. Mater.* 1994, *6*, 2023. (j) Nigam, A.; Shekharam, T.; Bharadwaj, T.; Giovanola, J.; Narang, S.; Malhotra, R. *J. Chem. Soc. Chem.*

Commun. 1995, 1547. (k) Nie, B.; Hasan, K.; Greaves, M. D.; Rotello, V. M. *Tetra. Let.* 1995, *36*, 3617 (l) Patil, A. O.; Schriver, G. W. *Macromol. Symp.* 1995, *91*, 73. (m) Anderson, H. L.; Boudon, C.; Diederrich, F.; Gisselbrecht, J.-P.; Gross, M.; Seiler, P. *Angew. Chem. Int. Ed. Engl.* 1994, *33*, 1628. (n) Bergbreiter, D. E.; Gray, H. N. *J. Chem. Soc. Chem. Commun.* 1993, 645.

147. (a) Olah, G. A.; Bucsi, I.; Lambert, C.; Aniszefeld, R.; Trivedi, N. J.; Sensharma, D. K.; Prakash, G. K. S. *J. Am. Chem. Soc.* 1991, *113*, 9387. (b) Loy, D. A.; Assink, R. A. *J. Am. Chem. Soc.* 1992, *114*, 3977. (c) Fedurco, M.; Costa, D. A.; Balch, A. L.; Fawcett, W. R. *Angew. Chem. Int. Ed. Engl.* 1995, *34*, 194. (d) Oszlanyi, G.; Forro, L. *Solid State Commun.* 1995, *93*, 265. (e) Zhao, Y. B.; Poirier, D. M.; Pechman, R. J.; Weaver, J. H. *Appl. Phys. Lett.* 1994, *64*, 577. (f) Nagashima, H.; Nakaoka, A.; Saito, Y.; Kato, M.; Kawanishi, T.; Itoh, K. *J. Chem. Soc. Chem. Commun.* 1992, 377.

148. (a) Cao, T.; Webber, S. E. *Macromolecules* 1995, *28*, 3741. (b) Cao, T.; Webber, S. E. *Macromolecules* 1996, *29*, 3826.

149. Bunker, C. E.; Lawson, G. E.; Sun, Y.-P. *Macromolecules* 1995, *28*, 3744.

150. Sun, Y.-P.; Lawson, G. E.; Bunker, C. E.; Johnson, R. A.; Ma, B.; Farmer, C.; Riggs, J. E.; Kitaygorodskiy, A. *Macromolecules*, 1996, *29*, 8441.

151. Camp, A. G.; Lary, A.; Ford, W. T. *Macromolecules* 1995, *28*, 7959.

152. Weis, C.; Friedrich, C.; Mülhaupt, R.; Frey, H. *Macromolecules* 1995, *28*, 408. (b) Loy, D. A.; Assink, R. A. *J. Am. Chem. Soc.* 1992, *114*, 3977.

153. Zhang, N.; Schricker, S. R.; Wudl, F.; Prato, M.; Maggini, M.; Scorrano, G. *Chem. Mater.* 1995, *7*, 441.

154. Samulski, E. T.; DeSimone, J. M.; Hunt, M. O., Jr.; Menceloglu, Y. Z.; Jarnagin, R. C.; York, G. A.; Labat, K. B.; Wang, H. *Chem. Mater.* 1992, *4*, 1153.

155. Hawker, C. J. *Macromolecules* 1994, *27*, 4836.

156. Liu, B.; Bunker, C. E.; Sun, Y.-P. *J. Chem. Soc. Chem. Commun.* 1996, 1241.

157. Patil, A. O.; Schriver, G. W.; Carstensen, B.; Lundberg, R. D. *Polym. Bull.* 1993, *30*, 187.

158. Lyons, K. B.; Hebard, A. F.; Innis, D.; Opila, R. L. Jr.; Carter, H. L. Jr.; Haddon, R. C. *J. Phys. Chem.* 1995, *99*, 16516.

159. Lawson, G. E.; Ma, B.; Rollins, H. W.; Hajduk, A. M.; Sun, Y.-P. *Res. Chem. Intermediates* 1996, in press.

160. Hebard, A. F.; Lyons, K. B.; Haddon, R. C. in *Fullerenes, Recent Advances in the Chemistry and Physics of Fullerenes and Related Materials;* Kadish, K. M.; Ruoff, R. S., Eds., Electrochemical Society, 1995, p. 11.

161. Cornett, D. S.; Amster, I. J.; Duncan, M. A.; Rao, A. M., Eklund, P. C. *J. Phys. Chem.* 1993, *97*, 5036.

162. Zhou, P.; Dong, Z.-H.; Rao, A. M.; Eklund, P. C. *Chem. Phys. Lett.* 1993, *211*, 337.

163. Wang, Y.; Holden, J. M.; Dong, Z.-H.; Bi, X.-X.; Eklund, P. C. *Chem. Phys. Lett.* 1993, *211*, 341.

164. Wang, Y.; Holden, J. M.; Bi, X.-X.; Eklund, P. C. *Chem. Phys. Lett.* 1994, *217*, 413.

165. Wang, Y.; Holden, J. M.; Rao, A. M.; Eklund, P. C.; Venkateswaran, U. D.; Eastwood, D.; Lidberg, R. L.; Dresselhaus, G.; Dresselhaus, M. S. *Phys. Rev. B.* 1995, *51*, 4547.

166. Menon, M.; Rao, A. M.; Subbaswamy, K. R.; Eklund, P. C. *Phys. Rev. B.* 1995, *51*, 800.

167. Rao, A. M.; Menon, M.; Wang, K.-A.; Eklund, P. C.; Subbaswamy, K. R.; Cornett, D. S.; Duncan, M. A.; Amster, I. J. *Chem. Phys. Lett.* 1994, *224*, 106.

168. Iwasa, Y.; Arima, T.; Fleming, R. M.; Siegrist, T.; Zhou, O.; Haddon, R. C.; Rothberg, L. J.; Lyons, K. B.; Carter, H. L. Jr.; Hebard, A. F.; Tycko, R.; Dabbagh, G.; Krajewski, J. J.; Thomas, G. A.; Yagi, T. *Science* 1994, *264*, 1570.

169. Lundin, A.; Sundqvist, B. *Phys. Rev. B* 1996, *53*, 8329.

170. Sundar, C. S.; Sahu, P. C.; Sastry, V. S.; Rao, G. V. N.; Sridharam, V.; Premila, M.; Bharati, A.; Hariharan, Y.; Radhakrishnan, T. S. *Phys. Rev. B* 1996, *53*, 8180.

171. Yamawaki, H.; Yoshida, M.; Kakudate, Y.; Usuba, S.; Yokoi, H.; Fujiwara, S.; Aoki, K.; Ruoff, R.; Malhotra, R.; Lorents, D. *J. Phys. Chem.* 1993, *97*, 11161.

172. Rao, C. N. R.; Govindaraj, A.; Aiyer, H. N.; Seshardi, R. *J. Phys. Chem.* 1995, *99*, 16814.

173. Stephens, P. W.; Bortel, G.; Faigel, G.; Tegze, M.; Jánossy, A.; Pekker, S.; Oszlanyi, G.; Forró, L. *Nature* 1994, *370*, 636.

174. Pekker, S.; Jánossy, A.; Mihaly, L.; Chauvet, O.; Carrard, M.; Forró, L. *Science* 1994, *265*, 1077.

175. Nunez-Regueiro, M.; Marques, L.; Hodeau, J. L.; Bethoux, O.; Perroux, M. *Phys. Rev. Lett.* 1995, *74*, 278.

176. Zhu, Q. *Phys. Rev. B* 1995, *52*, R723.

177. (a) Wang, Y. *Nature* 1992, *356*, 585. (b) Sariciftci, N. S.; Braun, D.; Zhang, C.; Srdanov, V. I.; Heeger, A. J.; Stucky, G.; Wudl, F. *Appl. Phys. Lett.* 1993, *62*, 585. (c) Smilowitz, L.; Sariciftci, N. S.; Wu, R.; Gettinger, C.; Heeger, A. J.; Wudl, F. *Phys. Rev. B* 1993, *47*, 13835. (d) Lee, C. H.; Yu, G.; Moses, D.; Pakbaz, K.; Zhang, C.; Sariciftci, N. S.; Heeger, A. J.; Wudl, F. *Phys. Rev. B* 1993, *48*, 15425. (e) Watanabe, A.; Ito, O. *J. Phys. Chem.* 1994, *98*, 7736. (f) Janssen, R. A. J.; Hummelen, J. C.; Lee, K.; Pakbaz, K.; Sariciftci, N. S.; Heeger, A. J.; Wudl, F. *J. Chem. Phys.* 1995, *103*, 788. (g) Janssen, R. A. J.; Christiaans, M. P. T.; Pakbaz, K.; Moses, D.; Hummelen, J. C.; Sacriciftci, N. C. *J. Chem. Phys.* 1995, *102*, 2628. (h) Kraabel, B.; Hummelen, J. C.; Vacar, D.; Moses, D.; Sariciftci, N. S.; Heeger, A. J.; Wudl, F. *J. Chem. Phys.* 1996, *104*, 4267. (i) Schlebusch, C.; Kessler, B.; Cramm, S.; Eberhardt, W. *Synth. Met.* 1996, *77*, 127. (j) Zakhidov, A. A.; Araki, H.; Tada, K.; Yoshino, K. *Synth. Met.* 1996, *77*, 127.

178. (a) Krusic, P. J.; Wasserman, E.; Keiser, P. N.; Morton, J. R.; Preston, K. F. *Science* 1991, *254*, 1183. (b) Krusic, P. J.; Wasserman, E.; Parkinson, B. A.; Malone, B.; Holler, E. R., Jr.; Keiser, P. N.; Morton, J. R.; Preston, K. F. *J.*

Am. Chem. Soc. 1991, *113*, 6274. (c) Morton, J. R.; Preston, K. F.; Krusic, P. J.; Hill, S. A.; Wasserman, E. *J. Phys. Chem.* 1992, *96*, 3576.

179. McEwen, C. N.; Mckay, R. G.; Larsen, B. S. *J. Am. Chem. Soc.* 1992, *114*, 4412.

180. Sun, Y.-P.; Lawson, G. E.; Ma, B.; Johnson, R. A. *Chem. Phys. Lett.* 1996, submitted.

181. Kamat, P. V.; Sun, Y.-P., unpublished results.

182. Riggs, J. E.; Lawson, G. E.; Sun, Y.-P., unpublished results.

10

Use of Photophysical Probes to Study Dynamic Processes in Supramolecular Structures

Mark H. Kleinman and Cornelia Bohne
University of Victoria, Victoria, British Columbia, Canada

I. INTRODUCTION

A. Supramolecular Systems

Supramolecular systems can be defined as a collection of molecular units which are held together by noncovalent bonds. Chemical and physical properties of supramolecular systems, which are different from the sum of the individual properties of each respective unit, are employed to modify chemical reactivity or perform complex functions.

Supramolecular systems can be classified into two broad categories: (1) systems formed from a relatively small number of components where interactions are rather specific, such as host–guest complexes, intercalation into DNA, or complexation to protein binding sites; (2) Systems which correspond to the self-assembly of a large number of components with similar structure. These systems, such as micelles and vesicles, can be frequently viewed as forming separate pseudo-phases from the bulk solvent. This categorization is simplistic and a continuum between the two extremes exists, such that a variety of systems that have intermediate properties between host–guest complexes and self-assembled aggregates have been observed. However, the distinction into categories is useful in order to discuss the properties of supramolecular function and the methodologies employed to characterize these systems.

The primary properties of supramolecular systems are to alter the reactivity of one or more of its components, or to perform a specific function. For example, host–guest systems have been designed to function as sensors [1–3], as catalysts [4], or for preferential solubilization [5,6]. Self-assemblies, such as micelles, vesicles, and biological membranes, have been explored to perform functions, such as preferential solubilization [7], delivery of drugs [8,9], or photoinduced charge separation [10–12].

The properties of supramolecular systems can be described at different levels of complexity. By definition, a supramolecular system involves the formation of structures which are held together by noncovalent bonds; therefore, equilibria are always involved. Any changes in experimental conditions may affect properties, such as the partitioning of solutes or the association/dissociation rate constants. These changes have significant implications on the function of the supramolecular systems.

Initial characterization is primarily concerned in establishing the structure of supramolecular systems. In the case of host–guest complexes, this includes determination of the structure of any synthesized host molecule and, preferentially, of the complexes formed. In the case of self-assembled systems, the structural characterization of each individual unit is generally straightforward, but parameters such as size, aggregation numbers, and polydispersity must be established. The next level of characterization is related to the physical properties of regions or complexation sites involved in supramolecular function. This includes determining the size and shape of complexation sites, the presence of specific interactions (e.g., hydrogen bonds), and the determination of local polarity and viscosity. Frequently, the importance of site geometry or specific interactions is examined by comparing equilibrium constants for similar molecules [13–15]. As a consequence, "improvement" in the design of "next-generation" supramolecular structures has sometimes been based on enhancement of equilibrium constants. Because supramolecular structures always involve some form of equilibria, it is important to characterize the relevant rate constants. For this reason, the next level of complexity in the characterization of organized systems is to understand how individual dynamic processes affect the function of the entire system. For example, a catalytic system has to bind the substrate efficiently, but products should be released rapidly to ensure a high turnover rate. To optimize catalysis, information on rate constants is necessary, as large equilibrium constants may possibly indicate a low turnover rate.

B.   Use of Photophysical Probes to Study Dynamics

The objective of this chapter is to address how photophysics can be instrumental in studying dynamic processes in supramolecular structures. Probe molecules report on the properties of their surroundings. Those molecules with finite lifetimes, such as excited states, will only report information regarding the volume that is explored before decaying back to their ground states. When analyzing mobility, the distance traveled is coupled to the amount of time taken for relocation to occur. One way of expressing this relationship is by considering the volume that a molecule can explore within a certain environment during a finite period of time. As the period of time decreases, less volume is explored. Thus, by choosing probes with different lifetimes, the volume being explored and the kind of information being recovered can be fine-tuned.

In order to study dynamic processes, the relationship between the lifetime of the probe and the time during which the dynamic process of interest occurs has to be considered. If the lifetime is much longer than the time for the completion of a dynamic process, no discrete information about the dynamics can be obtained, because the probe senses an average environment. However, if the probe has a much shorter lifetime than the dynamic process, it will experience a static environment, as changes do not occur during the lifetime of the reporter molecule. Thus, for dynamic information to be recovered, the lifetime of the probe should be of the same order of magnitude as the rates involved. It is important to compare rates because excited state lifetimes are unimolecular events, but dynamic processes may involve bimolecular reactions. The rate for the latter process can be controlled by varying concentrations of one of the reaction components. The advantage of using photophysical probe molecules is that fast dynamic processes in the nanosecond to microsecond range can be investigated. Although this review focuses on photophysical techniques, we should mention that other techniques are available. For example, ultrasonic relaxation gives access to the study of fast processes and has been occasionally employed for supramolecular systems, such as for the determination of the exchange rate of monomers in micelles [16–18] or the binding of ions to cyclodextrins (CDs) [19].

Movement of molecules can occur within different volumes (i.e., within a particular site of the organized system, between sites, and between the supramolecular structure and the homogeneous solvent). In the case of self-assemblies, the dissociation/association of individual monomers and the lifetime of the self-assembly are relevant for its function. The ability of molecules to move within the complexation/binding site is determined by the rigidity imposed by the surroundings and will influence chemical reactivity by defining the regioselectivity and/or stereoselectivity of reactions. The mobility within one site can be studied by microviscosity sensitive probes, and fluorescence polarization has been developed to examine this aspect of organized systems [20–22]. Mobility over larger distances involves the relocation of molecules between sites with different properties, as well as between the supramolecular structure and the homogeneous solvent. Both of these processes are conceptually the same. This review will be centered on this latter type of mobility—in particular, on the mobility between the supramolecular system and the homogeneous phase, which in most cases involves aqueous solutions.

C. Photophysical Techniques

Organic molecules in their excited singlet or triplet states have been employed as photophysical probes. The main advantage in measuring fluorescence is the sensitivity of the technique [23,24]. Thus, small concentrations of probes can be detected with high signal-to-noise ratios. Over the years, fluorescent probes with very specific properties have been developed, including those sensitive to polarity [25–29], pH [30], and specific ions [31], such as Ca^{2+} and labeling compounds for specific amino acids [32]. In addition to the detection of emission spectra, time-correlated or phase-modulated single photon counting (SPC) has been developed to measure the time dependence of the decay of excited singlet states. The large dynamic range for intensity measurements coupled with the statistical distribution of counts of this technique makes it possible to reliably separate multiexponential decays [24]. In homogeneous solution, the decay of excited singlet states is normally monoexponential, but multiexponential decays are frequently observed when probes are located in different binding sites of supramolecular structures [33,34].

Excited singlet lifetimes, also called fluorescence lifetimes, of organic molecules are normally smaller than 10 ns. Notable exceptions are polycyclic aromatic hydrocarbons, such as pyrene and naphthalene. Due to their short lifetimes, fluorescent probes can only explore a small volume; therefore, competition between the probe's decay to the ground state and dissociation from the supramolecular structure to the homogeneous phase occurs only infrequently. For example, in order to compete with the decay of the excited state, a probe with a 10 ns excited state lifetime must possess a dissociation rate constant from the supramolecular structure that is larger than 10^8 s^{-1}. For this reason, fluorescent probes are normally assumed not to relocate during their lifetimes; hence, explore a very limited volume.

Although the short lifetimes of fluorescent probes limits their use for direct relocation studies, the excited state can be employed as a marker for the access of other molecules to the site where the probe resides. Thus, quenching studies can yield information on the mobility of the quencher through the supramolecular structure. This is an indirect method, as the excited probe is employed to report on the mobility of an "invisible" molecule (i.e., the quencher). The quenching methodology assumes that electronically excited probes in different environments, such as those complexed to the supramolecular structure or in the bulk aqueous phase, are quenched with different efficiencies. The advantage of this selective quenching is that any differences in the lifetimes are generally enhanced for probes in different locations. The drawback of this methodology is that a mechanism for quenching has to be assumed, and the recovered rate constants are only relevant if the mechanism adequately describes the kinetic processes involved. To partially overcome this drawback and to test the validity of the mechanistic schemes, existing procedures for data analysis have been continuously refined and new ones have been introduced [33–49]. A further consideration is that for all quenching methods, it is assumed that the macromolecular structure being probed does not change with the incorporation of either the probe and/or quencher.

Organic molecules in their excited triplet state have long lifetimes, as the decay to the singlet ground state is a forbidden transition. The decay of triplet states can be followed by its phosphorescence or triplet–triplet absorption. Detection of phosphorence is more limited than observation of fluorescence, due to the smaller emission quantum yield for triplet states of organic molecules. For this reason, the use of laser flash photolysis (LFP), which measures the triplet–triplet absorption, is widely employed to study the decay of excited triplet states. Relative to singlet species, the longer lifetime of triplet states meets the requirement that the lifetime be comparable to the rate of the dynamic processes being studied. However, the use of triplet states has at least two drawbacks, the low signal-to-noise ratios of LFP when compared to SPC, and the sensitivity of the lifetimes to the presence of impurities. The latter characteristic is inherent to any transient with a long lifetime and can only be diminished by avoiding the presence of impurities. The signal-to-noise issue of LFP is, in part, a characteristic of the current technology, and we believe that in the future this issue will be addressed by employing stable and tunable sources as monitoring beams.

In the case of excited triplet states, the mobility of probe molecules can be determined either by direct kinetic studies or by employing the quenching methodology. Direct relocation measurements can be performed when the probe has an observable property that changes when located in different microenvironments.

We will restrict the focus of this chapter, as there are a large variety of supramolecular structures and photophysical probe molecules. Our choice is biased by our interest in systems with biological relevance and to systems which had a fundamental impact on the development of dynamic studies in organized structures. We will deal exclusively with systems in solution, mainly in the aqueous phase. The self-assembled systems being covered are micelles, due to their importance in the development of methodology, vesicles, which are mimetic systems for cell membranes, and bile salt aggregates. As an example of well-defined host–guest complexes, we will discuss the dynamics of probe complexation to cyclodextrins. Finally, we will also cover probe binding to DNA and proteins in order to illustrate the application of the concepts described to structures with biological relevance that cannot be classified either as self-assemblies or host–guest complexes. For all of the above systems, we are going to restrict our discussion to organic probe molecules. The principles being described can be equally applied to inorganic probe molecules. Indeed, the broader range of emission lifetimes and the greater number of excited state multiplicities available for inorganic probe molecules may prove important in the future to study larger organized systems and to understand different characteristics of supramolecular systems.

II. BASIC KINETIC SCHEME FOR PROBE AND/OR QUENCHER MOBILITY

This chapter centers on the mobility of molecules between supramolecular structures and the homogeneous phase. Mobility can be observed by following the relocation of the excited state or by measuring the reaction efficiency of the complexed excited state with quencher molecules. Figure 1 shows a scheme covering the simplest mechanistic outline to describe the dynamics of probe and/ or quencher mobility. Many reports in the literature have employed only a subset of this mechanistic scheme. The objective of presenting Fig. 1 is to provide a unified view of the dynamic processes, to qualitatively describe under which conditions different rate constants can be experimentally determined, and to state the important underlying assumptions.

The supramolecular (host) system is viewed as a confined space in which the probes and/or quenchers are located. Assumptions have to be made regarding the availability of complexation sites in each system. For host–guest complexes, such as cyclodextrins, stoichiometries of complexation are assumed or determined experimentally [50,51]. In the case of self-assembled systems, such as micelles or vesicles, the most common assumption is that at low probe/micelle ratios the distribution of probes follows a Poisson distribution [30]. This means that the probability of encountering a molecule in a particular micelle is independent of how many molecules are already in that micelle.

An equilibrium is established between the probe in its ground state and the supramolecular system. Equilibrium constants and complexation stoichiometries are established with a variety of techniques, including absorption and fluorescence measurements. Examples of these methodologies are the Benesi-Hildebrand [52] and Job plot [53,54] treatments. Equilibrium constants for the ground state molecules are established in order to know the fraction of probes

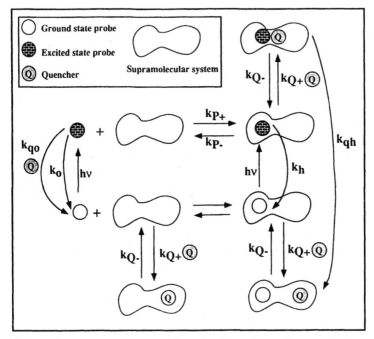

Figure 1 Schematic representation of probe and quencher association and dissociation processes, as well as excited state quenching in the presence of a supramolecular system.

that are complexed prior to their excitation. It is very important to realize that excited state molecules are electronic isomers of their ground states and that they have different properties, such as bond lengths and angles, pKa values, or dipole moments. The equilibrium constants for ground state complexation *cannot be equated* to the equilibrium constant for the excited state. For example, the values of k_{P+} or k_{P-} should not be derived from the equilibrium constants for the ground state and the knowledge of one of these rate constants for the excited state.

The excitation efficiency of bound and free ground state probe molecules has to be known in order to perform quantitative measurements. The easiest experimental approach is to work at an isosbestic point, so that the excitation efficiencies are equal. Unfortunately, this is not always possible when lasers are employed. When a system is being excited at a wavelength different from the isosbestic point, the ratio of free and complexed probe in the excited state will be different from that encountered in the ground state prior to excitation. The ratio in the excited state can be determined by taking into account the molar absorptivities of the free and complexed ground state probe at the excitation wavelength.

Excited probe molecules will decay to their ground state either through radiative or nonradiative processes. The decay rate constants in the homogeneous solvent is k_0, whereas k_h corresponds to the decay of the excited state complexed to the supramolecular structure. The former rate constant can be determined in independent experiments provided the probe is sufficiently soluble in the homogeneous phase. The value for k_h can frequently be estimated when working at high concentrations of host, such that all of the guest is complexed. If the excited state does not react with the supramolecular structure, the values for k_h are generally smaller than k_0, in part due to decreased mobility of the excited state or less accessibility of quenchers to the complexed probe.

Probe mobility can be measured in direct spectroscopic studies or by employing quenching methodologies. The association/dissociation rate constants obtained are always related to the excited state of the probe. In contrast, for the quenching methodology, the association/dissociation rate constants for the quencher are related to the ground state. Independent of the methodology, the photophysical probe molecule is always employed as a tag or marker. We will describe the most common scenarios for the mechanistic scheme in Fig. 1, with the objective of pointing out the potential advantages and limitations with respect to obtaining information on the dynamics of the probe and/or quencher dynamics. The arguments presented are valid for excited singlet as well as triplet states. One must always keep in mind that for the values for k_0 and k_h in the case of triplet states are smaller than for singlet states and that each excited state is better suited for different scenarios.

A. Probe Mobility

In the case of probe mobility different cases can be described.

Case 1: The dynamics is much faster than the excited state decay. For this condition to apply, the values of k_{P+} [H] (where H is the host) and k_{P-} have to be much larger than $k_0 + k_h$. Equilibration is not the rate-limiting step and will always occur before decay to the ground state. Thus, the excited state lifetime will only reflect the slowest of the decay pathways, either the decay in homogeneous solution, or in the supramolecular system, or both. It is worth noting that one of the rates being compared involves a bimolecular process. Because the association rate will depend on the host concentration, the rate-limiting step of the system can occasionally be fine-tuned (i.e., association can be transformed into the rate-limiting step when the host concentration is lowered). A second method that circumvents the condition above is to decrease the excited state lifetime with the addition of a quencher (cf. case 3).

In general, no information on the complexation dynamics of the probe can be obtained when the dynamics is fast compared to the excited state lifetime. An exception occurs when excitation leads to a significant change of the equilibrium constant between probe and host. This situation is analogous to that encountered in temperature or pressure jump experiments, in which equilibria are perturbed and the relaxation kinetics to the new equilibrium is observed [55–57]. The values for the association and dissociation rate constants can then be obtained from the relaxation kinetics, in which the pseudo-first-order experimental rate constant is measured at different host concentrations. For this methodology to be applicable, the observed property of the probe (i.e., absorption or fluorescence) has to be different when the probe is complexed to the host compared to the free probe found in the bulk solution.

Case 2: Decay of excited states is much faster than the dynamic processes. This situation is most frequently encountered for excited singlet states, and the condition applies when k_{P+} [H] and k_{P-} are much smaller than k_0 and k_h. The system can be viewed as having two populations of excited probe molecules which do not interchange during the excited state lifetime. Thus, no information on the dynamics of the probe can be obtained by directly following the decay of the excited states; rather, each population of probe reports on their respective environments.

Case 3: The rates of the dynamic processes are comparable to the decay rates. This condition leads to complex kinetics but allows the determination of the association and dissociation rate constants for probes with the supramolecular structure. Mobility can be directly monitored if the excited state probe molecule has a measurable property that depends on its location in the homogeneous phase or within the supramolecular binding site. Alternatively, quenching can be employed if such a property is not available. In this case, it is important that the quenching efficiency be different for the free and complexed probe. As the difference in quenching efficiency increases, so will the precision of the probe's association and dissociation rate constants. In general, a quencher is employed which mainly resides in the homogeneous phase and is very efficient in quenching the free excited states (large k_{qo}). With increasing quencher concentrations, a shortening of the lifetime for the free probe is achieved, and at sufficiently high quencher concentrations, the dissociation process (k_{P-}) becomes the rate-limiting step. Conversely, the association (k_{P+} [H]) can be the rate-limiting step when the quenching efficiency is higher for the complexed probe compared to the reaction with the excited probe in the aqueous phase ($k_{qh} > k_{qo}$). It is not necessary for the rate-limiting step to be completely achieved, as the association/dissociation rate constants can be recovered from the nonlinear dependence of the decay rate constant at different quencher concentrations.

In summary, the association and dissociation rate constants of probe molecules with supramolecular systems can only be determined if the decay rate constants are comparable or smaller than the rates of dynamic process, as described in cases 1 and 3. This situation is frequently achieved for excited trip-let states, but is rarely present for excited singlet states.

B. Quencher Mobility

Quencher association and dissociation rate constants are a direct measure of dynamics, whereas quenching rate constants can provide a measure of quencher accessibility when compared to quenching efficiencies in the homogeneous phase. The analysis of the quenching process is simplified if the assumption is made that the quencher association/dissociation rate constants are independent of the occupation of the supramolecular system with quencher and/or probe molecules. This is a valid assumption in the case of large self-assembled aggregates, such as micelles and vesicles, but is problematic in the case of host-guest complexes where defined stoichiometries are involved, because complexation of the quencher may preclude or significantly alter the complexation of the probe. However, accessibility of the quencher can still be inferred from the overall quenching rate constant.

Case 4: Only the quencher migrates. This is the most common situation with excited singlet state probes. The quenching efficiency will be determined by the rate of access of the quencher to the supramolecular structure ($k_{Q+}[H]$) and the efficiency of quenching within the supramolecular system (k_{qh}). The association to the supramolecular complex is a bimolecular process, whereas the quenching efficiency within the supramolecular structure is viewed as a unimolecular process. This picture is conceptually analogous to the formation of an encounter complex in solution before reaction, where the volume of the encounter complex is defined by the supramolecular structure. Thus, an overall effective quenching rate constant [$k_q(\text{eff})$] can be defined which takes into account the association process and the intrinsic reactivity:

$$k_q(\text{eff}) = \frac{k_{Q+}k_{qh}}{k_{Q-} + k_{qh}} \tag{1}$$

The unimolecular quenching rate constant, k_{qh}, defines the reaction efficiency when the probe and quencher are in the supramolecular structure. The value for the effective quenching rate constant can be obtained from conventional steady-state or time-resolved quenching experiments, provided all of the excited probe molecules are complexed to the supramolecular structure. This rate constant is an easy parameter to measure experimentally, but it only provides information on accessibility when compared to the quenching rate constant in homogeneous solution. This comparison assumes that the intrinsic reactivity (i.e., the reactivity after diffusion) is the same in the homogeneous solution and in the supramolecular system. This is a fair assumption when the reaction in solution is diffusion controlled, as this indicates that the rate constant for the intrinsic reactivity is much higher than the rate for diffusion, and any slight changes in the intrinsic rate constant does not affect the value of $k_q(\text{eff})$. The assumption of equal intrinsic reactivity may not be valid when the quenching reaction is not diffusion controlled. In this case, a change in the intrinsic reactivity might alter the quenching rate constant. For example, if the quenching mechanism involves charge transfer, a different intrinsic reactivity in water and in an apolar medium of a supramolecular complexation site may occur.

The value of k_{qh} is related to the reactivity within the supramolecular system. This rate constant will depend on the mobility of the quencher with respect to the probe inside the supramolecular structure, as well as on the chemical reactivity for the quenching process. Several models have been described for the mobility of quenchers with respect to probes in micelles [58–65]. Thus, the value of k_{qh} has a dynamic component to it, but it is not related to the association or dissociation processes of the quencher with the supramolecular system, which is the focus of this review. The values of k_{qh} will only be discussed when relevant to the association/dissociation studies.

The values for k_{Q+} and k_{Q-} can be determined for several mechanisms; the one represented in Fig. 1 is the simplest one. The probe's fluorescence decay generally corresponds to a first-order process in the absence of quencher when most of the probe molecules are bound to the supramolecular system. However, invariably a nonexponential decay is observed in the presence of quencher and the kinetics are fitted to equations containing several parameters. The relationship between these parameters and rate constants is established by an assumed mechanism. The values of k_{Q+} and k_{Q-} are recovered from the change in the fitting parameters with quencher concentration. The fitting equations are generally complex and the challenge of this methodology is to ensure that the mechanistic scheme accurately represents the processes involved.

Case 5: Probe and quencher migrate. Most initial fluorescence studies assumed that probes did not migrate during their excited state lifetimes. However, more detailed analyses of the quenching phenomenon, especially those that used more advanced curve-fitting techniques, have suggested that this assumption is not always correct [66–69]. It is especially questionable for long-lived excited singlet states such as pyrene and its derivatives. Probe migration introduces additional rate constants (k_{P+} and k_{P-}) to the equation employed to fit the excited state decay.

We want to emphasize that Fig. 1 is a minimum mechanistic scheme that describes the dynamic processes involved in probe and/or quencher mobility. Every supramolecular system can present additional complexities and has to be analyzed individually. Some examples of increased complexity are the possibility of probe and/or quencher migration by collision of aggregates [70–73], the lifetime of the aggregates when it is comparable to the dynamics of the probe/quencher, and the case when host–guest complexes have more than one complexation stoichiometry [74].

III. EXCITED SINGLET STATE PROBE MOLECULES

Excited singlet states are generally followed by their fluorescence, because an excellent signal-to-noise ratio can be achieved when measuring emission intensities. Fluorescence spectra and emission intensity measurements can be obtained with continuous irradiation. Quenching rate constants can then be inferred from Stern–Volmer constants obtained in quenching experiments (vide infra). However, time-resolved experiments are necessary for detailed analysis of the dynamics involved in the complexation of probes and/or quenchers to supramolecular systems. Time-correlated and phase-modulated single photon counting (SPC) are the most popular techniques to acquire fluorescence decay data with good signal-to-noise ratios. Several excellent reviews cover the details of these techniques [23,24,75,76], and it is sufficient to mention for the purpose of this chapter that the use of SPC has been essential for development of the methodology to study dynamics in organized systems.

A. Fitting Procedures

The fluorescence decay for probes being quenched in supramolecular systems typically involves multiexponential decays. To evaluate the "goodness of fit" of the experimental data to the equation derived from the mechanism proposed, it is essential that statistical parameters be analyzed and published. This enables the comparison of results obtained in different laboratories. Several statistical parameters are available for this analysis of the "goodness of fit" [23,24]. The reduced chi squared (χ^2) is the most common parameter employed, whereas the graphical analysis of the weighted residuals and of the autocorrelation function show any systematic deviation of the fitted curve when compared to the experimental data. Systematic deviations indicate that the proposed mechanism does not reveal the true intricacies of the system. Additional fitting parameters include the Durbin–Watson parameter, which differentiates among single, double, and higher-order exponentials, and the runs test which checks in mathematical form for the randomness of the residuals.

The most common fitting analysis employed is to fit the experimental data to nonlinear least squares regression analysis (NLLS). This analysis is employed for a single decay curve obtained at set experimental parameters, such as fixed excitation and emission wavelengths and fixed concentrations of probe, supramolecular structure, and quencher. The experimental data are fitted to equations derived from assumed mechanistic models. The fitting procedure, which is normally based on the Marquardt algorithm [77,78], minimizes χ^2. It is always necessary to deconvolute the instrument response function (IRF) from the experimental data, as the excitation pulse has a finite time dependence and the detection system can have a significant time dependence as well. The IRF has to be acquired with experimental conditions that match as close as possible the conditions for which the fluorescence decay was acquired. Artifacts, such as wavelength dependence of the detection system, drift in the excitation source, and time drifts in the electronics can significantly alter the IRF and lead to systematic deviations of the deconvoluted data. These problems are more severe when the time resolution of the system is in the picosecond time domain. To circumvent these systematic deviations, a reference convolution method can be employed [24,36,79]. In this method, the monoexponential decay of a compound with a known lifetime is employed for deconvolution. This compound is excited under the same conditions as the unknown sample.

With the increased use of equations with several fitting parameters, it has been necessary to develop more sophisticated fitting procedures. In recent years, many such procedures have been described [38,40,41,80–84]. Global analysis [33–37,43,45,48,78,79] and the maximum entropy method (MEM) [38,39] will be described briefly, as these methods are relevant for studies in supramolecular systems and have been developed sufficiently to be applied by most users.

Global analysis takes advantage of the fact that in complex mechanistic schemes, the fluorescence decay will depend on an experimental parameter that can be varied, such as the quencher concentration [33,34,36,79,85,86]. A nonlinear least squares (NLLS) fit is also employed, but the advantage over single-curve analysis is that linked parameters of the fitting equations are minimized to single values for several decays obtained under different experimental conditions. Comparisons between results obtained from fits to a single curve and when employing global analysis showed that the latter procedure unequivocally leads to smaller errors [63,87]. Under certain circumstances, global analysis can be employed to differentiate between competing mechanistic schemes, although unambiguous discrimination is not always possible [63,88–90]. This type of multicomponent analysis can be further advanced by using global compartmental analysis, in which more than one experimental condition is varied and all decays are fitted simultaneously. In addition, the species-associated emission spectrum can be constructed from global compartmental analysis [48,88,89,91].

The maximum entropy method (MEM) employs a distribution analysis of a series of exponentials to fit the fluorescence decay [38,92]. The amplitudes for the exponential series are reconstructed in such a manner as to maximize an entropy function related to these amplitudes. The difference between this method and those based on NLLS analysis is that no mechanistic scheme has to be applied a priori. Thus, the analysis is not initially biased by a mechanism, but care should be taken with potential artifacts [38,39,84,92–94]. Due to its independence from a mechanistic scheme, we believe that in the future, the maximum entropy method will be very useful when characterizing supramolecular systems for which the number and nature of binding sites is unknown. However, after recovering a series of lifetime distributions, it is imperative to establish their physical meaning.

B. Models to Study Dynamics

The accessibility of quenchers to probes in supramolecular structures can be inferred from bimolecular quenching rate constants. The necessary assumptions have been discussed in Section II.B. Quenching can be studied by the decrease of the fluorescence intensity with the addition of a quencher and the data are treated by employing the Stern–Volmer equation:

$$\frac{I_0}{I} = 1 + K_{sv}[Q] \tag{2}$$

The Stern–Volmer constant (K_{sv}) corresponds to the product of the overall quenching rate constant [$k_q(\text{eff})$] and the excited singlet lifetime (τ_0) in the absence of a quencher. The value for $k_q(\text{eff})$ can be calculated from K_{sv} when the lifetime in the absence of quencher is measured independently. The drawback of this method is that static quenching also leads to a decrease of the fluorescence emission intensity but does not correspond to a bimolecular process. Static quenching occurs when the probe and quencher are in close contact and quenching happens "instantaneously," leading to a decrease in the fluorescence intensity but not a decrease of the lifetime of the excited singlet state. In the homogeneous solution, this is only possible when the quencher and probe form a ground state complex. In contrast, in supramolecular structures, it is only necessary for the probe and quencher to reside in a very confined environment so that collisions between them occur in a time shorter than the time resolution employed to measure the decay. This latter process is not a classical static quenching mechanism because no complex is formed, but it experimentally appears as one. Because static quenching is always a real possibility when the probe and quencher are incorporated into supramolecular structures, any inference of mobility from steady-state fluorescence studies should be made with prudence.

Quenching rate constants can be directly measured in time-resolved experiments. For a fluorophore in homogenous solution, the observed decay rate constant at different quencher concentrations is given by

$$k_{obs} = k_0 + k_{q0}[Q] \tag{3}$$

where k_0 and k_{q0} are the decay rate constants in the absence of quencher and the quenching rate constant as defined in Fig. 1.

In the presence of a supramolecular system under conditions where *all* of the probe is bound, the effective quenching rate constant is obtained from

$$k_{obs} = k_h + k_q(eff) [Q] \qquad (4)$$

where k_h is the excited state decay rate constant in the absence of a quencher when the probe molecule is complexed to the supramolecular system, and $k_q(eff)$ is the overall quenching rate constant.

Frequently, the fluorescence decay of a probe in the presence of a supramolecular structure is not first order. When the decay is adequately fitted to the sum of two exponentials, it was assumed that one lifetime corresponds to the probe in a homogeneous solution and the second one to the probe in the supramolecular system. This assumption has been supported by the fact that the lifetime for the probe in the absence of the supramolecular system (i.e., in a homogenous solution corresponds to one of the two lifetimes recovered from the fit to the sum of two exponentials. In the presence of a quencher, the observed rate constant corresponding to the process in a homogenous solution is analyzed by Eq. (3), whereas the process corresponding to the complexed probe is analyzed by Eq. (4). The ratio of the two quenching rate constants $[k_{q0}/k_q(eff)]$ is a measure of the protection afforded by the supramolecular system due to decreased accessibility of the quencher to the probe molecule. The assumptions of this mechanistic scheme are that the probe does not interchange between the supramolecular system and the homogeneous phase during its excited state lifetime and that quenching within the supramolecular system is much faster than the exit of the quencher. The drawback of this analysis is that the values for k_{Q+} and k_{Q-} are not recovered.

Models with increasing sophistication for the analysis of dynamic processes in supramolecular systems, notably micelles, as well as for the determination of other parameters have been developed over the past two decades. The basic conceptual framework has been described early on [59,60,95,96] and has been classified into different cases which take into account the extent of quencher mobility and the mechanism of quenching [95]. Two of those cases lead to information about mobility and will be discussed. It is important to emphasize that this analysis is only applicable to self-assembled system such as micelles and vesicles; it cannot be applied to host–guest complexes. This model assumes that the probe is exclusively bound to the supramolecular system and that no probe migration occurs during its excited state lifetime. The distribution of probe and quencher has been modeled by different statistical distributions, but in most cases, data are consistent with a Poisson distribution. The Poisson distribution implies that the quencher association/dissociation rate constants to/from the supramolecular system does not depend on how many quencher molecules already occupy that system. It is this last assumption that makes it impossible to apply this model to host–guest complexes, as the limited environment for binding precludes the binding of more than one quencher molecule without altering the entry/exit processes.

The first case from which information about the dynamics of quencher mobility can be recovered occurs when the quencher is partially solubilized and where the quenching is much more efficient than the exit of the quencher from the self-assembly. The excited state decay is first order and the observed decay rate constant is given by

$$k_{obs} = k_0 + \frac{k_{Q+}[Q]}{1 + K_Q[H]}$$ (5)

$$K_Q = \frac{k_{Q+}}{k_{Q-}}$$ (6)

In the case of micelles, the host concentration is given by

$$[H] = [\text{micelle}] = \frac{[\text{monomer}] - \text{cmc}}{N}$$ (7)

where the cmc is defined as the critical micellar concentration, which corresponds to the concentration of monomers above which micelles are formed and N is the aggregation number, which corresponds to the number of surfactant monomers which form an individual micelle.

The second case from which dynamic information can be recovered is also the more general description of the model. The quencher is assumed to be mobile and the dissociation rate constant of the quencher from the supramolecular system (k_{Q-}) competes with the quenching process (k_{qh}). The mechanistic scheme of Fig. 1 is valid taking into account the general assumptions mentioned above. The fluorescence decay can be described by a function with four parameters [60,97]:

$$I(t) = A \exp\{-Bt - C[1 - \exp(-Dt)]\}$$ (8)

Each parameter can be described by the processes shown in Fig. 1. The expressions for K_Q and [H] are given in Eqs. (6) and (7), and the initial intensity is I_0:

$$A = I_0$$ (9)

$$B = k_0 + \left[\frac{k_{qh}k_{Q+}}{(k_{qh} + k_{Q-})(1 + K_Q[H])}\right][Q]$$ (10)

$$C = \left[\frac{k_{qh}^2 k_{Q+}}{(k_{qh} + k_{Q-})^2 k_{Q-}(1 + K_Q[\text{H}])} \right][Q] \tag{11}$$

$$D = k_{qh} + k_{Q-} \tag{12}$$

All the rate constants can be recovered by fitting the fluorescence decay curve to Eq. (8) at different quencher concentrations and employing Eqs. (9)–(12).

C. Overall Quenching Studies in Supramolecular Systems

1. Micelles

Micelles are self-assemblies (Fig. 2) formed from surfactants that have an alkyl chain and a hydrophilic headgroup that is frequently charged. These self-assemblies are very dynamic, with monomer exchange occurring in microseconds and with a lifetime of the self-assembly of milliseconds. Micelles have been modeled as spherical aggregates, but experimental evidence has shown that micelles have irregular surfaces and substantial water penetration [30]. Because these self-assemblies are fairly simple, most of the methodologies for using photophysical probes have been developed for the characterization of micellar systems.

New supramolecular structures are frequently characterized using fluorescent probes, and quenching can be employed in this initial characterization. For this reason, a large amount of quenching data is available. Because overall quenching studies do not provide detailed information on the entry/exit dynamics of the quenchers, we will not review all the data in the literature but will present selected examples. Overall quenching rate constants provide informa-

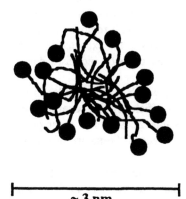

~ 3 nm

Figure 2 Schematic diagram of a micelle. The circles symbolize charged or polar headgroups, whereas the long tails are hydrocarbon chains.

tion on accessibility when the effective rate constant is compared to the quenching rate constant in the homogeneous solvent. For this reason, it is important to determine the latter quenching rate constant. This is not always trivial, as many studies are performed in aqueous solution and most probes are only sparingly soluble in water. If solubility is an issue, quenching rate constants should be determined in a closely matched solvent. In this respect, it is important to match the viscosity of the solvent when rate constants close to the diffusion controlled limit are expected. For example, in the case of water, ethanol is the solvent which has a similar viscosity and still maintains some hydrogen bonding capability.

Micelles can be formed from ionic, cationic, or neutral surfactants. The ionic micelles are the most common ones, and quenchers behave differently according to their charge. The charge on the surfactant is of no relevance for neutral quenchers. When the quencher has the same charge as the micelle monomer, it will be repelled, and only limited data are available for this situation. When the charge of the quencher is opposite that of the surfactant, the quencher will bind to the micelle; consequently, its behavior is more complex due to the attractive interaction with the surface potential.

Oxygen is a very important quencher not only because of its high quenching efficiency and small size but also because of its importance in biological systems. The quenching of micelle bound probes by oxygen was only slightly decreased and did not depend on the charge of the surfactant (Table 1). This indicated that the mobility of oxygen was very high. A very similar behavior was observed for nitromethane, whereas incorporation into sodium dodecyl sulfate (SDS) did not significantly change the quenching of excited singlet pyrene (Table 1). Acrylamide is a neutral but polar quencher. Its ability to quench different pyrene derivatives with and without negative charges in Cdtyl trimethyl ammonium chloride (CTAC) was similar for the different derivatives, indicating that the negative charge on the probe did not significantly change the location of the probe or that the quencher had access to all regions of the micelle (Table 1) [98]. An increase in temperature leads to the expected increase in k_q(eff) because the possibility of probe–quencher collision increases due to increased diffusional rates.

Carbon tetrachloride is an apolar quencher which mainly resides in the micellar phase. The decrease of the quenching efficiency for several probe molecules is compared in SDS and CTAC (Table 2) [99]. For most probe molecules, the decrease is moderate and independent of the nature of the micelle.

For ionic quenchers, it is expected that the overall quenching efficiency will decrease if the quencher has the same charge as the micelle. This was shown for the quenching of pyrene by iodide in SDS micelles where the value for the k_{qo}/k_q(eff) ratio was determined to be 1170 [100]. Quenchers with op-

Table 1 Quenching of Micelle-Solubilized Probes by Neutral Quenchers

Surfactant	Quencher	Probe	$k_q(\text{eff})$ $(10^9\ M^{-1}\ s^{-1})$	k_{q0} $(10^9\ M^{-1}\ s^{-1})$	$\dfrac{k_{q0}}{k_q(\text{eff})}$	Ref.
SDS	O_2	Pyrene	9.2^a	20^b	2.2	100
CTAB	O_2	Pyrene	8.3^a	20^b	2.4	100 and w/in
SDS	CH_3NO_2	Pyrene	3^a	3.6^b	1.2	100
CTAB	CH_3NO_2	Pyrene	1.8^a	3.6^b	2.0	100 and w/in
SDS	Triethylamine	Pyrene	0.25^a	0.23^b	0.92	100
CTAC	Acrylamide	1-Methyl Pyrene[c]	0.010^d	3.20	320	98
CTAC	Acrylamide	1-Pyrenesulfonate[c]	0.0012^d	0.24	200	98
CTAC	Acrylamide	Pyrenebutyric acid[c]	0.007^d	3.50	500	98
CTAC	Acrylamide	Pyrenebutyric acid[e]	0.020^d	3.90	195	98

[a] Obtained from time resolved (TR) experiments.
[b] Determined in methanol.
[c] Temp. = 20°C.
[d] Obtained from steady state (SS) experiments.
[e] Temp. = 50°C.

Table 2 Overall Quenching Rate Constants for Several Probe Molecules in SDS or CTAC by CCl_4

Probe	SDS $k_{q0}/k_q(\text{eff})$	CTAC $k_{q0}/k_q(\text{eff})$
Anthracene	5.5	5.2
Benzo(ghi)perylene	1.3	0.8
Benzo(a)anthracene	3.6	1.8
Benzo(a)pyrene	2.5	1.8
Biphenyl	7.7	13
2,3-Dimethylnaphthalene	4.1	4.3

Note: Quenching rate constants were determined from steady-state experiments and k_{q0} values were determined in ethanol.
Source: Ref. 99.

posite charge of the surfactants are expected to have an overall quenching efficiency that is higher than k_{q0}. The interpretation of the overall quenching rate constants derived from steady-state experiments or time-resolved experiments is difficult due to the effect of quencher mobility (*vide infra*). However, the quenching of pyrene emission in 0.1 M SDS by Ag^+ and Tl^+ was shown to follow pseudo-first-order behavior, with a $k_q(\text{eff})$ 1.3 and 1.8 times higher than in water, respectively [101], which indicated that a portion of these ions were located in the Stern layer or had increased access to the interior of the micelle. In contrast, the decay in the presence of Cu^{2+} was multiexponential and a more complex treatment had to be employed (*vide infra*) [101].

The quenching of pyrene derivatives in CTAC micelles by iodide was studied by time-resolved and steady-state measurements [98]. Some static quenching was observed, indicating that these experiments only provided an approximate picture of the quenching phenomenon. The values for the $k_{q0}/k_q(\text{eff})$ for the quenching of respectively 1-pyrenesulfonate and pyrenebutyric acid at 20°C were 0.26 and 0.09, suggesting that quenching was dramatically enhanced by the binding of iodide as a counterion.

Reversed micelles have three different environments: the water located in the interior of the micelle, the interface defined by a charged surfactant, and the organic solvent. A recent study employing pyrene derivatives, which are neutral, positively and negatively charged, and 2,2,6,6-tetramethyl-1-piperidinyloxy, free radical (TEMPO) derivatives, as quenchers showed that significant information [102], including measures of accessibility could be gained with simple quenching studies. Methylpyrene is predominantly solubilized in the heptane phase, and the quenching efficiency by TEMPO which is also mainly located in the organic phase remains unchanged (Table 3). However, 1,3,6,8-pyrenetetrasulfonate (PTS) will be exclusively located in the water pool, and

Table 3 Quenching of Pyrene Derivatives by TEMPO Derivatives Bearing Different Charges in AOT/Heptane Reversed Micelles

R [H$_2$O]/[AOT]	Quencher[a]	Probe[a]	k_q(eff) (10^9 M^{-1} s^{-1})	k_{q0} (10^9 M^{-1} s^{-1})	k_{q0}/k_q(eff)
4–27	TEMPO	1-Methyl Pyrene	13.2–14.7	15.5[b]	1.2–1.1
4	TEMPO	1-PyMe$^+$	5.3	6.8[c]	1.3
10	TEMPO	1-PyMe$^+$	3.8	6.8[c]	1.8
28	TEMPO	1-PyMe$^+$	4.0	6.8[c]	1.7
4–27	TEMPO	PTS	< 0.5–0.1	6.5[c]	High
4	TEMPOL	1-Methylpyrene	7.7	14.4[b]	1.9
10	TEMPOL	1-Methylpyrene	6.4	14.4[b]	2.3
28	TEMPOL	1-Methylpyrene	4.4	14.4[b]	3.3
5	TEMPAMINE	1-Methylpyrene	6.3		
28	TEMPAMINE	1-Methypyrene	2.8		
5–26	TEMPAMINE	PTS	5.0–5.2	74[d]	14–15

Note: Quenching rate constants obtained from steady-state and time-resolved data. [a]See Appendix for abbreviations.
[b]Determined in heptane.
[c]Determined in ethanol.
[d]Determined in water.
Source: Ref. 102.

almost complete protection from TEMPO quenching was observed. 1-pyrenyl(methyl)trimethylammonium iodide (PyMe$^+$) is solubilized at the interface because the positive charge is efficiently attracted to the negative charge of the surfactant. The difficulty in comparing k_q(eff) with the quenching rate constant in homogeneous solution is that k_{q0} for 1-methylpyrene varies from $7.4 \times 10^9\,M^{-1}\,s^{-1}$ in ethanol to $15.5 \times 10^9\,M^{-1}\,s^{-1}$ in heptane. As the polarity and viscosity at the interface are not known, it is difficult to adopt a value for k_{q0}. In fact, in addition to being more polar, ethanol is almost four times as viscous as heptane [103], which shows that polarity and viscosity effects must be separated. Nevertheless, when the lower value in ethanol is used, a decreased accessibility of TEMPO to the interface is observed (Table 3). TEMPOL, is a derivative of TEMPO with a hydroxy substituent which will lead to a partitioning of this quencher between the organic phase and the water phase. This partition led to a decreased ability of TEMPOL to quench methylpyrene. The quenching efficiency decreased further as either the R or the concentration of AOT was increased. Both of these observations can be explained by the partition of TEMPOL. Thus, the decreased quenching efficiency was not giving any information on the mobility of TEMPOL but could be employed to calculate other parameters, such as its partition coefficient. The quenching of PyMe$^+$ and PTS by TEMPOL gave curved Stern–Volmer plots indicating that some static quenching occurred. Finally, an amino TEMPO derivative (TEMPAMINE) that is positively charged was employed as a quencher. TEMPAMINE is believed to be bound to the interface through electrostatic interactions with the surfactant. The quenching of methylpyrene by this probe showed a marked decrease when R increased, suggesting that methylpyrene had less access to the quencher when the curvature of the interface decreased. As expected, no quenching was observed for PyMe$^+$, which might be attributed to charge repulsion. In the case of PTS, k_q(eff) was smaller than k_{q0} in water. This indicates that the accessibility of PTS to TEMPAMINE at the interface was hindered or that charge screening by the surfactant decreased the intrinsic reactivity.

In conclusion, the limited examples described show that quenching studies can provide some useful information on accessibility of quencher molecules to probes in micelles. However, the rate constants for association and dissociation are not recovered, and interpretation of the quenching results is not always straightforward.

2. Vesicles

Vesicles are self-assemblies formed by phospholipids or synthetic surfactants with two hydrophobic chains that form bilayers (Fig. 3). Vesicle systems are defined by three different regions: the internal water pool, the hydrophobic bilayer, and the homogeneous water phase. The size of vesicles depends on the method of preparation. Small vesicles have a diameter of the order of 30 nm,

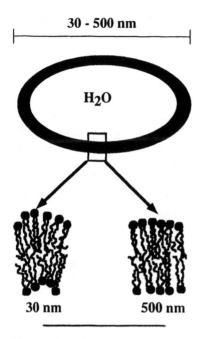

30 - 500 nm

H$_2$O

30 nm   500 nm

Figure 3 Schematic diagram of a unilamellar vesicle.

whereas large vesicles have diameters between 300 and 500 nm [104]. Smaller vesicles have more curvature in the bilayer. For this reason, larger vesicles are more representative of cell membranes. In contrast to micelles, which are quite mobile aggregates, the bilayer in vesicles is more structured and possesses a phase transition at a defined temperature. The curvature of the vesicular surfaces should be taken into consideration when discussing the exit and entry of solutes into the bilayer. As the size (volume) of the vesicle is reduced, the curvature of both the inner and outer leaflets increases. This packing tendency pushes the headgroups together while the lipophilic hydrocarbon tails encompass a larger volume than when dwelling in larger vesicles [104].

Dioctadecyldimethylammonium chloride (DODAC) is a popular synthetic amphiphile that forms vesicles. The quenching efficiency of pyrene derivatives by several quenchers has been studied [98,105]. When iodide is the quencher for the emission of 1-pyrenesulfonate, the k_{q0}/k_q(eff) ratios were respectively 0.016 and 0.023 at 20°C and 50°C. These values were smaller than observed for CTAC, suggesting that iodide was efficiently bound to the vesicle. It is known that water penetrates deep into the bilayer [106] and thus, I$^-$ may be located in these small water regions. It is also worth noting that an increase to a temperature above the transition temperature (33–37°C) [107] did not change the quenching efficiency by a large amount. In addition, a similar quenching ratio for CTAC and DODAC was observed for the quenching by acrylamide. These experiments seem to suggest that accessibility by neutral, aqueous quenchers to probes in CTAC micelles and DODAC vesicles is similar, whereas access for charged quenchers is affected by the number of quencher binding sites in the vesicle. In contrast, a different behavior was observed for CCl_4, for which mobility occurs within the hydrophobic region. The quenching efficiency for several pyrene derivatives in DODAC vesicles increased markedly at 50°C when compared to 20°C. The values for the quenching rate constants at 20°C were, in most cases, below those observed for CTAC micelles, whereas those at 50°C were higher by approximately an order of magnitude. This dependence on temperature for CCl_4 quenching indicated that the mobility within the hydrophobic phase was affected by the phase transition, but access from aqueous quenchers did not depend on the internal structural organization of the vesicles. In addition, the authors pointed out that other possibilities for the large increase in quenching rates exist, such as changes in the partition coefficient of the quencher or perturbation within the bilayer caused by binding of the quencher [98].

The access of quencher molecules to probes in vesicles can yield relevant information on the accessibility to molecules which are located in membranes in biological systems. The accessibility of iodide, which is an aqueous quencher to several anthracyclines incorporated into dimyristoylphosphatidylcholine (DMPC) vesicles has been studied [108]. The relative quenching efficiencies are shown in Table 4. As observed for DODAC, the accessibility of iodide decreased, but it is not markedly affected by the structure of the anthracycline, suggesting that reactivity of these anthracycline antibiotics with aqueous molecules does not depend significantly on the vesicle structure. According to the authors, quenching of the free drug took place by static and collisional mechanisms, whereas the bound probe was quenched predominantly by dynamic quenching. However, the decrease of iodide accessibility to the anthracyclines is significantly larger than observed for pyrene, pyrenemethyltrimethylammonium iodide, and 1-pyrenebutyltrimethylammonium iodide in dipalmitoylphosphphatidylcholine [DPPC, k_{q0}/k_q(eff) = 1.1–2.4] and in 80% DPPC/20% dipalmitoylphosphatidic acid [k_{q0}/k_q(eff) = 1.1–4.1] vesicles [109]. This result indicates that anthracycline fluorophores were buried deeper within the hydrophobic bilayer or that DMPC bilayers provided a larger barrier for iodide association. To resolve this question, the pyrene experiments would have to be performed in DMPC. This comparison demonstrates some of the difficulties associated with comparing experiments from different laboratories. No standard has been developed for quenching studies and in the case of vesicles a great variety of different experimental conditions has been employed.

Table 4 Overall Quenching Rate Constants for Anthracyclines Incorporated into DMPC vesicles

Probe	$k_q(\text{eff})$ $(10^9\ M^{-1}\ s^{-1})$	k_{q0} $(10^9\ M^{-1}\ s^{-1})$	$k_{q0}/k_q(\text{eff})$
Carminomycin	0.6	8.0	13
4-Dimethoxydaunomycin	0.7	8.1	12
Daunomycin	1.0	9.4	9.4
Rubidazone	1.0	8.8	8.8
Adriamycin	1.2	9.5	7.9
Daunomycinol	1.1	9.2	8.4
N,N-Dimethyldaunomycin	1.3	9.1	7
Carmimycinone	0.6	8.2	14
Daunomycinone	0.7	8.4	12
Adriamycinone	0.8	9.6	12
Daunomycinol aglycon	0.8	8.8	11
7-Deoxydaunomycinone	0.5	8.2	16
1,4-Dihydroxyanthraquinone	0.6	9.2	15

Note: Quenching rate constants were obtained from steady-state and time-resolved quenching experiments, k_{q0} was determined in phosphate buffer containing 8mM Na_2HPO_4 and 1mM KH_2PO_4 and at constant ionic strength (pH 7.4).
Source: Ref. 108.

3. Bile Salt Aggregates

Bile salts, such as sodium cholate (NaC) and sodium taurocholate (NaTC) are molecules with planar polarity. One face of the molecule is apolar, whereas the other face contains polar hydroxyl groups. Bile salts are natural solubilization agents, and besides their biological importance they have also been employed for analytical procedures [100,110–117]. Fluorescence has been used early on to characterize these aggregates which behave quite differently from conventional micelles. At low bile salt concentrations, a few molecules aggregate to form primary aggregates with hydrophobic binding sites. At higher concentrations, these small aggregates form larger structures (secondary aggregates), which contain hydrophilic binding sites (Fig. 4).

The values for overall quenching rate constants are shown in Table 5 for studies of probes in sodium taurocholate [100,118]. Only values that could be compared to quenching rate constants in homogeneous solution were included. The quenching by neutral molecules, such as nitromethane, oxygen, and triethylamine, was reduced to a much greater extent than was observed in the case of SDS micelles, indicating that bile salt aggregates provided a more rigid and protected environment. The reduction of the quencher accessibility was not

Hydrophobic
Site

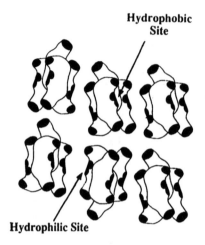

Hydrophilic Site

Figure 4 Schematic diagram of bile salt aggregates. The dark areas indicate the polar moieties of the bile salt monomer.

uniform, a larger reduction being observed for nitromethane followed by triethylamine. The aggregation of bile salts is known to be affected by the addition of sodium chloride. In the presence of 1 M NaCl, the accessibility of oxygen was significantly decreased when compared to quenching in the absence of NaCl. The monomers of NaTC are negatively charged. Nevertheless, cationic and anionic quenchers had decreased accessibility to incorporated probes. The similar behavior for quenchers with opposite charges is probably due to the low charge density of these aggregates when compared to conventional micelles. In addition, in the presence of 1 M NaCl efficient charge screening occurs. The addition of cations has also been shown to influence the aggregates [119,120] and could account for some of the changes in the overall quenching rate constants shown in Table 5.

The quenching of pyrene, naphthalene, and anthracene in sodium cholate (NaC) by iodide ions in the presence of 0.2 M NaCl was studied by steady-state and time-resolved fluorescence measurements [121]. Although iodide and sodium cholate are negatively charged some static quenching was observed. Static quenching was suggested to occur in the hydrophilic sites of the secondary aggregates. Dynamic quenching was measured in order to obtain the overall quenching rate constants for probes incorporated into the hydrophobic sites of NaC aggregates (Table 6). The protection is more efficient for naphthalene, followed by pyrene and anthracene. The significant dependence of protection ability by the bile salt aggregates on the shape of the probe indicates that the complexation site has a defined size.

Table 5 Overall Quenching Rate Constants for Pyrene Emission in Sodium Taurocholate (NaTC) obtained in Time-Resolved Experiments

[Monomer] (M)	Quencher	[NaCl] (M)	k_q(eff) ($10^9\ M^{-1}\ s^{-1}$)	k_{q0} ($10^9\ M^{-1}\ s^{-1}$)	k_{q0}/k_q(eff)	Ref.
10^{-2}	CH_3NO_2		0.12	3.6[a]	30	98
10^{-2}	O_2		4.6	20[a]	4.3	98
$(1–2) \times 10^{-2}$	O_2	1.0	1.5	11[b]	7.3	118
10^{-2}	Triethylamine		0.014	0.23[a]	16	100
10^{-2}	I^-		0.036	3.04[c]	84	100
$(1–2) \times 10^{-2}$	Cu^{2+}	1.0	0.29	5.3[b]	18	118
$(1–2) \times 10^{-2}$	MV^+	1.0	1.2	7.7[b]	6.4	118
$(1–2) \times 10^{-2}$	Tl^+	1.0	0.038	4.1[b]	110	118
$(1–2) \times 10^{-2}$	Eu^{3+}	1.0	0.042	1.6[b]	38	118

[a]Determined in methanol.
[b]Determined in water.
[c]Measured for 20 μM pyrenesulfonic acid in water.

Table 6 Overall Quenching Rate Constants by Iodide of Probes Solubilized in the Hydrophobic Site of NaC Aggregates

Probe	$k_q(\text{eff})$ $(10^9 \ M^{-1} \ s^{-1})$	k_{q0} $(10^9 \ M^{-1} \ s^{-1})$	$k_{q0}/k_q(\text{eff})$
Naphthalene	0.15	7.77 ± 0.07	51.8 ± 0.5
Pyrene	0.031 ± 0.001	1.11 ± 0.11[a]	36 ± 4
Anthracene	0.26 ± 0.06	5.27 ± 0.17[b]	20 ± 5

Note: [NaC] = 20 m*M* and [NaCl] = 0.2 *M*, time-resolved quenching studies.
[a]Determined in water.
[b]Determined in a mixture of 45/55 v/v ethanol/water.
Source: Data from Refs. 121 and 122.

In conclusion, bile salt aggregates are self-assembled systems with a more defined structure than conventional micelles, as they provide less access to molecules from the aqueous environment.

4. Cyclodextrins

Cyclodextrins (CDs) are cyclic oligosaccharides which contain 6, 7, and 8 glucose units (α-, β-, and γ-cyclodextrins). The size of the fairly hydrophobic cavity of these compounds is defined by the number of glucose units (Fig. 5) [123]. Cyclodextrins have been extensively studied as models for host–guest complexes with a defined binding site, as models for catalytic systems, and have found application in chromatography [124–127]. Photophysical probe molecules

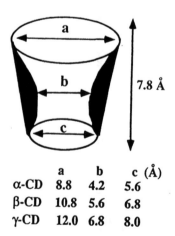

	a	b	c (Å)
α-CD	8.8	4.2	5.6
β-CD	10.8	5.6	6.8
γ-CD	12.0	6.8	8.0

Figure 5 Schematic representation of a cyclodextrin. Dimensions were obtained from Ref. 123.

have been used to characterize host–guest complexes with CDs, such as those involving nitrophenol [128], pyrene [129,130], pyrenesulfonate [131], naphthalene [129], substituted aminobenzoates [132], ANS [133], oxazine dyes [134], and dialkylaminobenzonitriles [135]. Fluorescence has been shown to be particularly suitable for determining equilibrium constants as well as the stoichiometry of complexation, such as for pyrene [74,136], cyanonaphthalene [137], acridine [138,139], fluorescein, erythrosin, and rose bengal [140]. CDs have shown a significant protective effect for quenching by aqueous molecules. It is remarkable that CD-complexed compounds are frequently phosphorescent at room temperature, as in the case of 1-bromonaphthalene [141] and arylpropiophenones [142].

Incorporation of probes into β-cyclodextrin protects them from quenching by aqueous quencher molecules regardless of their charge (Table 7) [132,143–145]. For example, the accessibility of oxygen and nitromethane was decreased substantially. The decrease for oxygen was larger than that observed for bile salt aggregates, whereas the decrease for nitromethane was somewhat smaller when compared to the decrease observed when pyrene was incorporated into sodium taurocholate aggregates. The small accessibility of oxygen to the CD cavity is responsible for the observation of phosphorescence in aerated solutions. For ionic quenchers, it was observed that the accessibility of cations can be reduced substantially. The decrease observed for anions was only moderate (Table 7).

The defined binding site for cyclodextrins provides the possibility that quenchers can be bound within the same cavity as the probe. This will lead to static quenching. Examples have been described in which "apparent" k_q(eff) values were calculated from K_{sv} values obtained in steady-state quenching experiments, such as the quenching of pyrene or naphthalene by several amines [129]. The apparent rate constant recovered showed an increase when compared to the quenching rate constant determined in homogeneous solution. The k_{q0}/k_q(eff) values varied from 0.0031 to 1.5, depending on both the size of the CD cavity and the quencher. Upon the addition of aliphatic amines, no dynamic quenching was observed with the use of time-resolved experiments; however, there was efficient static quenching within the cavity.

5. Proteins and DNA

In the case of DNA, no relevant quenching data for excited singlet probes have been reported. The fact that no extensive studies have been done with organic excited singlet states probably reflects the fact that nucleotides are efficient quenchers. However, inorganic probe molecules have been employed [146–151].

Proteins can have intrinsic fluorescent markers when their amino acid sequence contains one or more tyrosine or tryptophan residues. Although the

Table 7 Overall Quenching Rate Constants for Probes Bound to β-Cyclodextrins

Quencher	Probe	k_q(eff) ($10^9 M^{-1} s^{-1}$)	k_{q0} ($10^9 M^{-1} s^{-1}$)	k_{q0}/k_q(eff)	Ref.
Acrylamide	2,6-ANS	0.367[a]	1.96[c]	5.3	143
Acrylamide	Indole	2.0[b]	7.5[c]	3.8	144
CH_3NO_2	Pyrene	0.56[b]	8.1[c]	15	145
O_2	Pyrene	1.2[b]	11[c]	9.2	145
I^-	2,6-ANS	0.186[a]	0.24[c]	1.3	143
I^-	Methyl-2-aminobenzoate	0.83[a]	2.9	3.5	132
IO_3^-	Methyl-2-aminobenzoate	0.12[a]	3.0	25	132
Tl^+	Pyrene	0.02[b]	6.3	320	145
$C_2Pd^+Br^-$	Pyrene	0.07[b]	7.9	110	145
Cu^{2+}	Pyrene	0.17[b]	4.5	26	145
MV^{2+}	Pyrene	5[b]	7.8	1.6	145

[a] Steady-state measurements.
[b] Time-resolved measurements.
[c] Determined in water.

photophysics of tryptophan is quite complex [152,153], the advantage of using it as a marker is that its position is precisely known when the protein has been sequenced, and because the probe is intrinsic, the protein is in its natural state. When external probes are added, there is always the possibility of conformational changes. A variety of bimolecular quenching rate constants for very different proteins have been reported. These values were compared to quenching for N-acetyltryptophanamide (NATA). Oxygen was used as a quencher of tryptophan residues within protein matrixes (Table 8) [154–158]. The access of oxygen to indole within peptides was higher than observed in most proteins, and because the rate constant was almost diffusion controlled, it can be assumed that peptides provide no barrier for oxygen. The rate constant for most proteins was smaller than observed for peptides, suggesting that oxygen diffusion through the protein matrix was hindered. The addition of guanidine hydrochloride, which denatures the protein, led to an increase of the oxygen quenching efficiency by a factor of over 2. Results from different laboratories using steady-state or time-resolved methods can be compared for the quenching of indole in nuclease and RNase T_1. The agreement is very good for the latter, indicating that only dynamic quenching occurred. The agreement is not good for nuclease and the higher quenching rate constant recovered from the steady-state method could indicate the presence of some static quenching. The range of rate constants for acrylamide quenching (Table 9) [159–163] is larger than for oxygen. For example, acrylamide did not have access to indole incorporated into asparaginase, a protein that is known to have a buried tryptophan residue [155], whereas the quenching efficiency by oxygen was only diminished by a factor of 10 when compared to the peptides. This result indicates that the mobility of a small molecule such as oxygen cannot be affected by the protein matrix, whereas the mobility of the larger acrylamide can be dramatically affected by the protein's structure. Using RNase T_1, James et al. observed values of k_q(eff) from $(5.4 \pm 0.8) \times 10^6 \ M^{-1} \ s^{-1}$ for Cs^+ and $(7.6 \pm 0.8) \times 10^6 \ M^{-1} \ s^{-1}$ for I^- to $\sim 210 \times 10^6 \ M^{-1} \ s^{-1}$ for acrylamide and trichloroethanol [165]. This suggests that charged species have restricted access to some sites within the protein matrix.

D. Quencher Association and Dissociation Rate Constants

The values for the association (k_{Q+}) and dissociation (k_{Q-}) rate constants of the quenchers can be determined when the conditions discussed in Section II are met. Most of the studies in micelles have been based on the model that leads to Eq. (8), where rate constants are recovered from a four-parameter fit of the fluorescence decay in the presence of quencher. The assumptions of this basic model have been discussed in Section II. This model and inclusion of additional processes, such as probe and quencher migration, have been employed for over

Table 8 Quenching by Oxygen of Intrinsic Tryptophan Residues in Proteins and Polypeptides

Protein	Probe	$k_q(\text{eff})$ $(10^9\ M^{-1}\ s^{-1})$	$k_{q0}/k_q(\text{eff})^d$	Ref.
N-Acetyl-L-tryptophanamide	Indole (23°C)	13.6[a]	1	154
Gly-Trp-Gly	Indole (23°C)	10[a]	1.4	154
Leu-Trp-Leu	Indole (23'C)	10[a]	1.4	154
Glu-Trp-Glu	Indole (23°C)	9.8[a]	1.4	154
Lys-Trp-Lys	Indole (23°C)	10[a]	1.4	154
Aldolase	Indole (25°C)	2.7[a]	5.0	155
Asparaginase	Indole (25°C)	1.0	14	156
Azurin	Indole (25°C)	3.0[a]	4.5	155
BSA	Indole	2.4[a]	5.7	155
Carboxypeptidase A	Indole (20°C)	3.8[a]	3.6	155
Chorionic gonadotropin	Indole (25°C)	5.6[a]	2.4	154
α-Chymotrypsin	Indole	2.0[a]	6.8	155
Corticotropin	Indole (25°C)	8.5[b]	1.6	157
Cosyntropin	Indole (23°C)	7.5[a]	1.8	154
Edestin	Indole (25°C)	2.5[a]	5.4	155
Gastrin	Indole (23°C)	11[a]	1.2	154
Glucagon	Indole (23°C)	8.8[a]	1.5	154
HSA	Indole (30°C)	2.3[a]	5.9	154
HSA, [Guanidine Hydrochloride] = 6M	Indole (30°C)	5.2[a,c]	2.6	154
IgG	Indole (25°C)	4.6[a]	3.0	155
LADH	TRP-314 (25°C)	0.5	30	158
Lysozyme	Indole (25°C)	4.8[a]	2.8	155
Melittin, [NaCl] = 0 M (monomer)	Indole (25°C)	11.0[a]	1.2	154
Melittin, [NaCl] = 0.15 M	Indole (25°C)	10.8[a]	1.3	154
Melittin, [NaCl] = 2.4 M (tetramer)	Indole (25°C)	8.2[a]	1.7	154
Monellin	Indole (25°C)	7.7[a]	1.8	154
Monellin	Indole (25°C)	5.4[b]	2.5	156
Myelin basic protein	Indole (20°C)	5.9[a]	2.3	154
Nuclease	Indole (25°C)	4.2[a]	3.2	154
Nuclease	Indole (20°C)	2.7[b]	5.0	156
Pepsin	Indole (25°C)	5.2[a]	2.6	155
Phage fd	Indole (25°C)	2.5[b]	5.4	156

Table 8 Continued

Protein	Probe	$k_q(\text{eff})$ $(10^9\,M^{-1}\,s^{-1})$	$k_{q0}/k_q(\text{eff})^d$	Ref.
Parvalbumin (Cod)	Indole (25°C)	2.7[b]	5.0	156
Ribonuclease	Indole (25°C)	7.1[a]	1.9	155
RNase T_1	Indole (25°C)	2.1[a]	6.5	154
RNase T_1	Indole (25°C)	1.9[b]	7.2	156
Trypsin	Indole (25°C)	5.4[a]	2.5	155
Trypsinogen	Indole (25°C)	4.3[a]	3.2	155

[a]Steady-state measurements.
[b]Time-resolved measurements.
[c]Corrected for the effect of Guanidine hydrochloride on oxygen quenching [155].
[d]Assuming aqueous tryptophan has the same oxygen quenching rate constant as NATA.

a decade and several reviews have been written [63,64]. Our emphasis will be on reviewing the data related to the association and dissociation processes of quenchers from micelles. There are a few examples in which Eq. (8) has been applied to other supramolecular systems, and these will be mentioned at the end of this section.

Equations (9) – (12) describe the parameters for the fluorescence fit with respect to the mechanistic scheme shown in Fig. 1. However, fairly early, it was observed that parameter D varied with micelle concentrations [65,167], a fact that is not consistent with the mechanism shown in Fig. 1. In our view, the interpretation of this particular dependence is not yet resolved, and at least two possibilities to explain these results exist.

One mechanism proposed was that, in addition to the exit of the quencher into the aqueous phase, the quencher could change micelles through the collisions between two micelles. As this process is bimolecular and involves the collision of micelles, the rate of exchange increases linearly with micelle concentration. In order to quantify this exchange, an additional term has to be added for the equation defining parameter D which includes the exchange rate constant (k_e) and concentration of micelles ([H]) [167]:

$$D = k_{qh} + k_{Q-} + k_e[H] \qquad (13)$$

The exchange mechanism has been occasionally disputed [168], and recently a competing model which claims that the change in parameter D is due to changes in the micellar surface potential has been presented [169,170]. The dynamics of N-ethylpyridinium ion and Cu^{2+} used as quenchers for pyrene in SDS micelles was studied for various salt and micelle concentrations [169]. Data were analyzed using Eq. (8). The dissociation rate constant for the quenchers were shown to be strongly dependent on the salt concentration and moderately

Table 9 Quenching of Indole Residues of Proteins by Acrylamide or Iodide

Quencher	Protein	k_q(eff) $(10^9\ M^{-1}\ s^{-1})$	k_{q0}/k_q(eff)[d]	Ref.
Acrylamide	ACTH	3.5,[b] 4.2	1.7, 1.4	159, 160
Acrylamide	Asparaginase	≤0.01	High	163
Acrylamide	HSA	0.6	9.7	163
Acrylamide	LADH (TRP-15)	1.2	4.8	161
Acrylamide	LADH (TRP-314)	~0.01	~600	161
Acrylamide	Melittin (monomer)	3.2[c]	1.8	159
Acrylamide	Melittin (tetramer)	1.3[c]	4.5	159
Acrylamide	Nuclease	0.8[a]	7.3	159
Acrylamide	RNase T$_1$	0.2[a]	30	159, 165
Acrylamide	Glucagon	3.7	1.6	162
Acrylamide	Monellin	2.0	2.9	162
Acrylamide	β-Trypsin	1.1	5.3	162
I$^-$	Carboxypeptidase A	0.32[a]	6.3	155
I$^-$	Carbonic anhydrase	0.19[a]	10	155
I$^-$	Pepsin	2.1–1.4[a,b]	1.1–1.4	155
I$^-$	Trypsinogen	0.34[a]	5.9	155
I$^-$	LADH (TRP-15)	1.7	1.2	161
I$^-$	LADH (TRP-314)	~0	High	161
I$^-$	RNase T$_1$	0.0076 ± 0.0008[c]	260	165
Succinimide	ACTH	1.8[c]	2.2	159
Succinimide	Glucagon	1.7[c]	2.3	159
Succinimide	Melittin (monomer)	1.9[c]	2.1	159
Succinimide	Melittin (tetramer)	0.48[c]	8.1	159
Succinimide	Monellin	0.94[a]	4.1	159
Succinimide	Nuclease	0.22[a]	18	159
Succinimide	Parvalbumin	0.02[c]	200	159
Succinimide	Phage fd	0.27[a]	14	159
Succinimide	RNase T$_1$	0.04[c]	100	159

[a]Steady-state experiments.
[b]Downward curvature in Stern–Volmer plot.
[c]Time-resolved experiment.
[d]Calculated assuming k_q(eff) = 5.8 × 10^9 M^{-1} s^{-1} and 3.9 × 10^9 M^{-1} s^{-1} for NATA in aqueous solution and quenched by acrylamide and succinimide, respectively [159], or calculated assuming k_q(eff) = 2 × 10^9 M^{-1} s^{-1} for iodide quenching [164].

dependent on the micelle concentration. The salt dependence led the authors to propose that k_{Q-} is dependent on the micellar surface potential and that exchange is not important. The same experimental approach and conclusions have been reported for CTAC micelles [170].

Because the two competing mechanisms exist, without a complete resolution of the issue of why parameter D varies with micelle and salt concentration, we have adopted a strategy which allows us to compare k_{Q+} and k_{Q-} values with a certain confidence. We compared rate constants obtained at low surfactant concentrations. These experimental conditions should yield k_{Q-} values for which the value of parameter D was not significantly affected by exchange or changes in the micellar potential. The values for the exchange rate constants have been included in the tables when this mechanism was assumed. All the data being presented are derived from the application of Eq. (8) and the mechanism in Fig. 1, including cases where exchange between micelles was considered. Our discussion centers on the association and dissociation processes, and further details or more extensive discussion on the applicability of the model can be found in another review [63] and the references cited within.

Quenching processes involving micelles formed from ionic surfactants can be divided into groups with quenchers that are neutral and those that have the same or opposite charge relative to the surfactant. For any of these quenchers, k_{Q+} and k_{Q-} values can only be obtained if the quencher partitions between the micellar and aqueous phases. If the exit of the quencher expressed as k_{Q-} is much smaller than k_h, only the intramicellar quenching rate constant (k_{qh}) can be recovered. To ensure a partition between the two phases, most neutral quenchers are polar molecules. Representative examples are shown in Table 10 [60,61,171,172]. For all quenchers, the association process was diffusion controlled. The values for k_{Q+} and k_{Q-} agreed well for different reports, but the same was not true for k_{qh}. When 1-methylpyrene was used as a probe, the recovered k_{Q+} and k_{Q-} values were the same as for pyrene. This indicates that both pyrene derivatives were located in the same environment in the micelle, since the accessibility of the quencher was not altered. The binding site for pyrene derivatives is believed to be the Stern layer of the micelles [29,30]. Changing the quencher from m-DCB to p-cyanotoluene did not alter the entry or exit rate constants (Table 10). An increase in k_{Q-} was observed with increasing micelle or salt concentrations [171]. In the latter case, it was known that the micelles increase in size. In contrast to the dissociation process, the values of k_{Q+} decreased with an increase in micellar size, as expected for a diffusion-controlled process. In line with micellar growth, the values for k_{qh}, the intramicellar quenching rate constant, decreased with the addition of NaCl.

Alkyl iodides with different chain lengths were employed as quenchers for pyrene fluorescence in SDS (Table 10) [172]. There was a moderate effect

Table 10 Dynamics of Neutral Quenchers in SDS Micelles[a]

[SDS] (M)	Probe	Quencher	k_{qh} (10^9 s^{-1})	k_{Q+} (10^9 M^{-1} s^{-1})	k_{Q-} (10^6 s^{-1})	Ref.
0.02–0.08	Pyrene	m-DCB	0.027	11.5	7.6	61
0.02–0.08	Pyrene	m-DCB	0.044[b]	9.8	6.3	171
0.02–0.08	1-Methylpyrene	m-DCB	0.035[b]	10.5	5	171
	Pyrene	p-Cyanotoluene	0.075		6.1	61
0.040	Pyrene	Ethyl iodide	2.9	9.7	8.3	172
0.040	Pyrene	Butyl iodide	4.8	8.8	1.4	172
0.040	Pyrene	Hexyl iodide	5.4	7.8	0.75	172
0.040	Pyrene	Octyl iodide	5.9	6.6	0.4	172
0.1	Pyrene	Methylene iodide	0.75[c]	25	9.5	60

[a]No k_e was determined.
[b]30% errors.
[c]25% errors.

of the chain length on the association rate constant, which was close to a diffusion-controlled process. In contrast, the dissociation rate constant of alkyl iodides decreased by over an order of magnitude between ethyl and octyl iodide, suggesting that the interaction of the alkyl chain with the surfactants decreased the ability to exit into the aqueous phase. It is interesting to note that for chain lengths longer than four, the intramicellar mobility was fairly constant. Methylene iodide was a much more mobile molecule than most alkyl iodides with the exception of ethyliodide (Table 10) [60]. This indicates that the overall size of the quencher and the presence of hydrophobic portions that can interact with the surfactants are dominant factors that define the dissociation rate constants.

The association and dissociation rate constants for m-DCB from CTAC micelles ([CTAC] = 0.031 M) probed by the quenching of 1-methylpyrene were respectively 1.4×10^{10} M^{-1} s^{-1} and 8×10^{6} s^{-1} [173]. These values were similar to those observed for SDS, showing that the location of pyrenes and the mobility of neutral quenchers was not dramatically affected by the different charges of these surfactants.

The quenching methodology was initially developed to study the quenching behavior of counterions and has been used extensively in this area. In the case of SDS mono-, di-, and trivalent cations were employed (Table 11) [167,169,174]. Monovalent metal cations are very mobile, the association rate constants being diffusion controlled and the dissociation constant being at least one order of magnitude higher than those observed for neutral quenchers. This is probably the reason for the observation of pseudo-first-order rate constants in the presence of monovalent metal cations. The dissociation of ethylpyridinium chloride was slower than the same process for Ag^{+}, due to the interaction of the hydrophobic portion of this molecule with the surfactants. The mobility of most divalent cations were similar. The quenching by Cu^{2+} has been extensively studied and most of the model developments have been tested using this quencher. For constant and low SDS concentrations (~ 0.1 M or 0.025 M in Table 11; for additional values, see references cited within the table), the values recovered for k_{Q-} were similar and did not depend on whether pyrene or 1-methylpyrene was used as the fluorescent probe, or if quencher exchange was taken into account. It is important that the values determined in different laboratories, which are being employed to substantiate different models, are the same for similar experimental conditions. Trivalent cations are tightly bound to SDS micelles, and the dissociation rate constant was much smaller than the decay rate constant for excited singlet pyrene. For this reason, the fluorescence quenching methodology cannot be employed to determine the dynamics of these ions.

Iodide was used as a quencher that has an opposite charge to the alkyltrimethylammonium chloride surfactants. The dissociation rate constants for sur-

Table 11 Mobility of Cationic Quenchers in SDS Micelles

[Monomer] (M)	Quencher	Probe	k_{qh} (10^9 s^{-1})	k_{Q+} (10^9 M^{-1} s^{-1})	k_{Q-} (10^6 s^{-1})	k_e (10^9 M^{-1} s^{-1})	Ref.
0.1	Ag$^+$	Pyrene	0.016[a]	≤16	8.0	8.0[b]	167
0.080, [NaCl] = 0 M	N-Ethylpyridinium chloride	Pyrene	0.056		1.7		169
0.082, [NaCl] = 0.082 M	N-Ethylpyridinium chloride	Pyrene	0.040		3.9		169
0.1	Co^{2+}	Pyrene	0.0054[a]	1	0.1	0.7	167
0.025	Cu^{2+}	Pyrene	0.0349 ± 0.0006		0.07 ± 0.17		174
0.100	Cu^{2+}	Pyrene	0.0284 ± 0.0007		0.70 ± 0.06		174
0.021	Cu^{2+}	Pyrene	0.044		0.1		169
0.085	Cu^{2+}	Pyrene	0.043		1.1		169
0.1	Cu^{2+}	1-Me Pyrene	0.027[a]	1.2	0.12	0.6	167
0.1	Ni^{2+}	Pyrene	0.0089[a]	1	0.1	0.5	167
0.1	Pb^{2+}	Pyrene	0.0091[a]	7	0.3	0.3[b]	167
0.1	Cr^{3+}	Pyrene	0.0098[a]	<1	<0.1	<0.1	167
0.1	Eu^{3+}	Pyrene	0.016[a]	<1	<0.1	<0.1	167

[a]25% errors.
[b]Large errors.

factants which differ on the length of the alkyl chain are presented in Table 12 [175,176]. The dissociation rate constants for iodide from dodecyltrimethyl-ammonium chloride (DTAC) or tetradecyltrimethylammonium chloride (TTAC) were somewhat higher than those observed for cupric ions from SDS micelles [101,169] in the same micellar concentration range [$(3.1 - 7) \times 10^{-4}$ M in Table 12]. The dissociation rate constants are significantly lower for the larger micelle, CTAC, suggesting that the interaction of the counterions was dependent on the micellar size. This comparison should be made with caution, because the experiments were not obtained for exactly the same experimental conditions. However, judging by the good agreement found for data obtained in different laboratories using SDS micelles, we feel that the difference observed for DTAC/TTAC and CTAC is relevant.

The quenching of pyrene derivatives in alkyltrimethylammonium chloride micelles by alkylpyridinium chloride is an example where the surfactant and quencher have the same charge (Table 13) [177]. When the alkyl chain was longer or equal to 14 carbons, the quencher did not exit the micelle during the lifetime of excited singlet pyrene. This fact was observed for surfactants with different chain lengths. The experiments with DTAC and TTAC were done using 1-methylpyrene as the probe, whereas pyrene was employed to study CTAC. Based on the observations that the quenching of pyrene and 1-methyl-pyrene yield the same quencher dissociation/association rate constants, for SDS as well as CTAC, we assume that the results reported in Table 13 can be compared. The intramicellar quenching rate constant decreased with the increased size of the micelle. This observation was expected because the volume explored by the probe and quencher before encountering a collisional partner increased with an increase in micellar volume. As the alkyl chain of the surfactant increased, both the values for k_{Q+} and k_{Q-} decreased to approximately the same extent, suggesting that the partitioning coefficient did not change substantially for all cases but TTAC and $C_{12}Py^+$.

Alcohol molecules such as 1-butanol are water soluble and can interact with the Stern layer of the micelle. The alcohol can replace a water molecule in the Stern layer and hence decrease the amount of shielding between the charged headgroups. The increased electrostatic repulsion leads to a greater area per headgroup. For this reason, as 1-butanol is added to DTAC, the aggregation number decreases from 48 to 24 with the addition of 0.85 M alcohol [64]. Using cationic C_nPyC type quenchers, the effect of 1-butanol was observed in both SDS and DTAC micelles. At 1-butanol concentrations higher than 1 M, migration of the cationic quenchers ($C_{10}PyC$ and $C_{14}PyC$) was observed in SDS, whereas at lower alcohol concentrations, no mobility was seen. At large concentrations of alcohol (> 2 M), the exit rates of quenchers from both SDS and DTAC micelles were similar. Thus, the fragmentation–coagulation mechanism of transfer was postulated [64,73]. This mechanism allows the probe or quencher to be transferred between micelles without any dependence on the

Table 12 Dynamics of Iodide Used as a Quencher in Alkyltrimethylammonium Chloride Micelles

Surfactant	[Monomer] (M)	Probe	$k_{\phi h}$ $(10^9\ s^{-1})$	k_{Q+} $(10^9\ M^{-1}\ s^{-1})$	k_{Q-} $(10^6\ s^{-1})$	k_e $(10^9\ M^{-1}\ s^{-1})$	Ref.
DTAC[a]	0.05	Pyrene	0.0053	50	2.4	0.61–0.94	175
TTAC[b]	0.052	Me-pyrene	0.0161		2.9	0[d]	176
CTAC[c]	0.040	Me-pyrene	0.0077		0.98	0[d]	176

[a]cmc = 0.021 M.
[b]cmc = 0.005 M.
[c]cmc = 0.001 M.
[d]Analysis precluded exchange of quencher by micellar mechanisms.

Table 13 Mobility of Alkylpyridinium Chloride Used as a Quencher in Alkyltrimethylammonium Chloride Micelles

Surfactant	[Monomer] (M)	Probe	Quencher	k_{qh} (10^9 s^-1)	k_{Q+} (10^9 M^-1 s^-1)	k_{Q-} (10^6 s^-1)
DTAC	0.1	Me-pyrene	$C_{10}Py^+$	0.043	8.9	3.09
TTAC	0.1	Me-pyrene	$C_{10}Py^+$	0.0221	8.0	2.42
CTAC	0.361	Pyrene	$C_{10}Py^+$	0.018	5.8	1.81
DTAC	0.1	Me-pyrene	$C_{12}Py^+$	0.0416	2.9	0.49
TTAC	0.1	Me-pyrene	$C_{12}Py^+$	0.0226	1.7	0.17
CTAC	0.2	Pyrene	$C_{12}Py^+$	0.0114	1.4	0.25

Source: Ref. 177.

nature of the probe or quencher. It was postulated that a micelle could split up into submicelles which might contain a probe or quencher molecule and then recombine with other micelles.

The possibility of quencher migration has been taken into account in several of the experiments described above. However, for long-lived excited singlet states such as pyrene, the possibility of probe migration between micelles has also been investigated. The possibility of probe migration between CTAC micelles has been tested for 1-methylpyrene and pyrenesulfonate [67]. The quencher was the immobile tetradecylpyridinium chloride. The fluorescence decays were fitted to equations derived for competing models and global compartmental analysis with reference convolution was employed to analyze the fluorescence decay. No migration was observed for the more hydrophobic methylpyrene, whereas probe migration, which showed a dependence on micelle concentration, was observed for pyrenesulfonate. The same concept has also been applied to N-benzyl-N,N-dimethyltetradecylammonium chloride reverse micelles, and the migration rate was dependent on the water/surfactant ratio (R) [68].

The model described by Eq. (8) has also been applied for the quenching of pyrene incorporated into sodium taurocholate aggregates in the presence of 1.0 M NaCl [118]. The recovered association and dissociation rate constants for N,N-dimethylaniline were respectively 1.4×10^9 M^{-1} s^{-1} and 3.8×10^6 s^{-1}. These values are of the same order of magnitude as obtained for quenching of pyrene in SDS at the same concentration of salt. These results suggested that the access to hydrophobic probes in bile salt aggregates was similar to that in conventional micelles.

The association and dissociation rate constants for hydronium ions to cyclodextrin cavities has been studied by measuring the fluorescence of 2-naphthol [178]. The pKa of the excited single state of this molecule is much lower than for the ground state and the deprotonated species has different absorption and emission spectra. This spectroscopic signature was explored to characterize the mobility of H$^+$. Upon complexation to the cyclodextrin, the acidity of the naphthol decreased. In addition, excited state deprotonation was decreased in the presence of α-cyclodextrin due to the formation of a complex with 1 : 2 naphthol : cyclodextrin stoichiometry. Thus, the cyclodextrin environment precluded the exit of the hydronium ion from the cavity.

For proteins, the fluorescence of porphyrin moieties with myoglobin and hemoglobin can be employed to obtain information on oxygen mobility. The oxygen dissociation rate constants were found to be $(1.1 \pm 0.2) \times 10^8$ s^{-1} for myoglobin, and $(2.6 \pm 0.5) \times 10^7$ s^{-1} for hemoglobin. The association rate constants determined using phase-modulation spectroscopy were $(3.8 \pm 0.2) \times 10^8$ M^{-1} s^{-1} and $(6.7 \pm 0.2) \times 10^8$ M^{-1} s^{-1} for myoglobin and hemoglobin, respectively [166].

E. Conclusions

Fluorescence probe molecules have found widespread use in the characterization of supramolecular systems because fluorescence is an established technique which provides excellent signal-to-noise ratios. Even the time-resolved single photon counting systems can now be seen as routine, at least in the nanosecond time domain, and, for this reason, are accessible in many research laboratories. Fluorescent probe molecules have been in the forefront of the development of methodology to characterize supramolecular systems, and the understanding of association and dissociation processes has been no exception. The current discussion on suitable models to describe mobility in micelles and further development of more sophisticated models attests that fluorescence still has a major role to play in the field of supramolecular chemistry, and we expect that it will continue to be employed extensively. The intrinsic characteristic of excited singlet states of organic molecules is that they have relatively short lifetimes, in the range of tenths of nanoseconds. This is an advantage when it is desired that the excited probe should not move from its complexation site during its excited state lifetime. This property is very useful when characterizing the local microenvironment, but with respect to the association/dissociation process, it has the disadvantage that no direct studies of mobility can be performed. Thus, dynamics is inferred from quenching studies that are always based on the assumption of mechanisms.

IV. EXCITED TRIPLET STATE PROBE MOLECULES

The use of excited triplet states as probes occurred in concert with the development of fluorescent probe methodologies but have not been employed to the same extent. Phosphorescence quantum yields in solution are generally much lower than fluorescence quantum yields, and when the emission of triplet states is measured, the signal-to-noise advantage of the fluorescent probes is lost. In most cases, triplet excited states are followed by their absorption spectra using laser flash photolysis. A second reason for the scarcer reports on the use of excited triplet states probes in supramolecular systems is the fact that the laser flash photolysis technique has not been as widespread as fluorescence techniques. This situation has changed over the past decade and we expect that the number of studies which employ excited triplet states will increase.

The advantage of using excited triplet states to probe dynamic processes in supramolecular structures is that their longer lifetimes make it possible to monitor the mobility of probes by direct spectroscopic methods. In addition, the mathematical treatments for analysis of quenching experiments are simpler than those used for excited singlet species. The main disadvantage of using triplet states is the difficulty of performing precise measurements and the ease with which quenching by impurities occurs.

A. Direct Spectroscopic Measurements of Association/Dissociation of Probes to/from Supramolecular Systems

The conditions for probes to be suitable for direct spectroscopic studies have been discussed in Section II.A. The mobility of the probe between two compartments (i.e., the homogenous phase and the complexation site in the supramolecular system) competes with the decay of the excited triplet states in those compartments. This competition is only possible if the probe's excited state lifetime is either very much longer or of the order of magnitude of the rates for association and dissociation from the supramolecular system. Conceptually, the mechanism is analogous to the formation of excimers and the observation of monomer and excimer emission. The solution for the kinetic equations has been described for excimers several decades ago [179] and can be directly applied to the decay for triplet states in the presence of an organized system. This method assumes that at a particular wavelength the molar absorptivity of the free (P) and complexed triplet state (HP) are different. The absorbance is thus related to the sum of the probe concentrations:

$$\Delta A \; \alpha \; [P*] \; + \; [HP*] \tag{14}$$

The decay is fitted to the sum of two exponentials where the preexponential factors a_1 and a_2 are related to the initial concentrations of P* and HP* and to k_{P+} and k_{P-}.

$$\Delta A = a_1 \exp(-\gamma_1 t) + a_2 \exp(-\gamma_2 t) \tag{15}$$

When using laser flash photolysis, the preexponential factors are generally not analyzed as a function of concentrations and rate constants. One of the difficulties in obtaining quantitative data from the preexponential factors is that the molar absorptivities for P* and HP* have to be known. The latter parameter is very difficult to determine unless the equilibrium constants for the excited state is known or all of the excited state probe can be bound to the supramolecular system.

The exponential factors, γ_1 and γ_2, are given by an equation in which the three parameters A, B, and C are defined by the rate constants shown in Fig. 1:

$$\gamma_{1,2} = -\frac{1}{2}\Big[(A + B) \pm \sqrt{(A - B)^2 + 4C}\Big] \tag{16}$$

$$A = k_0 + k_{P+}\,[H] \tag{17}$$

$$B = k_h + k_{P-}$$ (18)

$$C = k_{P-} \, k_{P+} \, [H]$$ (19)

The values of the exponential factors ($\gamma_{1,2}$) are determined at different concentrations of the supramolecular system ([H]), and the rate constants can be obtained from the dependence of the sum and the products of the exponential factors γ, which are given by [180]

$$\gamma_1 + \gamma_2 = k_0 + k_h + k_{P-} + k_{P+} \, [H]$$ (20)

$$\gamma_1 \, \gamma_2 = k_0 \, (k_{P-} + k_h) + k_h \, k_{P+} \, [H]$$ (21)

Equations (20) and (21) are applicable when the rate constants for the decay are of the same order of magnitude (within a factor of 10) as the rates for the association and dissociation processes.

When the decay of bound and free excited probe is much slower than the association and dissociation processes, equations for $\gamma_{1,2}$ can be simplified. For this condition, the inequalities $k_0 \ll k_{P+} \, [H]$ and $k_h \ll k_{P-}$ are valid, and the expressions for γ_1 and γ_2 are reduced to

$$\gamma_1 = k_{P-} + k_{P+} \, [H]$$ (22)

$$\gamma_2 = 0$$ (23)

Under these conditions, a decay or growth is observed that levels off at a constant value. For this leveling off to occur, the values of k_0 and k_h should be smaller by a factor of at least 10^3 in the inequalities shown above.

There is a regime which falls in between the two extreme possibilities just discussed. This regime results when the decay rate constants are not small enough to lead to a leveling off, but the difference between k_0 and k_h with respect to k_{P-} and $k_{P+} \, [H]$ is sufficiently large (about a factor of 100) that the conditions $\gamma_1 \gg \gamma_2$ and $k_{P-} + k_{P+} \, [H] \gg k_0 + k_h$ apply. In this case, the decay data are fitted to the sum of two exponentials, and γ_1 can be expressed by Eq. (22). This is obviously an approximate method and is only applicable when γ_2 is of the order of magnitude of the errors associated with γ_1.

In most cases when laser flash photolysis has been employed to follow the decay of triplets, the data were analyzed for single decay curves. The precision of the recovered values for γ_1 and γ_2 as well as a detailed analysis of the preexponential factors in Eq. (15) would probably improve if more complex analysis procedures, such as global compartmental analysis, would be employed. However, it is important to note that the dynamic range of absorbance measurements for laser flash photolysis experiments is much smaller than can be obtained for single photon counting, thereby limiting the improvement that will be achieved with more sophisticated analysis procedures.

B. Quenching Methodologies

Quenching of excited triplet states that exchange between the homogeneous phase and the supramolecular systems is an indirect method employed to measure the association and dissociation rate constants for the excited probe molecule. The discussion regarding overall quenching rate constants in Section III.B is valid for excited triplet states as well as singlet species. The only relevant aspect that differentiates molecules in their excited triplet from their excited singlet states is their lifetimes. The longer lifetimes of triplet states makes it more difficult to encounter situations where the triplet does not migrate during its excited state lifetime between the homogeneous phase and the supramolecular system.

In addition to comparing overall quenching rate constants, it is also possible to recover the values of the quencher association and dissociation rate constants from quenching experiments. The same model that was employed for fluorescent probes can be employed. This model considered that the probe was immobile. The general solution to this model is given by Eq. (8), which has four parameters defined by the rate constants for the processes described in Fig. 1. However, the experimental results showed that the triplet state decayed by pseudo-first-order kinetics, suggesting that once the quenchers enter the supramolecular system, quenching occurred with an efficiency of unity. Under these conditions, Eq. (5) can be applied. In addition, if the condition that k_{Q+} [H] $>> k_{Q-}$ holds, Eq. (5) can be reduced to

$$k_{obs} = k_0 + \frac{k_{Q-}[Q]}{[H]} \qquad (24)$$

A second model leads to the recovery of the dynamic parameters for the probe. This model requires different quenching efficiencies for the excited probe in the homogeneous phase and the supramolecular structure and was initially developed for the mobility of arenes in micelles [62]. This method has been modified for studies with cyclodextrins in order to include the possibility of quenching of the probe within the supramolecular structure [181]. In the original derivation, it was assumed that the quenching of the complexed probe was negligible [k_q(eff) = 0]. However, for this method to be applicable, it is only necessary that the quenching efficiencies for the probe free in solution (k_{q0}) and complexed to the supramolecular system [k_q(eff)] are different. In most cases described in the literature, aqueous quenchers are employed leading to a higher quenching efficiency in the homogeneous phase. However, the reverse situation should also be possible, although additional complications such as static quenching could be introduced. The key assumption for this method is that the concentration of free probe ([P*]) is small when compared to the concentration of complexed probe [HP*]). With this assumption, it is possible to apply the steady-state condition to the rate law equations, and an expression for the observed pseudo-first-order rate constant can be obtained:

$$k_{\mathrm{obs}} = k_h + k_{P_-} + k_q(\mathrm{eff})[Q] - \frac{k_{P_-}k_{P_+}[H]}{k_{P_+}[H] + k_0 + k_{qo}[Q]} \tag{25}$$

When using this method, it is important that the triplet decay follows pseudo-first-order conditions for all the quencher concentrations employed. If the decay is multiexponential, this method is not applicable.

At high quencher concentrations, the decay of the excited state in the aqueous phase will be very fast and the exit from the supramolecular system will be rate limiting. Mathematically, at high quencher concentrations, the last term in Eq. (25) is negligible, and the observed rate constant is either a constant value corresponding to the sum of k_h and k_{P_-} when $k_q(\mathrm{eff})$ is zero, or a linear dependence with increasing quencher concentration is observed:

$$k_{\mathrm{obs}} = k_h + k_{P_-} + k_q(\mathrm{eff})\ [Q] \tag{26}$$

The values for the various rate constants are recovered from the fit of the experimental data to Eq. (25). The value for [H] is known and the values for k_0, k_h, and k_{q0} can be obtained in independent experiments. Thus, in the fit of k_{obs} versus [Q], three parameters are determined. This method has been employed in several studies involving excited triplet states in micelles [62,182], bile salts [183], and CDs [141,145,181]. However, no detailed discussion regarding the variation of several of the parameters in Eq. (25) has been provided. Figures 6–8 provide simulated results for Eq. (25) when the values for $k_q(\mathrm{eff})$, [H], k_{P_+}, and k_{P_-} are varied. It is important to analyze the data as rate constants and not lifetimes, as in the latter case, the errors are not weighted properly due to the inverse relationship of these variables. An increase in $k_q(\mathrm{eff})$ relative to k_{q0} decreases the curvature at moderate quencher concentrations (Fig. 6), with a corresponding increase in the errors of the recovered parameters. In principle, the value for $k_h + k_{P_-}$ can be determined from the intercept of the extrapolation of the linear region at high quencher concentrations [cf. Eq. (26)]. When collecting data, it is important to use high enough quencher concentrations to define this linear region well. It is worth noting that when the difference between the quenching efficiencies is about 1000, one can consider that the quenching for the probe complexed to the supramolecular system is negligible.

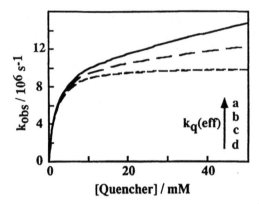

Figure 6 Simulation of the dependence of k_{obs} as defined in Eq. (25) with different $k_q(eff)$ values. The parameters kept constant for Eq. (25) are $k_h = 2.5 \times 10^4$ s^{-1}, $k_0 = 4 \times 10^4$ s^{-1}, $k_{q0} = 5 \times 10^9$ M^{-1} s^{-1}, $k_{P+} = 6 \times 10^8$ M^{-1} s^{-1}, $k_{P-} = 1 \times 10^7$ s^{-1}, and [H] = 0.01 M. The values for $k_q(eff)$ are (a) 1×10^8 M^{-1} s^{-1}, (b) 5×10^7 M^{-1} s^{-1}, (c) 1×10^6 M^{-1} s^{-1}, and (d) zero.

This model assumes that the concentration of probe is small when compared to that of the supramolecular system. At increasing concentrations of the supramolecular system, the initial slope decreases since a higher fraction of probe molecules will be complexed (Fig. 7). Analysis of Fig. 7 shows that in order to obtain data for the linear region at high quencher concentrations and

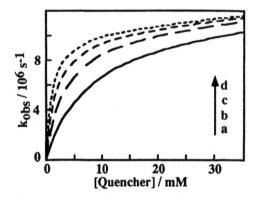

Figure 7 Simulation of the dependence of k_{obs} as defined in Eq. (25) with different concentrations of host. The parameters kept constant for Eq. (25) are $k_h = 2.5 \times 10^4$ s^{-1}, $k_0 = 4 \times 10^4$ s^{-1}, $k_{q0} = 5 \times 10^9$ M^{-1} s^{-1}, $k_q(eff) = 5 \times 10^7$ M^{-1} s^{-1}, $k_{P+} = 6 \times 10^8$ M^{-1} s^{-1}, and $k_{P-} = 1 \times 10^7$ s^{-1}. The values for [H] are (a) 0.05 M, (b) 0.02 M, (c) 0.01 M, and (d) 0.005 M.

enough points in the curved dependence of k_{obs} with quencher concentration, it is not always ideal to work at the highest concentration of the supramolecular system.

Figure 8 shows the effect of varying the values for k_{P+} and k_{P-}. These parameters are intrinsic to the system being studied and cannot be adjusted by changing experimental conditions as in the case for $k_q(eff)$, by the choice of an appropriate quencher, or the concentration of the supramolecular system. However, this Fig. 8 shows the limitations of this method. When the association rate is much faster than the dissociation process, information on k_{p-} is lost (Fig. 8A). For low k_{P+} values, the quencher concentration can be lowered and the parameters related to dynamics can still be recovered. The determination of the association rate constant can to some extent be fine-tuned by changing the concentration of the supramolecular system, but the condition [H] >> [P] always

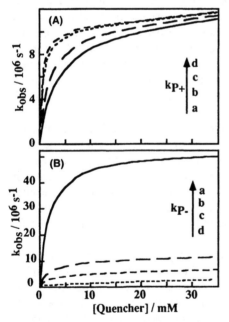

Figure 8 Simulation of the dependence of k_{obs} as defined in Eq. (25) with different (A) k_{P+} and (B) k_{P-} values. The parameters kept constant for Eq. (25) are $k_h = 2.5 \times 10^4$ s^{-1}, $k_0 = 4 \times 10^4$ s^{-1}, $k_{q0} = 5 \times 10^9$ M^{-1} s^{-1}, $k_q(eff) = 5 \times 10^7$ M^{-1} s^{-1}, and [H] $= 0.01$ M. The values for k_{P+} and k_{P-} are as follows: (A) $k_{P-} = 1 \times 10^7$ s^{-1} and k_{P+} values are (a) 1.2×10^9 M^{-1} s^{-1}, (b) 6×10^8 M^{-1} s^{-1}, (c) 1.2×10^8 M^{-1} s^{-1}, and (d) 6×10^7 M^{-1} s^{-1}; (B) $k_{P+} = 6 \times 10^8$ M^{-1} s^{-1}, and k_{P-} values are (a) 5×10^7 s^{-1}, (b) 1×10^7 s^{-1}, (c) 5×10^6 s^{-1}, and (d) 1×10^6 s^{-1}.

has to hold. An increased exit rate constant means that the residence time of the probe inside the supramolecular system is reduced when compared to the residence time in the homogeneous phase. Thus, the chances for quenching in the latter phase increases (Fig. 8B). Because the exit rate constant corresponds to a unimolecular process, it will not be affected by changes in concentrations of reactants.

In comparison to singlet states, much simpler models and fitting techniques have been employed for the studies involving excited triplet states. This is undoubtedly a reflection of the techniques being employed and the quality of the data, but it is also related to the fact that the longer lifetime of the triplet state makes it possible to introduce simplifications in the mechanistic scheme.

C. Direct Spectroscopic Studies

The conditions for direct spectroscopic studies have been discussed in Sections II.A and IV.A. These studies are preferred over the quenching methodology because the relocation of the probe can be followed using spectroscopy, and no mechanism has to be assumed a priori. The two conditions to be met for these studies are that the probe should have a property that depends on its location and, upon excitation, a driving force for relocation has to be created. Due to the rarity of the synchronicity of these phenomena, only xanthone has been used to explore the dynamics of probe relocation by a direct spectroscopic study.

The excited triplet state of xanthone has a π,π^* configuration in most solvents which makes this excited state fairly unreactive. In addition, the maximum of the triplet–triplet absorption spectra is dependent on solvent polarity [184]. In carbon tetrachloride, the maximum occurs at 655 nm, whereas in water, it is blue shifted to 580 nm [184,185]. The π,π^* nature of the excited triplet state of xanthone leads to a higher dipole moment for the excited triplet state when compared to the ground state. This is the impetus required for relocation.

Excited triplet xanthone was initially employed as a probe molecule to study the dynamics of surfactant aggregated to polyelectrolytes [186]. Triplet–triplet absorption spectra were obtained 40 ns and 1 μs after the laser pulse, and a blue shift was observed. These two spectra were ascribed respectively to xanthone within the macroassembly and in the aqueous phase. Quenching experiments were carried out where the probe residing in the aqueous phase was quenched, whereas that in the self-assembly was protected. The xanthone dissociation rate constant from the aggregates was 1.1×10^7 s^{-1} [186].

Cyclodextrins form complexes with ground state xanthone. The strength of the complex depends on the cavity size and was largest for β-CD, with an equilibrium constant of 1100 M^{-1} [185,187]. A spectral change was observed when the xanthone/β-CD complex was excited. At 30 ns after the laser pulse, the triplet–triplet absorption maximum was observed at 602 nm, but at 500 ns, the maximum shifted to 580 nm [185]. This shift was assigned to the relocation of xanthone from the CD cavity to the aqueous phase, due to the higher dipole moment of the excited state. When monitoring the kinetics of triplet xanthone, a fast decay or growth was observed respectively at the red or blue edge of the triplet–triplet absorption spectrum. This fast component was absent when triplet xanthone was formed in water in the absence of CDs and corresponds to the relocation process. In the early experiments, this fast decay was assigned to the dissociation rate constant and its value was determined to be $(1.0 \pm 0.1) \times 10^7$ s^{-1} [185]. This assignment was based on the comparison with the rate constant obtained from quenching studies with cupric ions, where, within experimental error, the same limiting rate constant was observed at high quencher concentrations. However, recent results uncovered a dependence of the rate constant for this fast process with CD concentration and showed that the coincidence of the observed decay for the fast process and the limiting rate constant at high quencher concentrations was fortuitous [187]. A linear dependence of k_{obs} for the fast decay with CD concentrations was observed. Equation (22) was employed to analyze the data, as the decay for triplet xanthone in water was slower by about two orders of magnitude when compared to the association/dissociation dynamics. The association rate constant was much higher for β-CD than for γ-CD, suggesting that the latter CD was too large to accommodate triplet xanthone (Table 14). Hydroxypropyl-β-cyclodextrin (Hp-β-CD) has several of the hydroxyl groups transformed into ether groups. This creates a more hydrophobic cavity, and, consequently, the association rate constant for triplet xanthone was smaller than for β-CD. The dissociation rate constants did not depend on the CD cavity size [187]. The equilibrium constant for triplet xanthone with the CDs can be calculated from the values for k_{P+} and k_{P-} (Table 14). These values are significantly smaller than the equilibrium constants for ground state xanthone, which were determined by fluorescence. This example shows that the equilibrium constants for excited and ground states can vary by

Table 14 Determination of Dissociation and Association Parameters for Xanthone–Cyclodextrin Complexes by a Direct Spectroscopic Technique

CD	K_G $(10^3 \ M^{-1})$	K_T $(10^3 \ M^{-1})$	k_{P-} $(10^6 \ M^{-1})$	k_{P+} $(10^8 \ M^{-1})$
β-	1.1 ± 0.2	0.048 ± 0.013	8.4 ± 0.7	4 ± 1
γ-	0.22 ± 0.03	<0.004	7.3 ± 0.5	$<0.3 \pm 0.3$
Hp-β-	1.8 ± 0.1	0.020 ± 0.004	7.1 ± 0.7	1.4 ± 0.3

Source: From Ref. 187.

orders of magnitude. Of course, xanthone may be an extreme example in which the change in dipole moment could be larger than for other probes, but any extrapolation of excited state behavior by using parameters obtained for the ground state should be avoided.

Quenching studies were also employed to determine the dissociation rate constant of triplet xanthone from CD cavities. These apparent rate constants were obtained by measuring the decay of triplet xanthone at high concentrations of cupric ions. The values for k_{P_-} were respectively $(9 \pm 2) \times 10^6$ s^{-1} and $(15 \pm 2) \times 10^6$ s^{-1} for β- and γ-CDs [185]. The higher value obtained in the quenching experiments for γ-CD indicated that the quencher had access to the probe within the CD cavity. Unfortunately, no detailed analysis employing Eq. (25) was performed and, for this reason, a comprehensive comparison between the two methods is not possible.

Alcohols and alkyl sulfates have been shown to affect the equilibrium constants of probe molecules with CDs [136,139,188–190]. When xanthone was used as a probe, the addition of alcohols led to the formation of "ternary" complexes that had ground state equilibrium constants for β-CD and γ-CD which were respectively significantly smaller and slightly larger than the binary xanthone/CD complexes [190]. In the case of γ-CD, the dissociation rate constants were obtained by directly following the decay of triplet xanthone, taking advantage of the fact that the association process is slow at the CD concentrations employed. The decay was fitted to the sum of three exponentials, which corresponded to the exit from the CD cavity without alcohols (k_{P_-}), the exit from cavities with alcohols ($k_{P_-(ter)}$) and the decay in water (k_0). The dissociation rate constants of triplet xanthone from CD cavities containing alcohols were much smaller than in the absence of alcohols (Table 15). For γ-CD, the same values for $k_{P_-(ter)}$ were obtained when the quenching methodology was employed. In the case of β-CD, only the quenching methodology could be used.

Table 15 Alcohol Effect on Ratios of Exit Rate Constants in Xanthone–CD Complexes

Alcohol	γ-CD $k_{P_-}/k_{P_-(ter)}$	β-CD[a] $k_{P_-}/k_{P_-(ter)}$
1-Butanol	21 ± 2	1.0 ± 0.1
2-Butanol	12 ± 2	13 ± 5
tert-Butanol	17 ± 5	5.9 ± 0.7
1-Pentanol	25 ± 5	21 ± 11
Cyclopentanol	12 ± 2	5.6 ± 0.3
Cyclohexanol	11 ± 1	10 ± 3

[a]Determined from quenching studies.
Source: From Ref. 190.

Although the effect of alcohol addition on the ground state equilibrium constant for β-CD was opposite to that observed for γ-CD, a decrease of the dissociation rate constant for triplet xanthone was also observed. The slow down of the dissociation process has been tentatively attributed to the preferential solvation of the CD cavity entrances by alcohols [190].

In conclusion, these experiments with xanthone as a probe show the potential of using direct spectroscopic studies. However, at this point, this methodology is limited to the use of xanthone and only when it is incorporated into an environment that leads to significant shifts of its triplet–triplet absorption spectrum. This methodology will be more broadly applicable when other probes which fulfill the requirements for direct studies are discovered and characterized.

D. Quenching Studies

Most of the reported data involving triplet states were obtained using quenching processes to recover the association/dissociation rate constants for the excited state triplet probes and/or quenchers. As in the case of singlets, overall quenching rate constants can yield limited information on accessibility, and representative examples have been included where relevant.

1. Micelles

Phosphorescence was employed to study the dynamics of polycyclic aromatic hydrocarbons with SDS micelles [62]. Equation (25) was applied with the assumption that k_{qh} is negligible. In the case of 1-bromonaphthalene, the val-

Table 16 Exit Rate Constants for Polycyclic Aromatic Hydrocarbon Probes Bound to SDS Micelles

Probe	Quencher	$k_{P_-}{}^a$ (10^6 s^{-1})	$k_{P_-}{}^b$ (10^6 s^{-1})	Ref.
Benzene	NO_2^-	4.4		62
Toluene	NO_2^-	1.3		62
p-Xylene	NO_2^-	0.44		62
Naphthalene	NO_2^-	0.25	>0.05	62
Anthracene	NO_2^-	0.017		62
1-Methylnaphthalene	NO_2^-		>0.05	62
1-Bromonaphthalene[c]	NO_2^-	0.033	0.025	62
1-Chloronaphthalene	NO_2^-	0.043		191
Pyrene	NO_2^-	0.0041		62

[a]Calculated assuming $k_{Q+} = 7 \times 10^9 \ M^{-1} \text{ s}^{-1}$.
[b]Determined from curve fit of phosphorescence data.
[c] $k_{P+} = (5–8) \times 10^9 \ M^{-1} \text{ s}^{-1}$.

ues of k_{P+} and k_{P-} could be recovered by using Eq. (25) (Table 16). However, for most other compounds only a lower limit of the dissociation rate constant could be obtained. Estimates for this parameter were obtained by assuming a diffusional association process. The value of k_{P-} was then calculated from the equilibrium constant for the ground state. Because similar values for k_{P-} were obtained in the case of 1-bromonaphthalene, the use of the equilibrium constant for the ground state was adequate in this case. The exit rate constant for the aromatic compounds decreased as the amount of substitution and annulation increased, suggesting a correlation with the hydrophobicity of the probe. The dissociation rate constant for 1-chloronaphthalene was determined in a similar experiment using triplet–triplet annihilation as the quenching process [191]. The exit rate constant is larger for the chloro derivative compared to bromona-phthalene, due to the larger polarity of the chloro compound as well as its smaller size.

Exit rate constants from CTAB micelles were assessed using the same quenching methodology [Eq. (25)]. Most exit rate constants of probes from CTAB are between five and eight times slower than those observed for the dissociation from SDS, but the general trend that the exit rate constants paral-lel solubility in the micellar phase still exists [62]. The slower exit rates could be caused by both the presence of a larger headgroup in CTAB monomers and the larger size of these micelles.

A variation of the quenching method leading to Eq. (25) is to follow the decay of the triplet state from the emission of a luminescent quencher, which is much more intense than the phosphorescence of the probe [192]. In this case, the observed rate constant corresponds to the growth and subsequent decay in the emission profile of the quencher. However, an additional rate constant, corresponding to the emission lifetime of the quencher, has to be included in Eq. (25). The exit rate constant was determined to be 2.5×10^4 s^{-1} for 1-bromonaphthalene when quenched by either Eu^{3+} or Tb^{3+}. This value is the same (Table 16) as that determined using the anionic quencher, NO$_2^-$ [62], showing that this method is useful. However, care should be taken to eliminate the possibility of reverse energy transfer.

The entry/exit rate constants of ketones with SDS micelles was studied by using either a micellar quencher such as γ-methylvalerophenone or nitrite ion as an aqueous quencher (Table 17) [193,194]. Using a micellar quencher, the values of k_{P-} were determined by employing Eq. (27), which incorporates the fraction of micelles containing a probe molecule (F_{occ}):

$$k_{obs} = k_0 + k_{P-} \, F_{occ} \tag{27}$$

In this model, the quencher was assumed to be immobile relative to the mobility of the probe. Quenching was observed to be dependent on the frac-tion of the micelles containing bound γ-methylvalerophenone. Therefore, the probe must enter a micelle that contains at least one quencher molecule. However, it is important that this type of experiment is conducted under conditions of varying micelle concentration because it provides a means to identify static quenching when both probe and quencher reside in the same micelle prior to excitation.

When nitrite was used as a quencher, the values for k_{P+} and k_{P-} were obtained from

$$\frac{1}{k_{obs} - k_0} = \frac{1}{k_{P-}} + \frac{k_{P+}[M]}{k_{P-}k_{P+}[Q]} \tag{28}$$

This equation, as a double reciprocal plot, is similar to Eq. (5) but is applied to probe mobility, not quencher dynamics. The same values for the dissociation rate constants (Table 17) were obtained when employing these two different methods with quenchers in different phases, suggesting that the underlying assumptions for the derivation of Eqs. (27) and (28) were reasonable. The entry rate constants were diffusion controlled, and the exit rate constants varied by a factor of ~ 3. In analogous fashion to the polycyclic aromatic hydrocarbon probes, the exit rate constants were faster for the more polar ketones p-methoxyacetophenone and acetophenone compared to isobutyrophenone or propiophenone [193,194].

The mobility of quencher molecules, instead of probe molecules, has been studied by using a probe, phenanthrene, that resides only in the micelle. To this

Table 17 Overall Mobility Rate Constants for Ketone Probes Bound to SDS

Probe	Quencher	k_{P+} $(10^9\ M^{-1}\ s^{-1})$	k_{P-} $(10^6\ s^{-1})$	Ref.
Acetophenone	γ-methyl-valerophenone		7.7	193
Acetophenone	NO_2^-	16	7.8	194
Benzophenone	NO_2^-	52[a]	2[a]	194
Isobutyrophenone	γ-methyl-valerophenone		1.6	193
Isobutyrophenone	NO_2^-	12	1.6	194
Propiophenone	γ-methyl-valerophenone		3.0	193
Propiophenone	NO_2^-	14	3.0	194
p-Methoxy-acetophenone	NO_2^-	28[b]	9.2[b]	194
Xanthone	NO_2^-	17	2.7	194

[a]Error $\sim$ factor of 2.
[b]Intercept in quenching plot ill defined.

end, several dienes were employed (Table 18) [195]. Pseudo-first-order decays were observed for phenanthrene and the data were treated according to Eq. (24) at several micelle concentrations. The use of Eq. (5) enabled the determination of k_{Q+} [195]. As the number of carbon atoms in the quencher increases, the exit rate constant decreases; the entry rate constant remains approximately constant. This dependence of the exit rate constants can be again explained by the hydrophobicity of the quencher.

Phosphorescent probes have been used as tags on detergent molecules to determine the rate of exchange of micelle monomers [182]. A comparison of the exit rate constant and its dependence on temperature is shown in Table 19 for the probe 10-(4-bromo-1-naphthoyl)decylsulfate. The phosphorescence was quenched by $Fe(CN)_3$ and the data were fitted to Eq. (25). As the monomer chain length increased, the exit rate constant decreased. In addition, an increase in temperature expanded the difference observed between the dissociation rate constants for the probes with different chain lengths. This demonstrates that the activation energy for the exit rate constant was larger for micelles of greater size, because the micellar volume increases at higher temperature. CTAC micelles were also investigated and it was shown that for labeled detergents with different alkyl chain lengths (5, 8, 10, and 11 carbon atoms in the chains), the exit rate constant varied between 6.8×10^2 s^{-1} to values greater than 5.0×10^4 s^{-1} for the longest and shortest probes, respectively. A direct comparison between DTAC and CTAC revealed that the exit rate constant decreased as the chain length of the monomer increased [182]. This is analogous to the process observed for STS and SDS (Table 19). In contrast to previous discussions, the exit rate constants in DTAC or CTAC were larger than for SDS or STS.

Table 18 Overall Mobility Rate Constants for Diene Quenchers of Triplet Phenanthrene in SDS

Quencher	k_{Q+} (10^9 M^{-1} s^{-1})	k_{Q-} (10^6 s^{-1})
1,3-Octadiene		0.13
2,5-Dimethyl-2,4-hexadiene		0.19
1,3-Cyclooctadiene		0.35
2,4-Dimethylpentadiene		0.37
cis-2-trans-4-Hexadiene	2.0	1.0
1,3-Cycloheptadiene		1.3
1,3-Hexadiene	0.83	2.3
1,3-Cyclohexadiene	1.3	5.6
trans-1,3-Pentadiene	0.95	6.9
cis-1,3-Pentadiene	1.2	8.9

Source: Ref. 195.

Table 19 Comparison of Exit Rate Constants from Alkyl Sulfate Micelles Using 10-(4-Bromo-1-naphthoyl)Decylsulfate as Probe and Fe(CN)$_3$ as Quencher

Temperature (°C)	Surfactant	k_{P_-} (10^6 s^{-1})
20	SC$_{10}$S	0.0057[a]
20	SC$_{12}$S	0.0014[a]
20	SC$_{14}$S	0.0007[a]
30	SC$_{10}$S	0.0098[a]
30	SC$_{12}$S	0.0025[a]
30	SC$_{14}$S	0.0008[a]

[a]10% error.
Source: Ref. 182.

However, this comparison was limited to the probe with 10 carbon atoms in the chain length and the relationship may not be general.

There has been very little work completed in reverse micelles. This may be due to the increased complexity of these systems, where probe migration and micelle clustering occurs on time scales similar to triplet decay. In addition, little work has been carried out in vesicles or liposomes. With the advent of diffuse reflectance laser flash photolysis, the turbid solutions observed for large vesicles no longer represent a stumbling block [196–198]. Thus, we believe that, in the future, much more work will be available in vesicular systems that will be able to complement the studies already completed in micelles.

2. Bile Salt Aggregates

Polycyclic aromatic hydrocarbons as well as dyes and ketones have been used to characterize bile salt aggregates [100,183,199]. For sodium taurocholate aggregates, the amount of protection afforded by the aggregates varied depending on the structure of the probe and its solubilization site. For quenching by nitrite, the ratio of k_{q0}/k_q(eff) for naphthalene was approximately 100, whereas it was 7 for xanthone [183]. The dissociation rate constants were determined from the fit of k_{obs} at different quencher concentrations to Eq. (25). No values for the association rate constants were obtained because the aggregation number for the primary and secondary aggregates were not known with certainty. The dissociation rate constant for triplet xanthone was $(25 \pm 8) \times 10^6$ s^{-1}, whereas for naphthalene under identical experimental conditions, it was $(2.2 \pm 0.5) \times 10^6$ s^{-1}. Naphthalene, with its smaller dissociation rate constant, was assigned to be incorporated into the primary aggregates, whereas xanthone was believed to be incorporated into the more hydrophilic sites of the secondary aggregates

[183]. These results show that bile salt aggregates have two sites that affect the mobility of the probe very differently. Because xanthone was incorporated into a fairly hydrophilic site, the direct spectroscopic methodology could not be applied, as no significant spectral shift could be observed. It is worth noting that the errors for the recovered k_{P-} values were much higher for xanthone than naphthalene [183]. This was a reflection of the smaller difference between k_{q0} and k_q(eff) for xanthone than for naphthalene, and the linear region at high quencher concentrations was never achieved in the case of xanthone quenching [183].

Rose Bengal has also been used to probe taurocholate aggregates. It was estimated that the exit rate constant was greater than 0.1×10^6 s^{-1} [199]. This measurement was limited by the time resolution of the equipment employed. In addition, the exit rate constant of anthracene from taurocholate aggregates has been estimated from the kinetics of the triplet–triplet annihilation process of anthracene [100]. The dissociation rate constant was estimated to be 2.8×10^3 s^{-1}, a value which is much lower than that observed for naphthalene. Because not much detail was given on the methodology employed, it was difficult to determine if this lower rate constant was related to the size of the probe molecule.

The dissociation rate constants for naphthalene and xanthone have been determined for sodium cholate and deoxycholate aggregates. In the case of sodium cholate, a factor greater than 30 separated the quenching efficiencies by nitrite for the two probes. The exit rate constant for xanthone was $(5 \pm 4) \times 10^6$ s^{-1}, whereas for naphthalene, the value was reduced by a factor of 5 [183]. In deoxycholate aggregates, a much larger degree of protection for quenching of triplet naphthalene by nitrite was afforded (~ 300), compared to taurocholate or cholate (110 or 160, respectively), suggesting that these aggregates are more rigid. However, the naphthalene exit rate constants were of the same order of magnitude as for the other bile salt aggregates.

3. Cyclodextrins

The amount of data obtained for this supramolecular system with excited triplet state probe molecules is much larger than with excited singlet states, showing that triplet states are very suitable to study the dynamics of host–guest complexes. Bromonaphthoyl derivatives [n-(4-bromo-1-naphthoyl)alkyl]trimethylammonium bromide (BNK-n) have been used to establish the effect of chain length on complexation to CD. No evidence was observed for a complex to α-cyclodextrin; however, complexes were observed for β- and γ-cyclodextrins. This suggests that the bromonaphthoyl moiety was too large to fit into the α-CD cavity but was of an appropriate size for inclusion into β- and γ-CD [181]. The association and dissociation rate constants were determined by using Co^{3+} as a quencher. The data were analyzed with Eq. (25), and in this case, some quenching was observed for the complexed probe molecule. In the case of β-CD, the entry rate constants for the bromonaphthalene derivatives were somewhat higher for the probe with the shortest alkyl chain. An even smaller dependence on chain length was observed for the dissociation rate constants (Table 20). In the case of γ-CD, the size of the probe had a marked effect on the magnitude of the exit rate constants, whereas the entry rate constant remained fairly similar [181]. The dissociation rate constant decreased by a factor of 3.5 when the alkyl chain contained 10 carbon atoms relative to the shortest chained probe. These results were attributed to the inclusion of the alkyl chain as well as the chromophore inside the CD cavity and were substantiated by molecular modeling experiments [181]. Furthermore, quenching of BNK-10 by nitrite yielded identical values within experimental error for the association/dissociation rate constants [181]. This showed that the exit rate of the probe was not dependent on the quencher employed.

   p-Methoxy-β-phenylpropiophenone has been employed to study the intramolecular mobility of the β-phenyl ring. The excited triplet state of this molecule was deactivated by interaction with the β-phenyl ring. In the case of α- and β-CD, only one of the rings can be included in the CD cavity, and protection from Cu^{2+} quenching was moderate. In the case of γ-CD, the larger CD cavity allowed both rings to be included. This led to a marked increase of the triplet lifetime. This excited state was better protected from quenching, showing that for γ-CD, the complete inclusion of the probe decreased the access of Cu^{2+} to the cavity [200]. This contrasts to the case when triplet xanthone was the included probe, as Cu^{2+} had some access to the probe within the cavity [185]. These examples show that accessibility of aqueous molecules to a probe within the CD cavity is very dependent on the structure of the probe

Table 20 Overall Quenching Rate Constants for Triplet Probes Bound to Cyclodextrins

CD	Probe	Quencher	k_{q0}/k_q(eff)	k_{P+} $(10^9\ M^{-1}\ s^{-1})$	k_{P-} $(10^6\ s^{-1})$
β	BNK-1[a]	Co^{3+}	9	0.035	0.05
β	BNK-5[a]	Co^{3+}	20	0.022	0.042
β	BNK-10[a]	Co^{3+}	24	0.024	0.040
γ	BNK-1[a]	Co^{3+}	8	0.015	0.042
γ	BNK-5[a]	Co^{3+}	20	0.019	0.043
γ	BNK-10[a]	Co^{3+}	24	0.015	0.012

Note: Error is not larger than 20%.
[a]n-([4-Bromo-1-naphthoyl)alkyl]trimethylammonium bromide; n = 1, 5, or 10.
Source: Ref. 181.

molecule. A higher degree of protection was afforded when the fit of the host-guest complex was tight.

Frequently, co-solvents are added to aqueous cyclodextrin solutions; in most cases, these co-solvents are employed to solubilize the probes. In Section IV.C, the decrease of the exit rate constant of triplet xanthone from CDs with the addition of alcohols was described. This effect was also apparent when studying the dynamics of 1-halonaphthalenes with β-CD in the presence of acetonitrile [141]. When the nitrite ion was used as quencher, the association rate constants decreased in the presence of the organic solvent while the dissociation rate constant increased (Table 21). The main rationalization to explain the change in mobility properties was that acetonitrile was small enough to co-include inside the cavity; a small amount of acetonitrile could preferentially solvate the entrances of the CD thereby leading to a different environment for the probe.

In the case of cyclodextrins, both the direct spectroscopic methodology and quenching studies have been employed to obtain information on the dynamics of guest complexation. Comparisons of the results indicate that when care is taken regarding the assumptions made, both methodologies yield similar rate constants. It is also worth noting that compared to micelles a greater variation of the association rate constants is observed for CD complexation.

4. Proteins and DNA

Several reviews have been written concerning the phosphorescence of intrinsic indole probes [201] and the use of triplet excited states to probe proteins [202]. Triplet tryptophan can be used to probe proteins, but its photophysics is complex, because in a homogeneous solution, the decays of the excited triplet tryptophan or indole derivatives do not usually follow first-order kinetics, due to protonation, triplet–triplet annihilation, or self-quenching [203,204]. In ad-

Table 21 Quenching of Nitrite of 1-Halonaphthalene in Mixed Solvents Containing β-Cyclodextrin

Probe	Solvent	k_{p+} $(10^9 \ M^{-1} \ s^{-1})$	k_{p-} $(10^6 \ M^{-1} \ s^{-1})$
1-Bromonaphthalene	Water[a]	0.07 ± 0.04	0.004 ± 0.03
1-Bromonaphthalene	10% ACN[b]	0.08	0.01[c]
1-Bromonaphthalene	2% ACN[b]	0.02	0.02[c]
1-Chloronaphthalene	2% ACN[b]	0.01	0.01[c]

[a]Temp. 15°C.
[b]ACN: Acetonitrile.
[c]± 10%.
Source: Ref. 141.

dition, most proteins being studied have multiple tryptophan residues, and there are numerous environments that are being probed simultaneously. Therefore, care must be taken when choosing a protein to investigate. Alternatively, a few studies with extrinsic probes have been reported [203,205,206].

Acrylamide is a polar, uncharged quencher that resides mainly in hydrophilic regions and efficiently quenches triplet probes by collisional processes [160]. Proteins have varied ability to inhibit the access of this quencher to indole residues that are buried within the matrix (Table 22) [163]. The absolute values for the quenching rate constants of excited singlet indoles in the same proteins were larger than for triplet quenching (cf. Tables 9 and 22), suggesting that factors in addition to accessibility contribute to the magnitude of the quenching rate constants. For example, these factors could be changes in the intrinsic quenching reactivity of the triplet with the polarity of the environment or a dependence of the quenching efficiency with viscosity [163].

Oxygen also quenches triplet tryptophan, but because quenching was at least partially due to energy transfer, a spin statistical factor which reduces the quenching rate constant has to be taken into account. For this reason, it was not surprising that the absolute values for k_q(eff) for triplet quenching were smaller than for singlet quenching (Table 23) [157,207–209]. However, the relative rate constants for different proteins were similar, indicating that the oxygen quenching efficiency can be used as an accessibility indicator. For example, the indole in the melittin tetramer was protected 2.5 times better than the monomer. It is well established that RNase T_1 contains a moderately buried indole residue [160,210,211], and so all quenching efficiencies were compared to it.

There are only a few examples of the use of extrinsic probes to investigate accessibility of quenchers to protein-complexed excited states. Polycyclic aromatic hydrocarbons were used by Geacintov et al. to probe bovine serum albumin (BSA) [205]. The accessibility of oxygen to the probe molecules was

Table 22 Quenching of Intrinsic Probes in Proteins by Acrylamide

Protein	Quencher	Probe	k_q(eff) $(10^9 \, M^{-1} \, s^{-1})$
ACTH	Acrylamide	Indole	0.78
Asparaginase	Acrylamide	Indole	≤0.00008
HSA	Acrylamide	Indole	0.009
Melittin (monomer)	Acrylamide	Indole	0.33
Melittin (tetramer)	Acrylamide	Indole	0.06
Nuclease	Acrylamide	Indole	0.04
RNase T_1	Acrylamide	Indole	0.0002

Source: Ref. 163.

Table 23 Quenching of Intrinsic Indole Residues in Proteins by Oxygen

Protein	Quencher	Probe	$k_q(\text{eff})$ $(10^9\ M^{-1}\ s^{-1})$	$k_q(\text{eff})/$ $k_q(\text{eff})_0$[a]	Ref.
Asparaginase	O_2	Indole (25°C)	0.1	0.3	157
Corticotropin	O_2	Indole (25°C)	5.1	20	157
HSA	O_2	Indole (25°C)	0.3, 0.09	1–3	157, 207
LADH	O_2	Indole (25°C)	0.1, 0.14	0.3–0.5	209, 208
Melittin (monomer)	O_2	Indole (25°C)	3.0	10	157
Melittin (tetramer)	O_2	Indole (25°C)	1.1	4	157
Monellin	O_2	Indole (25°C)	1.5	5	157
Nuclease	O_2	Indole (25°C)	0.9	3	157
Phase fd	O_2	Indole (25°C)	0.6	2	157
Parvalbumin (Cod)	O_2	Indole (25°C)	0.5	2	157
RNase T_1	O_2	Indole (25°C)	0.3	1	157

[a]Normalized relative to RNase T_1.

hindered by the protein matrix. Quenching rate constants that were one to two orders of magnitude smaller than for a diffusion-controlled process were observed (Table 24). Other studies have used intramolecular (intraprotein) quenching to study the accessibility from various different environments within the protein. Heme accessibility was determined using 6-bromo-2-naphthyl sulfate as a probe (Table 24). Probes in proteins that contained no heme moiety, such as lysozyme and apomyoglobin, were quenched only by self-quenching processes, whereas probes in proteins containing heme were quenched by self-quenching and extrinsic quenching mechanisms. The results indicated that the phosphorescence of the probe was quenched only upon contact with the heme portion; therefore, this methodology can be carried out successfully [203].

DNA is an interesting supramolecular structure, as it provides different complexation sites for molecules. Planar and relatively small molecules intercalate between base pairs, whereas larger molecules can bind to the minor or major grooves of the DNA helix. Interaction can be rather specific due to the chirality of the helix [147,150]. In addition, the negatively charged phosphate backbone is able to anchor or bind cationic probes or quenchers. Most studies have been performed with calf thymus DNA in its native double helical form.

Excited triplet organic probe molecules are more useful than excited singlet states to study DNA. In addition, it should be mentioned that inorganic metal complexes have been extensively employed to study DNA complexation [147,148,150,212–214], but they will not be covered in this chapter. Most quenching studies involved the mobility of oxygen with respect to DNA. This supramolecular structure affords a greater than 10-fold protection from oxygen quenching relative to quenching in homogeneous solvent and for probes such as acridine orange [215], ethidium bromide [216], and methylene blue [217] (Table 25). In addition, exit rate constants were estimated for acridine orange by assuming a diffusion-controlled association processes for Mn^{2+} and Ag^+. The values for the exit rate constants of the quencher from the helix were determined to be $\geq 1 \times 10^6$ s^{-1} for Mn^{2+} and $\geq 0.0005 \times 10^6$ s^{-1} for Ag^+ [219]. Using Eq. (5), the exit of acridine orange was estimated to be $\geq 1 \times 10^3$ s^{-1} [220]. Although DNA may be a complex macromolecule, mobility studies using triplet excited states as probes can offer valuable information.

E. Conclusions

Excited triplet probe molecules were shown to be useful in studying the dynamics of probe complexation to a variety of supramolecular structures, because the triplet lifetimes are frequently of the same order of magnitude as the entry/exit processes. Undoubtedly, the biggest advantage of excited triplet probes is that they can be used in direct spectroscopic measurements. However, the full potential of these direct studies will depend on the discovery of more probes which

Table 24 Quenching of Extrinsic Triplet State Probes Complexed to Proteins

Protein	Probe	Quencher	k_q(eff) $(10^9 M^{-1} s^{-1})$	Ref.
BSA	Benzo[a]pyrene	O_2	0.13 ± 0.005	205
BSA	Pyrene	O_2	0.095 ± 0.005	205
BSA	Benz[a]anthracene	O_2	0.17 ± 0.01	205
Apomyoglobin	6-Bromo-2-naphthyl sulfate	Apomyoglobin	0.004 ± 0.004	203
Lysozyme	6-Bromo-2-naphthyl sulfate	Lysozyme	0.010 ± 0.005	203
Cytochrome C	6-Bromo-2-naphthyl sulfate	Cytochrome C (oxidized)	1.8 ± 0.2	203
Cytochrome C	6-Bromo-2-naphthyl sulfate	Cytochrome C (reduced)	1.1 ± 0.1	203
Microperoxidase-11	6-Bromo-2-naphthyl sulfate	Microperoxidase-11	2.1 ± 0.2	203
Myoglobin	6-Bromo-2-naphthyl sulfate	Myoglobin	1.7 ± 0.2	203
Protoporphyrin IX	6-Bromo-2-naphthyl sulfate	Protoporphyrin IX	6.2 ± 0.5	203

Table 25 Overall Quenching Efficiency in DNA

Probe	Quencher	k_q(eff) $(10^9\ M^{-1}\ s^{-1})$	k_q^0 $(10^9\ M^{-1}\ s^{-1})$	k_q^0/k_q(eff)	Ref.
Benzo[a]pyrene	O_2	0.20 ± 0.02	2–3[a]	10	w/in 215
Benzo[e]pyrene	O_2	0.16 ± 0.02	2–3[a]	12	w/in 215
Benzo[a]anthracene	O_2	0.16 ± 0.02	2–3[a]	12	w/in 215
7,12-Dimethylbenz[a]anthracene	O_2	0.16 ± 0.02	2–3[a]	12	w/in 215
Acridine Orange	O_2	0.08 ± 0.01			215
Ethidium bromide	O_2	0.038 ± 0.006			216
Methylene blue[c]	O_2	0.24 ± 0.05	3.0 ± .01[b]	13	217
Methylene blue[d]	O_2	0.064 ± 0.010	3.0 ± .01[b]	470	217
Rose Bengal	DNA bases	<0.0004			218

[a] Hexane as solvent.
[b] Water as solvent.
[c] Experiment carried out in dAdT.
[d] Experiment carried out in dGdC.

have photophysical characteristics similar to xanthone. In the case of the quenching methodology, the mathematical treatments employed to recover the dissociation/association rate constants are frequently simpler than in the case of the use of excited singlet states. This is, again, a reflection of the longer triplet lifetime which makes it possible to introduce simplifications in the mathematical treatment. Triplet states have also proven to be more useful than excited singlet states to study the dynamics in host–guest complexes.

V. OTHER REACTIVE INTERMEDIATES

In addition to excited states, reactive intermediates can also be employed as probe molecules to study the mobility in supramolecular systems. The discussion regarding the relationship between lifetime of the probe and the explored volume is applicable. In principle, the equations described in this chapter for excited states could be used for other reactive intermediates. In fact, radicals and radical ions have been employed to explore the dynamics in micelles [221–223], vesicles [196], and host–guest complexes with CDs [224–226]. Although this chapter does not cover reactive intermediates, we will briefly mention a few examples that use methodologies that are unique to radicals. Radical dissociation rate constants can be studied when magnetic fields are employed to slow down the competitive geminate reaction in radical pairs [227]. For example, the exit rate constant for the benzophenone ketyl radicals from SDS micelles was determined to be 4.4×10^6 s^{-1} [228]. Techniques specific to radicals, such as chemically induced dynamic nuclear polarization, were employed to probe the mobility of benzyl radicals in micellar systems [229]. Quenching by the anionic, stable nitroxide, Fremy's salt, led to the exit rate values of $(1.4 \pm 0.3) \times 10^6$ s^{-1}, $(2.4 \pm 0.4) \times 10^6$ s^{-1}, and $(3.0 \pm 0.6) \times 10^6$ s^{-1} for benzyl, 4-methylbenzyl, and *sec*-phenethyl radicals, respectively. These values were identical to those obtained using LFP. These two examples show that not only can reactive intermediates be employed for probing mobility but there is also a potential to broaden the techniques being employed in dynamic studies.

VI. CONCLUDING REMARKS

This chapter covered the use of excited states as probes for dynamic studies involving supramolecular structures. It was shown that excited singlet and triplet states provide complementary capabilities for these studies. Micelles are the best characterized structures, and most of the methodology currently in use was developed when studying these supramolecular systems. Other self-assemblies, such as vesicles and bile salt aggregates, have been studied to some extent, but the dynamics of guest complexation has not been fully described and will undoubtedly be explored in the future. Cyclodextrins are widely employed as models for host–guest complexes, and mobility studies are no exception. We expect that the methodology employed for cyclodextrins will be easily applicable to other synthetic host compounds. The examples described for proteins and DNA showed that dynamic studies are still in their early stages, with overall quenching rate constants providing most of the information available to date. These biomolecules are examples of systems to which the methodologies described in this chapter will be applicable in future.

ACKNOWLEDGMENT

The support of the Natural Sciences and Engineering Research Council of Canada for our research in supramolecular systems is gratefully acknowledged.

APPENDIX: LIST OF ABBREVIATIONS

ACTH	Adrenocorticotropin
AOT	Sodium bis(2-ethylhexyl)sulfosuccinate
ANS	Anilinonaphthalene sulfonate
BNK-n	n-[(4-bromo-1-naphthoyl)alkyl]trimethylammonium bromide
BSA	Bovine serum albumin
CD	Cyclodextrin
C_nPyC	n-Alkylpyridinium chloride
cmc	Critical micelle concentration
CTAB	Cetyl trimethylammonium bromide
CTAC	Cetyl trimethylammonium chloride
DMPC	Dimyristoylphosphatidylcholine
DNA	Deoxyribonucleic acid
DODAC	Dioctadecyldimethylammonium chloride
DPPC	Dipalmitoylphosphatidylcholine
DTAC	Dodecyltrimethylammonium chloride
H	Host
HP*	Complexed host and excited probe
HSA	Human serum albumin
I_0	Initial fluorescence intensity
IRF	Instrument response factor
k_e	Rate constant for exchange of solute through micellar processes
k_h	Rate constant for decay of excited probe in the absence of quenching in a supramolecular system
k_0	Rate constant for decay of excited state in the absence of quencher

k_{obs}	Observed rate constant
k_{P-}	Rate constant for probe exit
k_{P+}	Rate constant for probe entry
$k_{P-(ter)}$	Rate constant for probe exit from ternary complexes
k_{Q-}	Rate constant for quencher exit
k_{Q+}	Rate constant for quencher entry
$k_q(eff)$	Overall quenching rate constant for the excited state bound to a supramolecular system
k_{qh}	Rate constant for quenching of the excited state when the quencher is bound to supramolecular system
k_{q0}	Rate constant for quenching of excited state in a homogeneous solution
K_Q	Equilibrium constant for quencher exit/entry
K_{sv}	Rate constant for Stern–Volmer quenching (equal to $k_q\tau_0$)
LADH	Liver alcohol dehydrogenase
LFP	Laser flash photolysis
m-DCB	*Meta*-dicyanobenzene
MEM	Maximum entropy method
MV^{2+}	Methyl viologen (Paraquat dihydrochloride)
N	Aggregation number
NaC	Sodium cholate
NATA	*N*-Acetyltryptophanamide
NaTC	Sodium taurocholate
NLLS	Nonlinear least squares
P	Probe
P*	Excited probe
PTS	1,3,6,8-pyrenetetrasulfonate
$PyMe^+$	1-Pyrenyl(methyl)trimethylammonium salt
Q	Quencher
SDS	Sodium dodecyl sulfate
SPC	Single photon counting (single photon timing)
SS	Steady state
STS	Sodium tetradecyl sulfate
TEMPAMINE	4-Amino-TEMPO (ammonium salt)
TEMPO	2,2,6,6-tetramethyl-1-piperidinyloxy, free radical
TEMPOL	4-Hydroxy-TEMPO
TR	Time resolved
TTAC	Tetradecyltrimethylammonium chloride

REFERENCES

1. Hamasaki, K.; Ueno, A.; Toda, F. *J. Chem. Soc., Chem. Commun.* 1993, 331–333.
2. Bates, P. S.; Kataky, R.; Parker, D. *J. Chem. Soc. Perkin. Trans.* 2 1994, 669–675.
3. Ueno, A.; Suzuki, I.; Osa, T. *Anal. Chem.* 1990, 62, 2461–2466.
4. Breslow, R.; Graff, A. *J. Am. Chem. Soc.* 1993, 115, 10988–10989.
5. Andersson, T.; Nilsson, K.; Sundahl, M.; Westman, G.; Wennerstrom, O. *J. Chem. Soc., Chem. Commun.* 1992, 604–606.
6. Armstrong, D. W.; Ward, T. J.; Armstrong, R. D.; Beeley, T. E. *Science* 1986, 232, 1132–1135.
7. Eicke, H.; Shepherd, J. C. W.; Steinnemann, A. *J. Colloid Interface Sci.* 1976, 56, 168–176.
8. Litzinger, D. C.; Huang, L. *Biochim. Biophys. Acta* 1992, 1113, 201–207.
9. Margalit, R.; Okon, M.; Yerushalmi N.; Avidor, E. *J. Controlled Release* 1992, 19, 275–287.
10. Atik, S. S.; Thomas, J. K. *J. Am. Chem. Soc.* 1981, 103, 3543–3550.
11. Fyles, T. M.; Hansen, S. P. *Can. J. Chem.* 1988, 66, 1445–1453
12. Aikawa, M.; Turro, N. J.; Ishiguro, K. *Chem. Phys. Lett.* 1994, 222, 197–203.
13. Kalsbeck, W. A.;Thorp, H. H. *J. Am. Chem. Soc.* 1993, 115, 7146–7151.
14. Nabiev, I.; Chourpa, I.; Manfiat, M. *J. Phys. Chem.* 1994, 98, 1344–1350.
15. Wakelin, L. P. G.; Creasy, T. S.;Waring, M. J. *FEBS Lett.* 1979, 104, 261–265.
16. Frindi, M.; Michels, B.; Zana, R. *J. Phys. Chem.* 1991, 95, 4832–4837.
17. Frindi, M.; Michels, B.; Zana, R. *J. Phys. Chem.* 1992, 96, 6095–6102.
18. Frindi, M.; Michels, B.; Zana, R. *J. Phys. Chem.* 1992, 96, 8137–8141.
19. Rohrbach, R. P.; Rodriguez, L. J.; Eyring, E. M.; Wojcik, J. F. *J. Phys. Chem.* 1977, 81, 944–948.
20. Lentz, B. R. *Chem. Phys. Lipids* 1993, 64, 99–116.
21. Toptygin, D.; Svobodova, J.; Konopasek, I.; Brand, L. *J. Chem. Phys.* 1992, 96, 7919–7930.
22. Zannoni, C.; Arcioni, A.; Cavatorta, P. *Chem. Phys. Lipids* 1983, 32, 179–250.
23. Lakowicz, J. R. *Principles of Fluorescence Spectoscopy;* Plenum Press: New York, 1983.
24. Eaton, D. F. *Pure Appl. Chem.* 1990, 62, 1631–1648.
25. Dong, D. C.; Winnik, M. A. *Photochem. Photobiol.* 1982, 35, 17–21.
26. Kalyanasundaram, K.; Thomas, J. K. *J. Phys. Chem.* 1977, 81, 2176–2180.
27. Kalyanasundaram, K.; Thomas, J. K. *J. Am. Chem. Soc.* 1977, 99, 2039–2044.
28. Parasassi, T.; De Stasio, G.; d'Ubaldo, A.; Gratton, E. *Biophys. J.* 1990, 57, 1179–1186.
29. Ramamurthy,V. *Photochemistry in Organized and Constrained Media;* VCH Publishers: New York, 1991.
30. Kalyanasundaram, K. *Photochemistry in Microheterogeneous Systems*; Academic Press: Orlando, FL, 1987.

31. Saavedra-Molina, A.; Uribe, S.; Devlin, T. M. *Biochem. Biophys. Res. Commun.* 1990, *167*, 148-153.
32. Kang, J. J.; Tarcsafalvi, A.; Carlos, A. D.; Fujimoto, E.; Shahrokh, Z.; Thevenin, B. J. M.; Shohet, S. B.; Ikemoto, N. *Biochemistry* 1992, *31*, 3288-3293.
33. Knutson, J. R.; Beecham, J. M.; Brand, L. *Chem. Phys. Lett.* 1983, *102*, 501-507.
34. Beecham, J. M.; Ameloot, M.; Brand, L. *Chem. Phys. Lett.* 1985, *120*, 466-472.
35. Eisenfeld, J.; Ford, C. C. *Biophys. J.* 1979, *26*, 73-84.
36. Boens, N.; Malliaris, A.; Van der Auweraer, M.; Luo, H.; De Schryver, F. C. *Chem. Phys.* 1988, *121*, 199-209.
37. Boens, N.; Luo, H.; Van der Auweraer, M.; Reekmans, S.; De Schryver, F. C.; Malliaris, A.*Chem. Phys. Lett.* 1988, *146*, 337-342.
38. Siemiarczuk, A.; Wagner, B. D.; Ware, W. R. *J. Phys. Chem.* 1990, *94*, 1661-1666.
39. Siemiarczuk, A.; Ware, W. R.; Liu, Y. S. *J. Phys. Chem.* 1993, *97*, 8082-8091.
40. Sugar, I. P. *J. Phys. Chem.* 1991, *95*, 7508-7515.
41. Sugar, I. P.; Zeng, J.; Vauhkonen, M.; Somerharju, P.; Chong, P. L.-G. *J. Phys. Chem.* 1991, *95*, 7516-7523.
42. Knutson, J. R. In *Methods in Enzymology;* L. Brand and M. L. Johnson, Ed.; Academic Press: New York, 1992; Vol 210; pp. 357-373.
43. Badea, M. G.; Brand, L. In *Methods in Enzymology;* C. H. W. Hirs and S. N. Timasheff, Ed.; Academic Press: New York, 1979; Vol. 61; pp. 378-424.
44. Bajzer, Z.; Prendergast, F. G. In *Methods in Enzymology;* L. Brand and M. L. Johnson, Ed.; Academic Press: New York, 1992; Vol. 210; pp. 200-236.
45. Beechem, J. M. In *Methods in Enzymology;* L. Brand and M. L. Johnson, Ed.; Academic Press: New York, 1992; Vol. 210; pp. 37-53.
46. Small, E. W. In *Methods in Enzymology;* L. Brand and M. L. Johnson, Ed.; Academic Press: New York, 1992; Vol. 210; pp. 237-278.
47. Ameloot, M. In *Methods in Enzymology;* L. Brand and M. L., Ed.; Academic Press: NewYork, 1992; Vol 210; pp. 279-304.
48. Ameloot, M.; Boens, N.; Andriessen, R.; Van den Bergh, V.; De Schryver, F. C. In *Methods in Enzymology;* L. Brand and M. L. Johnson, Ed.; Academic Press: New York, 1992; Vol. 210; pp. 314-339.
49. Faunt, L. M.; Johnson, M. L. In *Methods in Enzymology;* L. Brand and M. L. Johnson, Ed.; Academic Press: New York, 1992; Vol. 210; pp. 340-356.
50. Bright, F. V.; Catena, G. C.; Huang, J.; Zabrobelny, J.; Zhang, J. *Adv. Multidimen. Lumin.* 1993, *2*, 85-99.
51. Bortolus, P.; Grabner, G.; Kohler, G.; Monti, S. *Coord. Chem. Rev.* 1993, *125*, 261-268.
52. Benesi, H. A.; Hildebrand, J. H. *J Am. Chem. Soc.* 1949, *71*, 2703-2707.
53. Hill, Z. D.; MacCarthy, P. *J. Chem. Educ.* 1986, *63*, 162-167.
54. Gill, V. M. S.; Oliveira, N. C. *J. Chem. Educ.* 1990, *67*, 473-478.
55. Holzwarth, J. F.; Eck, V.; Genz, A. In *Spectroscopy and the Dynamics of Molecular Biological Systems* Academic Press: London, 1985; pp: 351-377.

56. Holzwarth, J. F. *NATO ASI Ser.*, *Ser. A* 1989, *178*, 383-412.
57. Laidler, K. J. *Chemical Kinetics*, 3rd ed.; Harper and Row Publishers: New York, 1987.
58. Barzykin, A. V.; Tachiya, M. *Chem. Phys. Lett.* 1993, *216*, 575-578.
59. Tachiya, M. *Chem. Phys. Lett.* 1975, *33*, 289-292.
60. Infelta, P. P.; Gratzel, M.; Thomas, J. K. *J. Phys. Chem.* 1974, *78*, 190-195.
61. Van der Auweraer, M.; Dederen, C.; Palmans-Windels, C.; De Schryver, F. C. *J. Am. Chem. Soc.* 1982, *104*, 1800-1804.
62. Almgren, M.; Grieser, F.; Thomas, J. K. *J. Am. Chem. Soc.* 1979, *101*, 279-291.
63. Gehlen, M. H.; De Schryver, F. C. *Chem. Rev.* 1993, *93*, 199-221.
64. Reekmans, S.; De Schryver, F. C. In *Frontiers in Supramolecular Organic Chemistry and Photochemistry;* H.-J. Schneider and H. Dürr, Ed.; VCH Verlagsgesellschaft: Weinheim, 1991; pp. 287-310.
65. Dederen, J. C.; Van der Auweraer, M.; De Schryver, F. C. *Chem. Phys. Lett.* 1979, *68*, 451-454.
66. Gehlen, M. H.; Van der Auweraer, M.; Reekmans, S.; Neuman, M. G.; De Schryver, F. C. *J. Phys. Chem.* 1991, *95*, 5684-5689.
67. Gehlen, M. H.; Boens, N.; De Schryver, F. C.; Van der Auweraer, M.; Reekmans, S. *J. Phys. Chem.* 1992, *96*, 5592-5601.
68. Gehlen, M. H.; De Schryver, F. C.; Bhaskar Dutt, G.; van Stam, J.; Boens, N.; Van der Auweraer, M. *J. Phys. Chem.* 1995, *99*, 14407-14413.
69. Gehlen, M. H. *Chem. Phys.* 1994, *186*, 317-322.
70. Almgren, M.; Löfroth, J.; van Stam, J. *J. Phys. Chem.* 1986, *90*, 4431-4437.
71. Almgren, M. *Adv. Colloid Interface Sci.* 1992, *41*, 9-32.
72. Jóhannsson, R.; Almgren, M.; Alsins, J. *J. Phys. Chem.* 1991, *95*, 3819-3823.
73. Malliaris, A.; Lang, J.; Zana, R. *J. Phys. Chem.* 1986, *90*, 655-660.
74. Muñoz de la Pena, A.; Ndou, T.; Zung, J. B.; Warner, I. M. *J. Phys. Chem.* 1991, *95*, 3330-3334.
75. Birch, D. J. S.; Imhof, R. E. In *Topics in Fluorescence Spectroscopy;* J. Lakowicaz, Ed.; Plenum Press: New York, 1991; Vol. 1; pp. 1-95.
76. O'Connor, D. V.; Phillips, D. *Time-Correlated Single Photon Counting;* Academic Press: Orlando, FL, 1984.
77. Marquardt, D. W. *SIAM J. Appl. Math* 1963, 14, 1176.
78. Johnson, M. L. In *Methods in Enzymology;* L. Brand and M. L. Johnson, Ed.; Academic Press: New York, 1992; Vol. 210; pp. 1-36.
79. Janssens, L. D.; Boens, N.; Ameloot, M.; DeSchryver, F. C. *J. Phys. Chem.* 1990, *94*, 3564-3576.
80. Duportail, G.; Lianos, P. *Chem. Phys. Lett.* 1988, *149*, 73-78.
81. Lianos, P. *J. Chem. Phys.* 1988, *89*, 5237-5241.
82. Duportail, G.; Lianos, P. *Chem. Phys. Lett.* 1990, *165*, 35-40.
83. Sugar, I. P.; Zeng, J.; Chong, P. L.-G. *J. Phys. Chem.* 1991, *95*, 7524-7534.
84. Andre, J. C.; Vincent, L. M.; O'Connor, D.; Ware, W. R. *J. Phys. Chem.* 1979, *83*, 2285-2294.
85. Luo, H.; Boens, N.; Van der Auweraer, M.; De Schryver, F. C.; Malliaris, A. *J. Phys. Chem.* 1989, *93*, 3244-3250.

86. Flom, S. R.; Fendler, J. H. *J. Phys. Chem.* 1988, *92*, 5908-5913.
87. Andriessen, R.; Boens, N.; Ameloot, M.; De Schryver, F. C. *J. Phys. Chem.* 1991, *95*, 2047-2058.
88. Boens, N.; Anriessen, R.; Ameloot, M.; Dommelen, L. V.; De Schryver, F. C. *J. Phys. Chem.* 1992, *96*, 6331-6342.
89. Boens, N.; Ameloot, M.; Hermans, B.; De Schryver, F. C.; Andriessen, R. *J. Phys. Chem.* 1993, *97*, 799-808.
90. Kowalczyk, A.; Meuwis, K.; Boens, N.; De Schryver, F. C. *J. Phys. Chem.* 1995, *99*, 17349-17353.
91. Ameloot, M.; Boens, N.; Andriessen, R.; Van den Bergh, V.; De Schryver, F. C. *J. Phys. Chem.* 1991, *95*, 2041-2047.
92. Siemiarczuk, A.; Ware, W. R. *Chem. Phys. Lett.* 1989, *160*, 285-290.
93. Shaver, J. M.; McGown, L. B. *Anal. Chem.* 1996, *68*, 611-620.
94. Shaver, J. M.; McGown, L. B. *Anal. Chem.* 1996, 68, 9-17.
95. Yekta, A.; Aikawa, M.; Turro, N. J. *Chem. Phys. Lett.* 1979, *63*, 543-548.
96. Turro, N. J.; Aikawa, M.; Yekta, A. *J. Am. Chem. Soc.* 1979, *101*, 772-774.
97. Infelta, P. P. *Chem. Phys. Lett.* 1979, *61*, 88-91.
98. Lissi, E. A.; Gallardo, S.; Sepulveda, P. *J. Colloid Interface Sci.* 1992, *152*, 104-113.
99. Encinas, M. V.; Rubio, M. A.; Lissi, E. *Photochem. Photobiol.* 1983, *37*, 125-130.
100. Chen, M.; Grätzel, M.; Thomas, J. K. *J. Am. Chem. Soc.* 1975, *97*, 2052-2057.
101. Grieser, F.; Tausch-Treml, R. *J. Am. Chem. Soc.* 1980, *102*, 7258-7264.
102. Alvarez, J.; Lissi, E. A.; Encinas, M. V. *Langmuir* 1996, *12*, 1738-1743.
103. Murov, S. L.; Carmichael, I,: Hug; G. L. *Handbook of Photochemistry;* 2nd. ed.; Marcel Dekker: New York, 1993.
104. Gennis, R. B. *Biomembranes: Molecular Structure and Function;* Springer-Verlag: New York, 1989.
105. Abuin, E.; Lissi, E.; Aravena, D.; Zanocco, A.; Macuer, M. *J. Colloid Interface Sci.* 1988, *122*, 201-208.
106. Pérochon, E.; Lopez, A.; Tocanne, J. F. *Biochemistry* 1992, *31*, 7672-7682.
107. Carmona Ribeiro, A. M.; Chaimovich, H. *Biochim. Biophys. Acta* 1983, *733*, 172-179.
108. Burke, T. G.; Tritton, T. R. *Biochemistry* 1985, *24*, 5972-5980.
109. Montich, G. G.; Cosa, J. J. *Bol. Soc. Chil. Quim.* 1990, *35*, 97-103.
110. Li, G.; McGown, L. B. *J. Phys. Chem.* 1993, *97*, 6745-6752.
111. Li, G.; McGown, L. B. *J. Phys. Chem.* 1994, *98*, 13711-13719.
112. Meyerhoffer, S. M.; McGown, L. B. *J. Am. Chem. Soc.* 1991, *113*, 2146-2149.
113. Meyerhoffer, S. M.; McGown, L. B. *Anal. Chem.* 1991, *63*, 2082-2086.
114. Chen, M.; Grätzel, M.; Thomas, J. K. *Chem. Phys. Lett.* 1974, *24*, 65-68.
115. Hinze, W.L.; Srinivasav, N.; Smith, T. K. In *Multidimensional Luminescence;* I. M. Warner and L. B. McGown, Ed.; JAI Press; Greenwich, CT, 1990.
116. Hertz, P. M. R.; McGown, L. B. *Anal. Chem.* 1992, *64*, 2920-2928.
117. Zana, R.; Guveli, D. *J. Phys. Chem.* 1985, *89*, 1687-1690.
118. Hashimoto, S.; Thomas, J. K. *J. Colloid Interface Sci.* 1984, *102*, 152-163.
119. Hofmann, A. F.; Small, D. M. *Annu. Rev. Med.* 1967, *18*, 333-376.

120. Nithipatikom, K.; McGown, L. B. *Anal. Chem.* 1988, *60*, 1043-1045.
121. Ju, C.; Bohne, C. *Photochem. Photobiol.* 1996, *63*, 60-67.
122. Ju, C. M. Sc. Thesis, University of Victoria, 1995.
123. Tabushi, I. *Acc. Chem. Res.* 1982, *15*, 66-72.
124. Guttman, A.; Paulus, A.; Cohen, A. S.; Grinberg, N.; Karger, B. L. *J. Chrom.* 1988, *20*, 553.
125. Hinze, W. L.; Riehl, T. E. *Anal. Chem.* 1985, *57*, 237-242.
126. Köhler, J. E. H.; Hohla, M.; Richters, M.; Konig, W. A. *Angew. Chem. Int. Ed. Engl.* 1992, *31*, 319-320.
127. Paleologou, M. L.; Purdy, W. C. *Can. J. Chem.* 1990, *68*, 1208-1214.
128. Cramer, F.; Saenger, W.; Spatz, H. C. *J. Am. Chem. Soc.* 1967, *89*, 14-20.
129. Kano, K.; Takenoshita, I.; Ogawa, T. *J. Phys. Chem.* 1982, *86*, 1833-1838.
130. Yorozu, T.; Hoshino, M.; Imamura, M. *J. Phys. Chem.* 1982, *86*, 4426-4429.
131. Kobashi, H.; Takahashi, M.; Muramatsu, Y.; Morita, T. *Bull. Chem. Soc. Jpn.* 1981, *54*, 2815-2816.
132. De Korte, A.; Langlois, R.; Cantor, C. R. *Biopolymers* 1980, *19*, 1281-1288.
133. Park, J. W.; Song, H. J. *J. Phys. Chem.* 1989, *93*, 6454-6458.
134. Herkstroeter, W. G.; Martic, P. A.; Farid, S. *J. Am. Chem. Soc.* 1990, *112*, 3583-3589.
135. Cox, G. S.; Hauptman, P. J.: Turro, N. J. *Photochem. Photobiol.* 1984, *39*, 597-601.
136. Zung, J. B.; Muñoz de La Peña, A. M.; Ndou, T.T.; Warner, I. M. *J. Phys. Chem.* 1991, *95*, 6701-6706.
137. Hamai, S. *J. Phys. Chem.* 1990, *94*, 2595-2600.
138. Schuette, J. M.; Ndou, T. T.; Munoz de la Peña, A.; Greene, K. L.; Williamson, C. K.; Warner, I. M. *J. Phys. Chem.* 1991, *95*, 4897-4902.
139. Schuette, J. M.; Ndou, T. T.; Munoz de la Peña, A.; Mukundan Jr., S.; Warner, I. M. *J. Am. Chem. Soc.* 1993, *115*, 292-298.
140. Flamigni, L. *J. Phys. Chem.* 1993, *97*, 9566-9572.
141. Turro, N.J.; Bolt, J. D.; Kuroda, Y.; Tabushi, I. *Photochem. Photobiol.* 1982, *35*, 69-72.
142. Casal, H. L.; Netto-Ferreira, J. C.; Scaiano, J. C. *J. Incl. Phenom.* 1985, *3*, 395-402.
143. Huang, J.; Bright, F. V. *J. Phys. Chem.* 1990, *94*, 8457-8463.
144. Örstan, A.; Ross, J. B. A. *J. Phys. Chem.* 1987, *91*, 2739-2745.
145. Hashimoto, S.; Thomas, J. K. *J. Am. Chem. Soc.* 1985, *107*, 4655-4662.
146. Kumar, C. V.; Barton, J. K.; Turro, N. J. *J. Am. Chem. Soc.* 1985, *107*, 5518-5523.
147. Barton, J. K.; Goldberg, J. M.; Kumar, C. V.; Turro, N. J. *J. Am. Chem. Soc.* 1986, *108*, 2081-2088.
148. Pyle, A. M.; Rehmann, J. P.; Kumar, C. V.; Turro, N. J.; Barton, J. K. *J. Am. Chem. Soc.* 1989, *111*, 3051-3058.
149. Kirsch-De Mesmaeker, A.; Barton, J. K.; Turro, N. J. *Photochem. Photobiol.* 1990, *52*, 461-472.
150. Turro, N. J.; Barton, J. K.; Tomalia, D. A. *Acc. Chem. Res.* 1991, *24*, 332-340.

151. Stemp, E. D.; Arkin, M. R.; Barton, J. K. *J. Am. Chem. Soc.* 1995, *117*, 2375–2376.
152. Willis, K. J.; Szabo, A. G.; Zuker, M.; Ridgeway, J. M.; Alpert, B. *Biochemistry* 1990, *29*, 5270–5275.
153. Szabo, A. G.; Willis, K. J.; Krajcarski, D. T.; Alpert, B. *Chem. Phys. Lett.* 1989, *163*, 565–570.
154. Lakowicz, J. R.; Maliwl, B. P.; Cherek, H.; Balter, A. *Biochemistry* 1983, *22*, 1741–1751.
155. Lakowicz, J. R.; Weber, G. *Biochemistry* 1973, *12*, 4171–4179.
156. Eftink, M. R.; Ghiron, C. A. *Biophys. J.* 1988, *53*, 290a.
157. Ghiron, C.; Bazin, M.; Santus, R. *Biochim. Biophys. Acta* 1988, *957*, 207–216.
158. Hagaman, K. A.; Eftik, M. R. *Biophys. Chem.* 1984, *20*, 201–207.
159. Eftink, M. R.; Ghiron, C. A. *Biochemistry* 1984, *23*, 3891–3899.
160. Eftink, M. R.; Ghiron, C. A. *Biochemistry* 1976, *15*, 672–680.
161. Eftink, M. R.; Selvidge, L. A. *Biochemistry* 1982, *21*, 117–125.
162. Eftink, M. R.; Ghiron, C. A. *Biochemistry* 1977, *16*, 5546–5551.
163. Ghiron, C.; Bazin, M.; Santus, R. *Photochem. Photobiol.* 1988, *48*, 539–543.
164. Calhoun, D. B.; Vanderkooi, J. M.; Englander, S. W. *Biochemistry* 1983, *22*, 1533–1539.
165. James, D. R.; Demmer, D. R.; Steer, R.P.; Verrall, R. E. *Biochemistry* 1985, *24*, 5517–5526.
166. Jameson, D. M.; Gratton, E.; Weber, G.; Alpert, B. *Biophys. J.* 1984, *45*, 795–803.
167. Dederen, J. C.; Van der Auweraer, M.; De Schryver, F. C. *J. Phys. Chem.* 1981, *85*, 1198–1202.
168. Malliaris, A.; Lang, J.; Zana, R. *J. Chem. Soc. Faraday Trans.* 1986, *82*, 109–118.
169. Alonso, E. O.; Quina, F. H. *Langmuir* 1995, *11*, 2459–2463.
170. Alonso, E. O.; Quina, F. H. *J. Braz. Chem. Soc.* 1995, *6*, 155–159.
171. Croonen, Y.; Geladé, E.; Van der Zegel, M.; Van der Auweraer, M.; Vandendriessche, H.; De Schryver, F. C.; Almgren, M. *J. Phys. Chem.* 1983, *87*, 1426–1431.
172. Löfroth, J.; Almgren, M. *J. Phys. Chem.* 1982, *86*, 1626–1641.
173. Roelants, E.; De Schryver, F. C. *Langmuir* 1987, *3*, 209–214.
174. Bales, B. L.; Almgren, M. *J. Phys. Chem.* 1995, *99*, 15153–15162.
175. Grieser, F. *Chem. Phys. Lett.* 1981, *83*, 59–64.
176. Roelants, E.; Geladé, E.; Smid, J.; De Schryver, F. C. *J. Colloid Interface Sci.* 1985, *107*, 337–344.
177. Malliaris, A.; Boens, N.; Luo, H.; Van der Auweraer, M.; De Schryver, F. C.; Reekmans, S. *Chem. Phys. Lett.* 1989, *155*, 587–592.
178. Park, H.-R.; Mayer, B.; Wolschann, P.; Köhler, G. *J. Phys. Chem.* 1994, *98*, 6158–6166.
179. Birks, J. B. *Photophysics of Aromatic Molecules*; Wiley-Interscience: London, 1970.
180. Cheung, S. T.; Ware, W. R. *J. Phys. Chem.* 1983, *87*, 466–473.
181. Turro, N. J.; Okubo, T.; Chung, C.-J. *J. Am. Chem. Soc.* 1982, *104*, 1789–1794.

182. Bolt, J. D.; Turro, N. J. *J. Phys. Chem.* 1981, *85*, 4029–4033.
183. Ju, C.; Bohne, C. *J. Phys. Chem.* 1996, *100*, 3847–3854.
184. Scaiano, J. C. *J. Am. Chem. Soc.* 1980, *102*, 7747–7753.
185. Barra, M.; Bohne, C.; Scaiano, J. C. *J. Am. Chem. Soc.* 1990, *112*, 8075–8079.
186. Abuin, E. B.; Scaiano, J. C. *J. Am. Chem. Soc.* 1984, *106*, 6274–6283.
187. Liao, Y.; Frank, J.; Holzwarth, J. F.; Bohne, C. *J. Chem. Soc., Chem. Commun.* 1995, 199–200.
188. Muñoz de la Peña, A.; Ndou, T. T.; Zung, J. B.; Greene, K. L.; Live, D. H.; Warner, I. M. *J. Am. Chem. Soc.* 1991, *113*, 1572–1577.
189. Edwards, H. E.; Thomas, J. K. *Carbohydrate Res.* 1978, *65*, 173–182.
190. Liao, Y.; Bohne, C. *J. Phys. Chem.* 1996, *100*, 734–743.
191. Turro, N. J.; Aikawa, M. *J. Am. Chem. Soc.* 1980, *102*, 4866–4870.
192. Almgren, M.; Grieser, F.; Thomas, J. K. *J. Am. Chem. Soc.* 1979, *101*, 2021–2026.
193. Scaiano, J. C.; Selwyn, J. C. *Photochem. Photobiol.* 1981, *34*, 29–32.
194. Scaiano, J. C.; Selwyn, J. C. *Can. J. Chem.* 1981, *59*, 2368–2372.
195. Selwyn, J. C.; Scaiano, J. C. *Can. J. Chem.* 1981, *59*, 663–668.
196. Barra, M.; Bohne, C.; Zanocco, A.; Scaiano, J. C. *Langmuir* 1992, *8*, 2390–2395.
197. Wilkinson, F. *J. Chem. Soc., Faraday Trans.* 1986, *82*, 2073–2081.
198. Kessler, R. W.; Wilkinson, F. *J. Chem. Soc., Faraday Trans.* 1981, *77*, 309–320.
199. Seret, A.; Van de Vorst, A. *J. Photochem. Photobiol. B: Biol.* 1993, *17*, 47–56.
200. Netto-Ferreira, J. C.; Scaiano, J. C. *J. Photochem. Photobiol. A: Chem,* 1988, *45*, 109–116.
201. Papp, S.; Vanderkooi, J. M. *Photochem. Photobiol.* 1989, *49*, 775–784.
202. Geacintov, N. E.; Brenner, H. C. *Photochem. Photobiol.* 1989, *50*, 841–858.
203. Bayles, S. W.; Beckman, S.; Leidig, P. R.; Montrem. A.; Taylor, M. L.; Wright, T. M.; Wu, Y.; Schuh, M. D. *Photochem. Photobiol.* 1991, *54*, 175–181.
204. Boens, N.; Van Dommelen, L.; De Schryver, F. C.; Ameloot, M. *SPIE* 1994, *2137*, 400–411.
205. Geacintov, N. E.; Flamer, T. J.; Prusik, T.; Khosrofian, J. M. *Biochem. Biophys. Res. Commun.* 1975, *64*, 1245–1252.
206. Gioannini, T. L.; Campbell, P. *Biochem. Biophys. Res. Commun.* 1980, *96*, 106–111.
207. Hicks, B.; White, M.; Ghiron, C. A.; Kuntz, R. R.; Volkert, W. A. *Proc. Natl. Acad. Sci. USA* 1978, *75*, 1172–1175.
208. Barboy, N.; Feitelson, J. *Photochem. Photobiol.* 1985, *41*, 9–13.
209. Barboy, N.; Feitelson, J. *Biochemistry* 1987, *26*, 3240–3244.
210. Eftink, M. R.; Ghiron, C. A. *Proc. Natl. Acad. Sci. USA* 1975, *72*, 3290–3294.
211. Gryczynski, I.; Eftink, M.; Lakowicz, J. R. *Biochim. Biophys. Acta* 1988, *954*, 244–252.
212. Friedman, A. E.; Chambron, J. C.; Sauvage, J. P.; Turro, N. J.; Barton, J. K. *J. Am. Chem. Soc.* 1990, *112*, 4960–4962.

213. Krotz, A. H.; Kuo, L. Y.; Shields, T. P.; Barton, J. K. *J. Am. Chem. Soc.* 1993, *115*, 3877–3882.

214. Tysoe, S. A.; Morgan, R. J.; Baker, A. D.; Strekas, T. C. *J. Phys. Chem.* 1993, *97*, 1707–1711.

215. Geacintov, N. E.; Prusik, T.; Khosrofian, J. M. *J. Am. Chem. Soc.* 1976, *98*, 6444–6452.

216. Atherton, S. J.; Beaumont, P. C. *J. Phys. Chem.* 1987, *91*. 3993–3997.

217. Berkoff, B.; Hogan, M.; Legrange, J.; Austin, R. *Biopolymers* 1986, *25*, 307–316.

218. Lee, P. C. C.; Rodgers, M. A. J. *Photochem. Photobiol.* 1987, *45*, 79–86.

219. Prusik, T.; Geacintov, N. E. *FEBS Lett.* 1976, *71*, 236–240.

220. Geacintov, N. E.; Waldmeyer, J.; Kuzmin, V.A.; Kolubayev, T. *J. Phys. Chem.* 1981, *85*, 3608–3613.

221. Turro, N. J.; Kraeutler, B.; Anderson, D. R. *J. Am. Chem. Soc.* 1979, *101*, 7435–7437.

222. Turro, N. J.; Chow, M.; Chung, C.; Weed, G. C.; Kraeutler, B. *J. Am. Chem. Soc.* 1980, *102*, 4843–4845.

223. Turro, N. J.; Wu, C.-H. *J. Am. Chem. Soc.* 1995, *117*, 11031–11032.

224. Shafirovich, V. Y.; Batova, E. E.; Levin, P. P. *Chem. Phys. Lett.* 1993, *210*, 101–106.

225. Rao, V. P.; Zimmt, M. B.; Turro, N. J. *J. Photochem. Photobiol. A: Chem* 1991, *60*, 355–360.

226. Yonemura, H.; Nakamura, H.; Matsuo, T. *Chem. Phys. Lett.* 1989, *155*, 157–161.

227. Turro, N. J.; Grätzel, M.; Braun, A. M. *Angew. Chem. Int. Ed. Engl.* 1980, *19*, 675–696.

228. Scaiano, J. C.; Abuin, E. B. *Chem. Phys. Lett.* 1981, *81*, 209–213.

229. Turro, N. J.; Zimmt, M. B.; Gould, I. R. *J. Am. Chem. Soc.* 1983, *105*, 6347–6449.

11

Photophysical and Photochemical Properties of Squaraines in Homogeneous and Heterogeneous Media

Suresh Das, K. George Thomas, and Manapurathu V. George
Council of Scientific and Industrial Research (CSIR), Trivandrum, India

I. INTRODUCTION

Squaraine dyes form a class of organic photoconducting materials, which, along with other organic photoconductors, such as phthalocyanines, napthaquinones, cyanines, and azopigments, have been extensively investigated for a variety of technical applications. Squaraines exhibit sharp and intense absorption in the visible region which becomes broad and red-shifted in the solid state due to strong intermolecular donor–acceptor interactions [1,2]. These properties, combined with their ability to photoconduct, have made squaraines ideally suited for applications in xerographic photoreceptors [3], organic solar cells [4,5], and optical recording media [6]. Unlike most other organic photoconductors, reports on the spectroscopic and photochemical properties of squaraines were however rather limited till recently. In this chapter we have summarized the recent studies on the spectroscopic and photochemical properties of squaraines in homogeneous and heterogeneous media as well as the effect of aggregation on these properties. Studies on various other aspects of squaraines, such as photosensitizing and nonlinear optical properties, potential applications as fluorophores in biological systems, and in the design of conducting polymers, are also discussed.

II. SYNTHESIS

Squaric acid undergoes condensation reactions with a variety of nucleophiles to form 1,3-disubstituted products possessing intense absorption in the visible and near-infrared region [7–11]. Schmidt [12] has proposed the widely accepted name squaraine for this class of dyes which was first reported by Treibs and Jacobs [7]. A major part of the earlier work on the synthesis of squaraines was carried out by the former group [7,13] and by Sprenger and Ziegenbein [14–16]. Later, Law and coworkers have contributed extensively to the synthesis of unsymmetric squaraines [17–20]. More recently, Nakazumi et al. [21] have reported a new class of cationic squaraines.

A. Symmetric Squaraines

Synthesis of symmetric squaraines is generally achieved by the condensation of one equivalent of squaric acid, with two equivalents of the respective nucleophiles, under reflux in an azeotropic solvent. In this extremely versatile one-pot synthesis, same substituents are attached on the 1,3-position of squaric acid to yield "symmetric squaraines." Nucleophiles like pyrroles [7], phenols [7], azulenes [14] and N,N-dialkylanilines [15] condense with squaric acid, and a representative example is given in Scheme 1. Nucleophiles obtained by the elimination of HI from 2-methyl substituted quinolium, benzothiazolium, and benzoselenazolium iodide also react with squaric acid to yield 1,3-disubstituted squaraines [16], as shown in Scheme 2 for the synthesis of 5.

    Farnum et al. [22] have synthesized squaraines using an alternative method involving [2+2] cycloaddition reactions of ketenes. Law et al. [23,24] observed that the synthesis of squaraines using alkyl squarates instead of squaric acid yields dyes with improved xerographic properties. This was attributed to the formation of squaraines in smaller particle size in this process. A new type of newer infrared absorbing squaraines containing 2,3-dihydroperimidine as terminal groups has been reported [25]. A series of symmetric bis(stilbenyl) squaraines with extended conjugation, possessing absorption in the longer wavelength (~ 720 nm) have also been reported recently [26].

    Squaric acid diesters react with 2- or 4-substituted pyridinium, quinolinium, and benzothiazolium iodides to yield 1,2-disubstituted dyes [16] (Structure 1). 1,2-Disubstituted dyes were also synthesized using Friedel–Crafts reaction of squaraine dichloride with various nucleophiles as a key step [27].

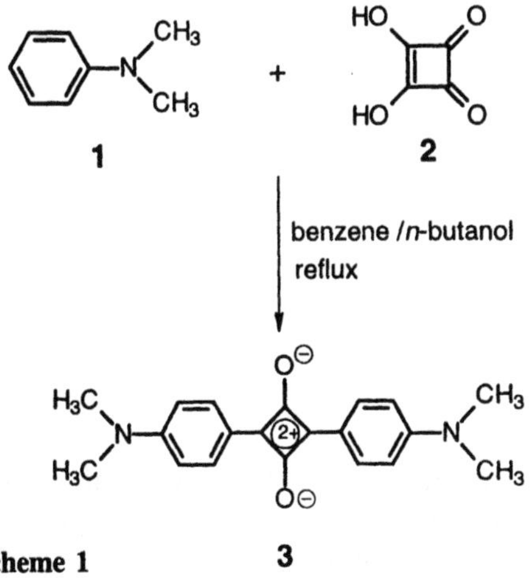

Scheme 1

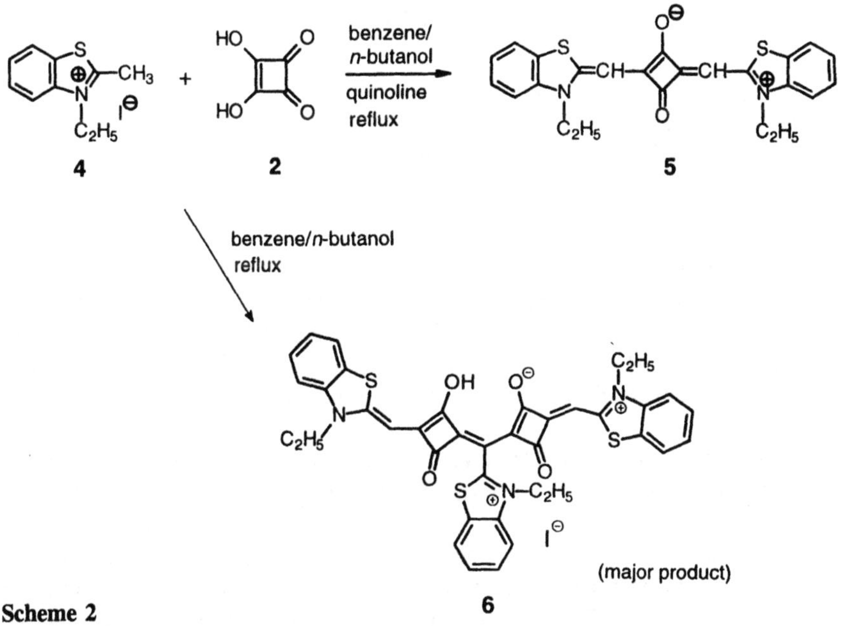

Scheme 2

B. Cationic Squaraines

Nakazumi et al. [21] have recently synthesized alkyl substituted cationic squaraines (for example, 6 in Scheme 2) possessing two squaric acid units and three benzothiazole units in an extended π-electron system. The x-ray structural analysis of these dyes indicated an intramolecular hydrogen bond between two squaric acid units.

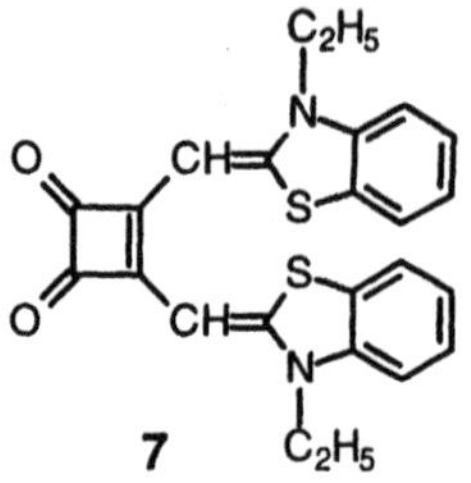

7

Structure 1

C. Unsymmetrical Squaraines

Of the several strategies reported in the literature for the synthesis of unsymmetrical squaraines, [17–20,27–34] two methods are discussed here. The first method, which was developed by Bellus [33] and improved upon by Law [17–20], involves the use of a cycloaddition reaction of an appropriate ketene with tetraethoxyethylene to yield 8 and 9 (Structure 2).

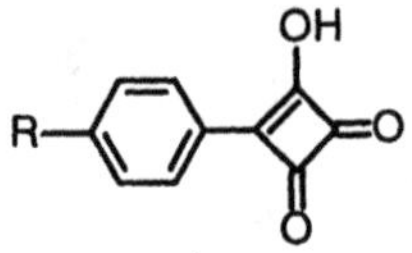

8, R = OCH$_3$

9, R = N(CH$_3$)$_2$

Structure 2

The reaction sequence for the synthesis of 1-(p-methoxyphenyl)-2-hydroxycyclobutene-3,4-dione, 8 [17,19,20] is illustrated in Scheme 3. The second strategy, involving the steps outlined in Scheme 4, was developed by West and coworkers [27] and by Law and Bailey [20]. The monoadducts 8 and 9 can react with various nucleophiles to yield unsymmetrical squaraines [11,17–20,35–39].

Adopting these methods, several symmetric and unsymmetric squaraines have been synthesized by various groups for specific applications such as molecular probes [28–30,40,41], trace metal ion detection [42], second harmonic generation [35,36,43], sensitizers for semiconductors [44,45], singlet oxygen generation [46], organic photoconductive materials for xerographic applications [11,34,47,48], optical recording devices [49], laser protective eyewear [50], electroluminescent materials [51], and photopolymerization initiators [52].

III. ABSORPTION AND EMISSION SPECTRA

Squaraine dyes can be generally described as compounds containing two donor moieties (D) connected to a central C$_4$O$_2$ electron withdrawing group (A)

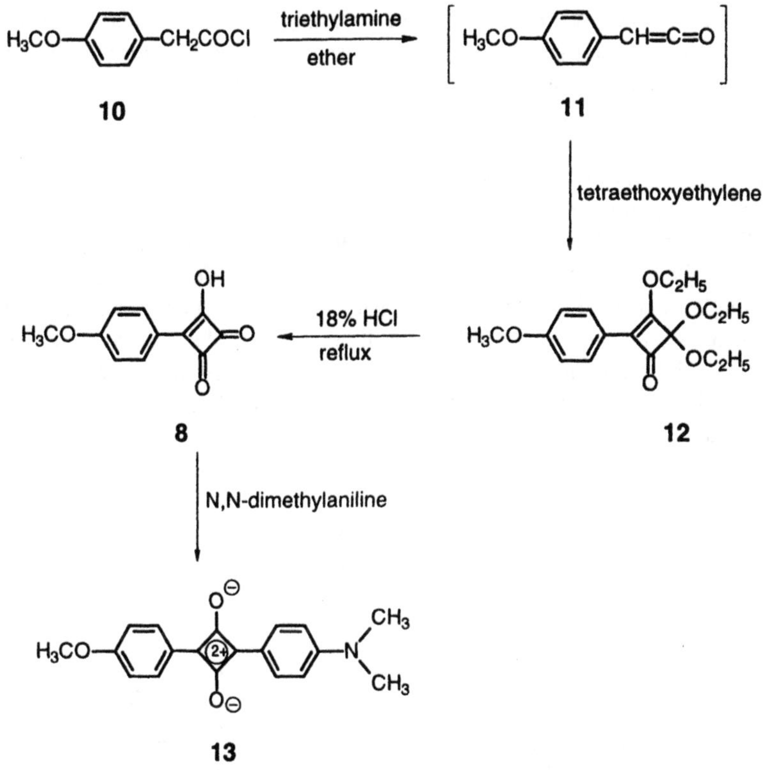

Scheme 3

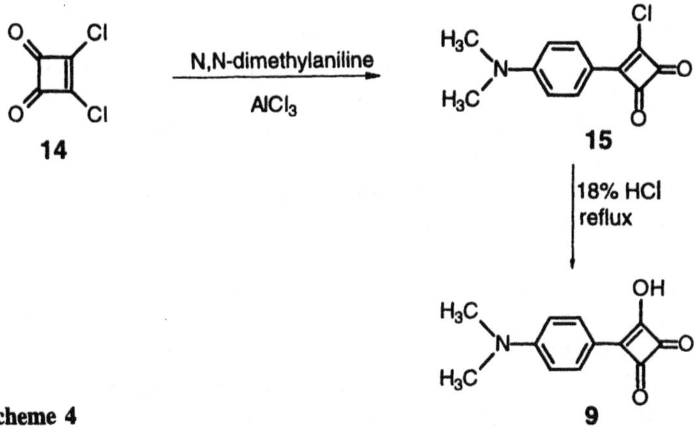

Scheme 4

(Structure 3). The optical properties of squaraine are very similar to those of polymethine cyanine dyes, suggesting a cross-conjugated structure as shown in Structure 4. In such a structure, the bond lengths between the conjugated carbons are expected to be equal, with values between those of a single and a double bond. X-ray crystallographic studies suggest that the average C–C bond length is about 1.414 Å, indicating a considerable amount of double bond character and suggestive of extensive delocalization throughout the molecule [43,53–55]. Whereas the C_4–C_5 bond length of 1.366 Å is indicative of a quinoidlike structure, the C–C bond length of the cyclobutane ring deviates closest to that of a single bond [43]. Thus while the net molecular carbon–carbon bond order suggests a cross-conjugated cyanine like structure, the squaraine also possesses cyclobutadienylium dication character. The true resonance structure must therefore be a complex mixture of the two resonance forms, shown in Structure 3 and 4.

3, R = CH_3

16, R = C_4H_9

Structure 3

Structure 4

MNDO and CNDO semiemperical calculations have also confirmed that the squaraines possess a distinct quinoid character [56]. These calculations have shown that both the ground and the excited states of squaraines are intramolecular D–A–D charge transfer (CT) states. The S_0–S_1 electronic excitation involves a charge transfer process that is primarily confined to the central cyclobutane ring, from each oxygen atom to the four-membered ring with a small degree of CT from the anilino moiety to the central C_4O_2 unit. The intramolecular CT character of this transition, combined with an extended conjugated π-electron donor network, gives rise to the sharp and intense bands in the visible region observed for squaraines [56].

The effect of substituents and solvents on the ground state absorption and emission properties of bis[4-(dimethylamino)phenyl]squaraine 3 (Structure 3) and its derivatives have been studied in detail by Law [57]. Increase in the chain length of the N-alkyl group as well as substitution of the C_4 and C_5 positions brought about bathochromic shifts in the absorption spectra of these squaraines.

These effects could partly be attributed to the minor involvement of the donor groups in the S_0-S_1 excitation. The enhancement in the D–A–D CT character brought about by the substituents can lead to a stabilization of the polarized charges. Law [57] has suggested that the bathochromic shifts can mainly be attributed to improved formation of solute–solvent complexes with increase in the stabilization of these charges. Evidence for the formation of solute–solvent complexes was obtained from solvent effect studies on the fluorescence spectra of these dyes [57,58]. These squaraines exhibit three emission bands, designated as α, β, and γ in the order of decreasing energy, which have been attributed to emissions from the excited state of squaraine, the excited state of the solute–solvent complex, and a relaxed twisted excited state, respectively. Studies on the fluorescence lifetimes of bis[4-(dibutylamino)phenyl]squaraine 16 (Structure 3) in various solvents indicated that the lifetime of the excited state squaraine is independent of solvent (2–3 $\pm$ 0.1 ns), whereas the lifetimes of the excited solute–solvent complex and the relaxed twisted excited state are solvent sensitive, varying from 0.6 to 3.5 ns and from 0.7 to 2.9 ns, respectively [59].

The solute–solvent complexation of squaraines has also been studied by NMR spectroscopy using 16 as a model compound [59]. Based on the correlation between the differences in the chemical shifts of the C_4 (α) and C_5 (β) protons (δ_α–δ_β) and the quantum yield of fluorescence of 16, it was proposed that increase in the solute–solvent complex leads to increased nonplanarity of the squaraine. From the NMR and fluorescence lifetime studies it was proposed that the main nonradiative decay process involves rotation of the C–C bond between the C_4O_2 unit and the phenyl group.

The absorption and emission properties of [4-methoxyphenyl-4'-(dimethylamino)phenyl]squaraine 13 (Scheme 3) and its derivatives, a class of unsymmetrical D'–A–D compounds, show interesting differences from those of the symmetrical squaraines [60,61]. Their absorption spectra are blue-shifted relative to those of the symmetrical squaraines, due to the presence of the less electron donating anisole ring. Introduction of asymmetry through the anisole ring enhances vibronic coupling during electronic transition, producing vibrational fine structure in both absorption and fluorescence spectra. The multiple fluorescence observed for these compounds has been shown to be the sum of vibronic bands of the unsymmetrical squaraine and its solvent complex. The fluorescence quantum yields of the unsymmetrical squaraines are a factor of 30 lower than those of symmetrical squaraines, and these differences have been attributed to radiationless decay, due to rotation of the C–C bond between the anisole ring and the central four-membered ring. Due to the anisole ring being a weaker electron donor than the dimethylaminophenyl group, the C–C bond between the anisole ring and the four-membered ring will have a considerable amount of single bond character, which can result in a lowering of the barrier for C–C bond rotation, facilitating rotational relaxation of unsymmetrical squaraines.

For the symmetrical bis[4-(dimethylamino)phenyl]squaraine 3 and its derivatives, complexation with alcohol solvents was shown to bring about bathochromic shifts in their absorption spectra [58]. For the bis(benzothiazolylidene)squaraine, such as 5 (Scheme 2), however, complexation with alcohol solvents brought about marked hypsochromic shifts in its absorption and emission maxima [62] (Figs. 1A and B).

The hypsochromic shifts have been attributed to the formation of hydrogen bonds between the alcohols and the oxygen atoms of the central cyclobutane ring, which can hinder the intramolecular charge-transfer process. A strong correlation was observed between the extent of the hypsochromic shift and the hydrogen bond donating strength of the alcohol. Such hypsochromic shifts have also been observed for similar squaraine dyes [36]. The differences in the behavior of the [4-(dialkylamino)phenyl]squaraines and bis(benzothiazolylidene)squaraines suggest that the quinoid like structure may be more pronounced in the latter.

IV. ACID–BASE PROPERTIES

There are very few studies related to the acid–base properties of squaraines. Based on the spectral changes, on addition of trifluoroacetic acid (TFA) to solutions of 3 in dichloromethane [63], stepwise protonation of the oxygen and nitrogen atoms as shown in Scheme 5 has been proposed.

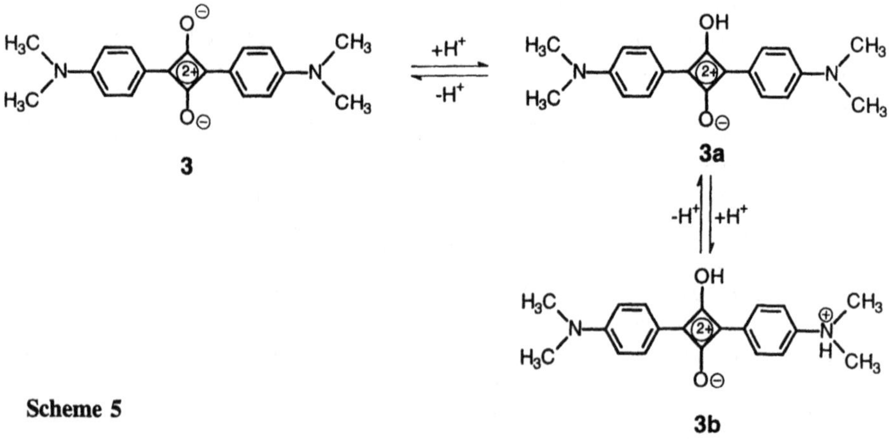

Scheme 5

The first of these protonations does not significantly affect the conjugation and brings about only minor changes in the absorption spectrum. The

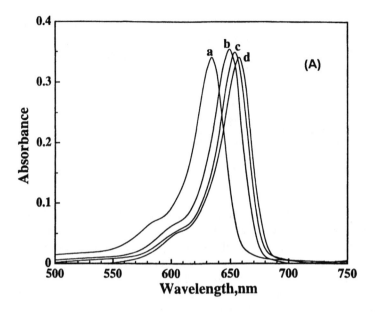

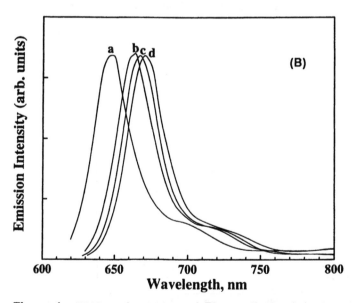

Figure 1 (A)Absorption spectra and (B) normalized emission spectra of 5 in hydrogen-bonding solvents: (a) trifluoroethanol, (b) methanol, (c) ethanol, and (d) 2-methoxyethanol. (From Ref. 62.)

addition of TFA in higher concentrations, which leads to additional protonation of one of the nitrogens, impedes conjugation in the molecule, bringing
about drastic changes in the UV spectrum. The 636 nm band decreases in intensity, and new bands appear at shorter wavelengths [63].

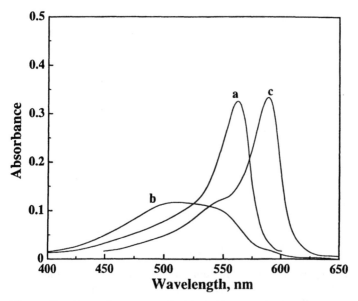

17, X = H ;   Y = H

18, X = OH; Y = H

19, X = OH; Y = Br

20, X = OH; Y = I

Structure 5

The acid–base properties of bis(2,4-dihydroxyphenyl)squaraine, 17 [64]
and bis(2,4,6-trihydroxyphenyl)squaraine 18 [65] (Structure 5) have been studied by following the changes in the absorption spectra as a function of pH.
Figure 2 shows the absorption spectra of 18 at different pH [65]. Four distinct
species could be detected (Scheme 6), and the pKa values, as measured from
the changes in absorption, were 3.5, 7.0, and 9.5. Above pH 12, the dye was
unstable and underwent irreversible chemical change. Similarly, two different
protonation equilibria with pKa values of 3.7 and 7.4 were observed for 17
[64].

Figure 2   Absorption spectra of 18 in 30% v/v methanol/water solutions: (a) pH 3.1,
(b) pH 5.5, and (c) pH 9.0. (From Ref. 65.)

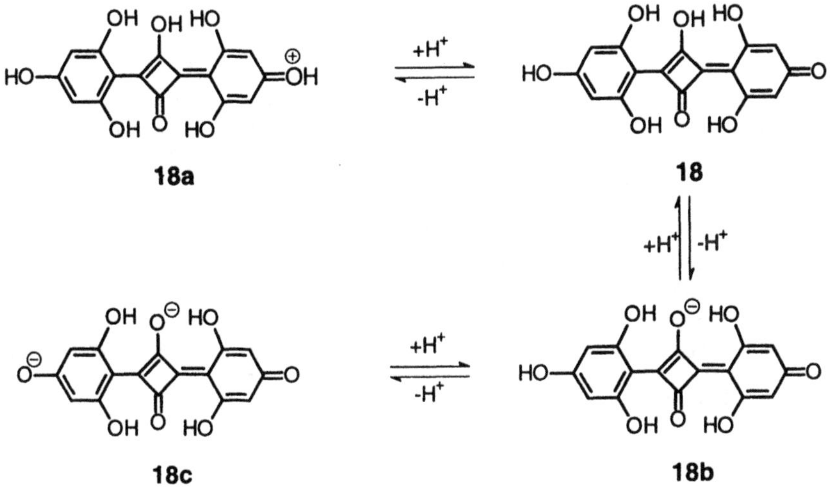

Scheme 6

The singly deprotonated forms of 17 and 18 were the most fluorescent species for both dyes. Similarly, acid–base properties of bis(3,5-dibromo-2,4,6-trihydroxyphenyl)squaraine 19 and bis(3,5-diiodo-2,4,6-trihydroxyphenyl) squaraine 20 (Structure 5) were also investigated [46]. The pKa values measured for the bromo derivative 19 were 2.3, 5.1, and 6.7, and the corresponding values were nearly the same for the iodo derivative. The comparison of the pKa of halogenated squaraines 19 and 20 with that of the parent dye 18 shows that the halogenated squaraines are more acidic, and this is attributed to the electron withdrawing nature of the bromo and iodo substituents.

The bis(benzothiazolylidene)squaraine derivatives show a sharp band around 645 nm, which became broad with a maximum around 515 nm on addition of acid. The acid–base equilibrium for the dye with a pKa of 6.4 was attributed to protonation/deprotonation of the nitrogen in the benzothiazole group [66].

V. EXCITED AND REDOX STATE PROPERTIES

A. Bis[4-(dimethylamino)phenyl]squaraine and Its Derivatives

The properties of the excited states and the reduced and oxidized forms of bis[4-(dimethylamino)phenyl]squaraine (3), bis[(4-(dimethylamino-2-hydroxy-phenyl)]squaraine (21), and aza-crown ether squaraine derivatives 22–24 (Structure 6) have been investigated by picosecond and nanosecond laser flash photolysis [67,68], and the results are summarized in Table 1.

Table 1 Excited State Properties of Squaraine Dyes

Dye	λ_{max} Abs. (nm)	λ_{max} Em. (nm)	Φ_f	λ_{max} S_1–S_n (nm)	τ_S (ns)	Φ_T	λ_{max} T_1–T_n (nm)	$E°$ D/D$^+$ V vs SCE	$\lambda_{max,}$ Dye$^+$ (nm)	$\lambda_{max,}$ Dye$^+$ (nm)
3	628	654	0.45	480	1.50	$<10^{-3}$	540	0.305	670	400
21	636	658	0.84	475	3.00	$<10^{-3}$	565	0.365	550	–
22	636	660	0.48	480	1.25		570	0.713	680	
23	634	657	0.50	480	1.20		570	0.792	680	
24	636	660	0.43	480	1.45		570	0.715	680	
25	568	587	0.08	442	0.22	0.007	420	0.976	505	435

Sources: Refs. 42, 67, 68, 70.

3, X = H
21, X = OH

22, n = 1
23, n = 2
24, n = 3

Structure 6

1. Excited Singlet and Triplet States

The transient absorption spectra, recorded at different time intervals, follow-ing 532 nm laser pulse (18 ps) excitation of 21 (Fig. 3) shown an absorption maximum around 480 nm [67]. The lifetime of the excited singlet state and the fluorescence quantum yield of 21 are about twice that of 3. This has been at-tributed to hydrogen bonding between the OH group in the phenyl ring and the CO group in the central cyclobutane ring, which can restrict the rotational relaxation of the excited state of 21, as proposed earlier by Law [57]. The excited singlet state of monoaza-crown ether derivatives of squaraines 22–24 also absorb in the 480 nm region with lifetimes ranging from 2–3 ns in ben-zene to ~100 ps in water [68].

The intersystem crossing efficiencies of 3 and 21 were very small (1%). The triplet–triplet spectra of these dyes were obtained using the triplet–triplet energy transfer method, employing 9,10-dibromoanthracene (DBA) as the sen-sitizer (Fig. 4). The transient absorption spectrum, recorded immediately af-

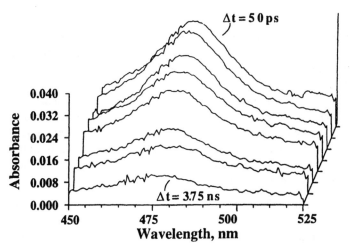

Figure 3 Transient singlet excited spectra of 21, recorded following 532 nm laser pulse excitation of 10 μM of 21 in CH$_2$Cl$_2$ at time intervals of 0.05, 0.15, 0.25, 0.5, 0.75, 1.75, 2.75, and 3.75 ns. (From Ref. 67.)

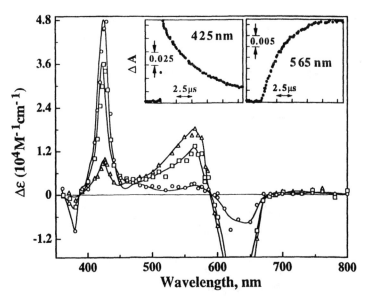

Figure 4 Energy transfer from triplet excited state of DBA to 21 in CH$_2$Cl$_2$. Transient absorption spectra were recorded, following 355 nm laser pulse excitation of a solution containing DBA (0.1 mM) and 21 (10 μM) at time intervals (O) 0 μs, (□) 1.3 μs, and (Δ) 5.6 μs. The absorption time profiles in the insets show decay of ^{3}DBA* at 425 nm and the formation of ^{3}21* at 565 nm. (From Ref. 67.)

ter laser pulse excitation (λ_{max} 425 nm) corresponds to the sensitizer triplet, while those recorded at time intervals greater than 10 μs correspond to the triplet state of 3. The triplet lifetimes of 3 and 21 were identical, indicating that the triplet excited states, unlike the singlet excited states, are insensitive to the presence of OH groups on the phenyl ring.

The triplet excited state characteristics of the crown-ether squaraines, determined by energy transfer sensitization, are also summarized in Table 1. The triplet states of the crown ether squaraine derivatives underwent a self-quenching process with bimolecular quenching rate constants varying from 0.5 $\times$ 10^9 to 1.9 $\times$ 10^9 M^{-1}s^{-1} with the ground state of the dyes to yield the radical cation and anion (Reactions 1 and 2).

$$^3(dye)^* + dye(S_0) \rightarrow 2\ dye(S_0) \tag{1}$$

$$^3(dye)^* + dye(S_0) \rightarrow dye^{\pm} + dye^{\pm} \tag{2}$$

2. Cationic and Anionic Radicals

Excitation using high-intensity 532 nm laser pulses led to photoionization of these dyes [67]. The transient absorption recorded following the laser flash excitation of 3 is shown in Fig. 5. The linear dependence of the absorption at

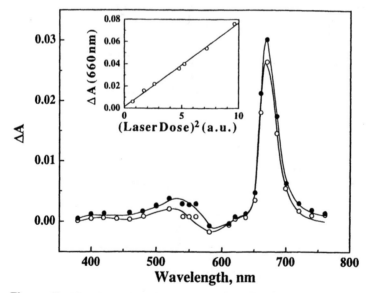

Figure 5 Photoionization of 3. The difference absorption spectra were recorded ($\bullet$) 0 μs and ($\circ$) 50 μs after 532 nm laser pulse excitation of 3 (10 μM) in CH$_2$Cl$_2$. The inset shows the dependence of A at 660 nm on the square of the laser intensity. (From Ref. 67.)

660 nm on square of the laser dose (inset, Fig. 5) indicated that the transient formation occurs via a biphotonic process. The transient intermediate was identified by comparison of the spectrum of the same species produced via pulse radiolytic oxidation in oxygen saturated methylene chloride solutions. The low oxidation potentials of these dyes facilitated their oxidation by the oxidative $RCl^\bullet$ radicals, produced in the radiolysis of chlorinated hydrocarbon solvents [69].

Direct excitation of the monoaza-crown ether derivative 22 [68] by 532 nm pulses gave rise to a transient absorption spectrum with maxima at 540 nm and 680 nm (Fig. 6). The transient absorption at 540 nm was quenched by oxygen and was attributed to the excited triplet state, by comparison with the triplet sensitization studies. The transient absorption at 680 nm was attributed to the radical cation and anion of the dye, formed via Reactions 3 and 4.

$$22 \xrightarrow{\ h\nu\ } 22^* \xrightarrow{\ h\nu\ } 22^{\overset{+}{\cdot}} + e^-_{solv} \tag{3}$$

$$e^-_{solv} + 22 \longrightarrow 22^{\overset{\cdot}{-}} \tag{4}$$

In air saturated solutions, the formation of the radical anion of 22 was suppressed due to scavenging of the solvated electrons by oxygen and a reduction in intensity of the 680 nm transient observed, suggesting that the radical cation and anion absorb in the same region.

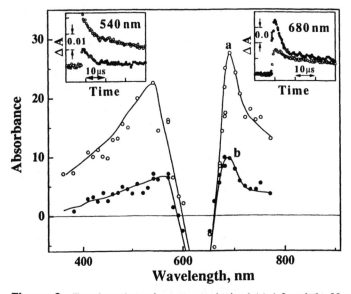

Figure 6 Transient absorption spectra obtained (a) 1.2 and (b) 22 μs after 532 nm laser pulse excitation of 22 (20 μM) in deaerated acetonitrile. Insets show absorption time profile at 540 and 680 nm of (○) N_2 saturated and (●) O_2 saturated solutions. (From Ref. 68.)

The EPR and UV-Vis spectral properties of radical cations from bis[4-(dimethylamino)phenyl]squaraine 3 generated by thermal oxidation has also been studied [63]. Oxidation of 3 in dichloromethane by tris(4-bromophenyl)-aminium (TBPA)$^+$ hexachloroantimonate led to a depletion of the intense 626 nm band and formation of a 668 nm band ($\varepsilon \approx 1.97 \times 10^5$ M^{-1} cm^{-1}) due to the formation of the one-electron oxidized species. This agrees well with the absorption spectra of the radical cation recorded by laser flash photolysis [67]. Although the oxidation of the radical cation of 3 to its dication form occurs at a potential of 1.0 V vs NHE (normal hydrogen electrode) [2] and TBPA$^+$ has a redox potential of 1.30 V, this conversion could not be observed even on addition of a 20-fold excess of TBPA$^+$. On addition of trifluoroacetic acid to a solution of 3, which had been oxidized by one equivalent of TBPA$^+$, the UV-Vis spectrum changed drastically, with the band with λ_{max} at 668 nm being replaced by one with λ_{max} at 628 nm. This was also reflected by significant changes in the EPR spectrum. The species formed under these conditions has been suggested to be the doubly protonated radical cation of the dye.

3. Cyclic Voltammetric Studies

The bis[4-(dialkylamino)phenyl]squaraine dyes exhibit two reversible oxidation peaks [2]. The first oxidation peak, as determined by cyclic voltammetry, was 0.35 and 0.41 V vs. Ag/AgCl for 3 and 21, respectively. Dyes 22 and 23 also exhibit two distinct reversible oxidations in methylene chloride. The first oxidation occurs in the 0.71–0.79 V vs. SCE potential range, and the second oxidation occurs in the 1.06–1.17 V vs. SCE potential range [68]. Both of these potentials are significantly greater than the corresponding oxidation potentials of bis[4-(dimethylamino)phenyl]squaraine 3 and its derivatives. Figure 7 shows the absorption spectra of 23, recorded in a spectro electrochemical cell at an applied potential of +0.8 V, which ensured selective, one-electron oxidation of the squaraine dye. The difference absorption spectrum exhibits a strong absorption band in the red region with a maximum at 680 nm and a bleach at 640 nm. The intensity of both bands increases with time as more of the dye gets oxidized on the electrode surface (inset, Fig. 7). Similar results were obtained for 22 and 24. The 680 nm absorption, obtained for the one-electron oxidized species is similar to the absorption spectra of radical cations of bis [4-(dialkylamino)phenyl]squaraines produced by laser flash photolysis [67] and thermal oxidation [63].

B. Bis(3-acetyl-2,4-dimethylpyrrole)squaraine

Unlike the bis[4-(dimethylamino)phenyl]squaraine derivatives, the pyrrole derivative 25 (Structure 7) is relatively nonfluorescent ($\phi_f = 0.08$) with an extremely short excited singlet state lifetime ($\tau_f = 222$ ps) [70]. The intersystem

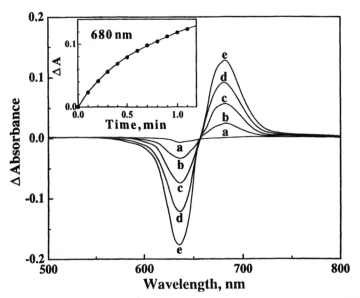

Figure 7 Difference absorbance spectra obtained by applying a potential of 0.8 V vs. Ag/AgCl in CH_2Cl_2 containing 50 μM of 23 and 0.1 M TBAP. Inset shows the increase in absorbance at 680 nm with time. (From Ref. 68.)

25

Structure 7

crossing efficiency was also very low (Table 1). The triplet excited state of the dye was characterized by both direct excitation and triplet sensitization using 9,10-dibromoanthracene as the sensitizer. Compound 25 undergoes irreversible oxidation around 0.98 V vs. SCE. The radical cation of 25, produced by pulse radiolysis of methylene chloride solution of 25, showed an absorption maximum at 505 nm and a bleach at 535 nm corresponding to depletion of the ground state.

VI. PHOTOPHYSICAL PROPERTIES OF SQUARAINES IN HETEROGENEOUS MEDIA

The intramolecular charge-transfer nature of the electronic transitions of squaraines makes their photophysical properties highly sensitive to the properties of the surrounding medium. In addition, the ability of these dyes to form solute–solvent complexes and their sensitivity to pH, polarity, and hydrogen bonding ability of solvents can make these dyes useful as probes for assessing the microstructures of organized assemblies and of polymers. Squaraines also have a strong tendency to form aggregates, and this property is largely dependent upon the nature of the surrounding medium.

A. Effect of β-cyclodextrin

Addition of β-cyclodextrin (β-CD) to aqueous solutions of 18b (Scheme 6) brought about a gradual red shift in the absorption band [71]. These changes were accompanied by a red shift and enhancement in fluorescence yield (Fig. 8). At the highest concentration of β-CD studied, the fluorescence yield of 18b (ϕ_f = 0.18) was nearly 90-fold of that observed in water. Benesi–Hildebrand analysis of the dependence of fluorescence yield of 18b on β-CD concentration indicated a 2:1 complex formation between β-CD and 18b. This complex has been envisioned as two β-CD molecules encapsulating the two phenyl rings of 18b as shown in Scheme 7.

　　　Encapsulation by β-CD molecules can enhance the microviscosity around the dye anion, which can restrict the free rotation of the two phenyl groups. Additionally, the hydrogen bonding of 18b with the solvent molecules will also

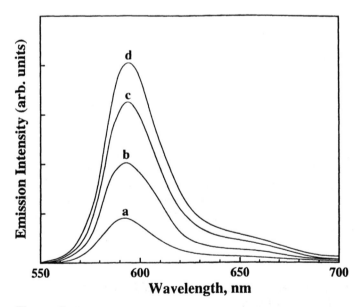

Figure 8　Influence of addition of β-CD concentrations on the emission spectrum of 0.35 μM 18 in aqueous solution at pH =8.6 [β-CD]: (a) 0.06, (b) 0.3, (c) 1.2, and (d) 2.4 mM. Excitation wavelength 560 nm. (From Ref. 71.)

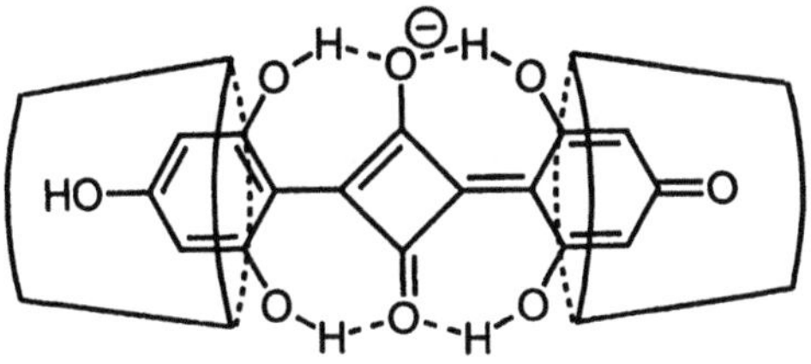

Scheme 7

be restricted. Due to a combination of these effects, the nonradiative decay routes become far less efficient, leading to the observed enhancement in fluorescence yield. From picosecond laser flash photolysis studies, the rate constant for nonradiative internal conversion was estimated as 1.18×10^{10} s^{-1} and 6.6×10^8 s^{-1} in the absence and presence of β-CD, respectively. Spectral properties of the excited singlet state of 18 in the presence and absence of β-CD obtained by laser flash photolysis are summarized in Table 2.

Complexation by β-CD also brought about significant enhancement in the chemical stability of the dye anion, making it insensitive to air oxidation in aqueous solutions. Formation of 1:1 complexes of α- and β-CD with squaraines [37,42] and 1:2 γ-CD squaraine complexes have also been reported [37].

B. Effect of Polymers

In basic methanolic solutions of 18, addition of poly(4-vinylpyridine) (P4VP) brought about a marked red shift in the absorption band of the dye anion [65], which was accompanied by a significant enhancement in the fluorescence quantum yield (Fig. 9, curve b). In neutral methanolic solutions, where the dye is predominantly in its neutral form (18) with minor amounts of the anionic form (18b) being present, addition of P4VP brought about a selective complexation of the anionic form, resulting in a slight enhancement in the fluorescence yields (Fig. 9, curve a). At the low concentrations of P4VP employed in these studies, the change in macroviscosity would not be sufficient to affect the fluorescence yields significantly. It was proposed that entrapment of the dye anion by the polymer macrocages leads to the observed effects, with the driving force for the entrapment being hydrophobic interactions between P4VP and the dye anion. The absorption maxima and lifetimes of the singlet excited states of the neutral and anionic forms of 18 are summarized in Table 2. For both forms, enhancement in lifetime was observed in P4VP-containing solutions, although the effect was much larger for the anionic form.

The molecular interactions between bis[4-(dimethylamino)phenyl]-squaraine 3 and polyvinylbutyral have been studied by fluorescence spectroscopy

Table 2 Absorption and Emission Characteristics of bis(2,4,6-trihydroxyphenyl)-squaraine (18) in the Presence and Absence of (a) β-cyclodextrin [β-CD] in water and (b) poly(4-vinylpyridine) [P4VP] in methanol

Solvent		λ_{max} (nm)		Φ_f	τ_S (ps)	Absorption max (S_1-S_n)(nm)
		Absorption	Emission			
	[β-CD]					
Water	0	584	595	0.002	85	—
Water	2.4	598	608	0.16	1200	—
	[P4VP]					
Methanol	0	510	597	<0.001	<30	414
Methanol	0.1	510,602	610	0.02	~100	
Methanol containing 3 mM KOH	0	588	601	0.02	185	420,490
Methanol containing 3 mM KOH	0.1	602	615	0.2	2000	425,495

Sources: Refs. 65, 71.

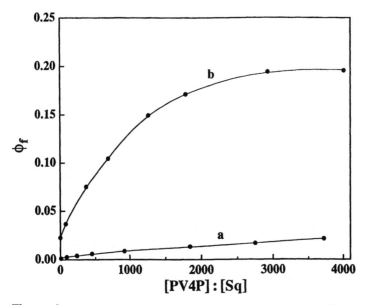

Figure 9 Dependence of fluorescence quantum yield of 18 on the concentration of P4VP: (a) neutral methanol, and (b) basic methanol ([KOH] = 3.0 mM). Excitation wavelength, 540 nm. (From Ref. 65.)

[72]. Based on changes in the fluorescence profile of the dye, complex formation between the dye and the hydroxyl groups of the polymer was proposed.

The absorption and emission properties of bis[4-(dimethylamino)-phenyl]squaraine (3), bis[4-(dimethylamino)-2-hydroxyphenyl]squaraine (21), bis(2,4,6-trihydroxyphenyl)squaraine (18), and an azulyl derivative of squaraine in poly(methyl methacrylate) and polystyrene have been reported [73]. The fluorescence lifetimes of new nanoseconds were observed for 3 and 21 in the polymer films, whereas for 18, the fluorescence lifetime was much shorter (10 ps). As discussed above, the neutral form of 18 is relatively nonfluorescent. A rapid excited state proton transfer tautomerization or deprotonation was proposed for the rapid deactivation of the excited state of 18.

C. Aggregation Properties of Squaraines

Photoconducting and other solid state properties of materials are dependent both upon the intrinsic molecular properties of the material as well as the intermolecular interactions that occur in the solid state. The sharp and intense bands of squaraines for example become broad and red-shifted in the solid state [2]. Studies on aggregates can help in developing a molecular level understanding of their solid state properties. In view of this, the aggregation behavior of squaraines in solutions [74–76], in vesicles [39,77], in thin films [37,38,78,79], in the solid state [80], and on semiconductor surfaces [44,45,64,81] has been extensively studied.

1. Aggregation in Solutions

The equilibrium constants and thermodynamic parameters for aggregate formation of bis(2,4-dihydroxyphenyl)squaraine (17) and bis(2,4,6-trihydroxyphenyl)squaraine (18) (Structure 5) have been studied by absorption spectroscopy [76]. In dry acetonitrile, at low concentrations, the dyes exhibited a broad absorption band centered around 480 nm. At higher concentrations, a sharp new band centered around 565 nm was observed (Fig. 10). Based on the dependence of the absorption spectral changes on dye concentration, the aggregated forms were determined to be dimeric species, and intermolecular hydrogen bonding has been proposed to be responsible for the dimerization process. This was confirmed by addition of hydrogen bond donating or accepting solvents to these solutions, which led to a decrease in the intensity of the dimer band and a concomitant increase in intensity of the monomer band.

According to exciton theory [82,83], the excited state energy level of the monomeric dye splits into two upon aggregation, one level being lower and the other higher in energy than the monomer excited state. The transition to the higher state is forbidden for head-to-tail (J-type) dimers, whereas the lower

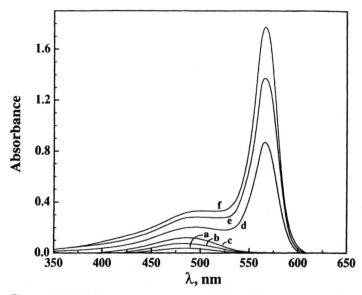

Figure 10 Visible absorption spectra of 17 at different concentrations in dry acetonitrile: (a) 0.5, (b) 1.1, (c) 1.6, (d) 8.1, (e) 9.6, and (f) 16.1 μM. (From Ref. 76.)

excited state is forbidden for the sandwich type (H-type) dimers. The sharp bands observed in the red region of the monomer bands indicate that J-type aggregates predominate in the case of these dyes.

Addition of iodine to acetonitrile solutions of 18 led to an increase in aggregate formation, and this has been attributed to the formation of charge-transfer complexes between iodine and 18. Squaraine dyes such as bis[4-(dimethylamino)phenyl]squaraine 3 have been observed to form charge transfer complexes with iodine [84].

The emission and excitation spectra of 17 in acetonitrile, shown in Fig. 11, suggest that the dimer aggregates tend to dissociate in the excited state [76]. At lower concentration, the dye is predominantly in the monomer state and exhibits emission in the 580 nm region (curve a'). The excitation spectrum (curve a) closely matches the absorption spectrum of the monomer. The large Stokes shift in emission is suggestive of deprotonation of the monomeric form in the excited state. The emission (curve b') and excitation (curve b) bands, measured in solution containing higher concentrations of the dye, indicate that the emission from the aggregate closely matches the monomer emission. This is suggestive of dissociation of the aggregate (Scheme 8) in the excited state, possibly involving proton exchange as well. Confirmation of dissociation of the

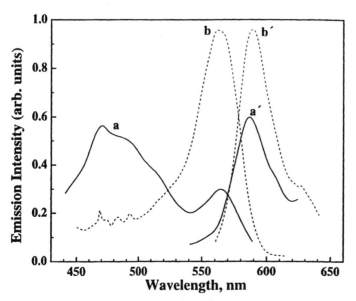

Figure 11 Excitation (a, b) and emission (a', b') spectra of 17 in acetonitrile at concentration levels of 3.7 µM (a, a') and 16.2 µM (b, b'). The excitation wavelengths were 520 and 540 nm for a' and b' and the emission wavelengths were 600 and 630 nm for a and b, respectively. Intensities were normalized for comparison. (From Ref. 76.)

dimer in the excited state was obtained from picosecond laser flash photolysis studies.

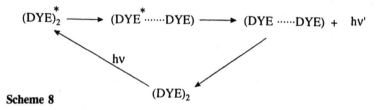

Scheme 8

The excited singlet states of 17 and 18 in the neutral form exhibited characteristic transient absorption in the 430–450 nm range, with lifetimes of 30 and 130 ps, respectively. Corresponding bleach and recovery of the ground state absorption was also observed (Fig. 12) [76]. At higher concentrations of the dye, bleaching of the dimer band was observed, and this was accompanied by the formation of a transient absorption band in the 430–450 nm region, indicating formation of the excited singlet state of the monomer. These results suggest that the squaraine aggregate dissociates in the excited state, as shown in Scheme 8. The decay time of the excited singlet state matched the recovery time of the dimer band. The rapid recovery of the dimer band suggests that

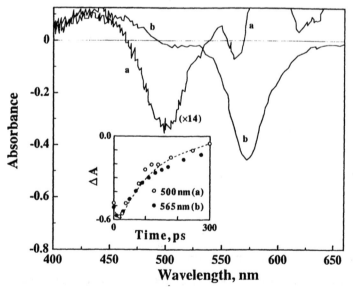

Figure 12 Transient absorption spectra, recorded immediately after 532 nm laser pulse excitation of 17 in acetonitrile. The concentrations of 17 were (a) 3 μM and (b) 15 μM. Inset shows the absorption time profiles at 500 and 565 nm. (From Ref. 76.)

dissociation of the dimer in the excited state leads to a solvent caged excited state/ground state dye pair.

The aggregation behavior of the squaraine dyes 26 and 27 (Structure 8) in water in the presence and absence of polyvinylpyrrolidone has been studied [66]. In aqueous solutions, these dyes show two absorption bands with maxima around 595 and 645 nm, which have been attributed to those of the aggregated and monomeric forms, respectively. The aggregated forms were identified as dimers from concentration dependent changes in the absorption spectra. The blue shift in absorption of the aggregated forms indicate the formation of H-type aggregates [82,83], unlike in the case of the bis(hydroxyphenyl)squaraine derivatives, where J-type aggregates were obtained [76].

26, R = CH$_2$—⟨ ⟩—CO$_2$H

27, R = CH$_2$CO$_2$H

Structure 8

Addition of polyvinylpyrrolidone (PVP) had interesting effect on the aggregation behavior of 27 [66]. At low concentrations of PVP (<0.25 mM), enhancement in aggregate formation was observed, whereas at higher concentrations of PVP (>0.25 mM), a decrease in the intensity of the aggregate form and formation of a fluorescent species with absorption maximum at 667 nm were observed. At low concentrations of PVP, hydrophobic interactions of 27 with the polymer can bring about a nonhomogeneous distribution of the dye, and the increased concentration of the dye around the polymer can enhance aggregate formation. At higher polymer concentrations, more sites become available for the dye to complex, which can lead to a breakup of the dye aggregate. Based on the similarities in the absorption and emission properties of the dye in aqueous solutions containing higher concentrations of PVP, with those observed in aprotic solvents such as DMSO and DMF, it was proposed that at higher polymer concentrations, encapsulation of the dye in polymer microcages protects it from hydrogen bonding interaction with the solvent water molecules.

UV-Vis spectroscopic studies showed that the squaraine dye 28 (Structure 9) can exist in the form of two aggregates, formed preferentially in different DMSO–water compositions [74]. In pure DMSO and DMSO–water mixtures containing more than 70% DMSO, 28 exists in the monomeric form that has a sharp absorption band with a maximum around 600 nm. In 70% v/v DMSO–water mixtures, 28 exists in a stable aggregated form that has a blue-shifted absorption band (λ_{max} = 530 nm, $\varepsilon \approx$ 80,000 M^{-1} cm^{-1}). In DMSO–water mixtures containing 20–50% of DMSO, a second type of aggregate form with a broad absorption band in the region 550–700 nm was observed. This species underwent a time dependent change to the aggregated species, observed in the DMSO-rich solvent mixture (Fig. 13). The rate of this change increased with increase in the DMSO content. Similar observations on the formation of thermodynamically and kinetically preferred aggregates were made for other bis[4-(dialkylamino)phenyl]squaraine derivatives [75].

28

Structure 9

2. Aggregation in Langmuir–Blodgett Films

The aggregate properties of surfactant squaraines 29–31 (Structure 10), which were designed to orient the "bricklike" squaraine chromophore in three different directions, when organized in monolayers and Langmuir–Blodgett films (LB films), have been reported [38].

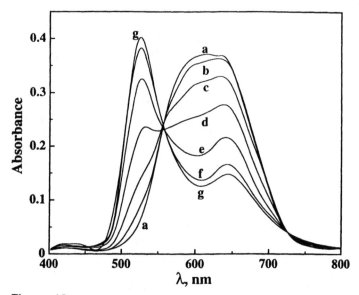

Figure 13 UV-Vis spectra of 28 in 30% v/v DMSO/water, monitored at 12 min intervals after sample preparation showing the dynamic conversion of aggregated form with spectrum (a) to the aggregated form with spectrum (g). (From Ref. 74.)

29, $R_1 = R_2 = CH_3$; $R_3 = R_4 = n\text{-}C_{18}H_{37}$

30, $R_1 = R_3 = CH_3$; $R_2 = R_4 = n\text{-}C_{18}H_{37}$

31, $R_1 = R_2 = R_3 = R_4 = n\text{-}C_{18}H_{37}$

Structure 10

Surface pressure–area isotherm studies show that the squaraine chromophores of 29, 30, and 31 orient as designed, namely being laid down vertically along the long axis, vertically along the short axis, and flat, respectively, on water. Thus the surface pressure–area isotherm of 29 on water surface exhibits only one transition with a limiting area of 52 Å²/molecule. From the single crystal x-ray structure of bis[4-methoxyphenyl]squaraine [55], x-ray powder diffraction patterns of bis [4-(dialkylamino)phenyl]squaraines [24], and molecular modeling studies, the dimensions of the squaraine chromophore have been estimated as 17 Å × 7 Å × 3.5 Å. Based on these values, the limiting area upon compression for 29, residing vertically on its long axis on water, would be 25 Å²/ molecule. However, depending upon the nature of the stearyl chain attachment, a spread of about 20–25 Å can be expected in these values [85]. Based on this and the nature of the intermolecular interaction of the aggregates, which involve C–O dipole–dipole interactions, it was proposed that 29 aligns vertically on its long molecular axis. The limiting area of the isotherm was 60 Å²/molecule for 30, which agrees well with the assumption that the molecule resides vertically along the short axis. Compound 31 showed two transitions of limiting areas of 126 and 85 Å²/molecule. Since the calculated molecular area of squaraine chromophore of 31 is 119 Å²/molecule, excluding the molecular chains, these results suggest that this dye lies flat on the water surface upon compression. Since the limiting area of the second transition is smaller than the molecular area of the squaraine chromophore, the pressure-isotherm curve suggests that the chromophores may become tilted and stack on each other at surface pressures higher than 10 mN/m.

The aggregation of 29 and 30 was studied by absorption spectroscopy. By comparison with the absorption spectra of model squaraine aggregates, it was concluded that 29 forms sandwich type dimers, which is essentially controlled by C–O dipole–dipole interactions. For 30, the aggregate band is red-shifted to that of the monomer, and this has been attributed to an aggregated form involving intermolecular charge-transfer interactions between the electron-donor and electron-acceptor groups of squaraines. The difference in aggregational behavior was attributed to orientational effects induced by the LB film technique.

More recently, it was observed that the blue-shifted aggregate of 29 could be converted to two different aggregates at higher temperatures from 65 to 100°C [38]. The absorption spectrum of a monolayer of 29 on a glass substrate (Fig. 14) showed an absorption peak at 530 nm that was attributed to the blue-shifted aggregate. On heating the film at 65°C under ambient conditions, the absorption peak at 530 nm decreases in intensity and an increase in absorption at around 660 nm was observed, indicating conversion of one aggregated form to another. Interestingly, the heat-generated, red-shifted aggregate in the LB film of 29 was similar to that obtained in freshly prepared LB film of 30. When the blue-shifted and the red-shifted aggregates of 29 were heated at 105°C for 10–20 min, absorption due to these species was replaced by that of a new species with a maximum at 695 nm (Fig. 15). This new species has been assigned to a J-aggregate of 29 based on similar characteristics of other squaraine dyes [76,86–88] and cyanine dyes [89,90]. When both the red-shifted and the J-aggregated films were treated with water vapor at 65°C, a gradual reversal in the aggregation process leading to regeneration of the blue-shifted aggregate was observed. The formation of J-aggregates was not observed for 30 and 31. Steam treatment of the red-shifted aggregates of these dyes did however lead to the

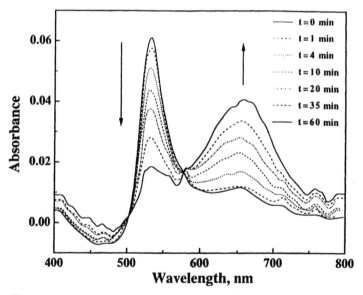

Figure 14 Absorption spectra of the LB film of 29 on glass as a function of heating time (65°C). (From Ref. 38.)

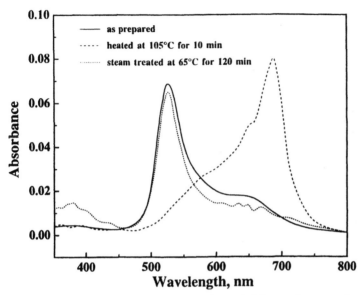

Figure 15 Absorption spectra of the LB film of 29 on glass as a function of heating and subsequent steam treatment. (From Ref. 38.)

formation of the blue-shifted aggregate. It was also observed that these dyes form blue-shifted aggregates in DMSO. On addition of the dye to DMSO–water mixtures of increasing water concentration, a decrease in the intensity of the 530 nm band and an increase in intensity of absorption at 650–680 nm were observed. Similar effects in DMSO–water mixtures were observed earlier for other bis[4-(dialkylamino)phenyl]squaraines by Buncel et al. and McKerrow et al. [74,75].

Based on these studies, Liang et al. [38] have proposed the molecular arrangement shown in Fig. 16 (a and b) for the blue-shifted and J-aggregates of 29 in LB films. The corresponding molecular arrangements proposed for the blue- and red-shifted aggregates of 30 are shown in Fig. 16 (c and d). In the LB films of 29, each squaraine molecule occupies 26 Å^2 in the vertical orientation, whereas its limiting molecular area in LB films was estimated as 51 Å^2 [38]. The excess free area enables tilting movement of the squaraine chromophores within the LB film on heating. Due to the relatively good match of the estimated areas occupied by the squaraines 30 and 31 to the limiting molecular areas in LB films, these films are less susceptible to rearrangement on heating.

The observed aggregational behavior has been rationalized in terms of an optimization between hydrophobic interactions (between hydrocarbon chains) and charge-transfer interactions (between the squaraine chromophores). In the

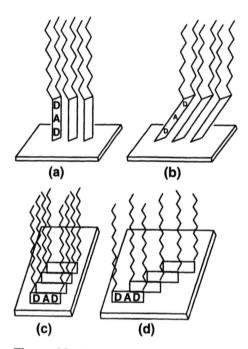

Figure 16   Schematics for the molecular arrangements of squaraine aggregates in LB films: (a) blue-shifted and (b) J-aggregates of 28; (c) blue-shifted and (d) red-shifted aggregates of 29. (Adapted from Ref. 38.)

air–water interface, due to predominance of the hydrophobic interactions, the card pack arrangement (blue-shifted aggregate) is favored. On being transferred to a glass substrate, this arrangement is maintained. Heating at 105°C however leads to the more stable J-aggregates, which are stabilized by the intermolecular charge transfer interactions. On exposure to steam, the hydrophobic interactions again become the dominating force, leading to reformation of the card-pack arrangement. These studies have recently been extended to several amphiphilic squaraines containing carboxylic units [79].

32

Structure 11

The aggregation of a surfactant squaraine, 4-[N-methyl,N-(carboxy-propylamino)phenyl]-4′-(N,N-dibutylamino)phenylsquaraine 32 (Structure 11) has been studied in a variety of media, including organic solvents, aqueous cyclodextrin solutions, vesicles, monolayers, LB films, and pure dry films [37].

In dilute aqueous solution, 32 has a strong absorption at 650 nm with a shoulder at 590 nm. At higher concentrations, the 594 nm band increases in relative intensity, and this band has been attributed to the H-dimer form of 32. The monomeric form fluoresces with a maximum at 678 nm, whereas the dimeric form is nonfluorescent. The structural assignment of the dimeric form was supported by studying the spectroscopic properties of 32 in cyclodextrin (CD) solutions. Compound 32 formed 1:1 inclusion complexes with α- and β-CD, which absorb at λ_{max} 650 nm and emit at λ_F ~672 nm with fluorescence quantum yields of 2–3 higher than in pure water. The formation of inclusion complexes of 32 with α- and β-CD was also supported by ^{1}H NMR studies, which indicate the squaraine chromophore to be in a relatively nonpolar environment. In γ-CD solution, due to the larger cavity size, a 2:1 complex between 32 and γ-CD was observed. The absorption and emission characteristics of the 2:1 complex were similar to those of the dimeric form. A further blue shift in absorption, λ_{max} 500–540 nm, was observed when 32 was incorporated in vesicles, monolayers, supported on LB films, and in pure solid dye film. Intermolecular interactions between squaraine chromophores, involving the C–O dipole–dipole interactions similar to those observed in microcrystals [55,80], has been proposed as the main driving force for the aggregations of these dyes.

3. Aggregation in Vesicles

The synthesis and aggregation behavior of several amphiphilic squaraines in aqueous and mixed aqueous–organic solutions, as well as in bilayer vesicles, have been reported [39,77]. In aqueous solutions containing the amphiphilic squaraines and phospholipids, clear solutions could be obtained upon sonication. Light scattering and membrane filtration studies indicate that the mixed vesicles were much larger than the corresponding pure phospholipid vesicles. At high concentrations of squaraine to phospholipid, the blue-shifted (520 nm) aggregate, similar to the spectra observed in LB films or in DMSO–water mixtures, was observed, whereas at lower dye/phospholipid ratios, increase in the intensity of the monomer band was observed. These squaraines were also capable of forming stable large vesicles upon probe sonication. Under these conditions only the blue-shifted aggregate band was observed.

The tendency for aggregation was found to increase with the increase in length of the alkyl chain substituent on nitrogen. Squaraines with quarternary ammonium head groups exhibited less tendency for aggregation probably due to electrostatic repulsion of the head groups. The aggregation number was determined by Benesi–Hildebrand type analysis to be ~4 for several of these squaraines. Based on the induced circular dichroism observed for the aggregate and Monte Carlo simulation results, Chen et al. [39,77] propose that the unit aggregate is a tetramer with chiral pinwheel structure.

4. Interaction with Semiconductors

The absorption, emission, and redox properties of squaraines make them highly suited for applications as photosensitizers. In view of this, the early studies on squaraines were focused on thin photovoltaic and semiconductor photosensitization properties [1,4,5,91–97]. Champ and Shattuck [98] first demonstrated that squaraines could photogenerate electron–hole (e–h) pairs in bilayer xerographic devices. Subsequently, extensive work has been carried out on the xerographic properties of squaraines [2,24,34,47,48,99,100], and these properties have been reviewed recently [11]. In an extensive study on the correlation's between cell performance and molecular structure in organic photovoltaic cells, squaraines were found to have much better solar energy conversion efficiencies than a variety of other merocyanine dyes [4,5].

The dynamics of charge injection from the excited state of squaraines to the conduction band of semiconductors have been investigated by picosecond laser spectroscopy [45,67,101]. The sharp absorption band of bis[4-dimethyl-amino-2-hydroxy)phenyl]squaraine 21 (Structure 6) with λ_{max} at 636 nm became broad and red-shifted (λ_{max} 670 nm) in the presence of colloidal TiO$_2$ (Fig. 17). The red shift of 35 nm in the absorption band has been attributed to a strong

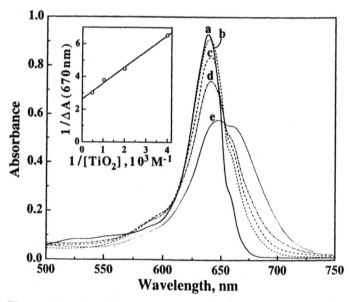

Figure 17   Absorption spectra of 21 (3 μM) in 2:3 CH$_2$Cl$_2$:CH$_3$CN containing (a) 0, (b) 0.25, (c) 0.5, (d) 1, and (e) 2 mM of colloidal TiO$_2$. The inset shows the fitting of the 670 nm absorption band to the Benesi–Hildebrand plot. (From Ref. 101.)

charge-transfer interaction of the dye with the TiO$_2$ surface. The apparent association constant for the complex formation between squaraine dye 21 and TiO$_2$ determined by the Benesi–Hildebrand analysis was 2667 M^{-1}.

On interaction of the squaraine dye 21 with TiO$_2$, the strong fluorescence of the dye (ϕ_f = 0.84 in CH$_2$Cl$_2$) was quenched, and this effect has been attributed to charge injection from the excited singlet of squaraine dye 21 to the conduction band of TiO$_2$. The oxidation potential of 21* (S$_1$), which is around 1.5 V versus NHE [2], and the energy level of the conduction band of TiO$_2$, which lies around –0.5 V in acetonitrile, provide favorable energetics for such charge transfer.

Picosecond laser flash photolysis of solutions of the dye with and without TiO$_2$ gave the transient spectra shown in Fig. 18. In the absence of TiO$_2$ (Fig. 18, curve a), a transient spectrum characteristic of the excited singlet state of 21 was observed. In the presence of TiO$_2$ (Fig. 18, curve b) a decrease in the intensity of the excited singlet state band and the formation of an additional band with λ_{max} at 580 nm, which was attributed to the dye radical cation, were observed. The absorption maximum of the radical cation was red-shifted ($\sim$20 nm) compared to that observed in solution, and this shift has been attributed to adsorption of the dye radical cation on the TiO$_2$ surface. The charge injec-

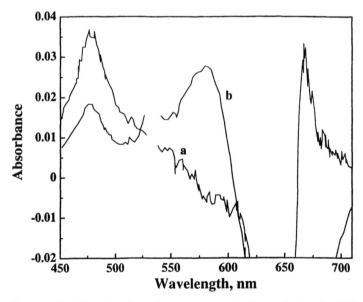

Figure 18 Transient absorption spectra recorded immediately after 532 nm laser pulse of 21 in 1:1 CH$_2$Cl$_2$:CH$_3$CN (a) without TiO$_2$ and (b) with 2.5 mM colloidal TiO$_2$. (From Ref. 101.)

tion occurred within the laser pulse duration ($k > 5 \times 10^{10}$ s^{-1}). The transient radical cation spectrum decayed by a first-order process ($k = 3.7 \times 10^9$ s^{-1}), and this has been attributed to the charge-recombination process (Eq. (5)).

$$21^{\overset{+}{\cdot}} + TiO_2 \, (e^-) \rightarrow 21 + TiO_2 \tag{5}$$

The photosensitization of TiO$_2$ particulate films by squaraine dyes 17 and 18 (Structure 5) has been investigated [64,81]. Highly porous, thin semiconductor films can be cast on optically transparent electrodes (OTE), which exhibit excellent photoelectrochemical properties [102–111]. The photosensitization of such TiO$_2$ and ZnO semiconductor films, using dyes, has been investigated [107–109]. The sensitizing behavior of 17 was probed by adsorbing the dye on a TiO$_2$-colloid coated OTE (OTE/TiO$_2$) electrode, using Pt as a counter electrode and aqueous 1.5 M KCl as electrolyte. Upon illumination of this electrode by visible light (500 nm), a rise in the photovoltage was observed, which remained steady as long as the irradiation was continued and dropped back to the original dark value when the lamp was turned off (Fig. 19) [64]. The photoelectrochemical effect at the OTE/TiO$_2$/17 was reproducible over several cycles of illumination. The photovoltage increased with an increase in the intensity of the lamp, and a maximum open-circuit photovoltage of 255 mV

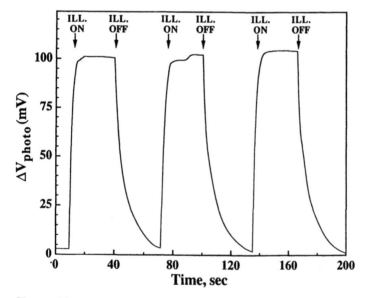

Figure 19 Photovoltage (open circuit) response of dye modified electrode (OTE/TiO$_2$/17) to illumination in a photoelectrochemical cell containing Pt foil as counterelectrode and aqueous 1.5 M KCl and electrolyte. (From Ref. 64.)

was observed. A close match was obtained between the absorption and the action spectra of the OTE/TiO$_2$/17 electrode (Fig. 20). A slight blue shift in the action spectra as well as a shoulder around 645 nm were attributed to aggregation of the adsorbed dye.

The J-type of dimer of 18 formed in dry acetonitrile solutions (see Sec. VI.C.1.) interacts strongly with TiO$_2$ [81]. The spectra recorded, immediately after addition of colloidal TiO$_2$ to acetonitrile solutions of 18, indicated a shift in the monomer–dimer equilibrium, in favor of the dimer (Fig. 21). These spectra underwent time dependent changes, leading to the formation of a new band in the 600 nm region (Fig. 22). This band has been attributed to charge-transfer interactions between the dimer and TiO$_2$, and the apparent association constant for this interaction was determined as 1600 M^{-1} by the Benesi–Hildebrand analysis of the changes in absorption at 610 nm as a function of TiO$_2$ concentration (inset, Fig. 22). Instantaneous formation of charge transfer complexes between the electron donor and TiO$_2$ is usually observed in colloidal TiO$_2$ [101,113,114]. The reasons for the slow formation of the charge-transfer band between the dimeric form of 18 and TiO$_2$ is not clear. A probable reason could be a slow change in the geometry of the dimer on the surface of TiO$_2$ that may be required for the CT complexation to occur.

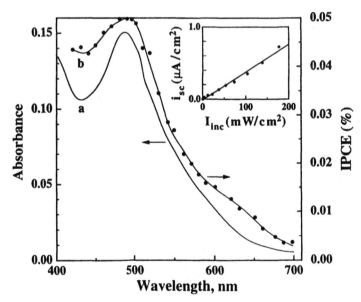

Figure 20 (a) Absorption spectrum of (a) 17 adsorbed on TiO$_2$ particulate film and (b) action spectrum of incident photon to current efficiency of OTE/TiO$_2$/17. Inset shows linear dependence of short circuit current on the incident light intensity. (From Ref. 64.)

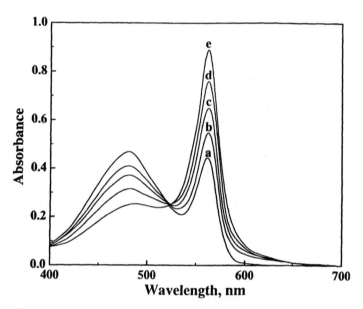

Figure 21 Absorption spectra of 9.8 μM of 18 in aceonitrile at various concentrations of colloidal TiO_2. The TiO_2 colloid concentrations were (a) 0, (b) 20, (c) 40, (d) 60, and (e) 100 μM. Spectra recorded immediately after addition of colloidal TiO_2 suspension. (From Ref. 81.)

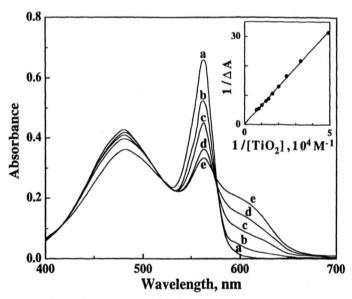

Figure 22 Absorption spectra of 11 μM of 18 in aceonitrile, measured 5 min after addition of colloidal TiO_2. TiO_2 colloid concentrations were (a) 0, (b) 20, (c) 50, (d) 81, and (e) 141 μM. Inset shows the dependence of reciprocal absorbance (610 nm) versus reciprocal concentration of TiO_2 colloid. (From Ref. 81.)

Nanocrystalline semiconductor films of TiO_2, ZnO, and SnO_2 containing 18 adsorbed on the surface indicated that both the monomeric and the aggregated forms participate in the charge-injection process.

Quite different results were obtained by Kim et al. [44] in the photocurrent generation studies on SnO_2 electrodes, modified by monolayers containing aggregates of bis[(4-dialkylamino)phenyl]squaraines] 29–31 (Structure 10) using the Langmuir–Blodgett technique. The 29-SnO_2 electrode showed an absorption maximum at 530 nm, which was significantly blue-shifted from its monomeric absorption (λ_{max} 633 nm) (Fig. 23), indicating the formation of squaraine aggregates on the SnO_2 electrode similar to those observed in LB films [38,78]. Under ambient conditions, a cathodic photocurrent was observed when the 29-SnO_2 electrode was illuminated and the action spectrum of the photocurrent generation was coincident with the absorption of the electrode (Fig. 24), indicating that the squaraine aggregate was responsible for the photocurrent. These observations suggest that electrons flow from the electrode through the LB film to the electrolyte solution where they are picked up by oxygen. Consequently a sharp decrease (>90%) in the photocurrent was observed in N_2 degassed solutions. Most electron donors were observed to attenuate the cathodic photocurrent. When hydroquinone (HQ) was used as donor, anodic photocurrents were observed in the absence of oxygen. The schematics of photocurrent generation under the two limiting conditions are shown in Fig. 25. In the case of the monomeric and J-aggregated forms of different squaraines

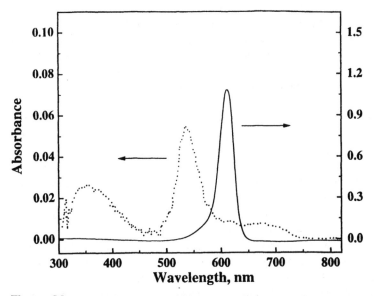

Figure 23 Absorption spectra of 29 (·····) on SnO_2 electrode and (——) dilute $CHCl_3$ solution. (From Ref. 44.)

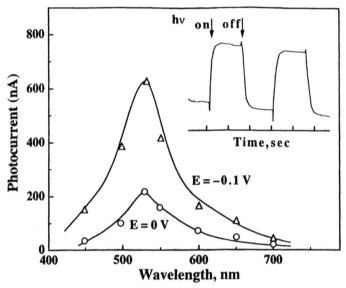

Figure 24 Spectral response curves of 29–SnO$_2$ electrode (E = bias voltage in volts, against Ag/AgCl). (From Ref. 44.)

[64,67,81,101], the photosensitization mechanism involved charge injection from the excited state of the dye or dye-aggregates to the conduction band of the semiconductor. Although such a process is energetically favorable for the LB films of 29 on SnO$_2$, facile electron transfer from the H-aggregate of this dye to oxygen and the absence of a hole scavenger in the electrolyte solution were proposed to be responsible for the observed cathodic photocurrent. Based on these studies, Kim et al. [44] suggest that oxygen may play an important role in the photogeneration mechanism of squaraine photoconductors in xerographic devices.

The photoelectrochemical reduction of 25, brought about by the band-gap excitation of TiO$_2$, has been investigated [70]. The conduction band of TiO$_2$ is -0.5 V, NHE, and photoelectrochemical reductions of several thiazine and oxazine dyes have been reported [114,115]. Selective excitation of TiO$_2$ by 308 nm laser pulses led to the formation of a transient absorption, attributed to that of the radical anion of 25 with a λ_{max} at about 435 nm. Steady-state irradiation of colloidal TiO$_2$ solution containing 25, using 325 nm, led to a loss in intensity of the ground-state absorption of the dye. The colorless product formed was stable in an inert atmosphere, and partial recovery of the dye absorption was observed in the presence of oxygen. The structure of the reduced product was suggested as 33 (Structure 12). Saturation of the carbon–carbon double bond between the central C$_4$O$_2$ unit and its neighboring group has been reported

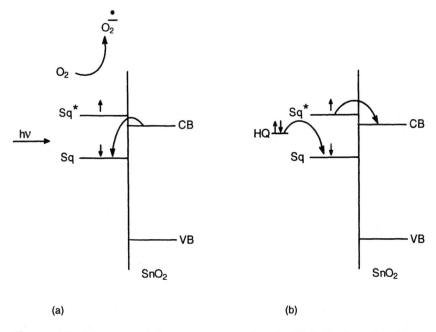

Figure 25 Schematics of photocurrent generations by 29–SnO$_2$ electrodes: (a) cathodic photocurrent under ambient conditions; (b) anodic photocurrent in degassed 1 M NaNO$_3$ in the presence of HQ. (From Ref. 44.)

33

Structure 12

to yield stable colorless products [22,27]. A radical disproportionation mechanism (Eqs. (8,9)) has been proposed for the formation of the reduced product.

$$TiO_2 \xrightarrow{\;hv\;} TiO_2(h^+) + TiO_2(e^-) \tag{6}$$

$$TiO_2(e^-) + 25 \rightarrow TiO_2 + 25^{\bullet-} \tag{7}$$

$$2(25)^{\bullet-} \rightarrow 25 + 25^{2-} \tag{8}$$

$$25^{2-} \xrightarrow{\;2H^+\;} 25\ H_2 \tag{9}$$

VII. BIOLOGICAL APPLICATIONS

There is considerable interest in the study of fluorescent dyes absorbing in the near infrared region (NIR) for a number of biological applications. Such dyes have applications in fluorescence lifetime sensing techniques for intracellular chemical imaging of analytes such as pH, O_2, K^+, or Ca^{2+} [116]. Soper et al. have recently reported detection sensitivity at the single molecule level in the near-infrared, using pulsed laser excitation and time-gated detection for the NIR fluorescent dye IR-132 [117,118]. IR-132 was observed to be superior to rhodamine 6G, in spite of its much lower quantum yield of fluorescence, for such applications, and this has been attributed primarily to the low background arising from fluorescent impurities in the solvent blank, when using most known NIR excitation [117]. However, very few NIR dyes possessing high quantum yields of fluorescence are known. Several squaraine dyes are known to absorb in the NIR region and also possess strong emission properties, which are highly sensitive to the nature of the surrounding medium. In view of this, the use of squaraines as fluorophores or fluoroionophores for biological applications is gaining increased attention. Terpetschnig et al. have synthesized and studied the biological applications of several new squaraine dyes [28–30,40,41]. It was found that indolenine-containing squaraines were the most photostable and that their lifetimes are significantly increased in the presence of bovine serum albumin (BSA). The synthesis of two amine reactive, N-hydroxysuccinimide esters of squaraines 34 and 35 (Structure 13) has been reported. The squaraine dye 35 is water solute due to the presence of a sulfobutyl group, and 34 was made water soluble by reacting it with taurine.

34, R = C_2H_5

Structure 13   **35**, R = $(CH_2)_4SO_3^{\ominus}$

The fluorescence quantum yields and lifetimes of the Sq-taurine derivative increased by 28-fold and 31-fold, respectively, on binding to BSA. The short lifetimes and low quantum yields of fluorescence in water of these squaraines increase significantly when bound to proteins. Also, their absorption maxima around 635 nm in water and 640 nm when bound to proteins allow excitation with diode lasers, making these dyes highly suited for applications as reactive fluorescent labels in immunochemical assays and biophysical studies of proteins.

The ability of monoaza-crown ether derivatives of squaraines 22–24 (Structure 6) to sense alkali metal ions in solutions by fluorimetric and electrochemical methods in solutions has been studied [42,119]. Binding of the alkali metal ions by the crown ether moiety of the dye brought about significant decrease in the quantum yield of fluorescence (Fig. 26), although the effects on the shape and position of the absorption and emission bands were negligible. The reduction in quantum yield of fluorescence was attributed to a reduction in the electron-donating ability of the nitrogen atoms brought about by complexation of the cation by the monoaza-crown ether moiety of the dye. The dye 22 exhibited two reversible oxidation peaks at 705 and 1030 mV versus Ag/AgCl in acetonitrile with tetrabutylammonium perchlorate (TBAP) as the electrolyte and could be subjected to several oxidation/reduction cycles without undergoing any significant chemical change. Complexation of the dye with alkali metal ions brought about a shift of these peaks to less positive potentials [119]. The oxidation peaks were sensitive to the presence of metal ions in solution. The dependence of the shift in the anodic peaks on the concentration

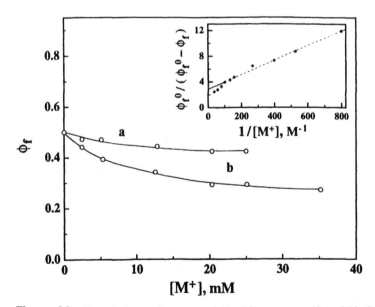

Figure 26 Plot of observed quantum yield of fluorescence (Φ_f) of 22 (3.0 μM) versus metal ion concentration in 30% v/v acetonitrile–toluene: (a) NaClO$_4$; (b) LiClO$_4$. Inset shows the plot of $1/(\Phi_f^0 - \Phi_f)$ versus reciprocal concentration of lithium ion concentration (Φ_f^0 = quantum yield fluorescence of uncomplexed dye). (From Ref. 42.)

of Na$^+$ ions is shown in Figure 27. Similar results were obtained for **23** and **24**. The solubility of these dyes in water and absorption in the NIR makes these dyes potentially useful for intracellular applications. Although the fluorescence quantum yields of these dyes in aqueous solutions are low, substantial enhancement on the fluorescence yields can occur in hydrophobic environments as observed on ß-cyclodextrin complexation of dyes in water. Other chromogenic squaraines have also been reported [120,121].

In spite of the strong absorption of several squaraines in the NIR region, their potential as photosensitizers in photodynamic therapy has not been explored. Due to the low intersystem crossing efficiency of most squaraines, the singlet oxygen generation efficiency of squaraines is expected to be low. Recently, heavy atom substituted squaraine derivatives **19** and **20** (Structure 5) have been synthesized, and their photophysical and photochemical properties have been explored [46]. Laser flash photolysis studies indicated triplet excited states to be the main transient intermediates for these dyes. The triplet quantum yields were found to be the highest for the neutral [ϕ_T = 0.12 (**19**); 0.24 (**20**)] and deprotonated forms [ϕ_T = 0.22 (**19**) and 0.5 (**20**)] of these dyes. Quantum yields of singlet oxygen generation by the anionic forms of **19** and

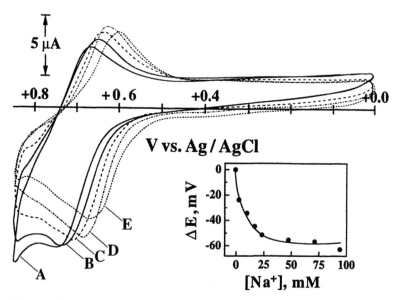

Figure 27 Electrochemical oxidation of **22** (5.7 mM) in CH$_3$CN containing 0.1 M TBAP. The concentrations of Na$^+$ were (A) 0, (B) 3.3, (C) 10, (D) 16.4, and (E) 93.7 mM. The inset shows the shift in the anodic peak vs. concentration of Na$^+$ ions. (From Ref. 119.)

20 were found to be 0.13 and 0.47, which are in good agreement with the triplet yields obtained in these systems. The photodynamic activity of these dyes have not been explored.

VIII. NONLINEAR OPTICAL PROPERTIES

Nonlinear optical (NLO) materials play an important role in modern technology, such as in image and information processing, telecommunications, integrated optics, and optical computing [122,123]. The origin of the second-order optical nonlinearities in organic conjugated molecules is fairly well explained by distortion of π-electrons induced by electron-donating (donor) and electron-accepting (acceptor) groups. In most such molecules however formation of centrosymmetric crystal structures is observed due to electrostatic interaction between adjacent molecules. One of the major problems to be addressed in these studies is the requirement of noncentrosymmetric alignment of molecules in crystals and thin films. Several squaraines have been designed and investigated for this purpose. Electric field induced second harmonic (EFISH) generation of several unsymmetrical (D'–A–D) squaraines in chloroform using 1.907 μm fundamental radiation indicated negative molecular first hyperpolarizabilities (β) with magnitudes comparable to that of 4-N,N-dimethylamino-4'-nitrostilbene (DANS) [36]. Negative values are uncommon and suggest that the dyes have larger ground-state than excited-state dipole, and this is consistent with negative solvatochromism observed for this kind of dye [36]. An unsymmetrical squaraine 36 (Structure 14) with extended conjugation was synthesized, and EFISH measurements in chloroform show large positive β values that are consistent with its observed positive solvatochromism [35]. The β value for this dye was about 8 times greater than that of DANS.

36

Structure 14

The large values reported for squaraines indicate that noncentrosymmetric crystals or films of these dyes should be capable of efficient second harmonic generation (SHG). A new class of chiral D'–A–D squaraines 37–44 obtained by 1,2-substitution of squaric acid showed second harmonic generation intensities of up to 64 times that of urea (Table 3) in the powder form [124,125]. X-ray crystal analysis indicated that two hydrogen bonds between adjacent mol-

Table 3 Relative SHG Intensities of 1,2-Disubstituted Squaraines 37–44

	X	Y	Z^a	λ_{max}^{b} nm	SHG^c
37	Me	H	OH	319	2.5
38	Me$_2$N	H	OH	379	0
39	Me$_2$N	H	NHCH$_2$CH(OH)Me	397	64
40	Me$_2$N	H	NHCH(Et)CH$_2$OH	396	6
41	Me$_2$N	H	NHCH(CH$_2$Ph)CH$_2$OH	394	8
42	Me$_2$N	H	$\overline{N(CH_2)_3C}HCO_2Bu\text{-}t$	396	8
43	Me$_2$N	Et	NHCH(Et)CH$_2$OH	397	26
44	MeO(CH$_2$)$_2$(Me)N	H	NHCH$_2$CH(OH)Me	397	58

[a]Chiral center is in italic type.
[b]In EtOH solution.
[c]Intensity relative to that of urea.
Source: Ref. 125.

ecules and the chirality of the molecule contribute to one-dimensional molecular alignment in spite of the strong dipole–dipole interactions, and that the conjugated system of the D′–A–D molecule extends from amino groups to cyclobutenedione ring to enhance the second harmonic nonlinearity of these crystals.

In a recent study, Ashwell et al. have observed that centrosymmetric squaraine dyes incorporated into LB monolayers showed second harmonic generation efficiencies, which compare favourably with the highest values hitherto reported for LB monolayers of noncentrosymmetric dyes [126]. Based on comparison of the absorption spectra and SHG characteristics of the LB films of squaraine 45 (Structure 15) as well as the effect of deposition pressure on these properties, it has been suggested that formation of noncentrosymmetric aggregates is responsible for these effects. Nonlinear optical studies have also shown that symmetric squaraines have quite large molecular second hyperpolarizabilities [127–131].

45

Structure 15

Vauthey et al. [132,133] have examined the hole-burning mechanisms of a few squaraine dyes in hydrogen-bonding and non-hydrogen-bonding polymers. In these cases, the spectral holes were not persistent, and decay with distribution rates ranging from 10^5 s^{-1} to about 1 s^{-1} [132,133].

IX. POLYSQUARAINES

Conjugated organic polymers possessing small band gaps are very important because they would show high intrinsic conductivity without doping and possibly better transpareny in the visible part of the spectrum. Havinga et al. [134,135] have synthesized a new class of organic polymer conductors containing squaraines 46 (Structure 16) and croconaines as the repeating unit. These polymers were observed to have a small band gap, down to 0.5 eV. It has been suggested that the small gap arises from the regular alteration of strong electron-donating and electron-accepting groups in a conjugated backbone. The band-gap was the smallest for a combination in which the electronegativity difference between donor and acceptor was greatest. Steric hindrance, which could prevent formation of a completely planar configuration, is found to counteract the decrease in band-gap energy. The Vis-NIR spectra of solutions of the polymers in chloroform were similar to the corresponding D–A–D monomeric squaraines, except for a marked red shift and a slight broadening. In solid state films, the red shift and broadening were enhanced.

46

Structure 16

Pyrrole and N-methylpyrrole were reported to form blue to blue-green colored insoluble polymers on condensation with squaric acid [7]. However, their absorption and conduction properties have not been investigated.

A systematic theoretical study of this class of polymers using Car-Parrinello techniques to optimize both the electronic and the geometrical structure of the polymer has been carried out [136,137]. These studies predict the possibility of polysquaraines with band-gaps as low as 0.2 eV.

X. CONCLUSIONS

The above-mentioned studies clearly illustrate the sensitivity of the photophysical and photochemical properties of squaraines to both substitutional and environmental changes. This tunability makes squaraines a highly versatile class of molecules. For example, the strong absorption in the NIR region and high fluorescence yields of these dyes make them potentially important for wide range of biological applications, which have not been fully exploited. Due to the large polarizability and the existence of a wide transparent window in the visible region of many squaraines, they are also ideally suited for second harmonic generation and other NLO applications. In addition, the studies on conducting polymers as well as the ability of squaraines to undergo reversible oxidation processes, both thermally and photochemically, combined with the synthetic flexibility of this class of molecules imply that they are ideally suited for application in a wide range of molecular photonic and electronic devices.

ACKNOWLEDGMENTS

The authors thank the Council of Scientific and Industrial Research and the Department of Science and Technology, Government of India.

We are thankful to Dr. Prashant V. Kamat of the Radiation Laboratory of the University of Notre Dame (USA) for his help in collaborative research and valuable discussions. This is contribution No. RRLT-PRU-80 from the Regional Research Laboratory (CSIR), Trivandrum. Copyright permission for all the figures, incorporated from published articles, have been obtained from the appropriate publishers.

REFERENCES

1. Loutfy, R. O.; Hsiao, C. K.; Kazmaier, P. M. *Photogr. Sci. Eng.* 1983, *27*, 5.
2. Law, K.-Y.; Facci, J. S.; Bailey, F. C.; Yanus, J. F. *J. Imaging Sci.* 1990, *34*, 31.
3. Tam, A. C. *Appl. Phys. Lett.* 1980, *37*, 978.
4. Morel, D. L.; Stogryn, E. L.; Ghosh, A. K.; Feng, T.; Purwin, P. E.; Shaw, R. F.; Fishman, C.; Bird, G. R.; Piechowski, A. P. *J. Phys. Chem.* 1984, *88*, 923.
5. Piechowski, A. P.; Bird, G. R.; Morel, D. L.; Stogryn, E. L. *J. Phys. Chem.* 1984, *28*, 934.
6. Emmelius, M.; Pawlowski, G.; Vollmann, H. W. *Angew. Chem. Int. Ed. Engl.* 1989, *28*, 1445.
7. Treibs, A.; Jacob, K., *Angew. Chem. Int. Ed. Engl.* 1965, *4*, 694.
8. Sprenger, H.-E.; Ziegenbein, W. *Angew. Chem. Int. Ed. Engl.* 1968, *7*, 530.
9. Schmidt, A. H. *Synthesis*, 1980, 961.
10. Seitz, G.; Imming, P. *Chem. Rev.* 1992, *92*, 1227.
11. Law, K. Y. *Chem. Rev.* 1993, *93*, 449.
12. Schmidt, A. H. In *Oxocarbons*; West, R. Ed.; Academic Press: New York, 1980, Chapter 10.
13. Treibs, A; Jacob, K. *Justus Liebigs Ann. Chem.* 1966, *699*, 153.
14. Ziegenbein, W.; Sprenger, H.-E. *Angew. Chem. Int. Ed. Engl.* 1966, *5*, 893.
15. Sprenger, H.-E., Ziegenbein, W. *Angew. Chem. Int. Ed. Engl.* 1966, *5*, 894.
16. Sprenger, H.-E., Ziegenbein, W. *Angew. Chem. Int. Ed. Engl.* 1967, *6*, 553.

17. Law, K.-Y.; Bailey, F. C. *J. Chem. Soc. Chem. Commun.* 1990, 863.
18. Law, K.-Y.; Bailey, F. C. *J. Chem. Soc. Chem. Commun.* 1991, 1156.
19. Law, K.-Y.; Bailey, F. C. *J. Org. Chem.* 1992, *57*, 3278.
20. Law, K.-Y.; Bailey, F. C. *Can. J. Chem.* 1993, 71, 494.
21. Nakazumi, H.; Natsukawa, K.; Nakai, K.; Isagawa, K. *Angew. Chem. Int. Ed. Engl.* 1994, *33*, 1001.
22. Farnum, D. G.; Webster, B.; Wolf, A. D. *Tetrahedron Lett.* 1968, *48*, 5003.
23. Law, K.-Y.; Bailey, F. C. *Can. J. Chem.* 1986, *64*, 2267.
24. Law, K.Y.; Bailey, F. C. *J. Imaging Sci.* 1987, *31*, 172.
25. Bello, K. A.; Corns, N. S.; Griffiths, J. *J. Chem. Soc. Chem. Commun.* 1993, 452.
26. Meier, H.; Dullweber, U. *Tetrahedron Lett.* 1996, *37*, 1191.
27. Wendling, L. A.; Koster, S. K.; Murray, J. E.; West, R. *J. Org. Chem.* 1977, *42*, 1126.
28. Terpetschnig, E.; Lakowicz, J. R. *Dyes Pigm.* 1993, *21*, 227.
29. Terpetschnig, E.; Szmacinski, H.; Lakowicz, J. R. *Anal. Chim. Acta.* 1993, *282*, 633.
30. Terpetschnig, E.; Szmacinski, H.; Ozinskas, A.; Lakowicz, J. R. *Anal. Biochem.* 1994, *217*, 197.
31. Pease, J.; Tarnowski, T. L.; Berger, D.; Chang, C. C.; Chuang, C. H. US Patent: *US 4,830,786 A*, 1989.
32. Treibs, A.; Jacob, K. *Justus Leibigs Ann. Chem.* 1968, *712*, 123.
33. Bellus, D. *J. Am. Chem. Soc.* 1978, *100*, 8026.
34. Kazmaier, P. M.; Burt, R.; Dipaola-Baranyi, G.; Hsiao, C.-K.; Loutfy R. O.; Martin, T. I.; Hamer, G. K.; Bluhm, T. L.; Taylor, M. G. *J. Imaging Sci.* 1988, *32*, 1.
35. Chen, C.-T.; Marder, S. R.; Cheng, L.-T. *J. Am. Chem. Soc.* 1994, *116*, 3117.
36. Chen, C.-T.; Marder, S. R.; Cheng, L.-T.; *J. Chem. Soc. Chem. Commun.* 1994, 259.
37. Chen, H.; Herkstroeter, W. G.; Perlstein, J.; Law, K.-Y.; Whitten, D. G. *J. Phys. Chem.* 1994, *98*, 5138.
38. Liang, K.; Law, K.-Y.; Whitten, D. G. *J. Phys. Chem.* 1994, *98*, 13379.
39. Chen, H.; Farahat, M. S.; Law, K.-Y.; Whitten, D. G. *J. Am. Chem. Soc.* 1996, 118, 2584.
40. Terpetschnig, E.; Szmacinski, H.; Lakowicz, J. R. *Proc. SPIE-Int. Soc. Opt. Eng.* 1994, *2137*, 608.
41. Terpetschnig, E.; Szmacinski, H.; Lakowicz, J. R. *J. Flouoresc.* 1993, *3*, 153.
42. Das, S.; Thomas, K. G.; Thomas, K. J.; George, M. V.; Kamat, P. V. *J. Phys. Chem.* 1994, *98*, 9291.
43. Dirk, C. W. Herndon, W. C.; Cervantes-Lee, P.; Selnau, H.; Martinez, S.; Kalamegham, P.; Tan, A.; Campos, G.; Velez, M.; Zyss, J.; Ledoux, I.; Cheng, L.-T. *J. Am. Chem. Soc.* 1995, *117*, 2214.
44. Kim, Y.-S.; Liang, K.; Law, K.-Y.; Whitten, D. G. *J. Phys. Chem.* 1994, *98*, 984.

45. Das, S.; Thomas K. G.; Kamat, P. V.; George, M. V. *Proc. Indian Acad. Sci. (Chem. Sci.)* 1993, *105*, 513.
46. Ramaiah, D.; Joy, A.; Chandrasekhar, N.; Eldo, N. V.; Das, S.; George, M. V. *Photochem. Photobiol.* 1997, (in press).
47. Law, K.-Y. U.S. Patent: *US 5,230,975A*, 1993.
48. Law, K.-Y. *J. Imaging Sci. Technol.* 1992, *36*, 567.
49. Murata, J.; Ozawa, T.; Kawana, M.; Urano, T. Patent: Jpn. Kokai Tokkyo Koho JP 05, 155, 142 [93, 155, 142], 1993; Jpn. Kokai Tokkyo Koho JP 05, 155, 143 [93,155,143], 1993; Jpn. Kokai Tokkyo Koho JP 05, 155, 144 [93,155,144], 1993.
50. Zepp, C. M. U. S. Patent: U.S. *US 5,211,885*, 1993.
51. Enokida, T.; Suda, Y. Patent: Jpn Kokai Tokkyo Koho JP 94 220,438 (JP 06, 220, 438), 1994.
52. Nagasaka, H.; Ota, K. Patent: Jpn. Kokai Tokkyo Koho JP 04, 106, 548, [92, 106, 548], 1992.
53. Farnum, D. G.; Neuman, M. A.; Suggs, W. T., Jr. *J. Cryst. Mol. Struct.* 1974, *4*, 199.
54. Kobayashi, Y.; Goto, M.; Kurahashi, M. *Bull. Chem. Soc. Jpn.* 1986, *59*, 311.
55. Bernstein, J.; Goldstein, E. *Mol. Cryst. Liq. Cryst.* 1988, *164*, 213.
56. Bigelow, R. W.; Freund, H.-J. *Chem. Phys.* 1986, *107*, 159.
57. Law, K.-Y. *J. Phys. Chem.* 1987, *91*, 5184.
58. Law, K.-Y. *J. Photochem. Photobiol. A: Chem.* 1994, *84*, 123.
59. Law, K.-Y. *J. Phys. Chem.* 1989, *93*, 5925.
60. Law, K.-Y. *J. Phys. Chem.* 1995, *99*, 9818.
61. Law, K.-Y. *Chem. Phys. Lett.* 1992, *200*, 121.
62. Das, S.; Thomas, K. G.; Ramanathan, R.; George, M. V.; Kamat, P. V. *J. Phys. Chem.* 1993, *97*, 13625.
63. Eberson, L. *J. Phys. Chem.* 1994, *98*, 752.
64. Kamat, P. V.; Hotchandani, S.; de Lind, M.; Thomas K. G.; Das, S.; George, M. V. *J. Chem. Soc. Faraday Trans.* 1993, *89*, 2397.
65. Das, S.; Kamat, P. V.; De la Barre, B.; Thomas, K. G.; Ajayaghosh, A.; George, M. V. *J. Phys. Chem.* 1992, *96*, 10327.
66. Das, S.; Thomas, K. G.; Thomas, K. J.; Madhavan, V.; Liu, D.; Kamat, P. V.; George, M. V. *J. Phys. Chem.* 1996, *100*, 17310.
67. Kamat, P. V.; Das, S.; Thomas, K. G.; George, M. V. *J. Phys. Chem.* 1992, *96*, 195.
68. Sauve, G.; Kamat, P. V.; Thomas, K. G.; Thomas, K. J.; Das, S.; George, M. V. *J. Phys. Chem.* 1996, *100*, 2117.
69. Ford, W. E.; Hiratsuka, H.; Kamat, P. V. *J. Phys. Chem.* 1989, *93*, 6692.
70. Patrick, B.; George, M. V.; Kamat, P. V.; Das, S.; Thomas, K. G. *J. Chem. Soc. Faraday Trans.* 1992, *88*, 671.
71. Das, S.; Thomas K. G.; George, M. V.; Kamat, P. V. *J. Chem. Soc. Faraday Trans.* 1992, *88*, 3419.
72. Law, K.-Y. *J. Imaging Sci.* 1990, *34*, 38.

73. Scott, G. W.; Tran, K. *J. Phys. Chem.*, 1994, *98*, 11563.
74. Buncel, E.; McKerrow, A. J.; Kazmaier, P. M. *J. Chem. Soc. Chem. Commun.* 1992, 1242.
75. McKerrow, A. J.; Buncel, E.; Kazmaier, P. M. *Can. J. Chem.* 1995, *73*, 1605.
76. Das, S.; Thanulingam, T. L.; Thomas, K. G.; Kamat, P. V.; George, M. V. *J. Phys. Chem.* 1993, *97*, 13620.
77. Chen, H.; Law, K.-Y., Perlstein, J.; Whitten, D. G. *J. Am. Chem. Soc.* 1995, *117*, 7257.
78. Law, K.-Y.; Chen, C. C. *J. Phys. Chem.* 1989, *93*, 2533.
79. Chen, H.; Law, K.-Y.; Whitten, D. G. *J. Phys. Chem.* 1996, *100*, 5949.
80. Law, K.-Y. *J. Phys. Chem.* 1988, *92*, 4226.
81. Hotchandani, S.; Das, S.; Thomas, K. G.; George, M. V.; Kamat, P. V. *Res. Chem. Intermed.* 1994, *20*, 927.
82. McRac, F. G.; Kasha, M. In *Physical Processes in Radiation Biology*; Augenstein, L.; Rosenberg, B.; Mason, S. F., Eds.; Academic Press: New York, 1963, pp. 23–42.
83. Kasha, M.; Rawls, H. R.; El- Bayoumi, A. *Pure Appl. Chem.* 1965, *11*, 371.
84. Das, S.; Thanulingam, T. L.; Thomas, K. G., unpublished results.
85. Nutting, G. C.; Harkin, W. D. *J. Am. Chem. Soc.* 1939, *61*, 1182.
86. Kim, S.; Furuki, M.; Pu, L. S.; Nakahara, H.; Fukuda, K. *J. Chem. Soc. Chem. Commun.* 1987, 1201.
87. Tanako, M.; Sekiguchi, T.; Matsumoto, M.; Nakamura, T.; Manda E.; Kawabata, Y. *Thin Solid Films* 1988, *160*, 299.
88. Iwamoto, M.; Majima, Y.; Hirayama, F.; Furuki, M.; Pu, L. S. *Chem. Phys. Lett.* 1992, *195*, 45.
89. Makio, S.; Kanamaru, N.; Tanaka, J. *Bull. Chem. Soc. Jpn.* 1980, *53*, 3120.
90. Hada, H.; Hanawa, R.; Haraguchi, A.; Yonezawa, Y. *J. Phys. Chem.* 1985, *89*, 560.
91. Kampfer, H. U.S. Patent *3,617,270*, 1971.
92. Morel, D. L. *Mol. Cryst. Liq. Cryst.* 1979, *50*, 127.
93. Ghosh, A. K.; Feng, T. *J. Appl. Phys.* 1978, *49*, 5982.
94. Morel, D. L.; Ghosh, A. K.; Feng, T.; Stogryn, E. L.; Purwi, P. E.; Shaw, R. F.; Fishman, C. *Appl. Phys. Lett.* 1978, *32*, 495.
95. Merritt, V. Y. *IBM J. Res. Develop.* 1978, *22*, 353.
96. Merritt, V. Y.; Hovel, H. *J. Appl. Phys. Lett.* 1976, *29*, 414.
97. Forster, M.; Hester, R. E. *J. Chem. Soc. Faraday Trans. I*, 1982, *78*, 1847.
98. Champ, R. B.; Shattuck, M. D. US Patent *3,824,099*, 1974.
99. Murti, D. K.; Kazmaier, P. M.; DiPaola-Baranyi, G.; Hsiao, C. K.; Ong, B. S. *J. Phys. D: Appl. Phys.* 1987, *20*, 1606.
100. Law, K. Y.; Bailey, F. C. *Dyes Pigm.* 1993, *21*, 1.
101. Kamat, P. V.; Das, S.; Thomas, K. G.; George, M. V. *Chem. Phys. Lett.* 1991, *178*, 75.
102. O'Regan, B.; Moser, J.; Grätzel, M.; Fitzmaurice, D. *Chem. Phys. Lett.* 1991, *183*, 89.

103. O'Regan, B.; Maser, J.; Grätzel, M.; Fitzmaurice, D. *J. Phys. Chem.* 1991, *95*, 10525.
104. Rothenberger, G.; Fitzmaurice, D.; Grätzel, M. *J. Phys. Chem.* 1992, *96*, 5983.
105. Rajeshwar, K. *Adv. Mater.* 1992, *4*, 23.
106 Hotchandani, S.; Kamat, P. V. *J. Electrochem. Soc.* 1992, *139*, 1630.
107. Hotchandani, S.; Kamat, P. V. *Chem. Phys. Lett.* 1992, *191*, 320.
108. Hotchandani, S.; Kamat, P. V. *J. Phys. Chem.* 1992, *96*, 6834.
109. O'Regan, B.; Grätzel, M. *Nature* (London), 1991, *353*, 737.
110. O'Regan, B.; Moser, J.; Anderson, M. Grätzel, M. *J. Phys. Chem.* 1990, *94*, 8720.
111. Vogel, R.; Pohl, K.; Weller, H. *Chem. Phys. Lett.* 1990, *174*, 241.
112. Ennaoui, A.; Fiechter, S.; Tributsch, H.; Giersig, M.; Vogel, R.; Weller, H. *J. Electrochem. Soc.* 1992, *139*, 2514.
113. Kamat, P. V. *J. Phys. Chem.* 1989, *93*, 859.
114. Kamat, P. V. *J. Photochem.* 1985, *28*, 513.
115. Kamat, P. V. *J. Chem. Soc. Faraday Trans. I*, 1985, *81*, 509.
116. Lackowicz, J. R. *Laser Focus World*, 1992, *28*, 60.
117. Soper, S. A.; Mattingly, Q. L. *J. Am. Chem. Soc.* 1994, *116*, 3744.
118. Soper, S. A.; Mattingly, Q. L.; Vengula, P. *Anal. Chem.* 1993, *65*, 740.
119. Das, S.; Thomas, K. G.; Thomas, K. J.; George, M. V.; Bedja, I.; Kamat, P. V. Anal. Proc. 1995, *32*, 213.
120. Yuan, D.-Q.; Fu, J.-S.; Xiao, S.; Xie, R.-G.; Zhao, H.-M. *Youji Huaxue*, 1994, *14*, 417.
121. Li, J.; Chen, Y.; Giang, Q.; Li, W.; Yu, L.; Zhao, H. *Youji Huaxue*, 1993, *13*, 608.
122. Ulrich, D. R. In *Organic Materials for Non-Linear Optics*; Hann, R. A.; Bloor, D., Eds.; Royal Society of Chemistry Proceedings, Special Publication No. 69, Royal Society of Chemistry; London, 1989, pp. 241–263.
123. Prasad, P. N.; Williams, D. J. *Introduction to Nonlinear Optical Effects in Molecules and Polymers*; John Wiley: New York, 1991.
124. Pu, L. S. In *Materials for Nonlinear Optics, Chemical Perspectives*; Marder, S. R.; Sohn, J. E.; Sluck, G. D., Eds.; ACS Symposium Series 455, American Chemical Society: Washington D.C., 1991, pp. 331–342.
125. Pu, L. S. *J. Chem. Soc. Chem. Commun.* 1991, 429.
126. Ashwell, G. J.; Jefferies, G.; Hamilton, D. G.; Lynch, D. E.; Roberts, M. P. S.; Bahra, G. S.; Brown, C. R. *Nature* (London), 1995, *375*, 385.
127. Scott, G. W.; Tran, K.; Funk, D. J.; Moore, D. S. *J. Mol. Struct.* 1995, *348*, 425.
128. Dirck, C. W.; Kuzyk, M. G. *Chem. Mater.* 1990, 2, 4.
129. Kuzuk, M. G.; Paek, U. C.; Dirk, C. W. *Appl. Phys. Lett.* 1991, *59*, 902.
130. Dirck, C. W.; Kuzyk, M. G. In *Materials for Nonlinear Optics: Chemical Perspectives*; Marder, S. R.; Sohen, J. E.; Stuckey, G. D.., Eds.; ACS Symp. Ser. No 455, ACS: Washington D.C., 1991, p. 687.
131. Dirk C. W.; Cheng, L.-T.; Kuzyk, M. G. *Int. J. Quantum Chem.* 1992, *43*, 27.

132. Vauthey, E.; Voss, J.; de Caro, C.; Renn, A.; Wild, U. P. *J. Lumin.* 1993, *56*, 61.
133. Vauthey, E.; Voss, J.; de Caro, C.; Renn, A.; Wild, U. P. *Chem. Phys.* 1994, *184*, 347.
134. Havinga, E. E.; ten Hoeve, W.; Wynberg, H. *Polym. Bull.* 1992, *29*, 119.
135. Havinga, E. E.; ten Hoeve, W.; Wynberg, H. *Synth. Met.* 1993, *299*, 55.
136. Brocks, G. *J. Chem. Phys.* 1995, *102*, 2522.
137. Brocks, G.; Tol, A. *J. Phys. Chem.* 1996, *100*, 1838.

12

Absorption, Fluorescence Emission, and Photophysics of Squaraines

Kock-Yee Law
Xerox Corporation, Webster, New York

I. INTRODUCTION

Squaraines are 1,3-disubstituted products synthesized by condensation of squaric acid with amines. When the amine is a tertiary amine, such as N,N-dimethylaniline and its derivatives, C-alkylation occurs [1]. The resulting squaraine products absorb strongly in the visible region. Because of the unique electronic structure, the nomenclature of these compounds has not been systematic. They have been named cyclotrimethine dyes [2], substituted 3-oxo-1-cyclobutenolates [1], cyclobutenediylium dyes [3], cyclobutenediylic dyes [4], and so forth. Schmidt proposed the name squaraine for these compounds in 1981 [5]. We find this nomenclature system very systematic; compounds of a variety of substituents, both at the nitrogen and in the phenyl ring, can be named unambiguously. MNDO and CNDO semiempirical molecular–orbital (MO) calculations by Bigelow and Freund [6] on bis(4-dimethylaminophenyl)squaraine indicated that both the ground and the excited states of squaraine are donor–acceptor–donor (D-A-D) charge transfer (CT) states. The anilino moieties and the oxygen atoms are the electron donors and the central four-membered ring is the electron acceptor. The calculation also showed that there is an increase in CT character during the S_0 to S_1 transition. Interestingly, the CT is primarily confined to the central C_4O_2 unit (80% from the oxygen atoms to the four-membered ring). The electronic transition of squaraine is undoubtedly very unique and is believed to be the origin of the very characteristic spectroscopic properties. Although the solution absorption of squaraine is sharp, their absorption in the microcrystalline solid state is very broad and panchromatic, covering most of the visible region and extending to the near-IR [7,8]. These optical characteristics, in conjunction to the known semiconductive and photoconductive properties, have made squaraines very attractive for a number of device applications (e.g., xerographic photoreceptors [9], organic solar cells [10–12], optical recording media [13,14], electroluminescence diodes [15], and, more recently, nonlinear optical devices [16–19].

By the nature of the squaric acid squaraine synthesis, squaraines are symmetrical. For a variety of reasons, unsymmetrical squaraines having two different donor groups were synthesized [20,21]. For convenience, we denote squaraines having two different aniline rings as psuedo-unsymmetrical squaraines and those bear one aniline ring and one anisole ring as "true" unsymmetrical squaraines. The intention of this chapter is to provide an overview on the spectroscopic properties and photophysics of both symmetrical and unsymmetrical squaraines. The effects of substituent, solvent, and ambient temperature on these properties are highlighted. The spectral properties of unsymmetrical squaraines are shown to be fundamentally different from those of symmetrical and pseudo-unsymmetrical squaraines. The origin is discussed.

II. ABSORPTION OF SYMMETRICAL SQUARAINES

Bis(4-dimethylaminophenyl)squaraine (Sq1) and its derivatives (Sq2–Sq26) all exhibit intense and shape absorption bands in the visible region. Their absorption maxima (λ_{max}) vary from 627 nm to 661 nm, depending on the substituent at the nitrogen and in the phenyl ring. The absorption spectral data of these compounds in CH_2Cl_2 are summarized in Table 1.

A. Effect of N-Alkyl Groups

The results in Table 1 show that λ_{max} of Sq1–Sq5 (X = H) shifts to the red by 12.4 nm as the length of the N-alkyl chain increases. The red-shift is accompanied with a very small increase in ε_{max}. This trend is also seen in three other series of squaraines, Sq7–Sq9 (X = OH), Sq10–Sq12 (X = CH_3), and Sq13–Sq15 (X = OCH_3), where red-shifts of 12.2, 13.6 and 12.0 nm, respectively, are observed. As noted in Section I, as both the ground and excited states of squaraine are intramolecular CT states, the observed red-shift may be attributable to stabilization of the CT states by the electron-releasing N-alkyl groups. Law [22], however, suggested that this interpretation is unlikely because if such

Table 1 Absorption Spectral Data of Symmetrical Squaraines in CH_2Cl_2

$$D — A — D$$

Squaraine	Substituent	λ_{max}[a]	log ε_{max}[b]	Ref.
Sq1	$R_1 = R_2 = CH_3$, $X = H$	627.6	5.49	22
Sq2	$R_1 = R_2 = C_2H_5$, $X = H$	634.1	5.51	22
Sq3	$R_1 = R_2 = C_3H_7$, $X = H$	638.8	5.53	22
Sq4	$R_1 = R_2 = C_4H_9$, $X = H$	640.0	5.53	22
Sq5	$R_1 = R_2 = C_{18}H_{37}$, $X = H$	641.8	5.52	22
Sq6	$R_1 = R_2 = CH_3$, $X = F$	630.0	5.09	22
Sq7	$R_1 = R_2 = CH_3$, $X = OH$	635.0	5.52	22
Sq8	$R_1 = R_2 = C_2H_5$, $X = OH$	641.1	5.57	22
Sq9	$R_1 = R_2 = C_4H_9$, $X = OH$	648.2	5.56	22
Sq10	$R_1 = R_2 = CH_3$, $X = CH_3$	643.5	5.42	22
Sq11	$R_1 = R_2 = C_2H_5$, $X = CH_3$	651.0	5.49	22
Sq12	$R_1 = R_2 = C_4H_9$, $X = CH_3$	657.1	5.47	22
Sq13	$R_1 = R_2 = CH_3$, $X = OCH_3$	631.8	5.40	22
Sq14	$R_1 = R_2 = C_2H_5$, $X = OCH_3$	638.8	5.48	22
Sq15	$R_1 = R_2 = C_4H_9$, $X = OCH_3$	643.5	5.40	22
Sq16	$R_1 = R_2 = CH_3$, $X = C_2H_5$	643.0	5.42	22
Sq17	$R_1 = CH_3$, $R_2 = CH_2C_6H_5$, $X = H$	628.2	5.56	23
Sq18	$R_1 = CH_3$, $R_2 = CH_2C_6H_5.F$, $X = H$	626.4	5.52	23
Sq19	$R_1 = CH_3$, $R_2 = CH_2C_6H_5.Cl$, $X = H$	626.4	5.55	23
Sq20	$R_1 = CH_3$, $R_2 = CH_2C_6H_5$, $X = F$	632.6	5.50	23
Sq21	$R_1 = CH_3$, $R_2 = CH_2C_6H_5.F$, $X = F$	630.9	5.49	23
Sq22	D =	634.6	5.49	24
Sq23	D =	638.4	5.46	24

(continued)

Table 1 Continued

	D = structure	λ_{max}[a]	ε[b]	Ref.
Sq24	D = (phenyl)-N(CH$_3$)$_2$ with ortho-C$_2$H$_5$	628.8	5.11	22
Sq25	D = (phenyl)-N(CH$_3$) (tetrahydroquinoline-type ring)	645.0	5.50	22
Sq26	D = (julolidine-type fused ring)-N	661.0	5.53	22

[a]Absorption maximum (in nm).
[b]Molar absorption coefficient (in cm^{-1} M^{-1}).

an effect is significant, one should observe a very large spectral redshift for Sq7–Sq9 relative to Sq1–Sq5, because OH is a much more powerful electron-donating group. The contrary is observed experimentally (Table 1).

Owing to the unique electronic structure, the four-membered ring of squaraine interacts with polar functionalities in the ground state. The interaction has been detected in the mass spectrometric analysis of squaraine [25] and also in proton NMR studies [26]. Because MO calculations suggested that the CT character of squaraine should increase when the polarized charges in the squaraine structure are stabilized, Law [22] then attributed the redshift induced by the N-alkyl group to the formation of the solute–solvent complex. As the length of the N-alkyl chain increases, the increased CT character shifts the equilibrium to the formation of the solute–solvent complex, which absorbs at a longer wavelength. This interpretation is consistent with the solvent effect data given below, as well as the fluorescence emission spectral data described in Sections III.B and V.A.

Additional support for this interpretation comes from the spectral data of N-benzyl squaraines (Sq17–Sq21) and N-pyrrolidino-substituted squaraines (Sq22–Sq23). The redshift induced by the N-pyrrolidino group is small relative to those of Sq1–Sq5 despite the fact that the N-pyrrolidino group is a much better electron-releasing group [24]. Again, the main driver for the red shift is the complexation process between squaraine and the solvent. In the case of N-benzyl groups, electron-withdrawing functionality in the benzyl ring is shown to produce a spectral blue-shift. The blue-shift may be attributable to the decrease in CT character, which disfavors the solute–solvent complexation process.

B. Effect of C2 Substituents

Substituents at C2 generally produce red-shifts on the λ_{max} (Table 1). The red-shifts are small (2–16 nm) and are consistent with MO calculation results simply indicating that the anilino moieties are not heavily involved in the electronic transition. The special feature of the results lies in the relative magnitude of the red-shift produced by these substituents. Because the excited state of squaraine is a CT state, the expected substituent effect would decrease in the following order: $CH_3O \sim OH > F > CH_3 \sim C_2H_5$. Experimentally, a reversed order is obtained. The red-shifts for Sq13 (CH_3O), Sq7 (OH), Sq6 (F), Sq10 (CH_3), and Sq16 (C_2H_5) (relative to Sq1) are 4.2, 8.3, 2.4, 15.9, and 15.4 nm, respectively.

In addition to the relatively large red-shift, a decrease in ε_{max} is also observed for Sq10–Sq12 and Sq16. As it is known in the literature that the absorption coefficient of a CT state would decrease when the CT interaction is decreased, the decrease in ε_{max} in Sq10–Sq12 and Sq16 suggests that there is a decrease in the D-A-D CT character in these squaraines. It was hypothesized that due to the steric repulsion between the C2 alkyl substituent and the phenyl ring in Sq10–Sq12 and Sq16, these squaraines become nonplanar [22]. The nonplanarity decreases the CT interactions and results in a nonplanar ground state, nonplanar excited state (lower energy) electronic transition [27,28]. The relative large bathochromic effect observed for these squaraines is thus a conformation effect. Subsequent proton NMR studies showed that these molecules are, indeed, nonplanar in the ground state [26].

The λ_{max} of OH-substituted squaraines (Sq7–Sq9) are red-shifted by ~ 8 nm from those of Sq1–Sq4 and the red-shift is about two times smaller than that induced by the methyl group. This observation again illustrates that CT stabilization through an induced effect is not the major contributor to the red-shift. The ε_{max} values for Sq7–Sq9 are in the range of $(3.3–3.5) \times 10^5$ cm^{-1} M^{-1} and are about 10% higher than those of Sq1–Sq4. The high ε_{max} values in Sq7–Sq9 can be attributed to the planarity of these molecules due to the intramolecular H-bonding between the C2 OH group and the C–O group in the four-membered ring [26].

The red-shift for methoxy-substituted squaraines (Sq13–Sq15) is small relative to that of Sq7–Sq9 despite the fact they both have very similar electron-releasing groups (OH versus OCH_3). This again indicates that CT stabilization is not the main driver for the red-shift. Because the ε_{max} values of Sq13–Sq15 are small relative to Sq1–Sq4 and Sq7–Sq9, Law [22] suggested that these squaraines are not planar, presumably due to steric replusion between the methoxy group in the C2 position and the C–O group in the four-membered ring. The twisted geometry decreases the CT interaction in the chromophore, leading to a smaller red-shift relative to OH-substituted squaraines.

Fluorine substitution at C2 is expected to give interesting results because of its σ electron-withdrawing and π electron-releasing character. The λ_{max} of

Sq6 is at 630 nm and is red-shifted from that of Sq1 by 2.4 nm. The red-shift, although small, nevertheless indicates that the F substituent increases the CT character of the chromphore and results in a spectral red-shift. Very similar small red-shifts are also obtained for fluorinated N-benzyl squaraines Sq20 and Sq21 and fluorinated N-pyrrolidino-substituted squaraine Sq23.

C. Effect of C3 Substituents

Very interesting results are obtained for C3-substituted squaraines. The λ_{max} of Sq24 is at 628.8 nm and is red-shifted by 1.2 nm from that of Sq1. The red-shift is > 12 times smaller than that caused by the C2 ethyl group (in Sq16). The ε_{max} of Sq24 is significantly smaller than that of Sq16 also: 1.28×10^5 cm^{-1} M^{-1}. Whereas the small red-shift in λ_{max} is not surprising because a substituent at C3 is not expected to produce any large CT stabilization effect anyway, the small ε_{max} for Sq24 may be due to another mechanism. For example, due to the strong steric replusion between the N,N-diemthylamino group and the C3 ethyl group, the lone pair electrons at the nitrogen may be twisted from the molecular plane of squaraine. The nonpolarity decreases hyperconjugation and CT interactions, resulting in a smaller ε_{max} value.

When the C3 substituent is structurally linked to the N,N-dialkylamino group, such as those in Sq25 and Sq26, relative large red-shifts are observed, 17.4 nm for Sq25 and 33.4 nm for Sq26. The ε_{max} values for these two compounds are also relative large: $(3.1–3.4) \times 10^5$ cm^{-1} M^{-1}. The overall results can be ascribable to the increase in CT character of the squaraine chromophore enabled by the rigid ring structures. For instance, contrary to Sq24, the CT character of the squaraine chromophores in Sq25 and Sq26 is enhanced due to delocalization of the lone pair electrons at the nitrogen to the π orbital of squaraine. The increased CT character facilitates strong solute–solvent complexation and results in the large red-shifts. This deduction is supported by the fluorescence emission spectra given below.

III. FLUORESCENCE EMISSION OF SYMMETRICAL SQUARAINES

A. Multiple Fluorescence Emission of Bis(4-dimethylaminophenyl)squaraine

Figure 1 shows the corrected fluorescence excitation and emission spectra of bis(4-dimethylaminophenyl)squaraine (Sq1) in CH$_2$Cl$_2$.* The excitation spectrum

---

*All the spectral data presented in this chapter were taken in a spectrofluorometer that was spectrally corrected from 510 to 750 nm by a quantum counter containing tetra-t-butyl metal-free phthalocyanine in 1,1,2-tricholoroethan ($\sim 1.2 \times 10^{-3}$ M). Details of the procedures have been described in Ref. 29.

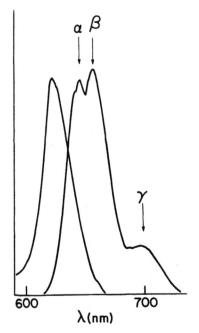

Figure 1   Corrected fluorescence excitation and emission spectrum of Sq1 in CH_2Cl_2 (conc. $\sim 3 \times 10^{-7}$ M).

was found to be identical to the absorption spectrum and is independent of the monitoring wavelength. Three emission bands at λ_F (646, 660, and ~ 702 nm) are observed. These bands are designated as α-, β-, and γ, in order of decreasing emission energy. All the symmetrical squaraines studied exhibit similar multiple emissions and the Stolks shifts for the α-, β-, and γ-bands are in the range 200–400, 530–780, and 1400–1700 cm^{-1} from their λ_{max}, respectively. The multiple emission band in Fig. 1 was shown to be an intrinsic emissive property of Sq1 because it was shown to be independent of concentration (10^{-6}–10^{-7} M) and was unaffected quantitatively by purification.

B.   Spectral Assignments and Photophysics

Probable causes for the multiple emission are (a) vibronic fine structure of the squaraine emission, (b) emission from a relaxed excited state, and (c) emission from an exciplex (with solvent) or the excited state of the squaraine–solvent complex. To differentiate these possibilities, the effects of solvent, temperature, and structural changes on the multiple emission have been studied. Sq4 was chosen as a model for these investigations because of its high solubility in organic solvents. In this subsection, we focus on results that aid the elucidation of the multiple emission. A more detailed discussion of the effects of solvent, temperature, and structural changes on the multiple emission will be given in Sections IV and V.

1. Solvent Effect Study in Alcohols: Evidence for the Solute–Solvent Complex

The results of the effect of alcoholic solvents on the absorption and fluorescence emission of Sq4 are summarized in Table 2 [30]. The λ_{max} of Sq4 is shown to be sensitive to the steric hindrance around the OH groups (e.g., λ_{max} is at ~ 634.1 in t-pentanol, ~ 639.4 nm in secondary alcohols, and 642.6 nm in primary alcohols). In the fluorescence emission spectra, Sq4 also exhibits multiple emission bands in alcohols and the spectral data are highlighted in Fig. 2. While the α- and β-bands are clearly observable in t-pentanol, the relative intensity of the α-emission decreases as the steric hindrance around the OH group decreases. The significance of the results here lies in its steric sensitivity on both λ_{max} and the composition of the emission. The steric sensitivity suggests that the observation is due to short-range interactions between Sq4 and the OH group in the alcoholic solvents. The specificity of the solvent effect in alcohols is also seen in the fluorescence yields (ϕ_F). The ϕ_F value is shown to increase systematically as the chain length in the alcohol increases (Table 2). Among all the pentanols, ϕ_F increases as the steric hindrance around the OH group increases. These results suggest that as the association between the squaraine and the solvent increases, the squaraine becomes increasingly nonplanar, which leads to lower ϕ_F. Law [22,30] then concluded that squaraine forms solute–solvent complexes in alcohols. Because of the polar nature of the four-membered ring, the sites for complexation are the OH group in the alcohol and the four-membered ring in Sq4 (Scheme 1).

Thus, excitation of squaraine in solution results in two excited states, namely the excited state of squaraine and the excited state of the solute–solvent complex. These two excited states emit and give rise the α- and β-emissions, respectively. The observation of a red-shift in λ_{max} accompanied by an increase in the β-emission intensity is found to be very general, including that in nonalcoholic solvents, the results which will be described in Section V.A.

Complexation between solvent molecules and compounds having intramolecular charge transfer states are known. For example, Wang [31] observed very similar solute–solvent complexation between the ground and excited states of polarized enones and solvent molecules. Wang also studied the exciplex formation between DMABN (p-dimethylaminobenzonitrile) and alkyl amines, where a steric effect on the exciplex formation was observed [32]. The steric effect reported in Wang's work is analogous to the Sq4–ROH complex. In a

Table 2 Absorption and Fluorescence Emission Data of Sq4 in Alcoholic Solvents

| Solvent | λ_{max}[a] | $\log \varepsilon$[b] | λ_F[c] | | | ϕ_F[d] | Ref. |
			α	β	γ		
Ethanol	642.7	5.46		667	703	0.14	30
1-Propanol	642.8	5.47		666	702	0.28	30
1-Butanol	642.8	5.47		665	704	0.39	30
1-Pentanol	642.7	5.51		665	~702	0.53	30
1-Hexanol	642.5	5.53		665	~704	0.57	30
1-Heptanol	642.4	5.50		665	~704	0.60	30
iso-Pentanol	642.2	5.52		~665	~703	0.49	30
2-Propanol	639.4	5.49		664	~705	0.37	30
2-Pentanol	639.3	5.53		664	~705	0.60	30
3-Pentanol	638.5	5.53		664	~703	0.68	30
t-Pentanol	634.1	5.55	651	661	~700	0.76	30

[a]Absorption maximum (in nm).
[b]Molar absorption coefficient (in $cm^{-1} M^{-1}$).
[c]Fluorescence wavelength (in nm).
[d]Fluorescence quantum yield

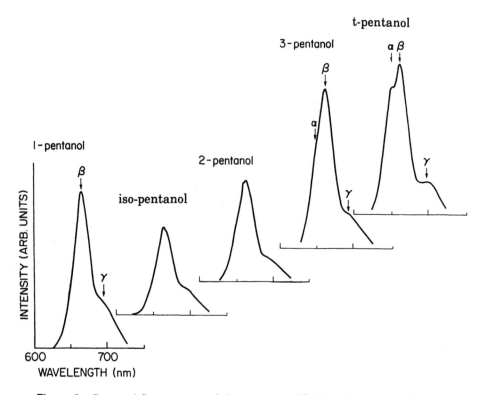

Figure 2 Corrected fluorescence emission spectrum of Sq4 in various pentanols (conc. $\sim 3 \times 10^{-7}$ M).

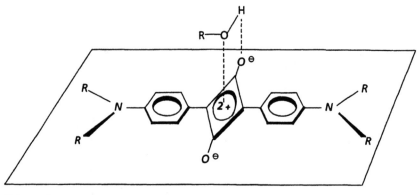

Scheme 1

similar context, Varma and co-workers [33–36] reported extensive evidence for the exciplex formation between DMABN and polar solvent molecules. In one of their articles, Visser and Varma [33] actually demonstrated the formation of a 1 : 1 exciplex between DMABN and alkylnitrile in cyclohexane, using a mixed-solvent experiment. The dipole moment for the excited state of DMABN is ~14D [37]. Assuming that the polarized charges in the excited state of DMABN are localized in the dimethylamino group and the cyano group, each of these groups should bear ~0.42e of opposite charges. The formation of an exciplex is presumably due to interactions between these polarized charges and dipoles of solvent molecules. The polarized charges in the ground state of Sq4 as estimated from the MO calculations of Sq1 [6] are approximately – 0.35e at the oxygen atoms, + 0.37e in the four-membered ring of squaraine, and + 0.4e at the nitrogen atoms. Thus, interactions between the polarized charges in the ground state of squaraine and solvent molecules are not irrational. In fact, complexation processes involving squaraine and solvents are not unprecedented. Merrit [38] reported the preparation of a high concentration (up to ~10^{-2} M) of Sq7 in ethylene diamine and propylamine. The amine solution was "straw colored" and did not show any normal squaraine absorption. Sq7 can be re-generated from the amine solution when the solvent is evaporated. Merrit concluded the high solubility of Sq7 in amine solvents (10^{-2} M as compared to ~10^{-5} M in organic solvents) to the formation of a solute–solvent complex. Complexation between squaraines and polar species have been observed in the vapor phase. Law and co-workers [25] reported a general complexation process between squaraines and the anilino fragments in a mass spectrometric study of squaraines. In that study, these authors studied the chemical ionization mass spectrum of Sq1 using CH_2Cl_2 as an ionization gas and detected the [Sq1 : Cl] complex. More concrete evidence for the occurrence of ground state complexation between squaraines and solvent molecules was subsequently obtained in proton NMR experiments [26].

2. Low-Temperature Experiments

Figures 3a and 3b show the corrected fluorescence excitation and emission spectra of Sq4 in diethyl ether at room temperature and at 77 K, respectively. At room temperature, Sq4 exhibits primarily an α-emission at λ_F=641 nm, attributable to the squaraine emission. The excitation maximum is at ~620 nm. The β-emission is very weak. At 77 K, only a single emission band at λ_F=664 nm is observed. The excitation maximum is at 654 nm and is red-shifted by 32 nm relative to the room-temperature λ_{max}. This observation is genuine for squaraine because similar spectral results were also obtained in 2-methyl-tetrahydrofuran matrix at 77 K [30]. The observation of spectral red-shifts in both absorption and emission spectra is certainly against general expectation from a vibronic fine structure, as lower vibronic bands are expected to be intensified and should result in spectral blue-shifts at low temperature [39]. The overall spectral results in Fig. 3 are therefore consistent with the solute–solvent complex model proposed above. For example, at 77 K, squaraine molecules all form complexes in the matrix due to the stabilization effect provided by the low temperature, and these complexes absorb at longer wavelengths. They emit at

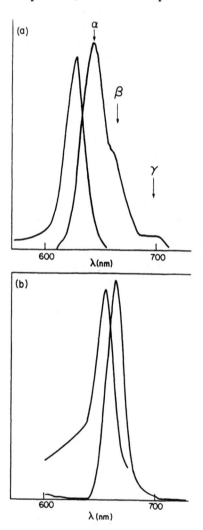

Figure 3 Corrected fluorescence excitation and emission spectrum of Sq4 in diethyl ether (a) at 298 K and (b) 77 K (conc. ~3 × 10^{-7} M).

$\lambda_F = 664$ nm. By comparison with the room-temperature spectrum, one can assign the emission band at 660 nm to the emission from the solute–solvent complex and the emission band at 650 nm to the emission from the excited squaraine itself.

Another significant finding in the low-temperature emission spectrum is the absence of the λ-emission at ~ 700 nm. As will be shown in Section III.B.4, the λ-emission is derived from a relaxed excited state which is formed by rotation of the C–C bond between the phenyl ring and the central four-membered ring of squaraine. Its absence simply indicates that the rotation motion is prohibited at low temperature.

3. Mixed-Solvent Experiments

The solute–solvent complex model proposed above is further supported by results obtained from mixed-solvent experiments. In diethyl ether, Sq4 exists primarily as free squaraine (Fig. 3a). The addition of a complexing solvent should drive the equilibrium for complexation. As a result, both λ_{max} and λ_F should shift to the red and the intensity of the β-emission should increase. To eliminate any complication due to the increase in dielectric constant of the mixed solvent during experimentation, the mixed-solvent experiment was first performed in a ternary system consisting of ether ($\varepsilon = 4.43$), chloroform ($\varepsilon = 4.7$), and n-hexane ($\varepsilon = 1.9$). The addition of n-hexane in the mixture is to keep the dielectric constant constant as the concentration of chloroform increases. The absorption and fluorescence spectral results are summarized in Figs. 4a and 4b, respectively. The data showed that λ_{max} and λ_F shift to the red and the intensity of the β-emission increases as [CHCl$_3$] increases at [CHCl$_3$] < 1.12 M. Simultaneously, an isosbestic point at ~ 625 nm and isoemissive point ~ 662 nm are observed in the absorption and fluorescence spectra, respectively. The observation of an isosbestic point in the absorption spectra and an isoemissive point in the emission spectra provide positive evidence that (1) Sq4 forms a complex with chloroform in the ethereal solutions, (2) the β-emission is the emission from the excited solute–solvent complex, and (3) the α-emission is from the excited squaraine. The stoichiometry of the complex at [CHCl$_3$] < 1.12 M is 1 : 1, as indicated in the plot of the intensity of the β-emission (I_β) versus [CHCl$_3$] (inset in Fig. 4b).

Further red-shifts on both λ_{max} and λ_F and a further increase in I_β are observed at [CHCl$_3$] > 2.24 M. The absorption and emission spectra no longer pass through the isosbestic and isoemissve points. The results can be attributable to the formation of the 1 : n solvent complexes between Sq4 and chloroform.

Complexation between squaraines and solvent molecules are very general. Analogous absorption and fluorescence spectral data have also been obtained

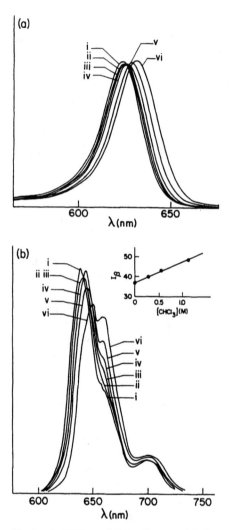

Figure 4 Effect of chloroform on (a) absorption and (b) corrected fluorescence spectra of Sq4 in diethyl ether ([CHCl$_3$] = (i) 0, (ii) 0.28, (iii) 0.56, (iv) 1.12, (v) 2.24, and (vi) 4.48 *M*).

in other solvent mixtures, such as mixtures from ether and *n*-butyl acetate, benzene, methylene chloride, DMF, and 2-propanol [30].

4. γ-Emission and Photophysics of Squaraines

One of the important findings in the low-temperature emission spectrum in Fig. 3 is the absence of the γ-emission at 77K. In recent work, Law extended the study to the fluorescence of Sq4 in a polystyrene matrix at room temperature (Fig. 5) [30]. Sq4 exhibits two emission bands at $\lambda_F = 650$ and 659 nm and there is very little emission beyond 700 nm. By comparison with the fluorescence emission of Sq4 in toluene (see later results in Table 4), one can assign the band at 650 nm to the α-emission and the band at 659 nm to the β-emission. The observation of the β-emission in polystyrene indicates that the styrenic chromophore in the polymer matrix is capable of forming complexes with Sq4. The major distinction between the emission spectrum in polystyrene and that in toluene is the absence of the γ-emission. Its absence supports the notion (from low-temperature experiments) that a rotational motion is required for the generation of the emitting state for the γ-band. In polystyrene, the rotational motion is presumably retarded by the rigid polymer matrix. As will be detailed in Sections IV.B and IV.C, there is strong evidence to suggest that the rotation occurs in the C–C bond between the phenyl ring and the four-membered ring of squaraine because the γ-emission is shown to increase significantly when the phenyl ring and the four-membered ring of squaraine is twisted from planarity. The photophysics of squaraine can accordingly be summarized (Scheme 2). Excitation of squaraine in solution leads to two excited states: the excited state of squaraine and the excited state of the solute–solvent complex. These two excited states emit to give rise to the α- and β-emission, respectively. They also undergo a molecular relaxation by rotation of the C–C bond between the phenyl ring and the four-membered ring. A twisted, relaxed excited state is generated as a result. This relaxed excited state can either emit to give rise to the γ-emission or undergo a radiationless decay by rotation of the C–C bond to the ground state.

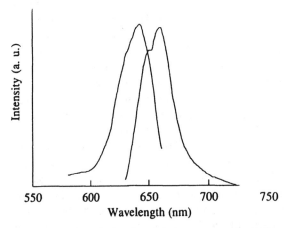

Figure 5 Corrected fluorescence excitation and emission spectra of Sq4 in polystyrene at room temperature (conc. $\sim 10^{-5}$ M).

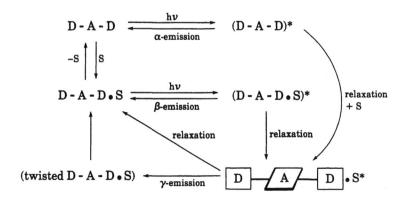

squaraine : D - A - D
solvent   : S

Scheme 2

IV. EFFECTS OF STRUCTURAL CHANGES ON THE FLUORESCENCE EMISSION OF SYMMETRICAL SQUARAINES

A. Effect of *N*-Alkyl Groups

Figure 6 shows the fluorescence excitation and emission spectra of Sq2–Sq5 in CH_2Cl_2. In each case, the excitation spectrum was found to be identical to the absorption spectrum and is independent of the monitoring wavelength. The spectral results are summarized in Table 3. Although the effect of chain length on λ_F may be small, it has a profound effect on the composition of the emission band. For example, for $N = CH_3$, the intensities of the α- and β-bands are about the same (Fig. 1). As the chain length is increased, the intensity of the α-band decreases whereas the opposite is observed for the β-band (Fig. 6). The gradual dominance of the β-emission indicates that the equilibrium constant for the solute–solvent complex increases as the chain length increases. This is actually consistent with the solute–solvent complex model discussed above. Namely as the CT D-A-D state of squaraine is stabilized (by the electron-releasing *N*-alkyl group), the tendency for complexation increases [6].

The ϕ_F of Sq1 to Sq5 are in the range of 0.65 to 0.74. After correction for the γ-emission, the quantum yields for the (α + β) band, $\phi_{\alpha\beta}$, range from 0.58–0.66. As outlined in Scheme 2, the major radiationless decay process for the excited states of squaraine is by rotation of the C-C bond between the phenyl ring and the four membered ring, the small increase in ϕ_F may be attributable to the decrease in rate of rotation of the *N,N*-dialkylanilino group as its size increases.

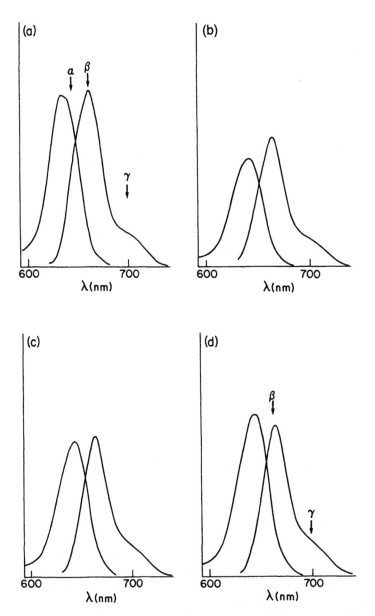

Figure 6 Corrected fluorescence excitation and emission spectra of (a) Sq2, (b) Sq3, (c) Sq4, and (d) Sq5 in CH_2Cl_2 (conc. ~3 × 10^{-7} M).

Table 3 Fluorescence Emission and Fluorescence Lifetime Data of Symmetrical Squaraines in CH_2Cl_2

Squaraine	λ_F (nm) a			ϕ_F[a]	ϕ_γ/ϕ_F[a,b]	τ (ns)[c]
	α	β	γ			
Sq1	~646	~660	~702	0.65	11%	1.3 (1.5)[d]
Sq2	650	660	~702	0.69	12%	
Sq3	651	662.5	~702	0.70	14%	
Sq4		664	~705	0.70	13%	1.8
Sq5		664.5	~705	0.74	11%	2.0
Sq6	652	661.5	~701	0.091	17%	0.27
Sq7	650	660	~704	0.86	11%	2.8 (3.0)[d]
Sq8		661.5	~705	0.86	12%	
Sq9		665	~702	0.83	12%	2.7
Sq10		669.5	~706	0.023	27%	0.18
Sq11		671.5	~705	0.036	34%	
Sq12		675	~705	0.054	40%	0.30
Sq13	650	661	~704	0.042	18%	0.28
Sq14		665	~702	0.049	29%	
Sq15		664	~704	0.26	17%	0.89
Sq16	638	650	~704	0.42	40%	
Sq17	~650	~660	~700	0.70	14%	1.7
Sq18	~647	~659	~700	0.68	10%	
Sq19	~647	~659	~700	0.67	12%	
Sq20	~650	~662	~704	0.19	18%	0.30
Sq21	~650	~661	~704	0.22	24%	
Sq22	649	657	~698	0.65	14%	2.2
Sq23	650	658	~700	0.66	14%	1.6
Sq24	638	650	~704	0.42	12%	1.0
Sq25		666.5	~701	0.45	10%	1.1
Sq26		680	~705	0.32	50%	1.0

[a]Data taken from Ref. 22 unless specified.
[b]Contribution of the γ-band to the total emission.
[c]New data reported in this work.
[d]From Ref. 40.

The γ-emission is resolvable from the α and the β-emission. The ϕ_γ/ϕ_F values for Sq1–Sq5 are found to be insensitive to the chain length of the N-alkyl group. Since rotation of the C-C bond between the phenyl ring and the four membered ring is involved in both the population and the de-population of the relaxed state (Scheme 2), the constant ϕ_γ/ϕ_F values observed for Sq1–Sq5 can be rationalized as a compensation effect.

The spectral results of Sq17–Sq19 and Sq22 in Table 3 suggest that the effects of N-benzyl and N-pyrrolidino substituents on the fluorescence (compared to Sq1), in terms of the emission wavelengths and ϕ_F values, are very small. While the small N-benzyl effect may be understandable, the small N-pyrrolidino effect is somewhat surprising in view of its strong electron-releasing character [24]. However, when one compares the fluorescence data between the C2 fluorinated squaraines, e.g., Sq6 and Sq23, the N-pyrrolidino effect is evident. Although the emission wavelengths between these two squaraines are similar, the ϕ_F for Sq23 is > 6 times higher than that of Sq6. We suggest the increase in ϕ_F for Sq23 to the increase in CT character enabled by the strong electron-releasing N-pyrrolidino group. The increased CT interaction probably increases the double bond character of the rotating C-C bond and results in a high ϕ_F value.

B. Effect of C2 Substituents

1. C2 Methyl Groups

Figures 7a, b and c show the fluorescence emission spectra of Sq10–Sq12 in CH_2Cl_2 and the spectral data are listed in Table 3. Only β- and λ-bands are observed in the emission spectra and they are shifted by 400–600 and 1034–1376 cm^{-1} from their λ_{max}, respectively. As in Sq1–Sq5, the N-alkyl effect on λ_F is very small. The ϕ_F values for Sq10–Sq12 are 0.023, 0.036 and 0.054, respectively. After subtracting the contributions from the γ-emissions, the ϕ_β values are 0.017, 0.024 and 0.032, respectively. These values are 20–30 times less than those of Sq1–Sq5 and show a strong dependence on the N-alkyl group. Since proton NMR results indicate that the phenyl ring and the four membered ring in these squaraines are twisted from planarity due to steric replusion between the methyl group and the C-O group [26], and since Rettig and Gleiter [41] reported earlier that the rate for the formation of a twisted intramolecular CT state can increase considerably as the twist angle increases, the about 30 times decrease in ϕ_F for Sq10–Sq12 relative to Sq1–Sq5 can readily be attributed to a relaxation process involving rotation of the C-C bond between the phenyl ring and the four membered ring. As shown in Scheme 3, excitation of these squaraines will generate a nonplanar excited state. The excited states can either emit to give the β-emission or they can undergo a rapid radiationless decay by rotation of the C-C bond between the phenyl ring and the four membered ring. The gradual increase in ϕ_F from Sq10 to Sq11 to Sq12 is consistent with this mechanism because the rate of rotating the N,N-dialkylanilino group should decrease as the size of the rotating group increases. A very similar rotational de-excitation mechanism has been reported in the excited states of stilbenes [42], rhodamines [43], coumarins [44], and p-dimethylamino-benzonitriles and derivatives [45].

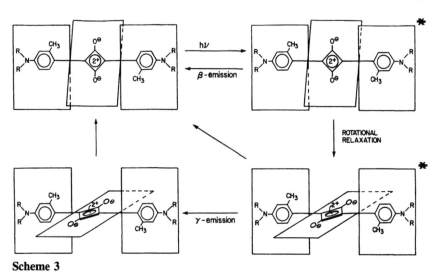

Scheme 3

The most significant finding in the emission spectra in Figs. 7a–7c is the relatively high λ-emission. The ϕ_γ/ϕ_F values for Sq10–Sq12 are 27%, 34%, and 40%, respectively. They are two to three times higher than those of Sq1–Sq5 and exhibit a strong dependence on the N-alkyl group. As seen in Scheme 3, the relatively high λ-emission intensity in these compounds may be due to the high rate of populating the emitting relaxed excited state. Because the rate of intersystem crossing for squaraine is very slow [40], one can actually estimate the efficiency of populating the relaxed excited state from the ϕ_β values, which is about 95–98%. The almost constant efficiency for populating the relaxed excited states indicates that the gradual increase in γ-band intensity is due to a steady decrease in the radiationless decay rate. We suggest that the so-gener- ated relaxed (twisted) excited state can emit to give rise to the low energy γ- emission or it can deexcite by rotating the C–C bond between the phenyl ring and the four-membered ring to the ground state (Scheme 3). The rate of rota- tion decreases as the size of the rotating group increases and this results in an N-alkyl group effect on the relative intensity of the γ-emission.

2.   C2 F, C_2H_5, OH, and OCH_3 Groups

The fluorescence emission spectrum of Sq6 is given in Fig. 7d. Three emis- sion bands (α, β, and λ) are observed and they are shifted by 536, 746, and 1608 cm^{-1}, respectively, from the λ_{max}. The ϕ_F for S16 is 0.091 and the ϕ_γ/ϕ_F value is ~17%. Both the ϕ_F and ϕ_γ/ϕ_F values for Sq6 is between Sq1 and Sq10. The intermediate values correlate well to the size of the fluorine atom, which suggests that the twisting induced by the fluorine atom is smaller than that induced by the methyl group.

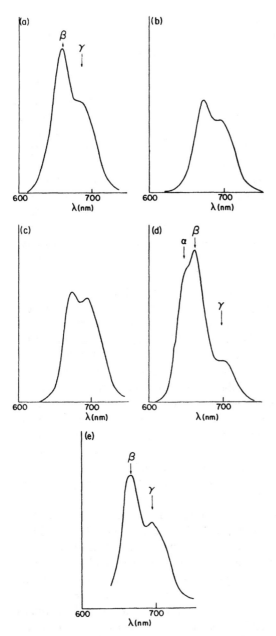

Figure 7 Corrected fluorescence emission spectra of (a) Sq10, (b) Sq11, (c) Sq12, (d) Sq6, and (e) Sq16 in CH_2Cl_2 (conc. ~3 × 10^{-7} M).

The fluorescence emission spectrum of Sq16 is given in Fig. 7e. As in the spectra of Sq10–Sq12, only β- and γ-bands are observed and they are shifted by 616 and 1269 cm^{-1}, respectively, from the λ_{max}. The ϕ_F for Sq16 is 0.024 and the ϕ_γ/ϕ_F value is ~40%. After correcting for the β-emission, ϕ_β becomes 0.014 and is smaller than that of Sq10. The decrease in ϕ_β is probably due to the larger twist angle between the phenyl ring and the four-membered ring in Sq16. As shown in Scheme 3, the twist geometry not only speeds up the rotational relaxation process and results in a low ϕ_F value, it also facilitates the generation of the twisted excited state. The large ϕ_γ/ϕ_F value for Sq16 may be attributable to the increased population of the twisted relaxed excited state.

The fluorescence emission spectra of Sq7 to Sq9 are given in Figs. 8a–8c. Three emission bands are observed for Sq7, and the emission becomes dominated by the β- and λ-emission in Sq8 and Sq9. This trend is identical to that seen in Sq1–Sq5. The gradual dominance of the β-emission in the emission spectra along with a small red-shift on λ_F as the chain length of the N-alkyl group increases suggest that as the D-A-D CT character in squaraine is enhanced, more solute–solvent complexes are formed.

The ϕ_F values for Sq7–Sq9 are ~0.83–0.86 and the ϕ_γ/ϕ_F values are ~12%. The constant quantum yield data are in contrast to those observed in Sq1–Sq5 and Sq10–Sq12, where both ϕ_F and ϕ_γ/ϕ_F are found to be sensitive to the length of the N-alkyl chain. Because proton NMR spectral data suggest that Sq7–Sq9 are planar [26] and a molecular model study shows that the intramolecular H-bonding between the C2 OH group and the C–O group in the four-membered ring of squaraine is feasible throughout a full rotation of the C–C bond, the high ϕ_F values and the independence on the N-alkyl group can be attributed to the H-bonding effect. Due to the dominant influence of the H-bonding on the photophysics, the ϕ_F and ϕ_γ/ϕ_F values become very similar.

The fluorescence emission spectra of Sq13–Sq15 are given in Figs. 8d–8f. Three emission bands are observed for Sq13 and the emission becomes dominated by the β- and γ-bands in Sq14 and Sq15. The N-alkyl group effect is identical to those observed for Sq1–Sq5 and Sq7–Sq9.

The ϕ_F values for Sq13, Sq14, and Sq15 are 0.042, 0.049, and 0.26, respectively. After correcting for the γ-emission, the $\phi_{\alpha\beta}$ values become 0.034, 0.035, and 0.22, respectively. These values are higher than those of Sq10–Sq12 but lower than those of Sq1–Sq4. Because rotational relaxation has been established as a major radiationless decay process for the excited state of squaraine, the intermediate ϕ_F values suggest that OCH$_3$-substituted squaraines are closer to planarity than CH$_3$-substituted squaraines. This observation can be attributed to the OCH$_3$ group in the C2 position. Although steric repulsion alone may have twisted the chromophore to nonplanarity, the electron-releasing OCH$_3$ group increases the CT character of the chromophore and may have flattened the squaraine structure.

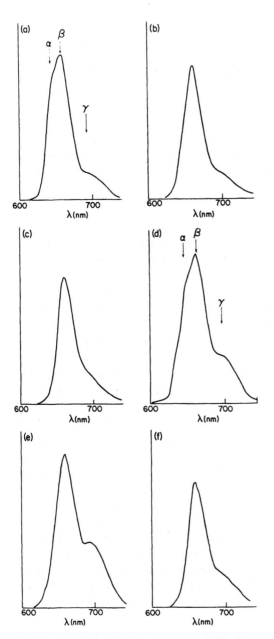

Figure 8 Corrected fluorescence emission spectra of (a) Sq7, (b) Sq8, (c) Sq9, (d) Sq13, (e) Sq14, and (f) Sq15 in CH_2Cl_2 (conc. ~3 × 10^{-7} M).

The ϕ_δ/ϕ_F values for Sq13–Sq15 are 18%, 29%, and 17%, respectively. The results are different from those of Sq10–Sq12, which increases as the chain length of the N-alkyl group increases. Although the increase in the ϕ_γ/ϕ_F value from Sq13 to Sq14 may be explainable by the free rotor mechanism, the small ϕ_γ/ϕ_F value for Sq15 is unexpected. The result may be attributable to the increased CT character relative to Sq13 and Sq14. The increased CT character may make Sq15 the most planar molecule among this series of squaraines. Thus, the rotational relaxation for the excited state of Sq15 becomes the slowest, resulting in higher ϕ_F and lower ϕ_γ/ϕ_F values. The deviation in photophysical behavior for Sq15 relative to Sq13 and Sq14 may be a conformation effect.

C. Effect of C3 Substituents

The fluorescence emission of Sq24 to Sq26 are shown in Figs. 9a–9c. Sq24 exhibits three emission bands, α, β, and γ, and they are shifted by 229, 518, and 1700 cm^{-1}, respectively, from the λ_{max}. The ϕ_F of Sq24 is 0.49 and is about ~30% lower than that of Sq1. Because the absorption spectral data suggest that the C–N bond in Sq24 is twisted due to replusion between the C3 ethyl group and the N,N-dimethylamino group, the decrease in ϕ_F for Sq24 may be due to molecular relaxation by rotation of the C–N bond. It is very important to note that the effect exerted by the C2 ethyl group in Sq16 is significantly larger than that in Sq24. The result thus supports the photophysical model proposed in Schemes 2 and 3; specifically, rotation of the C–C bond between the phenyl ring and the four-membered ring is the major radiationless decay process for the excited state of squaraine.

Only two emission bands, β and γ, are observed for Sq25 and Sq26. Their ϕ_F values are 0.42 and 0.32, respectively. After correction for the γ-emission, the ϕ_β values for Sq25 and Sq26 are 0.38 and 0.16, respectively. They are smaller than that of Sq1, even though the C–N bond in these two compounds are rigidized. The finding again supports the photophysical model given in Schemes 2 and 3 (e.g., rotation of the C–C bond between the phenyl ring and the four-membered ring is the major radiationless decay process for the excited state of squaraine). The fluorescence spectra in Figs. 9b and 9c show that these two compounds form solute–solvent complexes in CH$_2$Cl$_2$. As it is known that solute–solvent complexation induces nonplanarity in the squaraine chromophore [26], the decrease in ϕ_β for Sq25 to Sq26 can be considered as a conformation effect. The rotational relaxation rate increases as the chromophore is twisted from planarity.

The ϕ_γ/ϕ_F values for Sq24 and Sq25 are about 12% and are similar to those of Sq1 to Sq5. On the other hand, the ϕ_γ/ϕ_F value for Sq26 is anomalously high, ~50%. The increase in ϕ_γ/ϕ_F value can again be explained by the

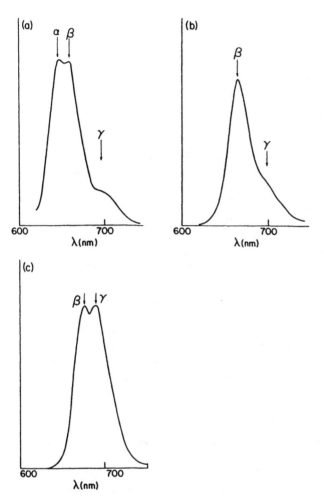

Figure 9 Corrected fluorescence emission spectra of (a) Sq24, (b) Sq25, and (c) Sq26 in CH_2Cl_2 (conc. ~3 × 10^{-7} M).

conformational effect. For instance, due to solute–solvent complexation, Sq26 is nonplanar. This nonplanarity decreases the ϕ_F value, at the same time populating the relaxed excited state (Scheme 3), leading to a high ϕ_γ/ϕ_F value.

D. Fluorescence Lifetimes

The fluorescence lifetime data of symmetrical squaraines are also summarized in Table 3. For each squaraine, monoexponential decays with similar lifetimes (within 0.1 ns, $\chi^2 < 1.4$) across the multiple emission band (from 645 to 705 nm) are observed. Because rotation of the C–C bond between the phenyl ring and the four-membered ring is the major radiationless decay process, the similar lifetimes across the multiple emission band suggests that the lifetimes between the squaraine and the complex are similar, at least in CH_2Cl_2.

1. Effect of N-Alkyl Groups

There is a steady increase in lifetime from Sq1 to Sq2 to Sq5 as the chain length increases. A similar chain length effect is also seen from Sq10 to Sq12, and Sq13 to Sq15. The increase in fluorescence lifetime as the chain length increases parallels the ϕ_F data and is consistent with the rotational deexcitation mechanism.

The chain length effect is not observed for OH-substituted squaraines Sq7 and Sq9. The lifetimes for these two squaraines are ~2.8 ns and are longer than those of Sq1 and Sq4. Similar to the ϕ_F data, the longer lifetimes for Sq7 and Sq9 can be ascribable to the intramolecular H-bonding between the C2 OH group and the C–O group in the four-membered ring. The intramolecular H-bonding slows down and dominates the rotational deexcitation process, resulting in a longer lifetime and the lack of a chain length effect.

The fluorescence lifetimes for Sq17 and Sq23 are about ~1.6 ns in CH_2Cl_2 and are longer than that of Sq1. The longer lifetimes for Sq17 and Sq23 can be attributed to the increase in size for the rotating group, namely N-methylbenzylanilino and N-pyrrolidinoanilino groups are larger in size than that of the N,N-dimethylanilino group.

2. Effect of C2 Substituents

The fluorescence lifetimes for C2-substituted squaraines (in CH_2Cl_2) decreases from Sq7 (OH, 2.8 ns) to Sq1(H, 1.3 ns) to Sq13 (OCH$_3$, 0.28 ns) ~Sq26 (F, 0.27 ns) to Sq10 (CH$_3$, 0.18 ns). This trend parallels approximately to the planarity of the squaraine chromophore, as revealed by proton NMR spectroscopy [26]. Because rotation of the C–C bond is the major radiationless decay process, the rates of C–C bond rotation in these squaraines can be estimated to be 1.1×10^8, 5.5×10^8, 33.3×10^8, 34.6×10^8, and 53.5×10^8 s^{-1}, respectively, asssuming that the rate of intersystem crossing is small relatively and the radiative decay rate of squaraine is ~2.44×10^8 s^{-1} [30]. The estimated C–C bond rotation rate increases as the nonplanarity of the squaraine chromophore or as the twist angle between the phenyl ring and four-membered ring increases.

3. Effect of C3 Substituents

The fluorescence lifetimes for Sq24–Sq26 are ~1.1 ns and are comparable to that of Sq1. The lack of a C3 substituent effect on the lifetime (e.g., compare

Sq12 to Sq24) once again demonstrates that C–C bond rotation, not C–N bond rotation, is responsible for the radiationless decay of excited squaraines.

V. EFFECTS OF SOLVENT AND TEMPERATURE ON THE ABSORPTION AND FLUORESCENCE EMISSION OF SYMMETRICAL SQUARAINES

A. Effect of Solvent

The absorption and fluorescence emission data of Sq4 in nonalcoholic solvents are summarized in Table 4. Small red-shifts are observed for both λ_{max} and λ_F. The small shifts suggest that there is very little solvent reorganizarion during the electronic transition. The observation is in agreement with the MO calculation results reported by Bigelow and Freund [6], which showed that both the charge distributions in the S_0 and S_1 states of Sq1 are similar and that the S_0 to S_1 transition is localized in the central four-membered ring of squaraine.

A number of attempts were made to correlate the solvent effects with different solvent parameters, such as the dielectric constant E_T [46], Z [47], δ [48], Py [49], π^* [50], and so forth. The relationships between λ_{max} and these solvent parameters are quite scattered except π^*. The plot of λ_{max} of Sq4 as a function of solvent parameter π^* is given in Fig. 10. Along with the red-shift on λ_{max}, a systematic and gradual change in the composition of the multiple emission band is observed (see insets in Fig. 10). Sq4 exhibits primarily α-emission in diethyl ether . As the solvent polarity increases, the intensity of the β-emission increases. The β-emission eventually dominates the fluorescence. Because the β-emission is the emission from the solute–solvent complex, the overall spectral results suggest that the solvent effect on λ_{max} may be due to the shift in equilibrium for the complex formation as π^* increases. For solvents with π^* ranging from 0.273 to 0.567, both α- and β-emission bands are discernible simultaneously. Assuming that the spectral bandwidths of these two bands are similar and that they are not sensitive to solvent, Law [30] has deconvoluted the contribution of the α- and β-bands in the multiple emissions. The relative intensity of these two bands can then be used to estimate the relative concentrations of the free squaraine and the complex. From the ratio of the α- and β-emissions and the molar concentration of the solvent, the equilibrium constants (K_{eq}) in these solvents are calculated. A plot of K_{eq} versus π^* is depicted in Fig. 11, and a linear plot is obtained. The result simply indicates that the equilibrium constant for solute–solvent complexation increases as π^* increases.

The ϕ_F of Sq4 is also found to be solvent sensitive (Table 4). It generally decreases as the solvent polarity increases, but there is no correlation with π^* or other solvent parameters. According to Schemes 2 and 3, ϕ_F should be sensitive to the twist angle between the phenyl ring and the four-membered ring.

Table 4 Absorption and Fluorescence Emission Data of Sq4 in Non-Alcoholic Solvents

| Solvent | π^{*a} | λ_{max}^b | $\log \varepsilon^b$ | λ_F^b | | | ϕ_F^b | ϕ_γ/ϕ_F^b |
				α	β	γ		
Ether	0.273	623.5		640	~660c	~700	0.86	12.7
p-Xylene	0.426	643.9	5.56	647	657	~701	0.90	11.4
n-Butyl acetate	0.460	631.6	5.54	~647	659	~700	0.81	11.6
1,1,1-Trichloroethane	0.490	633.4	5.52	~649	659	~701	0.91	11.9
Toluene	0.535	635.2	5.58	~649	659	~701	1.0	11.9
Ethyl acetate	0.545	631.6	5.52	~649	659	~700	0.84	13.6
1,4-Dioxane	0.553	633.8	5.50	~650c	~660	~700	0.85	14.0
Tetrahydrofuran	0.576	636.6	5.53	~650c	~660	~700	0.78	12.8
Benzene	0.588	637.0	5.53	~650c	659	~701	0.96	13.0
Methyl ethyl ketone	0.674	639.4	5.54		662	~701	0.47	14.6
Acetone	0.684	638.8	5.51		664	~702	0.39	15.9
Acetonitrile	0.713	640.7	5.48		665	~701	0.15	22.7
Chloroform	0.760	638.2	5.52		661	~702	0.92	14.5
Methylene chloride	0.802	640.0	5.53		664	~705	0.70	13.0
1,2-Dichloroethane	0.807	641.8	5.50		668	~701	0.75	19.0
1,1,2-Trichloroethane	0.892	642.3	5.50		665	~705	0.79	14.2
N,N-Dimethylformamide	0.875	650.0	5.40		670	~705	0.33	16.0

aSolvent parameter π^*, data taken from Ref. 50.
bNotations identical to that used in Tables 1–3.
cShoulder.

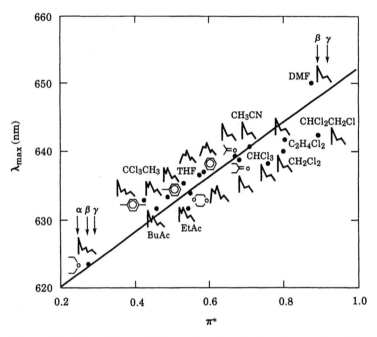

Figure 10   Plot of λ_{max} of Sq4 as a function of solvent parameter π^*.

The variation of ϕ_F in different solvents may be due to the different twist angles in different solute–solvent complexes. Fortunately, the geometry and the complexation process between squaraines and solvent molecules are accessible by proton NMR spectroscopy. Law [26] showed that the chemical shifts of the aromatic protons α and β to the four-membered ring (δ_α and δ_β, respectively) are very characteristic and they are at 8–9 and 6–7 ppm, respectively. The difference in chemical shift for these two protons ($\delta_\alpha - \delta_\beta$) was shown to be a useful measure for the planarity of the squaraine chromophore; specifically, the largest ($\delta_\alpha - \delta_\beta$) value is obtained when the squaraine is perfectly planar. Results on the correlations of ($\delta_\alpha - \delta_\beta$) with solvent parameter π^* and λ_{max} are depicted in Figs. 12 and 13, respectively. Linear relationships are obtained in these two plots, with the exception of aromatic solvents. The plot in Fig. 12 suggests that as the solvent-squaraine association increases, the planarity of the squaraine chromophore decreases. Because the site for complexation in squaraine is believed to be in the four-membered ring, it is reasonable to expect that as the association is increasing, the phenyl ring and the four-membered ring become increasingly distorted due to steric effect (Scheme 4a) [26]. Aromatic solvents (benzene and toluene) are exceptions. Although their interactions are strong, they do not distort the planarity of the squaraine chromophore be-

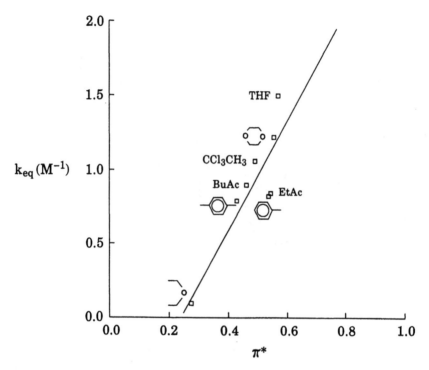

Figure 11 Plot of K_{eq} versus π^*.

cause of the π–π interactions (Scheme 4b) [26]. The ability of correlating NMR results with both π^* and λ_{max} (Figs. 12 and 13), along with a parallel increase in the intensity of the β-emission, lead to the conclusion that the solvent effects on both absorption and fluorescence of squaraine are all originated from the solute–solvent complexes.

Figure 14 shows that $(\delta_\alpha-\delta_\beta)$ correlates with ϕ_F too. This is very significant, as $(\delta_\alpha-\delta_\beta)$ is a probe of the planarity of the chromophore [26]. The correlation basically confirms the photophysical model presented in Schemes 2 and 3. Rotation of the C–C bond between the phenyl ring and the four-membered ring is the major radiationless decay process for the excited state of squaraine, and the rate of rotation increases as the twist angle around the rotating bond increases.

B. Effect of Temperature

Figures 15 and 16 show the effect of temperature on the absorption and fluorescence emission of Sq4 in THF and toluene, respectively. These two solvents

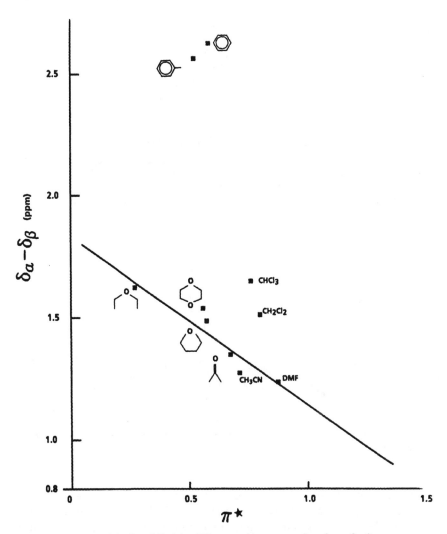

Figure 12 Plot of $\delta_\alpha - \delta_\beta$ of Sq4 in different solvents as a function of π^*.

were selected because they were found to form complexes with squaraine with different geometries [26]. Specifically, the solvent complex with THF should be increasingly nonplanar (Scheme 4a), whereas that with toluene should always be planar (Scheme 4b), as the solute–solvent association increases due to temperature lowering. The absorption results in Fig. 15a show that λ_{max} shifts to the red as the ambient temperature decreases; at the same time, an isosbestic point at ~ 637 nm is observed. In the same temperature range, the composi-

(a)

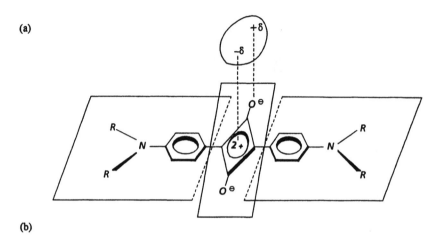

(b)

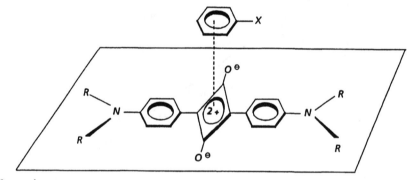

Scheme 4

tion of the fluorescence emission spectrum changes from having two emission bands, α and β, at 25°C to primarily β at ~-17°C (Fig. 15b). Similar absorption and fluorescence spectral results are also obtained in toluene. The results again can be interpretated using the solute–solvent complex model. As the ambient temperature is lowered, more solute–solvent complexes are formed, leading to a spectral red-shift in the absorption spectra and an increase in the β-emission in the fluorescence spectra.

In toluene, the α- and β-bands can be resolved reasonably well by making the assumption that the bandwidth between the squaraine and the complex emissions are similar. From the relative intensity of the α- and β-emissions, the enthalpy for the formation of the Sq4–toluene complex is calculated to be ~-0.7 kcal/mol^{-1} according to an Arrhenius analysis.

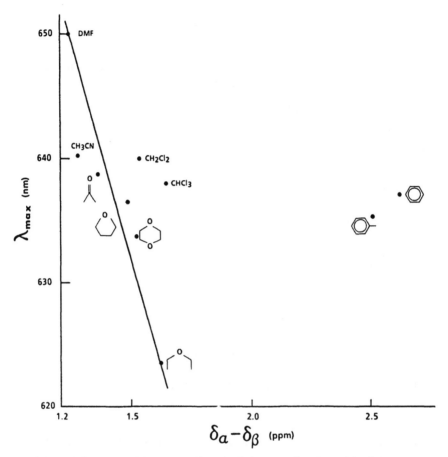

Figure 13 Plot of λ_{max} of Sq4 in different solvents as a function of $\delta_\alpha-\delta_\beta$.

C. Effects of Solvent and Temperature on the Fluorescence Lifetimes

1. Effect of Solvent

The fluorescence decays of Sq4 in various solvents have been studied at three wavelengths (640–645, 670–675, and 705 nm) and the results are summarized in Table 5. In most cases, the fluorescence decay is monoexponential. Exceptions are observed in toluene and ethyl acetate. In toluene, the decay of Sq4 is monoexponential at 645 nm with a lifetime of 2.4 ns. The decay becomes biexponential at 670 nm and lifetimes at 2.4 (80%) and 3.5 ns (20%) are recorded. The decay at 705 nm is monoexponential and the lifetime is 2.7 ns.

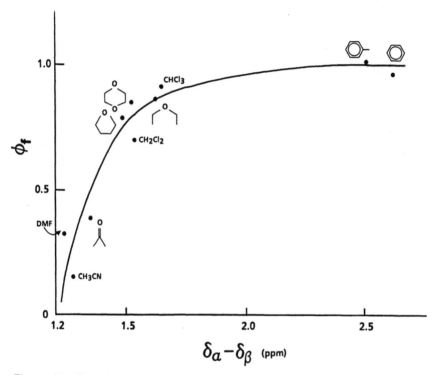

Figure 14 Plot of ϕ_F of Sq4 in different solvents as a function of $\delta_\alpha - \delta_\beta$.

According to Scheme 2, Law [30] suggested that three different excited states of Sq4 in toluene were detected. Because the decay at 645 nm is primarily from the free squaraine, the 2.4-ns decay was assigned to the excited Sq4, the 3.5-ns decay to the excited solute–solvent complex, and the 2.7-ns decay to the relaxed excited state. The fact that the solute–solvent complex has a longer lifetime is consistent with the geometry of the complex, which is shown to be rigidized upon complexation due to π–π interaction (Scheme 4b). The three lifetimes recorded for Sq4 in toluene provide kinetic evidence for the existence of three different excited states of squaraine in solution.

In ethyl acetate, biexponential decays of lifetimes of ~2.3 and ~1.3 ns were obtained at 645 as well as at 670 nm. Based on the percentage of contribution and the wavelength, Law [30] assigned the 2.3-ns decay to the S_1 of Sq4 and the 1.3-ns decay to the solute–solvent complex. In this case, the complex is shown to have a shorter lifetime. This, again, is consistent with the solute–solvent complex model because complexation in slightly polar solvents usually

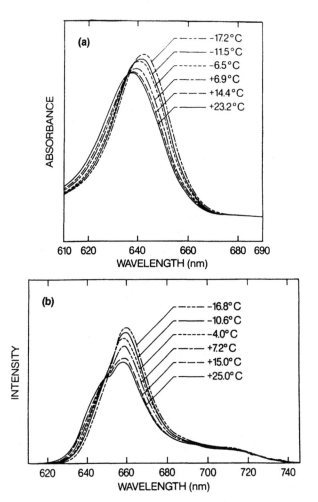

Figure 15 Effect of temperature on (a) the absorption (conc. ~2 × 10⁻⁶ M) and (b) the fluorescence emission of Sq4 (conc. ~5 × 10⁻⁷ M) in THF.

distorts the squaraine chromophore from planarity and leads to shorter lifetime (Scheme 4a). The lifetime at 705 nm is 2.3 ns and is attributable to the decay from the relaxed excited state. Again, three different fluorescence lifetimes are recorded.

Inspection of the results in Table 5 suggests that the lifetime for the S_1 of Sq4 is ~2.2 ns and is independent of solvent. Although the total ϕ_F for Sq4 in these solvents are high, they are definitely less than unity. As the radiative rate is not expected to be solvent-sensitive, the constant fluorescence lifetime

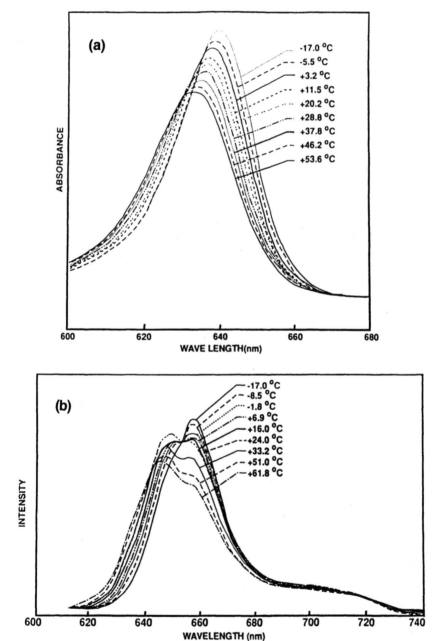

Figure 16 Effect of temperature on (a) the absorption (conc. ~2 × 10⁻⁶ *M*) and (b) the fluorescence emission of Sq4 (conc. ~5 × 10⁻⁷ *M*) in toluene.

Table 5 Effect of Solvent on the Fluorescence Lifetime of Sq4

Solvent	Lifetime (ns)			k_{rot} (s^{-1})	$\delta_\alpha - \delta_\beta$ (ppm)[d]
	α[a]	β[b]	γ[c]		
Ether	2.4	2.5	2.9	1.58×10^8	1.634
Benzene	2.4	2.4	2.5	1.74×10^8	2.631
Toluene	2.4	2.4 (80%) 3.5 (20%)	2.7	0.38×10^8	2.505
1,4-Dioxane	2.2	2.3	2.4	2.00×10^8	1.520
Ethyl acetate	1.3 (38%) 2.2 (62%)	1.4 (55%) 2.4 (45%)	2.3	4.96×10^8	
Tetrahydrofuran	2.1	2.2	2.4	2.10×10^8	1.492
Chloroform		2.5	2.6	1.64×10^8	1.652
Methylene chloride		1.8	1.9	3.18×10^8	1.526
Acetone		1.0	1.1	7.66×10^8	1.351
Acetonitrile		0.6	0.7	14.0×10^8	1.271
DMF		1.1	2.4	6.33×10^8	1.246

[a]Monitored at 645 nm.
[b]Monitored at 670–675 nm.
[c]Monitored at 705 nm.
[d]Data from Ref. 26.

indicates that there must exist a radiationless decay that is insensitive to solvent, at least within the solvent range considered (from ether to THF in Table 5). In earlier proton NMR study, Law [26] showed that one of the relaxation processes for Sq4 in the ground state is by rocking and wiggling around the C–C bond between the phenyl ring and the four-membered ring; it was hypothesized that these motions, which require very little free volume or solvent reorganization, may be responsible for the radiationless decay for the S_1 state of Sq4.

The results in Table 5 also show that the lifetime for the excited solute-solvent complex is solvent-sensitive, and that it decreases roughly as the solvent polarity increases. Because rotation of the C–C bond between the phenyl ring and the four-membered ring is the major radiationless decay process, one can estimate the rate of C–C bond rotation from the known radiative decay rate of Sq4 (2.44×10^8 s^{-1}, assuming it is not solvent-sensitive) [30]. These rotation rates are included in Table 5. Because the planarity of Sq4 in solvents can be estimated by the chemical shift difference between the two aromatic protons ($\delta_\alpha - \delta_\beta$), one can correlate the rotation rate of the C–C bond with the non-planarity of the chromophore or the twist angle between the phenyl ring and the four-membered ring. The plot is given in Fig. 17 showing that the C–C bond rotation rate increases considerably as the twist angle increases. The lifetime data in Fig. 17 complements well the steady-state results in Fig. 14.

The lifetime of the γ-emission is solvent-sensitive also, indicating that solvent molecules do play a role in the radiationless decay process. The variation is less systematic and not informative, however.

2. Effect of Temperature

Table 6 summarizes the results of the effect of temperature on the lifetimes of the excited states of Sq4 in toluene and THF. In toluene, the fluorescence emissions from the squaraine and the complex can be isolated by carefully choosing the measuring wavelengths. The lifetimes for the excited squaraine and the excited complex are found to be ~ 2.3 and ~ 3.3 ns, respectively, and are insensitive to temperature. The lack of a temperature effect suggests that the rotational relaxation process, which probably involves rocking and wiggling motions around the C–C bond [26], are insensitive to free-volume changes. On the other hand, as indicated by the longer lifetime, the complex is rigidized relative to the free squaraine due to π–π interaction with solvent molecules. Its lack of a temperature effect may be due to the already rigidized structure in the complex (Scheme 4b). At $-196°$C, however, the lifetime increases to 4.1 ns. The long lifetime indicates that all rocking and wiggling motions are prohibited at the liquid-nitrogen temperature.

In THF, the lifetimes of the β- and γ-emissions are insensitive to the ambient temperature. Because steady-state fluorescence spectra indicate that

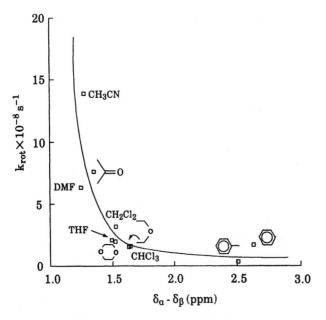

Figure 17 Plot of k_{rot} as a function of $\delta_\alpha-\delta_\beta$.

there is a shift in equilibrium for the solute–solvent complex as the temperature is lowered and because complexation is known to induced nonplanarity (Scheme 4a), the lack of a temperature effect in this case may be due to the compensation effect. Whereas temperature lowering is expected to decrease the

Table 6 Effect of Temperature on the Fluorescence Lifetime of Sq4[a]

Temp. (°C)	Lifetime in toluene (ns)			Lifetime in THF (ns)		
	α	β	γ	α	β	γ
~25	2.3	2.4 (80%)	2.7	2.1	2.2	2.4
		3.5 (20%)				
0	2.3	3.2	3.0		2.3	2.4
-23	2.2	3.3	3.3		2.4	2.5
-42	2.2	3.1	3.1		2.5	2.6
-61		3.0	3.2		2.5	
-77		3.1	3.2		2.5	
-100		3.3	3.4		2.4	
-196		4.1				

Note: Notations identical to that used in Tables 1–3.

rate of rotational relaxation and increase the lifetime, the formation of the solute–solvent complex twists the squaraine from planarity and reduces the lifetime.

VI. ABSORPTION AND FLUORESCENCE EMISSION OF PSEUDO-UNSYMMETRICAL SQUARAINES

A. Absorption

Many pseudo-unsymmetrical squaraines have been synthesized in the literature
for a variety of reasons. Table 7 summarizes a systematic investigation on the
effect of structural changes on the absorption, fluorescence emission, and lifetime of these compounds. A bathochromic effect on the absorption is observed
for N-alkyl substituted pseudo-unsymmetrical squaraines. The λ_{max} of USq1–
USq7 are red-shifted relative to Sq1. The magnitude of the red-shift is small,
however [e.g., the λ_{max} of USq7 is 8 nm red-shifted from that of Sq1, which
is about two times smaller than the red-shift observed from Sq5 to Sq1 (15
nm)]. We attribute the small red-shift to the smaller number of N-alkyl groups
in these unsymmetrical squaraines. The mechanism for the red-shift appears to
be identical to that described in Section II.A.

A similar reduction in bathochromic effect is also seen in C2-substituted
pseudo-unsymmetrical squaraines, USq8–USq12. The largest red-shift is observed for Usq11, ~9 nm. The red-shift for the symmetrical analog (Sq10) is
~16 nm. The smaller red-shift is attributable to the presence of only one C2
substituent in Usq8–USq12. What is important is the trend of the C2 substituent effect. The red-shift decreases from CH_3 to OCH_3 to OH to F to H, identical to that observed in symmetrical squaraines. The results suggest that the
same mechanism is operating in Usq8–USq12 and the effect is additive.

B. Fluorescence Emission and Lifetimes

Similar to symmetrical squaraines, pseudo-unsymmetrical squaraines also exhibit multiple fluorescence emission. Representative emission spectra are given
in Figs. 18 and 19. Spectral assignments can be made by comparing the Stokes
shifts and the spectral characteristics with those of symmetrical squaraines. The
fluorescence emission data, together with the measured lifetimes are included
in Table 7.

1. Effect of N-Alkyl Groups

Figures 18a–18c highlight the effect of a mono N-alkyl group on the multiple
fluorescence. A comparison with the spectrum of Sq1 in $CHCl_3$ [29] shows that

Table 7 Absorption and Fluorescence Emission Spectral Data of Pseudo-Unsymmetrical Squaraines (in Chloroform)

$$D — A — D'$$

| Squaraine | λ_{max} | $\lambda_F{}^a$ | | | $\phi_F{}^a$ $(\tau)^a$ |
		α	β	γ	
Sq1[b]	624	646	656	~695	0.80 (2.2)
USq1	626[c]	645	656	~695	0.84 (2.1)
USq2	628[c]	645	655	~695	0.77 (2.1)
USq3	630[c]	646	655	~695	0.82 (2.1)
USq4	630[c]	645	655	~695	0.82 (2.1)
USq5	630[c]	645	656	~695	0.79 (2.2)
USq6	630[c]	645	655	~695	0.84 (2.1)
USq7	632[d]	~645	656	~698	0.80 (2.1)

$Ar =$ structure with N, R_1, R_2

Sq1[b]: $R_1 = CH_3$, $R_2 = CH_3$
USq1: $R_1 = CH_3$, $R_2 = C_2H_5$
USq2: $R_1 = CH_3$, $R_2 = C_4H_9$
USq3: $R_1 = CH_3$, $R_2 = C_8H_{17}$
USq4: $R_1 = CH_3$, $R_2 = C_{12}H_{25}$
USq5: $R_1 = CH_3$, $R_2 = C_{16}H_{33}$
USq6: $R_1 = CH_3$, $R2 = C_{18}H_{37}$
USq7: $R_1 = R_2 = C_{18}H_{37}$

(continued)

Table 7 Continued

Ar =

Squaraine		λmax	λ_F^a			ϕ_F^a $(\tau)^a$
			α	β	γ	
USq8	D' 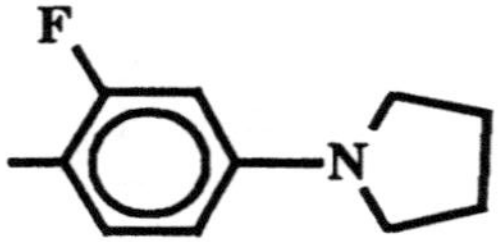 =	632[e]	647	656	~700	0.76 (2.0)
	D' 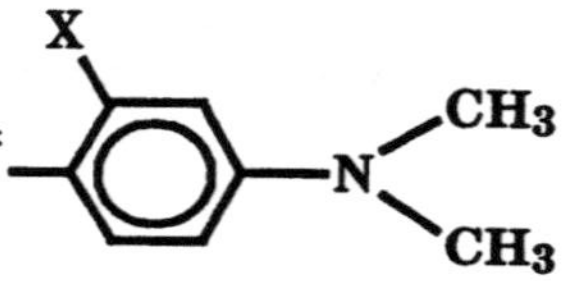 =					
USq9	X = F	625[f]	645	655	~700	0.70 (1.6)
USq10	OH	626[f]	647	656	~700	1.0 (2.4)
USq11	CH₃	633[f]	—	660	~700	0.081 (0.23)
USq12	OCH₃	627[f]	644	660	~700	0.097 (0.35)

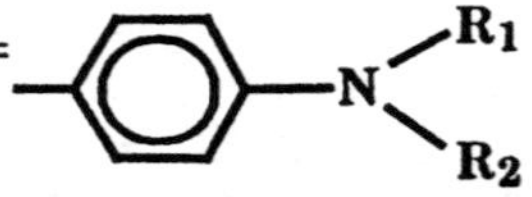

Note: Notations identical to that used in Tables 1–3.
[a]This work, unless specified.
[b]Ref. 29.
[c]Ref. 51.
[d]Ref. 52.
[e]Ref. 24.
[f]Ref. 20.

there is a gradual increase in the intensity of the β-emission as the length of the N-alkyl chain increases. The effect of the N-alkyl group on λ_F and ϕ_F is small. In any event, the increase in the intensity of the β-emission indicates that there is a shift in equilibrium for the complex formation as the chain length increases. The effect is identical to that observed in symmetrical squaraines (Section IV.A), simply confirming that there is a higher tendency for the squaraine to form complex as the D-A-D CT state is stabilized by the N-alkyl group. The magnitude of the N-alkyl group effect for USq1–USq7 is smaller than that in Table 3. The smaller effect is readily attributable to the presence of only one alkyl group in USq1–USq6 and two alkyl groups in USq7. The fluorescence spectra between USq7 and USq8 (Figs. 18c and 18d) are very similar. The result is consistent with the fact that the N-pyrrolidino group is a good electron-releasing group.

Table 7 also shows that the ϕ_F values and the fluorescence lifetimes among USq1–USq8 are very comparable. Considering the fact that the effect of N-alkyl

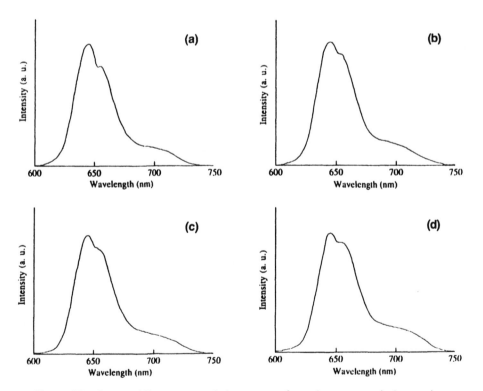

Figure 18 Corrected fluorecence emission spectra of pseudo-unsymmetrical squaraines in CHCl$_3$ [conc. ~3 × 10^{-7} M, (a) USq1, (b) USq2, (c) USq7, and (d) USq8].

group on ϕ_F and the lifetime is small to begin with (Table 3), the lack of an effect in Table 7 is not surprising

The spectral results showing the effect of C2 substituents on the fluoresecence of pseudo-unsymmetrical squaraines are summarized in Figs. 19a–19d. Three emission bands, α, β, and γ, are observed for USq9, USq10, and USq12; only two emission bands, β and γ, are observed for USq11. The substituent effect on the λ_F is expectedly small. Both the ϕ_F values and the lifetimes for these C2-substituted pseudo-unsymmetrical squaraines decrease as the substituent changes from OH to H to F to OCH_3 to CH_3. The trend is similar to that seen in symmetrical squaraines, indicating that the photophysical behavior of symmetrical and pseudo-unsymmetrical squaraines are the same. The effect observed in Sq9–Sq12 is smaller and is again attributable to the additive nature of the substituent effect.

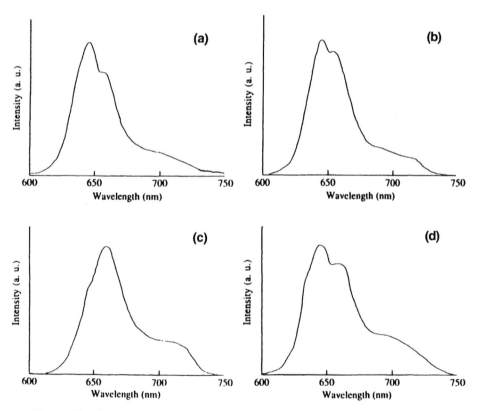

Figure 19 Corrected fluorescence emission spectra of pseudo-unsymmetrical squaraines in CHCl$_3$ [conc. ~3 × 10^{-7} M (a) USq9, (b) USq10, (c) USq11, and (d) USq12].

VII. ABSORPTION AND FLUORESCENCE EMISSION OF UNSYMMETRICAL SQUARAINES

A. Absorption

Similar to symmetrical and pseudo-unsymmetrical squaraines, 4-dimethyl-aminophenyl-4'-methoxyphenylsquaraine and its derivatives (USq13–USq26) also exhibit intense absorption in the visible region, and the data are summarized in Table 8. Comparison of the absorption maxima of these compounds with those in Tables 1 and 7 indicates that the λ_{max} of USq13 to USq26, which range from 562 to 593 nm, are blue-shifted from those of Sq1–Sq26 and USq1–USq12. Typical absorption spectra are depicted in Fig. 20 using USq13 and USq17 as examples. Because both the ground and excited states of squaraine are intramolecular CT states, the observed blue-shift is readily attributable to the substitution of one of the aniline rings by a less powerful, electron-releasing anisole ring in USq13 to USq26. The substitution decreases the CT character and results in higher-energy transitions.

Another notable distinction between the absorption of these unsymmetrical squaraines and those of symmetrical and pseudo-unsymmetrical squaraines is the absorption shoulders in the blue edge of the absorption band. As will be shown in Section VII.B, these are vibrational fine structures. In the case of USq13 in $CHCl_3$, these bands are at 580.1 (0,0), 546.3 (0,1), and 517.3 (0,2) (Fig. 20). The fine structures are found to be particularly pronounced when the aniline ring is substituted with a 2-OH group, such as in USq17 (Fig. 20).

The molar extinction coefficients for US13 to USq26 are in the range of $(1.2–2.5) \times 10^5$ cm^{-1} M^{-1}. Although these values are considered high for organic compounds, they are generally lower than those of symmetrical squaraines. We suggest that the slightly lower values may be due to the unsymmetrical structure. Substituents, both in the aniline ring and the anisole ring, are profoundly affecting the spectroscopic properties of unsymmetrical squaraines, and a discussion of the substituent effects will be given in Section VIII.C.

B. Fluorescence Emission

1. Multiple Fluorescence Emission and Spectral Assignments

Figure 21 shows the corrected fluorescence excitation and emission spectra of USq13 in $CHCl_3$. The excitation spectrum is similar to the absorption spectrum and is independent of the monitoring wavelength. In the emission spectrum, an emission maximum at 598 nm, a shoulder at ~606 nm, and a number of weaker emission bands from 620 to 740 nm are observed. Very similar multiple emission was also observed in toluene [53]. The major clue for the origin of the multiple emission bands comes from the low-temperature (77 K)

Table 8 Absorption, Fluorescence Emission, and Fluorescence Lifetime Data for Unsymmetrical Squaraines

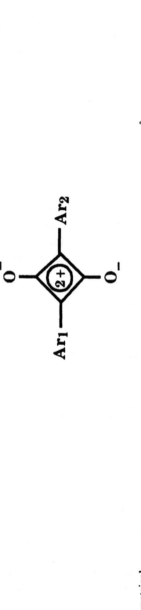

| Unsymmetrical squaraine | Ar₁ | Ar₂ | λ_{max} | $\log \varepsilon_{max}$ | λ_F | | ϕ_F | τ^a |
					α	β		
USq13	CH₃O—⟨⟩	⟨⟩—N(CH₃)₂	578.8	5.37	598	606 (s)	0.005	2.3
USq14	CH₃O—⟨⟩	F—⟨⟩—N(CH₃)₂	581.1	5.40	597	609 (s)	0.0014	2.4
USq15	CH₃O—⟨⟩	CH₃—⟨⟩—N(CH₃)₂	583.5	5.32	599	609 (s)	0.00011	2.3

ID	Structure (left)	Structure (right)					
USq16	CH₃O–C₆H₄–CH₃ (4-methoxytoluene)	3-CH₃O-4-CH₃-C₆H₃–N(CH₃)₂	583.6	5.23	597	607(s)	0.00092	2.4
USq17	CH₃O–C₆H₄–CH₃	3-HO-4-CH₃-C₆H₃–N(CH₃)₂	563.6	5.20	585	593	0.016	2.3 (54%) / 1.0 (46%)
USq18	3,4-(CH₃O)₂-toluene	4-CH₃-C₆H₄–N(CH₃)₂	587.0	5.34	601	606	0.011	2.5
USq19	3,4-(CH₃O)₂-toluene	3-F-4-CH₃-C₆H₃–N(CH₃)₂	590.6	5.32	602	607	0.0059	2.3
USq20	3,4-(CH₃O)₂-toluene	3-CH₃-4-CH₃-C₆H₃–N(CH₃)₂	592.4	5.35	602	609	0.0019	2.5
USq21	3,4-(CH₃O)₂-toluene	3-CH₃O-4-CH₃-C₆H₃–N(CH₃)₂	582.4	5.08	599	605(s)	0.0033	2.4

(continued)

Table 8 Continued

Unsymmetrical squaraine	Ar_1	Ar_2	λ_{max}	$\log \varepsilon_{max}$	λ_F α	λ_F β	ϕ_F	τ^a
USq22	(dimethoxyphenyl, CH_3O, CH_3O)	(N,N-dimethylamino phenyl; CH_3, CH_3; HO, CH_3)	572.1	5.20	—	598	0.030	2.2 (73%) 0.4 (27%)
USq23	(dimethoxyphenyl, CH_3O, CH_3O)	(N,N-dimethylamino phenyl; CH_3, CH_3)	583.1	5.39	602	610	0.0077	2.3
USq24	(dimethoxyphenyl, CH_3O, CH_3O)	(N,N-dimethylamino phenyl; CH_3, CH_3; HO, CH_3)	562.4	5.12	601	608	0.022	2.3 (21%) 0.79 (79%)
USq25	(methylenedioxyphenyl)	(N,N-dimethylamino phenyl; CH_3, CH_3)	585.6	5.26	601	609 (s)	0.005	2.4
USq26	(methylenedioxyphenyl)	(N,N-dimethylamino phenyl; CH_3, CH_3; HO)	569.1	5.06	—	595	0.019	2.5 (33%) 1.6 (67%)

Note: notation: identical to that used in Tables 1–3.
Source: Data from Ref. 54.
[a]At 77K in 2-MeTHF matrix.

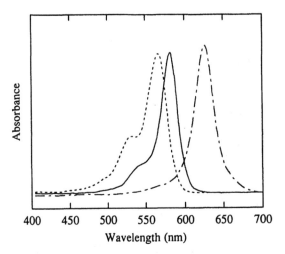

Figure 20   Absorption spectra of USq13 (——), USq17 (- - - -), and Sq1 (- - -) in CHCl₃.

fluorescence excitation and emission spectra in toluene (Fig. 22). The complex emission spectrum at room temperature was resolved into three emission bands centered at λ_F=609, 663, and 727 nm. The spacing between these bands is identical, ~1332 cm⁻¹. As they are in mirror-image relationship with the excitation spectrum, which exhibits maxima at 565 and 603 nm, the three emis-

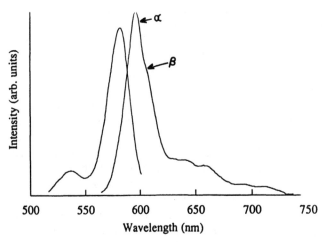

Figure 21   Corrected fluorescence excitation and emission spectra of USq13 in CHCl₃ (~3 × 10⁻⁷ *M*).

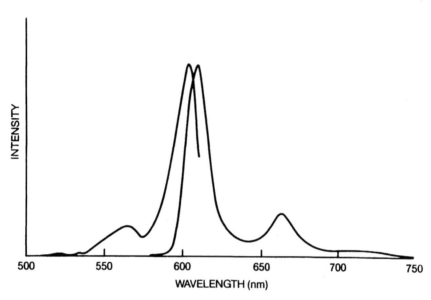

Figure 22   Corrected fluorescence excitation and emission spectra of USq13 in toluene at 77 K ($\sim 5 \times 10^{-7}$ *M*).

sion bands was then assigned to vibronic bands, (0,0), (0,1), and (0,2), of the fluorescence emission [53].

It is worthy noting that there is a spectral red-shift for both absorption and fluorescence spectra of USq13 when the recording temperature changes from room temperature to 77 K. For squaraine, this is very indicative of the formation of the solute–solvent complex [22,30]. This hypothesis is supported by variable-temperature spectral data [53]. Figures 23 and 24 show the effect of temperature on the absorption and fluorescence emission of USq13 in toluene, respectively. The λ_{max} shifts to the red, from 583.8 to 588.1, as the ambient temperature decreases from 50.8 to - 17.1°C (Fig. 23). At the same time, an isosbestic point at ~ 581 nm is observed. In the fluorescence spectra (Fig. 24), a bathochromic shift in λ_F, accomplished by a slight increase in the relative intensity of the emission shoulder at ~ 609 nm, and an isoemissive point at ~ 589 nm are observed. The spectral red-shifts, coupled with the observation of an isosbestic point and an isoemissive point in the absorption and fluorescence spectra, respectively, leave very little doubt that the long-wavelength emission shoulder at ~ 609 nm is from the excited state of the solute–solvent complex and the emission at 590 nm is from the excited USq13 itself [53].

Now if one assumes that the emissions from the solute–solvent complex and the squaraine are very similar, one can deconvolute the complex emission

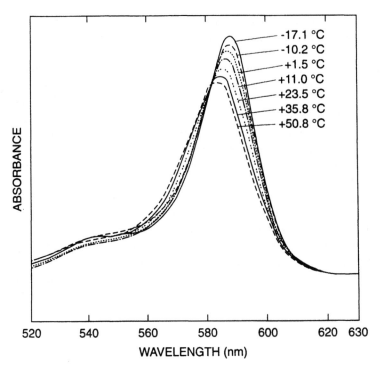

Figure 23 Effect of temperature on the absorption spectra of USq13 in toluene (conc. ~2 × 10⁻⁶ *M*).

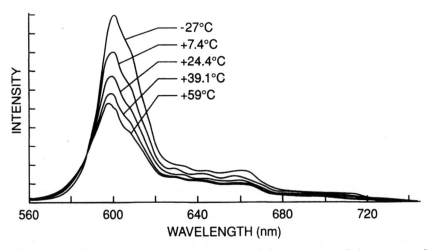

Figure 24 Effect of temperature on the corrected fluorescence emission spectra of USq13 in toluene (conc. ~3 × 10⁻⁷ *M*).

spectrum observed at room temperature (e.g., in Fig. 21) based on the spectral parameters (bandwidth, spacing, and shape) in Fig. 22. The calculated emission spectra for USq13 and its solvent complex are given in Fig. 25. The results show that the multiple emission bands can be resolved into six vibronic bands, attributable to the emissions from the S_1 state of USq13 at 597, 636, and 685 nm and from the S_1 state of the solute–solvent complex at 609, 663, and 706 nm. These bands are the (0,0), (0,1), and (0,2) vibronic bands of the respective fluorescence. The sum of the intensity of these six vibronic bands are in ∼98% agreement with the observed spectrum. The good agreement indicates that these six vibronic bands essentially account for all the fluorescence, and any emission from the relaxed excited states of USq13 should be insignificant. It is important to point out that the $I(0,1)/I(0,0)$ ratio for the room-temperature (calculated) fluorescence spectrum of the solute–solvent complex is much higher than that at 77 K. The result is consistent with intuition because one would expect to have a higher population of the zero vibrational level at 77 K [39].

Similar spectral analyses have also been carried out for several more unsymmetrical squaraines. The data consistently suggested that vibronic fine structures are the main contributors to the multiple emissions of these compounds [54].

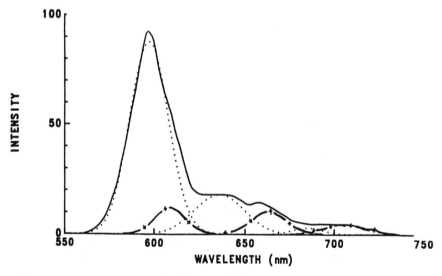

Figure 25 Observed and calculated emission spectra of USq13 in CHCl$_3$ at ∼25°C: (—) experimental data; (· · ·) calculated emission spectrum for USq13; (–+–) calculated emission spectrum for the squaraine–solvent complex.

2. Origin of the Vibronic Fine Structures

The spectral data so far indicate that the spectral characteristics of unsymmetrical squaraines are different from those of symmetrical and pseudo-unsymmetrical squaraines in terms of the origin of the multiple emissions. Law [54] suggested that although the electronic transition in squaraine is localized in the central four-membered ring, the two very different donor groups in USq13–USq26 have nevertheless perturbed the symmetry of the electronic transition. The perturbation enhances vibronic coupling, leading to vibronic fine structures in both absorption and emission spectra. The vibronic structures are quite pronounced for hydroxy-substituted unsymmetrical squaraines (e.g., USq17, USq22, USq24, and USq26). As will be discussed in Section VII.C, one of the predominating resonance structures for these hydroxy-substituted unsymmetrical squaraines is a tautomeric structure formed by transferring the proton in the OH group to the central C–O group. Law [54] suggested that this contributed resonance structure enhances the asymmetry of the electronic transition. The impact of H-bonding on the vibronic fine structure was tested by studying the absorption spectra of USq13 and USq17 in ethanol (Fig. 26). In comparison with the absorption spectra in CHCl₃ (Fig. 20), the intensities of the vibronic fine structures are indeed enhanced in both cases.

Aromatic molecules such as benzene, naphthalene, anthracene, and pyrene all exhibit vibrational fine structures in their absorption as well as fluorescence emission spectra [55–63]. Also, solvents are known to affect the relative inten-

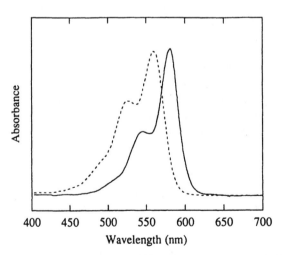

Figure 26   Absorption spectra of USq13 (——) and USq17 (- - - -) in ethanol.

sity of these vibronic bands. Notable examples are the effect of solvent on the absorption fine structure of benzene (Ham effect) [55] and the effect of solvent on the fluorescene emission of pyrene [56–59]. In both cases, interactions of the aromatic molecules with solvent molecules were reported to reduce the molecular symmetry, which decreases the forbiddance of certain vibronic coupling. Thus, the observation of vibrational fine structures in the electronic spectra of USq13–USq26, which is attributable to the asymmetric electronic distribution induced by the two very much different donor groups, is quite reasonable.

3. Photophysics of Unsymmetrical Squaraines

The fluorescence quantum yields of USq13–USq26 are in the ranges of 1×10^{-4} to 3×10^{-2} in chloroform (Table 8) and are a factor of >30 lower than those of symmetrical and unsymmetrical pseudo squaraines, whose fluorescence yields vary from 0.08 to 1.0 (Tables 3 and 7). Their measured lifetimes are <0.25 ns in fluid solution at room temperature. At 77K in 2-methyltetrahydrofuran glass, drastic increases in fluorescence intensities and lifetimes are observed. With the exception of hydroxy-substituted unsymmetrical squaraines, which show biexponential decays, all other compounds show monoexponential decays with a measured lifetime of ~ 2.4 ns (Table 8). The large temperature effect indicates that there exists a very efficient radiationless decay process at room temperature. Because the photophysics of pseudo-unsymmetrical squaraines are shown to be similar to those of symmetrical squaraines, Law [54] attributed the deviation in photophysical behavior for USq13–USq26 to the anisole ring. The proposition is not unprecedented. For example, Loutfy and Law [45] reported that the fluorescence quantum yield of p-N,N-dimethylaminobenzylidenemalononitrile (DMBM) is $\sim 10^{-3}$ at room temperature and the value increases to > 0.5 at 77 K in 2-methyltetrahydrafuran glass. Inoue and Itoh [64] showed that p-methoxybenzylidenemalononitrile (MBM) is nonfluorescent at room temperature and the quantum yield increases as the chromphore is rigidized or as the ambient temperature is lowered (Scheme 5). These reports suggest that ϕ_F can decrease by a factor of > 10 when the donor group in the CT molecule changes from an N,N-dimethyaminophenyl group to an anisole group. The small ϕ_F for USq13–USq26 relative to Sq1–Sq26 and USq1–USq12 is not unreasonable.

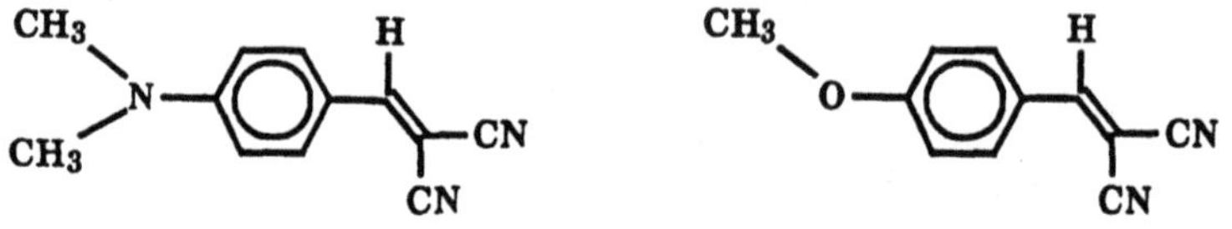

DMBM ($\phi_f \sim 10^{-3}$)   MBM ($\phi_f < 10^{-4}$)

Scheme 5

Probable mechanisms for the fast radiationless decay in USq13–USq26 that is associated with the anisole ring are (1) rotation of the C–C bond between the anisole ring and the four-membered ring and (2) rotation of the C–OCH$_3$ bond in the anisole ring. The contribution from the latter can be excluded because there is practically no increase in ϕ_F for Usq25 and USq26 when the C–OCH$_3$ bond is rigidized (Table 8). The rapid radiationless decay, on the other hand, can be rationalized based on the resonance structures shown in Scheme 6. Due to the difference in electron donicity, resonance structure II is expected to be predominant. As a result, the C–C bond between the anisole ring and the four-membered ring will consist of a considerable amount of single-bond character. It is believed that this single-bond character lowers the activation barrier for the C–C bond rotation process, facilitating fast rotational relaxation in the excited states of USq13–USq26.

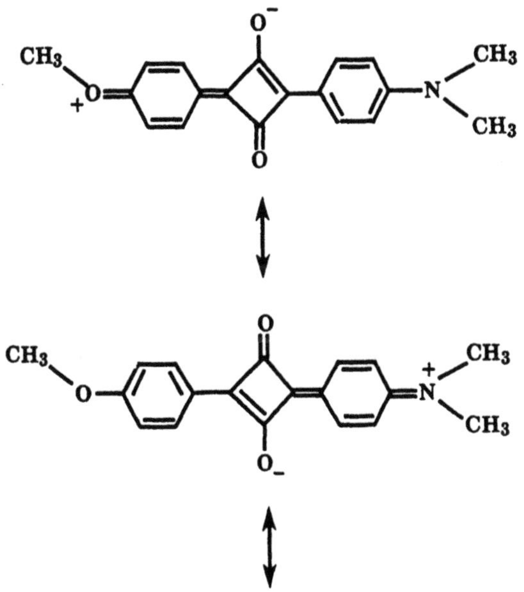

Scheme 6

All hydroxy-substituted unsymmetrical squaraines exhibit biexponential decays, one at ~2.4 ns and the other at < 1 ns. The fact that all other unsymmetrical squaraines show a monoexponential decay with lifetimes of ~2.4 ns suggests that the biexponential may be a conformational effect. For instance, hydroxy-substituted unsymmetrical squaraines may exist in two conformers: a planar conformer with a lifetime similar to other unsymmetrical squaraines (~2.4 ns), and a nonplanar conformer that is stabilized by the intramolecular H-bonding between the C2 OH group in the aniline ring and the C–O group in the four-membered ring. Since it is well-known that intramolecular proton transfers facilitate rapid radiationless decay, Law [54] hypothesized that an analogous proton transfer may be responsible for the subnanosecond lifetimes observed in these OH-substituted unsymmetrical squaraines [65–67].

C. Substituent Effects

Figures 27 and 28 show the fluorescence emission of a number of unsymmetrical squaraines. Multiple emissions similar to that of USq13 are observed, and the spectral data are tabulated in Table 8. Although the low ϕ_F values for USq13–USq26 have been rationalized, within a series of unsymmetrical structures, a systematic variation in ϕ_F is discernible. The results indicate that substituents, both in the aniline ring and the anisole ring, influence the photophysical behavior of these compounds.

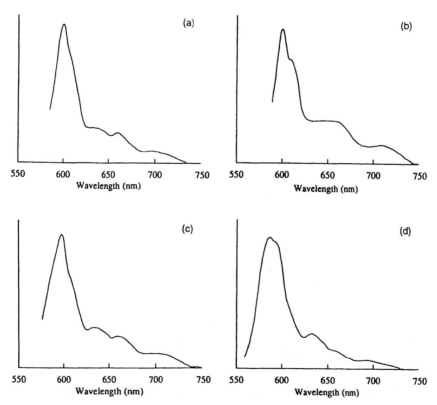

Figure 27 Fluorescence emission spectra of (a) Usq14, (b) Usq15, (c) Usq16, and (d) USQ17 in CHCl$_3$.

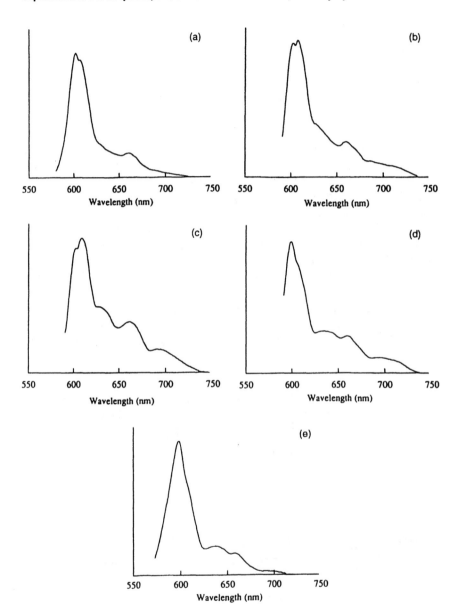

Figure 28 Fluorescence emission spectra of (a) USq18, (b) USq19, (c) USq20, (d) USq21, and (e) USq22 in CHCl$_3$.

1. Substituent Effects in the Aniline Ring

Substituent effect studies for symmetrical and pseudo-unsymmetrical squaraines (Tables 1, 3, and 7) show that the CT character of the squaraine chromophore increases as the C2 substituent in the aniline ring changes from F to H to CH_3 to OCH_3 ~ OH. This increased CT character shifts the equilibrium for the formation of the solute–solvent complex, leading to red-shifts in λ_{max} and λ_F and an increase in the intensity of the β-emission. The ϕ_F value should be sensitive to the planarity of the chromophore and intuitively should decrease as the substituent changes from OH to H to F to OCH_3 to CH_3. The substituent effect for USq13–USq15 and USq18–USq20 follows the same trend. Specifically, λ_{max} of these compounds shift to longer wavelengths and the intensity of the β-emission increases as the C2 substituent changes from H to F to CH_3. The ϕ_F values also decrease in the same order and is attributable to the nonplanarity of the squaraine chromophore due to steric replusion between the C2 substituent and the C–O group.

The most unusual substituent effect is that generated by the 2-OH group in the aniline ring. A blue-shift in λ_{max} and a decrease in ε_{max} are observed in USq17, USq22, USq24, and USq26 relative to USq13. The results are contrary to intuitively expected from the electron-releasing OH group. Law [54] rationalized the results based on the resonance structures in Scheme 7. In OH-sub-

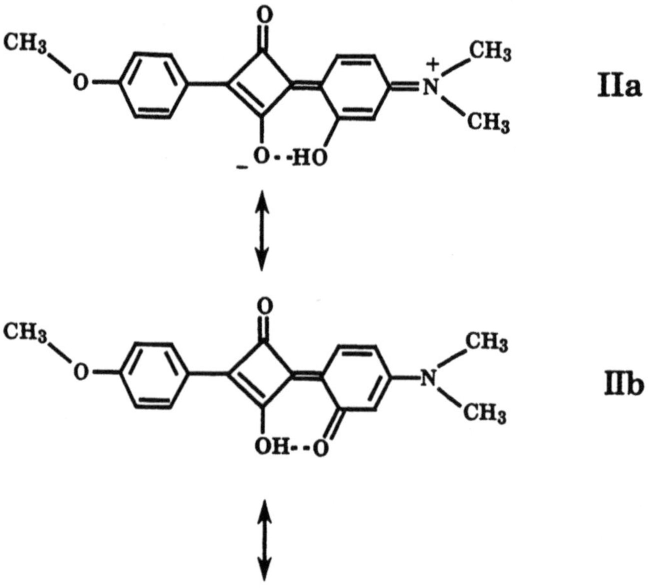

IIa

IIb

Scheme 7

stituted unsymmetrical squaraines, contribution from the tautomeric resonance structure IIb may be very important. Any significant contribution from IIb is expected to decrease the CT character of the electronic transition, which is supposed to be localized in the four-membered ring [6]. The decrease in CT character then leads to a blue-shift in λ_{max} and a decrease in ε_{max}. Moreover, it increases the asymmetry of the electronic distribution, providing a good rationale for the pronounced vibrational structures in the absorption and fluorescence emission spectra of these compounds. This explanation is supported by the spectral data obtained for USq16 and USq21. Presumably due to the contribution from similar resonance structures, these two compounds also show an anomalously short λ_{max} and low ε_{max}, despite having a strong electron-releasing substituent in OCH_3. The effects observed in USq16 and USq21 are smaller and is presumably due to the lack of transferrable protons in these compounds.

2. Substituent Effects in the Anisole Ring

A comparison of the spectral data of USq13, USq18, and USq23 (Table 8) suggests that there is a red-shift in λ_{max}, from 578.8 in USq13 to 587.0 nm in USq18 when the second OCH_3 group is introduced to the anisole ring. The red-shift can be ascribable to the increase in CT character of the chromophore due to an inductive effect provided by the second electron-releasing OCH_3 group. The increase in CT character is also revealed by the increase in ϕ_F from 0.005 for USq13 to 0.011 for Usq18. The increase in ϕ_F is presumably due to the increased contribution by the resonance structrure of type I in Scheme 6, as it would increase the double-bond character of the rotating C–C bond and lower the rate of the rotational deexcitation process. On the other hand, when the third OCH_3 is added, there appears to be a decrease in CT character. The λ_{max} of USq23 is at a shorter wavelength relative to USq18. A similar trend is also observed for USq22 and USq24. It was suggested that due to the steric replusion between the methoxy groups in USq23 and USq24, the CT character in these compounds is reduced (compare structures IIIa and IIIb in Scheme 8). The proposed steric effect in these squaraines is analogous to that observed for methoxy-substituted benzylidenemalononitriles [64].

The interpretation regarding the impact of the CT character on the electronic spectra is further supported by the results obtained in USq25 and USq26. Although the anisole ring is rigidized in these two compounds, there does not seem to be any increase in CT character, as indicated by the λ_{max}, ε_{max}, and ϕ_F values. The unexpected result was attributable to the unavailability of the lone pair electrons at the oxygen atoms for delocalization due to geometrical constraints in the rigidized ring structure. Therefore, the CT character in USq25 and USq26 is midway between USq13, USq17, and USq18 and USq22, respectively. The λ_{max}, ε_{max}, and ϕ_F data in Table 8 refect this point fairly well.

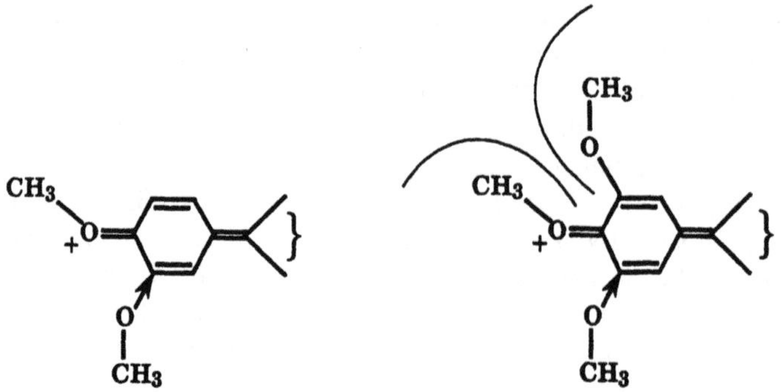

IIIa   **IIIb**

Scheme 8

VIII. SUMMARY AND REMARKS

This work summarizes systematic investigations of the spectroscopic properties and photophysics of symmetrical and unsymmetrical squaraines in the author's laboratory. Both the ground and excited states of squaraine are D-A-D intramolecular CT states. Owing to the unusual electronic structures and arrangements, the electronic transitions (S_0 to S_1) are localized in the four-membered ring [6]. This unique characteristic leads to the very narrow-bandwidth absorption with a very high extinction coefficient and intense fluorescence with small Stokes shifts. Also, the polarized charges in the central four-membered ring becomes a site of dipole–dipole or specific interactions with solvent molecules. Even though the interaction is not very strong (ΔH for the solute–solvent complex is about -0.7 kcal/mole^{-1} in toluene), it is still detectable because of the narrow fluorescence emission bandwidth. Excitation of squaraine in solution leads to two excited states: the excited state of squaraine and the excited state of the solvent complex. These two excited states fluoresce to give rise to the α- and β-emissions. They can undergo a rotational relaxation process to generate a relaxed excited state, which can fluoresce to give the γ-emission or relax by a rotational mechanism back to the ground state. The proposed photophysical model is supported by solvent effect, temperature effect, and mixed-solvent studies. The conclusion is further complemented by structural effect studies and fluorescence lifetime measurements.

When a small asymmetry is introduced to the electronic structure, such as in USq1–USq12, the spectroscopic properties and photophysics remain unchanged. The observation is readily attributed to the localization of the electronic transition, which is in the four-membered ring. On the other hand, when two significantly different donor groups are present in the same squaraine structure, such as in USq13–USq26, the symmetry of the electronic transition is broken. As a result, vibronic couplings are enhanced and vibrational fine structures are observable in both absorption and fluorescence spectra. Unsymmstrical squaraines also form solvent complexes with solvents and result in multiple emissions. Using low-temperature spectral data, we have been able to show that vibrational fine structures from the emissions of the unsymmetrical squaraine and the solvent complex are basic contributors to the multiple emissions.

As noted in Section I, a large number of squaraines have been synthesized lately for one reason or another. Although the spectroscopic properties of these compounds have not been evaluated in detail, they are actually rationalizable by the principles described herein. The structures of some of these squaraines along with their absorption and fluorescence emission data are given in Table 9 [68–75]. Sq27 absorbs at ~540 nm in CHCl$_3$ and the λ_{max} is significantly blue-shifted relative to that of Sq1. The blue-shift is attributable to the significant decrease in CT character due to the two anisole rings in the squaraine structure [6]. It is worthy pointing out that the absorption of Sq27 also does not show any significant vibronic fine structure [69]. The absence supports an earlier conclusion that an asymmetrical electronic distribution is the prerequisite for observing the vibronic fine structure. The fluorescence emission of Sq27 is very weak, probably due to the decreased CT character in the squaraine chromophore and other radiationless processes available in the anisole ring.

Sq28 and Sq29 are synthesized by condensation of squaric acid with phenol derivatives. They exhibit absorption at wavelengths comparable to that of Sq27. Their ε_{max} are low relatively [~(5–10) $\times$ 10^4 M^{-1} cm^{-1}]. The fluorescence is weak and the Stokes shift is relatively large compared to other squaraines [70–72]. The existence of multiple radiationless processes (e.g., proton transfer, tautomerization, etc.) may be the reason for the low ϕ_F [72], and systematic work is required to elucidtate the photophysics of these compounds.

On the other hand, Sq30–Sq32 exhibit absorptions with extinction coefficients very typical of squaraines, ~2 $\times$ 10^5 cm^{-1} M^{-1}. Sq30 absorbs at 568 nm and fluoresces at 586 nm [73]. The absorption and emission wavelengths are blue-shifted from that of Sq1 and are ascribable to the pyrrole ring structure (less conjugation). The ϕ_F of Sq30 is low and can be explained by a rotational deexcitation mechanism similar to that described in Scheme 3. Presumably due to the steric replusion between the C2 methyl group and the C–O group, the squaraine chromophore is twisted, resulting in a facile rotational relaxation for the excited state. The absorption and fluorescence wavelengths, the Stokes shifts, and the ϕ_F for Sq31 and Sq32 are comparable to the squaraines described in Tables 1, 3, and 7. The overall data suggest that the spectroscopic

Table 9 Absorption and Fluorescence Emission Spectral Data of Miscellaneous Symmetrical Squaraines

Squaraine	Ar	Solvent	λ_{max} (log ε_{max})	λ_F	ϕ_F	Ref.
Sq27		$CHCl_3$	540 (5.15)	—	—	68,69
Sq28	X = H	CH_3CN/H_2O	~530	582	0.037	70
Sq29	= OH	CH_3OH/H_2O	508	—	<0.0001	71
		C_2H_5OH	510 (5.11)	610	0.002	72

Compound		Solvent				Ref.
Sq30	CH₃, CH₃, CH₃, CH₃, C, O (pyrrole structure)	CH_2Cl_2	568 (5.27)	586	0.08	73
Sq31	CH, S, N—C_2H_5 (benzothiazole structure)	benzene	679 (5.43)	690	0.54	74
Sq32	CH₃, CH₃, CH, N—C_2H_5 (indolenine structure)	$CHCl_3$	635	645	0.09	75

Note: Notations identical to that used in Tables 1–3.

properties and photophysics of Sq31 and Sq32 are similar to those of Sq1–Sq26. The findings are not totally surprising because according to MO calculations, the electronic transition of squaraine is primarily taking place in the central four-membered ring [6]. The similarity suggests that the CT character in Sq31 and Sq32 should be comparable to those in Sq1–Sq26. In other words, the nucleophilicities of *N*-ethyl-2-methylbenzothiazolium and 2-methylene-1,3,3-trimethylindolenine should be comparable to those of *N,N*-dialkylanilines. Indeed, they are comparable synthetically [75,76]. In addition, unsymmetrical squaraines from *N*-ethyl-2-methylbenzothiazolium, 2-methylene-1,3,3-trimethylindolenine, and derivatives have also been prepared [76] and they are found to exhibit spectroscopic properties very similar to their symmetrical analogs [75]. The observation can again be attributed to the unique electronic transition of squaraine, which is localized in the central four-membered ring.

REFERENCES

1. Sprenger, H. E.; Ziegenbein, W. *Angew. Chem. Int. Ed. Engl.* 1966, *5*, 894.
2. Treibs, A.; Jacob, K. *Angew. Chem. Int. Ed. Engl.* 1965, *4*, 694.
3. Sprenger, H. E.; Ziegenbein, W. *Angew. Chem. Int. Ed. Engl.* 1967, *6*, 553; 1968, *7*, 530.
4. Rehak, V.; Israel, G. *Chem. Phys. Lett.* 1986, *132*, 236.
5. Schmidt, A. H. *Oxocarbons*, R. West, Ed.; Academic Press: New York, 1980.
6. Bigelow, R. W.; Freund, H. J. *Chem. Phys.* 1986, *107*, 159.
7. Law, K. Y.; Facci, J. S.; Bailey, F. C.; Yanus, J. F. *J. Imaging Sci.* 1990 *34*, 31.
8. Wingard, R. E. *IEEE Trans. Ind. Appl.* 1982, 1251.
9. Law, K. Y. *Chem. Rev.* 1993, *93*, 449 and references cited therein.
10. Loutfy, R. O.; Hsiao, C. K.; Kazmaier, P. M. *Photogr. Sci. Eng.* 1983, *27*, 5.
11. Morel, D. L; Stogryn, E. L.; Ghosh, A. K.; Feng, T.; Purwin, P. E.; Shaw, R. F.; Fishman, C.; Bird, G. R.; Piechowski, A. P. *J. Phys Chem.* 1984, *88*, 923.
12. Merritt, V. Y.; Hovel, H. J. *Appl. Phys. Lett.* 1976, *29*, 414.
13. Gravesteijn, D. J.; Steenbergen, C.; van der Veen, J. *Proc. SPIE* 1983, *420*, 327.
14. Jipson, V. P.; Jones, C. R. *J. Vac. Sci. Technol.* 1981, *18*, 105.
15. Mori, T.; Miyachi, K.; Kichimi, T.; Mizutani, T. *Jpn. J. Appl. Phys.* 1994; *33*, 6594.
16. Ashwell, G. J. *Adv. Mater.* 1996, *8*, 248.
17. Ashwell, G. J.; Jefferies, G.; Hamilton, D. G.; Lynch, D. E.; Roberts, M. P. S.; Bahra, G. S.; Brown, C. R. *Nature*, 1995, *275*, 385.
18. Andrews, J. H.; Khaydarov, J. D. V.; Singer, K. D.; Hull, D. L.; Chuang, K. C. *J. Opt. Soc. Am. (B)* 1995, *12*, 2360.
19. Chen, C. T.; Marder, S. R.; Cheng, L. T. *J. Am. Chem.Soc.* 1994, *116*, 3117.
20. Law, K. Y; Bailey, F. C. *Can. J. Chem.* 1993, *71*, 494.
21. Law, K. Y.; Bailey, F. C. *J. Org. Chem.* 1992, *57*, 3278.
22. Law, K. Y. *J. Phys. Chem.* 1987, *91*, 5184.

23. Law, K. Y.; Bailey, F. C. *Dyes Pigm.* 1988, *9*, 85.
24. Law, K. Y.; Bailey, F. C. *Dyes Pigm.* 1993, *21*, 1.
25. Law, K. Y.; Bailey, F. C.; Bluett, L. J. *Can. J. Chem* 1986, *64*, 1607.
26. Law, K. Y. *J. Phys. Chem.* 1989, *93*, 5925.
27. Emslie, P. H.; Foster, R.; Fyfe, C. A.; Harman, I. *Tetrahedron* 1965, *21*, 2843.
28. Mulliken, R. S. *J. Chem. Soc.* 1952, *74*, 811.
29. Law, K. Y. *Chem. Phys. Lett.* 1988, *150*, 357.
30. Law, K. Y. *J. Photochem. Photobiol. A: Chem.* 1994, *84*, 123.
31. Wang, Y. *J. Phys. Chem.* 1985, *89*, 3799.
32. Wang, Y. *Chem. Phys. Lett.* 1985, *116*, 286.
33. Visser, R. J.; Varma, C. A. G. O. *J. Chem. Soc. Faraday Trans.* 2 1980, *76*, 453.
34. Visser, R. J.; Weisenborn, P. C. M.; Varma, C. A. G. O. *Chem. Phys. Lett.* 1985, *113*, 330.
35. Visser, R. J.; Varma, C. A. G. O.; Konijnenberg, J.; Weisenborn, P. C. M. *J. Mol. Struct.* 1984, *114*, 105.
36. Visser, R. J.; Varma, C. A. G. O.; Konijnenberg, J. *J. Chem. Soc., Faraday Trans.* 2 1983, *79*, 347.
37. Grabowski, Z. R.; Rotkiewicz, K.; Siemiarczuk, A.; Cowley, D. J.; Baumann, W. *Nouv. J. Chim.* 1979, *3*, 443.
38. Merritt, V. Y. *IBM J. Res. Devel.* 1978, *22*, 353.
39. Becker, R. S. *Theory and Interpretation of Fluorescence and Phosphorescence*; Wiley: New York, 1969; p. 42.
40. Kamat, P. V.; Das, S.; Thomas, K. G.; George, M. V. *J. Phys. Chem.* 1992, *96*, 195.
41. Rettig, W.; Gleiter, R. *J. Phys. Chem.* 1985, *89*, 4676.
42. Hammond, G. S.; Saltiel, J.; Lamola, A. A.; Turro, N. J.; Bradshaw, J. S.; Cowan, D. O.; Counsell, R. C.; Vogt, V.; Dalton, C. *J. Am. Chem. Soc.* 1964, *86*, 3197.
43. Drexhage, K. H. *Top. Appl. Phys.* 1973, *1*, 147 and references therein.
44. Jones, G.; Jackson, W. R.; Choi, C.; Bergmark, W. R. *J. Phys. Chem.* 1985, *89*, 294.
45. Loutfy, R. O.; Law, K. Y. *J. Phys. Chem.* 1980, *84*, 2803.
46. Reichardt, C. *Angew. Chem. Int. Ed. Engl.* 1965, *4*, 29.
47. Kosower, E. M. *J. Am. Chem. Soc.* 1958, *80*, 3253.
48. Herbrandson, H. F.; Neufeld, F. R. *J. Org. Chem.* 1966, *31*, 1140.
49. Dong, D. C.; Winnik, M. A. *Photochem. Photobiol.* 1982, *35*, 17.
50. Kamlet, M. J.; Abboud, J. L.; Taft, R. W. *J. Am. Chem. Soc.* 1977, *99*, 6027.
51. Law, K. W.; Bailey, F. C. *Dyes Pigm.* 1992, *20*, 25.
52. Law, K. Y.; Chen, C. C. *J. Phys. Chem.* 1989, *93*, 2533.
53. Law, K. Y. *Chem. Phys. Lett.* 1992, *200*, 121.
54. Law, K. Y. *J. Phys. Chem.* 1995, *99*, 9818.
55. Ham, J. S. *J. Chem. Phys.* 1953, *21*, 756.
56. Nakajima, A. *Bull. Chem. Soc. Jpn.* 1971, *44*, 3272.
57. Nakajima, A. *J. Mol. Spectrosc.* 1976, *61*, 467.
58. Hara, K.; Ware, W. R. *J. Chem. Phys.* 1980, *51*, 61.

59. Dong, D. C.; Winnik, M. A. *Can. J. Chem.* 1984, *62*, 2560.
60. Durocher, G; Sandorfy, C. *Mol. Spectrosc.* 1964, *14*, 400; 1966, *20*, 410.
61. Koyanagi, M.; Kanda, Y. *Spectrochim. Acta.* 1964, *20*, 993.
62. Koyanagi, M. *J. Mol. Spectrosc.* 1968, *25*, 273.
63. Nakajima, A. *Spectrochim. Acta A* 1974, *30*, 860.
64. Inoue, K.; Itoh, M. *Bull. Chem. Soc. Jpn.* 1979, *52*, 45.
65. Hou, S. Y.; Hetherington, W. M.; Korenowski, G. M.; Eisenthal, K. B. *Chem. Phys. Lett.* 1979, *68*, 282.
66. Ford, D,; Thistlethwaite, P. J.; Woolfe, G. J. *Chem. Phys. Lett.* 1980, *69*, 246.
67. Law, K. Y.; Shoham, J. *J. Phys. Chem.* 1994, *98*, 3114.
68. Farnum, D. G.; Webster, B.; Wolf, A. D. *Tetrahedron Lett.* 1968, 5003.
69. Law, K. Y. *J. Phys. Chem.* 1988, *92*, 4226.
70. Kamat, P. V.; Hotchandani, S.; de Lind, M.; Thomas, K. G.; Das, S.; George, M. V. *J. Chem. Soc., Faraday Trans.* 1993, *89*, 2397.
71. Das, S.; Kamat, P. V.; De la Barre, B.; Thomas, K. G.; Ajayaghosh, A.; George, M. V. *J. Phys. Chem.* 1992, *96*, 10327.
72. Scott, G. W.; Tran, K. *J. Phys. Chem.* 1994, *98*, 11563.
73. Patrick, B.; George, M. V.; Kamat, P. V.; Das, S.; Thomas, K. G. *J. Chem. Soc., Faraday Trans.* 1992, *88*, 671.
74. Das, S., Thomas, K. G.; Ramanathan, R.; George, M. V.; Kamat, P. V. *J. Phys. Chem.* 1993, *97*, 13625.
75. Terpetschnig, E.; Szmacinski, H.; Lakowicz, J. R. *Anal. Chim. Acta* 1993, *282*, 633.
76. Terpetschnig, E.; Lakowicz, J. R. *Dyes Pigm.* 1993, *21*, 227.

Index